民國建築工程期刊匯編

MINGUO JIANZHU GONGCHENG QIKAN HUIBIAN

51

《民國建築工程期刊匯編》編寫組 編

廣西師範大學出版社
GUANGXI NORMAL UNIVERSITY PRESS
·桂林·

第五十一册目録

南大工程

25521

南大工程

桂銘敬

THE JOURNAL OF THE LINGNAN ENGINEERING ASSOCIATION

VOL. 1

[illegible]B. 1948

[illegible]刊

第壹期

[illegible]大學工程

中華[illegible]三十七年[illegible]十[illegible]

南大工程

嶺南大學工程學會會刊

鳴謝

本刊是次出版，經費無着，於民國卅六年春曾舉行電影籌欵，荷蒙校內外人士熱烈贊助，乃克完成。高誼隆情，至爲銘感。其次並得威行洋服。雄德公司各惠助本刊廣告費港幣八十元。李耀記，惠康公司，友誠行。國民漆廠暨天利行各惠助廣告費港幣四十五元。宏記及永安營造廠各惠助廣告費港幣二十五元，並蔚興印刷場惠助廣告費國幣六十三萬元。此外，更蒙張嘉會先生特別捐助港幣四十五元，暨漢生先生捐助港幣二十五元。本刊同人，至深紉感，特此一併鳴謝。

南大工程

康樂再刊 第一期

—目 錄—

再刊前言

吳秉俠

爲着繼承先者的耕耘，而拓展發揚於來日；爲着交遞我們的所知，而檢討本身之所學；爲着灌長研究的風尚，而擴寬智識之範疇。雖則荏苒處處，我們也努力於使這本刊册再度出版。

事實如此，智識之獲得，是學理與實驗的湊合。尤以我們在實科中鑽的一羣，更脫不了實習的證示。然而，考之今日我們學習的環境，不禁黙言難語。爲了國家科學的落後，學術的幼稚，人材的缺少，資源的拮据，以至我們不但在實驗的器材與場所方面給絆倒了。甚而在尋求學理之參考書本方面，也受到嚴緊的覊束。（尤其甚者，基於今日教育制度的不健全，一部同學連參閱課外書籍的時間也給剝削了。）在這種學理與實習皆趨於窒息的狀況下，要我們長久地只在課本中去吸吮狹小的智識，未免過於殘酷。爲着打破這近於沉暮的局面，多找機會去攫取智能，和大胆去檢討本身之所學，我們唯有寄托於這刊物上。

遠在十五年前（民國廿二年春），先進者便出版了這刊物，以後每年一期，直迄民國廿五年因戰事影响與學校遷港而停頓。及至學校復原於康樂後之第二年（民卅六年春），本會有另出版「工程學報」的意思，後卒因經費及枝節問題，未能實現。延至今季（卅六年秋），在教授與同學支持下，方出版了這刊子，名稱照舊緣用「南大工程」，因爲它是本會唯一會刊之故。假使能力所及，我們希望能達到每學期出一册的目的。不然，也力爲保持每年一期的出版，以免有負愛好本刊讀者之厚望。

在南中國這片荒蕪的學術園地裏，關於工程的刊物（除了工務關機的報告刊册外），更屬鳳毛麟角。這對於學習工科的我們，不啻爲一種諷刺。這本刊册，我們未敢斷言對大衆有何重大的貢獻。然而，最少希望能爭取一部份人士對工程方面的注視與興趣，至其收效，還需等候社會人士的判斷。

最後，我們渴望着得到讀者嚴正的批評與珍貴的指示。

轉　載

來湛鐵路

西南西北鐵路幹綫出口路綫之一段

• 桂銘敬 •

來湛鉄路乃自廣州灣之湛江市起而迄於廣西境內之來賓。爲西南西北鉄路幹綫出口路綫之一段，其對南路之交通，影响至鉅。作者桂銘敬先生爲湘桂鉄路局副局長兼副總工程司，並來湛段粵境工程處處長及本系系主任（在假）。本文曾刊載於本年一月一日之湛江市大光報。値此復員建設期間，本刊爲求社會人士對該段鉄路有所認識，爰爲轉載，俾讀者得窺其全也。　　編者誌

總理實業計劃中原有渝欽鐵路幹線之議，今已列爲戰後第一五年築路計劃之首選，並已積極準備分段實施，惟起迄地點畧有變更，計由西北之蘭州起，經天水，成都，貴陽，柳州以達湛江市之西營，全長共約三千二百七十餘公里，歷經甘，陝，川，黔，桂，粵六省成爲西南西北鐵路之主要幹線，本線除都匀至柳州，柳州至來賓一段經已完成外（現屬湘桂黔鐵路範圍）其餘天蘭，天成，成渝，隆筑，都筑，來湛六段全屬新工，刻已積極展開工作，來湛段由來賓至湛江經已定測完畢，查本線出口港灣原有欽州灣，北海，及廣州灣，三處之比較，結果以廣州灣（卽今之湛江）具有天然之優勢條件，允爲西南唯一之出口良港，且該港與海南島遙遙相對，而海南島礦產豐富，一經開發可成爲南中國重要工業區，又有富比台灣之說，爲謀開發海南寶藏使與大陸相貫通，則湛江港口之決定，至爲適當，關於路線起點之所以選定在蘭州市亦以蘭州爲我國眞正之中心點，欲謀開發西北，必先繁榮蘭州，欲謀繁榮蘭州，則非發展蘭州與沿海各省之交通不可，是選定蘭州市爲起點之決定，亦屬必然之事實。

本路線綿延六省，長達三千餘公里，貫通西南西北，允爲我國鐵路計劃中之大幹線，然如何完成此艱巨工程，則財力物力，與時間之估計，誠屬必要，查戰前每公里鐵路建築費約爲一二〇，〇〇〇元（以戰前銀元爲本位），現全線三二七二公里，計需欵3,272×120,000＝392, 640,000元約合四億銀元根據黔桂鐵路已成部份屬於材料資本支出之分析，橋工佔百分之九、五四，軌道佔百分之一一.三二，電話電報佔百分之二.〇六，

機件佔百分之〇.六二,車輛佔百分之七.一〇,合計佔全部資本支出百分之三〇.六四,此百份之三〇.六四之數,設減去工費,及其他雜費約爲百分之十一則餘者概爲購置材料之費約爲百分之二十,今西南西北幹線全部

湘桂黔鐵路路綫圖

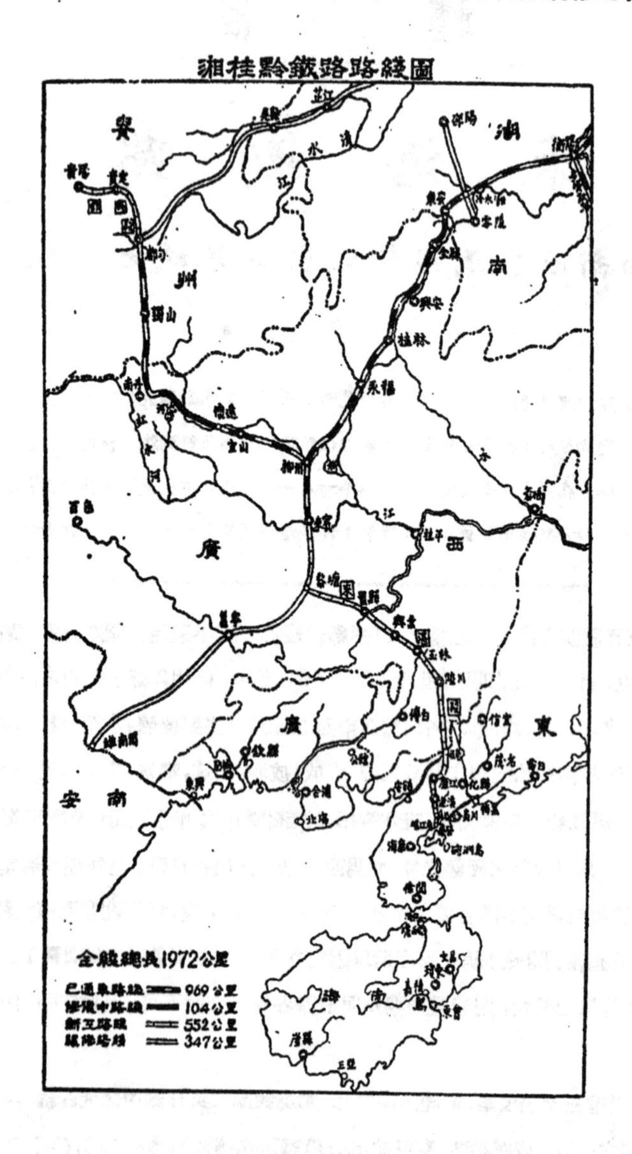

建築費設爲四億餘元,則用於購置材料者約爲八〇,〇〇〇,〇〇〇餘元,假定全部材料以鋼料爲代表,戰前鋼料價格約爲二百七十元一公噸,則全線需用材料約爲二百七十分之八〇,〇〇〇,〇〇〇等于二九六,二九

六公噸，粗計約合三十萬公噸此三十萬公噸之材料如何分別輸入沿綫使用，至成問題，查沿綫可能運輸之路綫大別有四：一爲隴海路，二爲長江水運，三爲粵漢湘桂綫，四即本段——來湛段，然目前隴海路正遭戰事影響，復軌通車至海州出海一時尚未能辦到，其次長江之三峽以上水道又不能航行大汽輪，運輸能力有限，粵漢路綫又曲折迤邐，頗不合算，然則來湛鐵路旣爲本路綫出口之一段且與湛江港口聯接，將爲本路唯一之主要運輸綫當無疑義，然若全綫趕工，以上指出運輸路綫及其他可能通達之運輸路綫仍須四方八面同時配合，使用方可克奏膚功也。

按照中央政府預定本幹綫完成日期，照分段施工計劃，四年內可從事本計劃之能否實現端賴國家財力是否充裕與乎地方人士之是否竭誠協助以爲斷，蓋一切築路材料，大都仰給外來，非有充裕之財力，不足以成大事，其他如征集就地人工與材料，則又非有地方人士之明瞭與協助不爲功也。

來湛段鐵路旣爲我國西南西北鐵路線出口路線之一段，且爲完成本幹線之主要運輸線，則其進行近狀與完成日期，與及配合之築港工作，自必爲地方人士所矚目與關懷，筆者奉命參與本段工程及建港工作，僅就所知，報道大畧，想爲社會人士所樂聞也。

來湛段鐵路爲施工方便起見現分桂境粵境兩工程處負責，目前均屬湘黔桂鐵路局管轄範圍，共長四百零一公里，桂境佔三百二十一公里，粵境因須兼辦港口工程，鐵路里程較短共長八十九公里，桂境工程處設在廣西境內之貴縣，現設九個工務總段二十七個分段及若干橋工處分別執行施工工作，大橋及一部份路基工程已開工。

粵境工程處現有三個工務總段九個分段負責鐵路部份工作，另先設一個港口工程所負責西營碼頭堤岸工程，將來港工展開，或再添設工程所一處，至內部組織除粵處加設一港工組外，其餘兩處均設有工務，總務，材料，會計各組及人事秘書两室，另設副處長二人協助處務，两處技術人員現已有四百餘名，將來全段展開工作時尙需添用技術人員一二百名，目前因國家財力未充，分配工欵甚少，且以治安關係，暫未曾展開工作，如沿線治安寧靜，工欵充足，本段四百公里鐵路工程預計一年半即可完成。

來湛段全段除大橋數處外，尙無其他艱巨工程，粵境綫內所得稱述者厥爲湛江總車站之建築查湛江總站設在西營，爲全路之終點站，蓋其位置重要上握蘭廣幹線之咽喉爲西南通內陸之門戶，故列爲一等站，爲配合港口發展，車站位置緊接海岸與市區，車站之東，隣接市區西部，其南聯接海港碼頭，倉庫與堤岸，西毗霞山村，北隣龍興村，又爲發展雷州半島工業，在站南正線之起端西出以接雷湛支線，總計全部用地除築港範圍及雷湛支線聯絡站未計外，計共征用土地二千四百餘市畝，內軌道佔地約爲全部面積百分之二十三，廠房佔地爲百分之三十，房屋佔地爲百分之二十八，貨物站場佔地爲百分之十五，材料場佔地爲百分之四。站內佈置如下：因軌道方向爲南北向，故靠市區，一面爲旅客站場，有四十公尺寬之交通大道直達市區近海岸方面爲貨物站場，站之東北端入口處附近爲軍用站場，所有各種站場均係在正線之右邊，站南靠市區一面分佈職員宿舍

25531

鐵路賓館、郵局、電信局、站園及機、工車、警段所辦公室等區域，以寬二十公尺之馬路縱橫貫通之，並留路口凡六處通達市區，以便交通。

本站場完成後具有極完備之發電廠與機械廠，全部計劃實現後，不獨可供鐵路及港口各項工程與行車之用，兼可有剩餘電力供給爲湛江市各種工業需求，此外關於湛江總站站房，亦已設計完妥，茲併將該站房透視圖附後，以明大概。

抑更有進言者，最近湘桂黔全段路線，幾經艱苦，在物資極度困難之時，能努力搶趕，已於本年十一月底，由柳州復軌通車至衡陽，如此沿粵漢鐵路，向北可達漢口南可通廣州香港，綜觀全國鐵路，在華北，在東北，無日不在共匪破壞及搶修之中，虛耗物力財力，不可以數計，而在華南，竟有此一段復軌通車之成就，展望交通建設之前途，自當感相當興奮也，且查華中之浙贛鐵路，正在加深復軌之中，一兩月後即可告竣，最近之將來，更可由浙贛，滬杭等路，而與京滬各地聯接，使華中與華南之交通動脈貫通，必更爲便利，至於由來賓至貴縣一段，路綫在戰時已築有路基，稍加修補，及建築來賓附近之紅水河大橋工程，設法再籌購鋼軌材料，則通車到達貴縣，亦非極難之事，由貴縣可沿西江水路經梧州以達廣州，如此粵桂兩省，息息相通，商旅往來，必更便利，路通貴縣之後，爲謀桂省土產之出口，更足以促成來湛鐵路之成功，而使華南之雷州半島上之經濟文化，均有極度之發展，國計民生，因之可由解決，五年十年之後，湛江建港，逐步實施，其經濟建設之優越地位，當可媲美香港，而成爲華南一重要港口，可無疑義也。

SPECIAL REPORT

A BRIEF REPORT ON THE DESIGN OF THE LEUNG KAM SHAN IRRIGATION PROJECT.

Leung Kin Hing

The author of this report, Mr. K. H. Leung, professor of the Engineering College, is an expert in hydraulic engineering. He was appointed the consulting engineer to supervise the designing work of the Leung Kam Shan Irrigation Project. This article was written as a report of his achievements there.

The whole project was initiated last November and is expected to be completed by the end of the coming March. Its completion will make possible the irrigation of 10,000 mows of arid land and will help to improve the crop production of the district. (The Editor)

I. Introduction.

With the present food shortage problem in Kwangtung Province, the United Nations Relief and Rehabilitation Administration (UNRRA) had brought in huge amount of foods such as flour, rice, oat meal etc., from food surplus countries such as Siam, Canada, the United States of America and others. Through China Relief and Rehabilitation Administration (CNRRA), Kwangtung International Relief Committee (KIRC), and other social affairs organisations these foods were distributed to the badly needed as a direct relief.

Besides those measures, UNRRA advised repair of river dikes, which were in a very bad shape due to war conditions. One of these works was shown in the reconstruction of the Tsin Yuen District dike which was completed at the end of 1946, under the supervision of KIRC with the help of CNRRA, Pearl River Water Conservancy Commission and the local authorities. That project was a direct work relief one. It brought fruit within a year because the 1947 flood had not done much damages to that area.

In order to solve the problem basically, increase of food production is the right answer. Irrigation, that is, proper control and distribution of water to the rice fields is one of the many solutions. Again through KIRC a typical project was planned and carried at the Leung Kam Shan Irrigation system at Hoi Ping Hsien.

II. General Survey.

Hoi Ping Hsien is one of the fourth Hsiens composing Sz Yip, the main native villages where overseas Chinese come from.

The valley to be irrigated by the Leung Kam Shan Project is owned approximately by 90 villages. Of the 90 villages about 70 belong to the Cheung Clan which form the "Soi Kon" Heung and the other 20 villages are divided up among six clans such as the Yu, Hui, Wong, Fung, Tsiah and Tam, which form the "Six Clan Heung" with a total population of approximately 40,000. That valley comprises approximately an area of 50,000 mows, about 40% of that area consists of high lands. The high lands are either non-cultivated or abandoned field. The low lands yield two crops of rice a year, and about 10% of these low lands yield an extra winter crop of garlics which are the main resources for the tenant farmers. Thus the limited amount of land available to the farmer averages to 1.2 per person, which means rice has to be imported to that area every year.

At present there are 20% of the local populace without employment or land on which to grow food to supply their needs.

The object of the irrigation project is to furnish water for the high lands for at least one crop of rice and also water for the low land for the winter crop of garlics. But due to the limit of the reservoir capacity, only approximately 10,000 mows of lands may benefit from the project.

III. Technical Problems.

The reservoir site with its watershed, as well as the valley to be irrigated was surveyed by the No. 6 Surveying Party of the Bureau of Reconstruction of the Kwangtung Provincial Government. The leader was Mr. S. T. Ting. The catchment area for that reservoir is two square miles with a capacity of 12,355,000 cubic feet. Mr. K. C. Lau, engineer of the said Bureau with consultation of the author made a preliminary plan for the project. The plan consists of an earth dam with core wall, and a culvert as an outlet and a spillway to take care of the flood flow. Since there were no drilling equipments available, only six open pits were made. It was found that a layer of limestone was laid at the bottom of the reservoir at a depth of 6 to 8 feet. The earth on the top of that layer of stone is identified as sandy loam with embéded gravels and boulders. The actual hydraulic gradient for water flowing through that material was found with two different open pits, again, it comes out between 1 to 1 to 1 to 2. The discrepancy of the results was due to the embedded gravels. The test consists of filling water into a 3 × 3 feet pit (about 6 feet deep) and measuring the drop of the water line at another brench 4 × 6 feet distant from the pit. The spillway to take care of the flood flow is to be constructed at "Yung Yick" the little elevated ground at the south-east side of the dam.

The earth dam was designed to have a height of 46 feet from the original ground level and 520 feet long on top with upstream slope of 1 to 3 and downstream slope of 1 to 2, top bank of 13 feet and concrete core wall. The upstream slope is protected from the action of the waves by 12 inches stone rip rap and the downstream slope will be sodded with grass. The free board is 8 feet. The total earth filling will be 1,200,000 cu. feet.

The core wall is of the reinforced concrete diaphram type, it intersects the upstream face at the highest water level and projects 3 ft. above and thus taking care of the wave action and protecting the embankment.

The upper 20 ft. of the core wall is 12″ thick, the lower portion is 18″. The core wall is reinforced near both sides with 3/8″ round bars space 19″ center to center both ways. Vertical expansion joints are placed 150 feet apart, painted with asphaltum.

A 3″ × 3′ × 235′ reinforced concrete culvert will be built near the center of the dam and at the base of the dam, which will supply water to two main irrigation canals.

A reinforced concrete control tower 4 feet in diameter and 45 feet will be built just next to the core wall at the upstream face, and that tower leads to 2-12″ steel gate valves of the retaining type' which regulate the flow of the water for irrigation through the culvert.

A concrete spillway, consisting of plain concrete slab embedded to hard pan, has a width of 120 feet wide, 1 weir, 3 aprous, 2 slope revetment trench, 1 back water pool to make a hydraulic jump which is necessary to prevent scouring due to the chute of water from flood flow.

At present a model study of that spillway is built at Lingnan University Hydraulic Laboratory in order to study the effect of scouring at different regions of the spillway.

The model scale is 1 to 20. The study will be made by Messrs. H. Y. Chiu, H. K. Wong and W. Y. Lui senior students of the Sciences and Engineering College as their thesis works.

IV. Local Problems.

The total cost of the project will amount to near $150,000 U.S. CNRRA had allocated to that project 164.1 tons of rice. The American Aids for China contributed nearly $3,000 U.S. and a special fund from Canadian Relief funds of $10,000 U.S. was donated to that project, local contributions were collected which amounted to $1,000 U.S. With the shortage of funds the project shall be divided into two stages of construction. Relief labor has to be recruited among the villages. The funds for the second stage of construction has to be collected from local people because once the project shows benifits to their lands.

Because the location of the reservoir is right in a local graves yard which is famous for its good "Foon Shui," a superstition practiced in China for more than thousand years, that is, if one buries his ancesters in a place known as a good "Foon Shui" terrain, he and his descendant will enjoy prosperity, and happiness. There were nearly 5,000 to 8,000 graves in that yard. The removal of these will cause trouble with the owners for that means they have to sacrifice their present conditions of prosperity to a certain period of times until they find another terrain as good or even better "Foon Shui" than the present one.

Because the dam site is located at a place not near to any highway so that materials could not be transported to the the dam site with motorised vehicles, hence a temporary highway has to be built to join the Sha Shui highway. That temporary highway requires some lands and those lands are at present under cultivation. The owners are not willing to give them up temporary, so social workers have to be sent down to show them the necessity for such sacrifice; it is a hard job for these workers because it is difficult to put up the vision of the future while they have to have "back luck" for the present.

If once the project is completed, it will be another hard job for the administration to water the different fields since most of the fields owners are small farmers, and at present a Commission is formed for that purpose. That Commission has representatives from all villages of that valley, and also appointed a few outside well known authorities to settle a compromise in case there will be dispute. Thus protecting and safeguarding in such a manner that the interests of the project do not fall into the hands and manupulation of the few privileged.

V. Conclusion.

The project, once functioned, will give a big example for other localities to start one of their own in order to solve their food shortage problems.

It also shows to the villagers that sciences in irrigation is worth to put more interest than their "Foon Shui."

Moreover, an international understanding can be put in effect even in such a remote area if that project is completed and also local dispute can be settled peacefully, thus preventing the clan fights which happened quite frequently some 50 years ago.

MEASUREMENT OF PHASE ANGLE BETWEEN FUNDAMENTAL COMPONENTS OF TWO NON-SINUSOIDAL PERIODIC WAVES.

by

Ping-Chuan Feng

I. **Introduction:** In many engineering problems, it is desired to measure the phase angle between the fundamental components of two periodic waves which are not necessarily sinusoidal. This kind of problem occurs very frequently in electrical engineering. This paper presents a simple method by which this phase angle can be directly measured on the screen of a cathode ray oscilloscope. If the two waves are electrical, the two voltages are applied either directly or indirectly, through a voltage divider, to the two pairs of deflecting plates of the cathode ray tube. If the waves are not electrical in nature, they can be translated into electrical voltages by means of electro-mechanical or electro-accoustical devices such as piezo-electric crystals, photo-cells, microphones, together with their auxiliary apparatus such as input circuits and amplifiers. It will be shown in the following that there is a definite relationship between this phase angle and the area enclosed by the open loop or loops of the Lissajous figure so formed on the screen of the oscilloscope. From this area, the phase angle can be readily measured.

II. **Relation between the Phase Angle and the Enclosed Area.**

Let us consider the general case in which neither one of the two waves is sinusoidal. Expressing them in the form of Fouriers series, we have,

$$x = X_0 + \sum_{n=1}^{\infty} X_n \cos(n\omega t - \alpha_n) \quad \text{.........................} \quad (1)$$

and

$$y = Y_0 + \Sigma\, Y_n \cos(n\omega t - \beta_n) \quad \text{.........................} \quad (2)$$

where X_0 and Y_0 are the constant terms, X_n and Y_n are amplitudes of the nth harmonic, and α_n and β_n the phase angles. If we consider the fundamental component of x as our reference, then $\alpha_1 = 0$ in equation (1). Let us assume that the Lissajous figure shown on the screen has the form shown in fig. (1) which has two loop-areas A_1 and A_2. Let the cross point P in fig. (1) be defined by,

$\omega t = \theta$ in the first half cycle

and $\omega t = \phi$ in the second half cycle.

25538

Note that at point O which corresponds to the instant $\omega t = 0$, x or y does not necessarily have a maximum value. The area of the loop is equal to,

$$\oint y dx = \oint y \cdot \frac{dx}{dt} dt \qquad \text{.........................} \qquad (3)$$

The areas A_1 and A_2 are given by,

$$A_1 = \int_{\omega t=0}^{\theta} y dx + \int_{\omega t=\phi}^{2\pi} y dx \quad ; \quad A_2 = \int_{\omega t=0}^{\pi} y dx + \int_{\omega t=\pi}^{\phi} y dx \qquad (4)$$

The total net area A is equal to,

$$A = A_1 + A_2 = \int_0^{2\pi} y \frac{dx}{dt} dt \qquad \text{....................} \qquad (5)$$

Differentiating equation (1) with respect to time and substituting in equation (5), we get,

$$A = -\int_0^{2\pi} Y_0 \sum_{n=1}^{\infty} n X_n \sin(n\omega t + \alpha_n) \, d\omega t$$

$$= -\int_0^{2\pi} \sum_{m=1}^{\infty} \sum_{n=1}^{\infty} n X_n Y_n \sin(n\omega t + \alpha_n) \cos(m\omega t + \beta_m) \, d\omega t$$

$$= -\int_0^{2\pi} \sum_{n=1}^{\infty} n X_n Y_n \sin(n\omega t + \alpha_n) \cos(n\omega t + \beta_n) \, d\omega t$$

Fig (1)

where $n = 1, 2, 3, 4, \ldots\ldots\ldots$ (6)

and $m = 1, 2, 3, 4, \ldots\ldots\ldots$ with $m \neq n$.

Performing the integration in equation (6) and taking the proper limits, it is obvious that the first and the second terms in equation (6) vanish, giving,

$$A = \sum_{n=1}^{\infty} n \pi X_n Y_n \sin(\beta_n - \alpha_n) \qquad \text{........................} \qquad (7)$$

Equation (7) is a very important result. It gives an expression for the total net area enclosed by the loops in terms of the amplitudes of the voltages and the phase angles. Although we used a double-loop figure in deriving equation (7), it is evident that the result we obtained is quite general and is applicable to any shape of path of operation.

III. Application of Equation (7) in an Electron-tube Amplifier.

As an illustration in a rather special case, let us consider a vacuum-tube amplifier where the input voltage is sinusoidal. Here, X corresponds to the grid voltage and Y the plate voltage. Under this condition, equation (7) becomes,

$$A = \pi X_1 Y_1 \sin \beta_1 \quad \ldots\ldots\ldots\ldots\ldots\ldots \quad (8)$$

We see from equation (8) that if $\beta = 0$ or π, the area A equals zero. And conversely, if the area A is zero, the phase angle must be either zero or 180 degrees. There are three possible cases:

Case 1: Path of operation is a straight line. This means not only that $A = 0$, but also that $A_1 = A_2 = 0$. The expressions for A_1 and A_2 may be obtained from equation (4) where θ now equals $\pi/_2$, and ϕ equals $3\pi/_2$. Substituting these values in equation (4) and integrating, we get,

$$\left.\begin{aligned} A_1 &= -\sum_{n=2}^{\infty} X_1 Y_n \left[\frac{\sin(n+1)\pi/_2}{n+1} - \frac{\sin(n-1)\pi/_2}{n-1}\right] \sin\beta_n + (X_1 Y_1 \sin\beta_1)\cdot \pi/_2 \\ A_2 &= +\sum_{n=2}^{\infty} X_1 Y_n \left[\frac{\sin(n+1)\pi/_2}{n+1} - \frac{\sin(n-1)\pi/_2}{n-1}\right] \sin\beta_n + (X_1 Y_1 \sin\beta_1)\cdot \pi/_2 \end{aligned}\right\} \quad (9)$$

It can be seen from equation (9) that if $A_1 = A_2 = 0$, either

$$\beta_1 = \beta_2 = \ldots\ldots\ldots = \beta_n = 0 \text{ or } \pi$$

or

$$\beta_1 = 0 \text{ or } \pi \text{ and } Y_2 = Y_3 = \ldots\ldots\ldots = 0$$

It is obvious that the first condition ($\beta_1 = \beta_2 = \ldots = 0$) is never realized in practice. We can conclude, therefore, that $A_1 = A_2 = 0$ is a necessary and sufficient condition for zero harmonic output and zero phase shift.

Case 2: Path of operation is a symmetrical figure of eight. In this case, $A = 0$ but $A_1 = A_2 \neq 0$. A study of equation (9) shows that the above conditions can be satisfied if,

$$\beta_1 = \pi \text{ and } Y_n \neq 0 \quad \ldots\ldots\ldots\ldots\ldots\ldots$$

Equation (9) may be rewritten as,

$$|A_1| = |A_2| = \sum_{n=2}^{\infty} X_1 Y_n \left[\frac{\sin(n+1)\pi/_2}{n+1} - \frac{\sin(n-1)\pi/_2}{n-1}\right] \sin\beta_n \quad (10)$$

If we neglect the harmonics higher than the second, the above equation reduces to,

$$|A_1| = |A_2| = \frac{4}{3} X_1 Y_2 \sin \beta_2 \quad \text{.........................} \quad (11)$$

It will be recalled from circuit theory that when the load circuit is tuned to resonance for the fundamental frequency, it offers almost a pure capacitive load to the higher harmonics. This means that sin β_2 is nearly equal to unity, and the above equation can be used as a rough check of the amount of the harmonic content.

Case 3: Path of operation is an open or an unsymmetrical figure. This is an indication that the circuit is not in tune, for it is seen in equation (9) that this happens only when $\beta_1 \neq 0$ or π.

IV. Application of Equation (7) in an Electron-tube Oscillator.

It has been shown previously that when neither the plate nor the grid voltage is sinusoidal, the total area of the loops is given by,

$$A = \pi \sum_{n=1}^{\infty} n X_n Y_n \sin(\beta_n - \alpha_n) \quad (7)$$

For the sake of simplicity, let us consider only the first two terms in equation (7), neglecting the effect of the higher harmonics. We have then,

$$A = \pi [X_1 Y_1 \sin \beta_1 + 2 X_2 Y_2 \sin(\beta_2 - \alpha_2)] \quad (12)$$

Imagine that the circuit of an oscillator has been so adjusted that the area A is zero. We want to know what information can be derived from equation (12). There are three possible ways to make A equal to zero in equation (12):

(1) $\beta_1 = 0$ or π, and $(\beta_2 - \alpha_2) = 0$ or π

(2) $\beta_1 \neq 0$ or π, and $(\beta_2 - \alpha_2) \neq 0$ or π, but $X_1 Y_1 \sin \beta_1 = -X_2 Y_2 \sin(\beta_2 - \alpha_2)$

(3) $\beta_1 = 0$ or π, and X_2 or $Y_2 = 0$

The first two are only mathematically possible and can not be realized physically. For if the fundamental components of the plate and grid voltages are 180 degrees out of phase, the second harmonic voltages, in general, will not have the same phase difference. It can also be easily shown that, in general, sin $(\beta_1 - \alpha_1)$ and sin $(\beta_2 - \alpha_2)$ will always have the same sign, that is, the angles $(\beta_1 - \alpha_1)$ and $(\beta_2 - \alpha_2)$ are in the same quadrant. Therefore the first two possibilities are ruled out leaving only the last. We can conclude therefore that if the condition $A = 0$ is satisfied, we know that first, the phase angle must be 180 degrees, and secondly, the plate or grid voltage must be free from harmonics. This is the best adjustment for an oscillator.

V. Measurement of Phase Angle from Cathode-ray Oscillograms.

If one of the voltages X and Y is sinusoidal, equation (7) reduces to,

$$A = \pi X_1 Y_1 \sin \beta_1$$

from which we may write,

$$\beta_1 = \sin^{-1} \frac{A}{\pi X_1 Y_1} \qquad \text{(13)}$$

By applying X and Y to the vertical and horizontal deflecting plates of a cathode-ray oscilloscope through proper high frequency voltage dividers, one can observe the path of operation on the screen of the oscilloscope and measure the quantities X, Y, and A. The area A is measured by a planimeter. The justification of the use of equation (13) depends upon the condition that either x or y is practically sinusoidal. This can be roughly checked by measuring the positive and negative peaks of the loop and determining if the amplitudes on both sides of the axis are the same. The percentage distortion can be easily shown to be given approximately,

$$\%\ \text{harmonic content} = \frac{\frac{1}{2}(D_a + D_b) - D_a}{\frac{1}{2}(D_a + D_b)} \times 100 \qquad \text{(14)}$$

where D_a is the maximum deflection above the axis and D_b is the maximum deflection below the axis.

The accuracy of equation (13) depends on how small is the harmonic content. If either x or y is free from harmonics, equation (13) is exact. If both x and y are non-sinusoidal, equation (13) is only approximate. Let us consider the effect of the second harmonic. Assuming, for example, that the second harmonic distortion is 3% in both the grid and plate voltages, and assuming furthur that the phase angle has a value that is most unfavorable, that is, when $\sin(\beta_2 - \alpha_2) = 1$, then the expression for A in this case is, from equation (12),

$$A = \pi X_1 Y_1 \sin \beta_1 + 2\pi X_2 Y_2 \sin(\beta_2 - \alpha_2)$$

Therefore, the phase angle is given by,

$$\beta_1 = \sin^{-1}\left[\frac{A}{\pi X_1 Y_1} - 0.0018\right]$$

For β_1 equal to about 2 degrees, the effect of the correction term is about 1/10 of a degree. For β_1 equal to about one degree, the correction term is about 5 minutes. It is obvious that for large values of β_1, the correction term has little effect. It follows from the above example that provided the harmonic content is not too large, equation (13) gives a very simple way of measuring the phase angle.

The author has used this method in measuring the phase angle between the grid and plate voltages of power oscillators and found it very convenient. Undoubtedly, it can be used for many other purposes.

COMMENTS

METERS vs YARDS.

Rev. Joseph A Hahn, M.M.

During my high school days, I recall how my teachers extolled the metric system and advocated its prescribed use for the whole world. "There must be something to it," I thought, and marvelled over the simplicity made possible by decimals throughout. Then I experienced the confusion resulting from the difference between a gram of mass and a gram of weight, and when I began to realize that the length of the meter was very arbitrary and depended upon the length of a certain piece of metal I began to lose interest.

Why did they choose such a length to be a meter? Why not pick out a length so that the acceleration due to gravity at a certain point on the earth's surface would be exactly 1000 cm per second per second? The gram would be slightly smaller, to be sure, but at least a lot of multiplying and dividing would be saved. If the choice was for an arbitrary length, why did they not make it equal to the English yard?

And what particular advantage has the Centigrade scale over the Fahrenheit? Why not base a scale on absolute temperatures and make the melting point of ice, say 100 degrees? You would avoid the confusion arising from plus and minus temperatures. That there are exactly 100 degrees between freezing and boiling of water under standard conditions is of no particular advantage.

"The metric system saves a lot of computation" say some of its proponents. This is nonsense. If you are working with centimeters you are unlikely to use kilometers. If you are working with millimicrons, you are not interested in centimeters. The confusion arising from the use of these Greek combinations is a result of oversimplification. If you are working with pounds you can use decimals just as easily as in the metric system, using tenths, hundredths, thousands, etc. of pounds. If you are working in ounces you can do likewise. Similarly for tons. What advantage is gained by being able to convert millimicrons into parts of a kilometer by moving the decimal point? You hardly ever have occasion to do it. In measuring distances, for example, how is the use of a figure like 75.483 kilometers any more advantageous than a figure like 75.483 miles. Aside from the basic unit there is no difference.

Now for something practical. Since a very great part of engineered imports throughout the world come from factories that use the English system, why saddle the distribution of these items by demanding their conversion into metrical equivalents? If engineers are taught from textbooks using the English system of units, why should they be forced to convert all their data into an unwieldy system which pretends to be superior?

If we are going to force the world to use one standard system, let us make it a good system to start with.

First of all change the arabic method of numbers. It is more natural to divide in two, and so the English system has the advantage of using halves, quarters and eighths. But for greater precision it is better to use decimals. Why not use twelve unit digits instead of ten so that numbers ten and eleven will be figures of but one digit. Then "one" "zero" (10) will really be twelve, and if you wanted to divide this new "10" by three you would get four. It will be twice factorable by two, and the need of an infinite decimal for numbers divided by three would disappear.

Secondly, choose a suitable standard for the meter, based on something which cannot be destroyed, such as a wavelength of light, acceleration due to gravity, etc. In effect this is done now since such things are exactly related to the present meter, but let us make the length of the meter something which will really save arithmetic.

Thirdly, while the gram and calorie will follow the change of the meter, change the temperature scale to start with absolute zero, so that the melting point of ice will be at some easy figure like 100 or 200 degrees.

In conclusion, I might point out that none of these suggestions will ever be taken up seriously, because we have become so used to the metric system. Therefore, metric system advocates ought not impose their system upon those people who have become so used to the English system.

ENGINEERING. ITS STUDY AND APPLICATION

by

Donald R. Bowman

(Mr. Donald R. Bowman is the first exchange student from the United States in our Department since the conclusion of the war. He joined our Department in Sept., 1947. The justification of the statements which he made in this article concerning conditions in China is left to the reader.—The Editor.)

Since the war, engineering has risen to be the most sought after of all college courses. Previously more students were enrolled in the School of Arts & Sciences, but the last year showed the enrollment in engineering to be the highest. In the United States there is, as there has always been, much discussion as to the best method for the education of engineers. I will, therefore, try as my first aim to list, as impartially as possible, the different methods used and experimented with in the United States as well as those I have observed in China. Secondly, I will try to show how this education has been used in the past and how it can be used in the future.

I find that there is very little material on this subject at hand, so this article will consist largely of my own opinions and observations. Few statistical books of this type are available, since the war in China has prevented the compiling of such statistics, and also the acquiring of books on this field of education in the United States. I want, therefore, to excuse myself from any misquoting or statement of disputable facts. I will try to refrain from giving what information I have as fact.

There has always been much controversy in the United States about the degree of specialization that should be taught to undergraduate engineers. This takes in both cultural education to enable the engineer to get along better in life and practical education to enable him to know exactly how the product he designs or builds is put together, how the workers have to operate, and how the product will be used.

A great many students in the United States, having been born in cities where they have everything provided, learn very little about the making of the articles they use. In China, mechanical devices are not widely used, so the average Chinese student probably knows even less about the mechanics of simple machines. This lacks of knowledges is partly compensated for by laboratory classes which are taught in connection with courses, but there is still the production shop that the students have little opportunity to learn about. In many colleges in the United States "shop courses" are required for all engineers. In these courses the student learns how to operate lathes, milling machines, steel planers, turret lathes, and a variety of tools used in the industrial shops and factories of today. They also learn the many different methods of welding, forging, heat treating, and the effect of alloying elements on different metals. All these things can be studied in books, but in doing the various operations with one's hands one can come to understand the problems of the worker who, to manufacture the product you desire, must work from the blue prints which you have drawn. Many of these have little effect on the civil engineer. However, in the forging and metal testing courses, the civil engineer learns by doing and understands why he must use a certain kind of steel for reinforcement on one job and a different variety on another. He learns such things as why one cable will "wear out" and render a bridge unsafe within a year while another of identical composition and properties will support the bridge for eternity. The difference lies only in the method of prefabrication of the steel that goes into the cables. Many colleges which did not offer such courses in the past are now installing them because the factory managers more readily hire engineers who have been well trained in their field and understand the problems of the workers.

Many an engineer, a few years after he gets out of school, finds himself in an executive position. He then wishes he had taken less engineering and more business and economic courses. Due to this, many colleges now require a certain number of courses along the business and economic line. Thus, if an engineer has a chance for a good managerial job, he has the education background to step in and take over with the assurance of being able to do his job well. Very few people in the engineering field or in any other field go through life without at some time having to resort to the law in order to retain their rights. Thus, an elementary course in law is added to those who considered desirable to a well-rounded engineering education.

One does not intend to spend all of his time at his job. He, therefore, should learn something of the so called "finer things" of life. A thorough knowledge of the language of his country is necessary, and it would be to his advantage to know something of the language of others. It is often very beneficial to know something of the history of the world as well. These courses have in many colleges in the United States been combined into what are called Man in the Social World, Man in the Cultural World, etc. In these courses, a little of various arts, histories, and geographies of all the world is taught. Many students dislike these courses, and they are still somewhat in the experimental stage, but they are the present solution to the cultural education of engineering students without taking too much of their time away from the subjects in their fields. Many schools formerly offering courses that were wholly technical are now realizing the importance of a liberal education and are adding such courses to their curriculum. Massachusetts Institute of Technology, recognized as one of the leading schools in the United States for technical education, has recently installed liberal arts courses along with its technical courses. Its educational leaders have found that students are much better prepared for life if they have a working knowledge of other fields as well as a mastery of their own.

The practical application of technical courses is also coming to be used more. At the University of Cincinnatia approximately half of the time is spent in class rooms, and the other half is spent working in the factories of the surrounding city. The students work on the actual production line as well as in the drawing rooms, the testing divisions, and the administrative offices. Thus, they learn to use their knowledge as they acquire it. This is a relatively new system in the United States, and I do not know of its being practiced in any other country. It will be an interesting observation to note its progress.

There are, of course, many trade schools where the entire education is obtained by just doing the actual work, but in schools such as this the students learn only the industrial processes and obtain no foundation for their knowledge. By applying some of the tradeschool methods to colleges, the students are taught how the theories that they study are applied. Thus, they will not have to take the time when they get out of school to learn to apply the many formulas to the actual work. This would be hard to do in China where industrialization has made little progress, but it is most important since they do not have the experience of others to draw upon.

The amount of theory teaching as compared with practical teaching is necessarily limited. Those who want to teach or enter into work in research need a complete theoretical knowledge of all courses, but the average engineer does not. On the other hand, every engineer needs some theoretical foundation or he cannot understand the principles with which he works. Each can borrow from the other and benefit.

In the present-day shrinking-world the ways of foreign countries are coming to be recognized as another source of information. I would say that there is no country in the world whose people do not regard themselves at least as good as the people of any other country of the world, if not better. In the past when transportation and communication was slow compared with what it is now, it was more difficult to study the engineering achievements of other countries. Also, people did not feel close to their neighbors and, therefore, were not concerned with what they did or how they did it. Now this situation is greatly changed and is changing more every day. People are coming to realize that their foreign neighbors are just as intelligent as they are themselves and in many cases are learning the horrible truth that they are even more intelligent. This has brought about an exchange of students between countries which, discontinued by the war, is being renewed now with hopes of allowing more students to study in foreign countries. This will bring about better ideas in all countries concerned. Many minds are better than one. In cooperation, untold progress can be brought about.

Engineering in China is young in comparison with other industrial countries. Roads, dams, skyscrapers, and other symbols of modern industrialization are not yet evident. The backbone of China is her agriculture. Yet good roads, dams, and industry have their place too. With a complete network of transportation, the food products can be transported faster to where they are most needed. Dams will help to control destructive floods. Industry will manufacture the raw products so that this form of income will stay within China rather than be gained by foreign countries. Since China does not have many roads for transporting materials, the beginning of engineering projects is rather slow. Roads of any size are very expensive and take a long period of time to build. It is, therefore, possible that many engineering projects will have to be carried out by transporting materials by old methods or, in the case of fairly compact articles, by air.

In the future of mountainous China, the air will undoubtly play a very great part. Airfields can be built at a fraction of the cost of a road. Already there are many airfields in China which were built during the war by Chinese labor under the direction of United States engineers. It is quite impractical to transport the raw products of heavy industry by air, but machinery, small articles, and other compact things can now be transported by air almost as cheaply as by roads, even in countries where roads are available. With the development of better planes and the possible utilization of glider trains, the field of air transport opens even wider.

The Chinese government has already inaugurated a program of building dams on many of the major rivers of China. This program would even now be under way if it were not for the civil war. Dams are not a complete answer to floods and soil erosion, but they do help. They also supply huge quantites of electrical power which could be used to many advantages in China. The Chinese farmer knows little of the use of electricity in the village, but in due time it could be utilized to give added comforts, efficiency, and more happiness to the Chinese peasant.

China ranks among the greatest potential producers of raw mineral products. It is true that in coal and iron it is not as well off as it might be, but in many other elements it leads the world. In the United States the use of aluminum has taken the place of steel in a great many cases. China has an abundance of aluminum, and by using it in every case possible, and using steel in only those places where it is not possible to use aluminum, she can probably produce enough raw products to build industries of her own with very little import of such materials from other nations. Coal is an essential in the reduction of iron ore to iron, but in the reduction of aluminum electricity is the prime factor. With a large system of dams, this basic need is taken care of. One large dam supplies huge quantities of electricity, and by building dams in the mountainous country where they are best needed, two problems are solved. Cities can be built where they won't occupy the valuable farm land; thus some of the strain of over-population can be relieved, and the utilization of valuable raw products can be realized.

While Chinese cities now have many of the benefits of the modern world, they do not realize the use to which these benefits can be put. Canton has an electrical power system, but it cannot be compared with the power systems of cities having its size in the United States. Good engineering could greatly improve it, and education of the people would bring about its more effective use. With the advent of further modernization, the fertilizer now used on the farms could be first processed. It would lose none of its utility as a plant food, and the spread of typhoid fever and other disease could be lessened. If China is to lead the world again as she once did, she must become concerned with the little man. The peasant who toils and dies without honor must be raised to his proper position and recognized. It is the millions of workers who build nations, not the statesmen. Engineering can play a great part in making life better and more comfortable for the Chinese peasants. The result will be less disease and death and more production and happiness.

Dams are the harness to a valuable resource found in all countries. China is especially fortunate in having much of this resource. Her rivers in the past have been her sorrow; in the future they can be her glory. Many rivers are navigable. Where dams are needed and are possible, locks can be built. Thus, use of rivers as a throughfare will be unhampered. A complete system of dams can be used to many more advantages than are evident at first. They furnish power, control floods, raise the water level in arid regions to where it can be easily used for irrigation, thus bringing more land into utilization, and hold water so that rainfall is more general.

It is true that only one dam will not control a huge river, but appropriate placement of dams according to a plan devised as a result of a complete study of the river system will give almost positive control. The use of dams is beneficial not only to countries lacking in other natural resources but are of sufficient value to be very useful even in industrialized countries that do have other means of power in quantity. Recently in the United States a plan was inaugurated to dam the entire Missouri and Mississippi River valleys. In this project the dams will number in the hundreds and the cost will be figured in billions, but the profit counted in saving crop land and the harnessing of power dwarfs the original cost. However, it would probably not be advisable to start such a project unless it is fully intended to finish it, for too few dams might well develop into a hazard instead of a help.

It is held be some in the United States that airplanes will soon replace the automobile. There are more, however, who disagree. In a country where there is already a fairly complete system of roads it is indeed doubtful if airplanes will even offer very serious competition in the very near future. It is far more practical to use a car or truck in the United States for almost any distance less than one hundred miles and in many cases for even longer distances. However, in China it often takes weeks to do one hundred miles even by the fastest means possible. By air, it takes only a few minutes. Air fields are expensive, but their cost in comparison to roads is very meager indeed. Also, air fields cover but a fraction of the land taken up by roads; thus, the valuable farm land is left to be used for the growing of food. It is true that for bulky articles air transport is expensive, but it is becoming cheaper as science discovers better engines, better wing design, and more efficient loading factors. Thus, by the time of a sufficient system of roads could be built, it is possible that air transport would be cheapest. Meanwhile, the initial cost is much less.

There must necessarily be some road building even in the advent of complete utilization of the airplane. Transport of raw materials to the refinery would be much better accomplished by the building of railroads. The transport of processed raw materials to the factory where they would be made into consumer goods would in many cases be more efficient if railroads or motor roads were built. In many places where the terrain is unfavourable for the building of roads, the airplane could still be utilized. Inter-village communications would also never be able to utilize the airplane efficiently. Village communication would not in most cases require the transport facilities of large cities. Goods could be flown to central cities, and from there systems of roads could spread out to serve the surrounding communities. In many cases with the erection of telephone or, better still, radio communication, the present village paths would suffice even in the modern industrial world. These forms of communication would come into use in the case of emergency, and the products that the villagers have for market could be brought into central points in the same way that they have been brought in the past. There would still be the airplane to drop supplies in the case of epidemic or other serious disaster. During the war a fairly efficient system of pickup was worked out, and where the need called for this system could be utilized in China. The Government could thus use these methods to insure the people against death from starvation, disease, and other natural phenomena which often occur without any forewarning. A case of this type arose only last year in the United States. A great portion of Colorado was cut off from communication by a snow storm. Supplies were flown in not only to save the lives of the starving and freezing people but to save livestock on the ranges as well. This valuable livestock was thus saved to feed the starvation of the world in the year to come. If the Government of China could guarantee such help to her people at all times it would undoubtly secure much greater allegiance.

Along with the advent of engineering achievements must necessarily come an educational program that will instruct the people as to their uses. People will gain no benefit from articles they do not know how to use. Already many of the things found in China which have been imported from foreign countries are not being utilized to the fullest. China is one of the greatest food-growing potentialities in the world. If she can put them to use to feed her overpopulated country she can rise to be one of the leading powers of the world. China's prolonged war of recent years can possibly be traced to economic difficulties, which, are partly due to overpopulation. She must devise a system whereby more of her people can be fed properly. Engineering can help this by utilizing more efficient production methods and by making it possible to transport food quickly to where it is most needed. Engineering will also offer more jobs to relieve the strain on present job shortage. It will not, it is true, offer enough jobs to solve the problem, and, in some cases, may even offer less jobs, but as a whole it will help the situation.

The government, too, must adopt a policy of helping the lower classes by creating more conveniences for them. The government, in its policy of promoting the establishment of industry, must also adopt a policy of following the advice of the engineer, and the engineer must live up to this trust.

連續有托梁感應線求法舉例

(Influence Lines for Continuous Haunched Beams)

●王銳鈞●

(一) 總論

感應線　當一單位豎向集中荷重經過一梁或桁架跨度時，該梁或桁架某一截面或部份之應力卽隨該單位豎向集中荷重之位置變動而改變其數值。用以表示梁或桁架某一截面或某一部份各種應力之直線或折線或曲線，稱爲該截面或該部份某種應力之感應線。感應線可用以表示某一纖維之單位應力，某截面之切力，某一截面內對重心之彎矩，支座上反力之一分力，或結構上任何一線之旋度，(rotation)。

感應線用途　感應線普通有兩種用途(一)藉感應線大致形狀用以指示求最大活荷重應力時活荷重在結構上應佔之位置。(二)用感應線以表示梁或桁架某一部份應力之值。前者祇須得其近似形狀後者則須詳計其數值。本文舉例乃屬後者。

(二) 例題

一連續式拋物線形有托洋灰梁(Haunched Beam)橋共三孔，中孔淨跨度爲10.92米，兩旁孔淨跨度爲10.98米，橋台兩端梁深爲70.5公分中部梁深爲56公分，兩橋墩處梁深爲143.00公分，求各截面彎矩及各支點切力感應線。

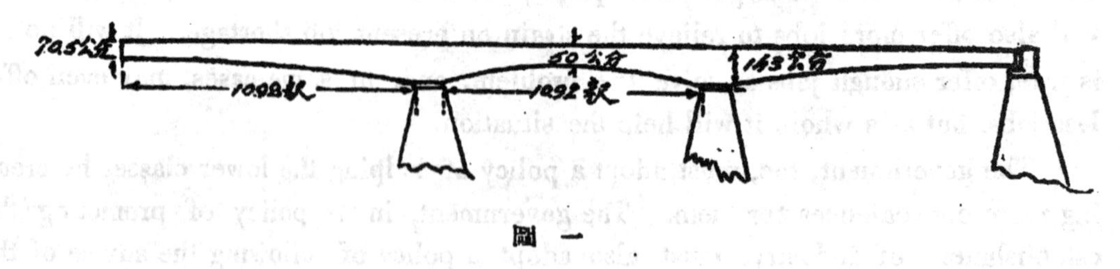

圖一

(一)　各截面彎矩感應線求法

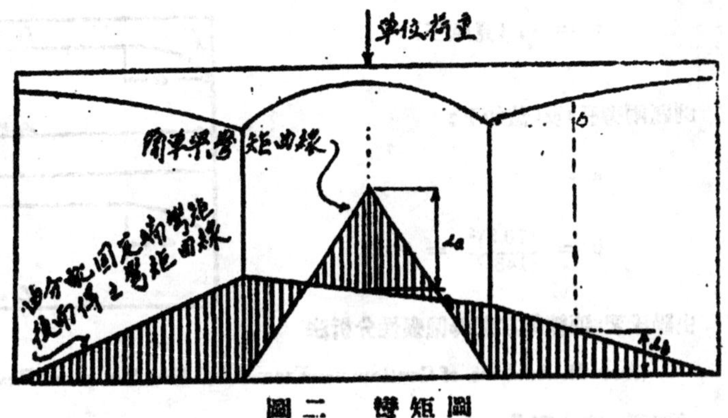

圖二　彎矩圖

應用原理說明　假設一單位集中荷重置於三孔連續梁橋 a 點上時其彎矩圖如圖二所示。

則 i_a ＝單位荷重在 a 點時，該點彎矩感應線於該點之縱座標值(Ordinate)。

i_b ＝單位荷重在 a 點時，b 點彎矩感應線於 a 點之縱座標值。

欲用此法繪一完全之感應線，須將單位荷重在各位置時之彎矩圖繪出。每一彎矩圖之各縱座標值即為單位荷重在某點時各截面彎矩感應線於該點之縱座標值。

由上述原理，感應線各值之求法，依下列步驟，似覺方便。

（甲）　繪出各跨度固定端彎矩 (Fixtd-end Moments) 感應線（見大圖 1/4）

（乙）　假設一單位未平衡彎矩 (unbalanced moment) 於各跨度之兩端，用力矩分配法 (momen- distribution) 求出各支點之彎矩。

（丙）　放置一單位荷重於一跨度任意一點上，用（甲）之感應線求出該跨度固定端彎矩。將此等固定端彎矩乘以（乙）算出之值即得各梁端最後彎矩值。

一單位荷重在某點時，任意一截面之彎矩等於該截面由簡單梁 (simple beam) 而生彎矩與由跨度兩端彎矩影響而生彎矩之和（見圖二畫影線部份）。應用此原理可繪出一單位荷重在一指定點時之彎矩曲線。

（丁）　當一單位荷重在不同各點放置時，由（丙）所得各彎矩曲線可得某一截面彎矩感應線之各值。

茲試求單位荷重在左旁孔點(6)時，中孔點(6)之彎矩感應值。

橋台兩端梁深 ＝ 70.5 公分

70.5^3 ＝ 350,402（隨性率因梁深之立方而變）

橋墩處梁深為 143 公分

143^3＝ 2,925,000

橋之中部梁深為 56 公分

56^3＝ 175, 616

假設 a 為有托部份 (Haunch) 長度與跨長之比（見圖三）

25555

$b = \frac{I}{I'}$ (見圖三)

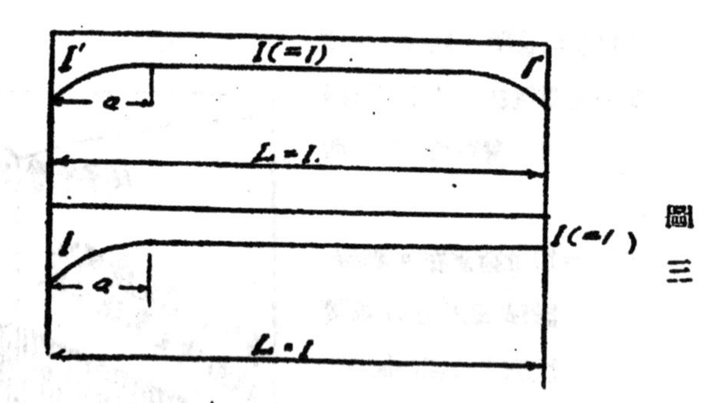

圖三

例題兩旁孔a及b值如下：

$a = 1.0$

$b = \frac{(70.5)^3}{(143)^3} = 0.12$

由羅氏著：連續架之束縛佩强性分析法

(Russell: Analysis of Continuous Frames by The Method of Restraining Stiffnesses)

第四項 表 I 可得

$K_{12} = 1.35 = K_{43}$

$K_{21} = 3.75 = K_{34}$

K 爲各端之佩强性係數

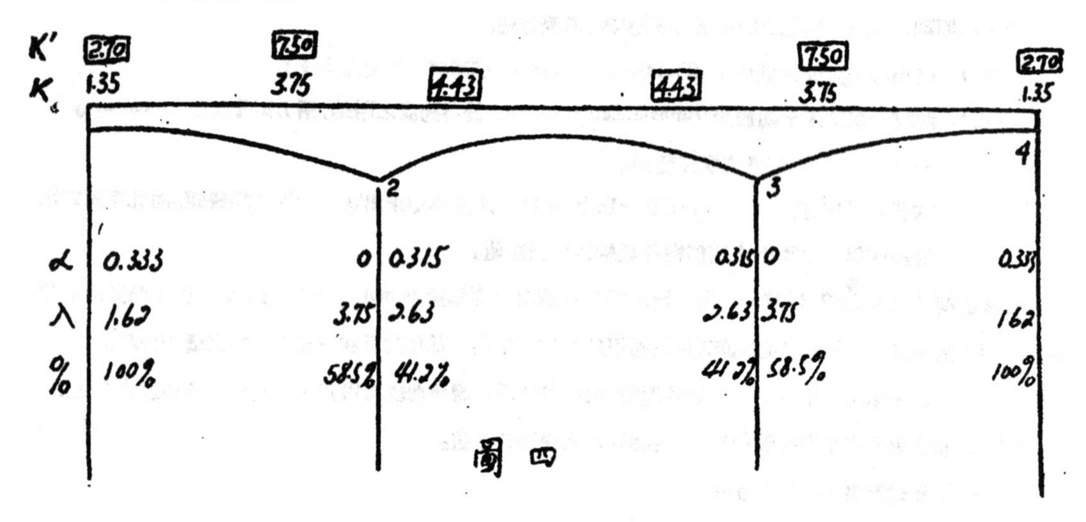

圖四

中孔 a 及 b 值如下

$a = 0.5$

$b = \frac{(56)^3}{(143)^3} = 0.06$

由第三項 表 I 可得

$K_{23} = K_{32} = 4.43$

25556

$K'_{12}\ K_{12} \times \frac{(70.5)^3}{(56)^3} = 2.70 = K'_{43}$

$K'_{21} = K_{21} \times \frac{(70.5)^3}{(56)^3} = 7.50 = K'_{34}$

有 ▭ 爲變更後相關偏強性值 (Modified Relative Stiffnesses)　(見圖四)

由精確力矩分配法 (The method of precise moment distribution)

$\lambda = \frac{K}{2-\alpha}$　　　　$\alpha = \frac{\lambda}{K+2\lambda}$

λ = 精確分配因數 (precise distribution factor)

α = 精確傳遞因數 (precise carry - over factor)

$\alpha_{21} = 0$

$\lambda_{21} = \frac{K_{21}}{2-\alpha_{21}} = \frac{7.50}{2-0} = 3.75$

$\alpha_{32} = \frac{\lambda_{21}}{K_{32}+2\lambda_{21}} = \frac{3.75}{4.43+2\times 3.75} = 0.315$

$\lambda_{32} = \frac{K_{32}}{2-\alpha_{32}} = \frac{4.43}{2-0.315} = 2.63$

$\alpha_{43} = \frac{\lambda_{32}}{K_{43}+2\lambda_{32}} = \frac{2.63}{2.70+2\times 2.63} = 0.333$

$\lambda_{43} = \frac{K_{43}}{2-\alpha_{43}} = \frac{2.70}{2-0.333} = 1.62$

$\frac{3.75}{3.75+2.63} \times 100 = 58.8\ \%$

$\frac{2.63}{3.75+2.63} \times 100 = 41.2\ \%$

假設單位集中荷重在左旁孔點〔6〕位置時，其彎矩分配如下：(固定端彎矩見大圖1/4)

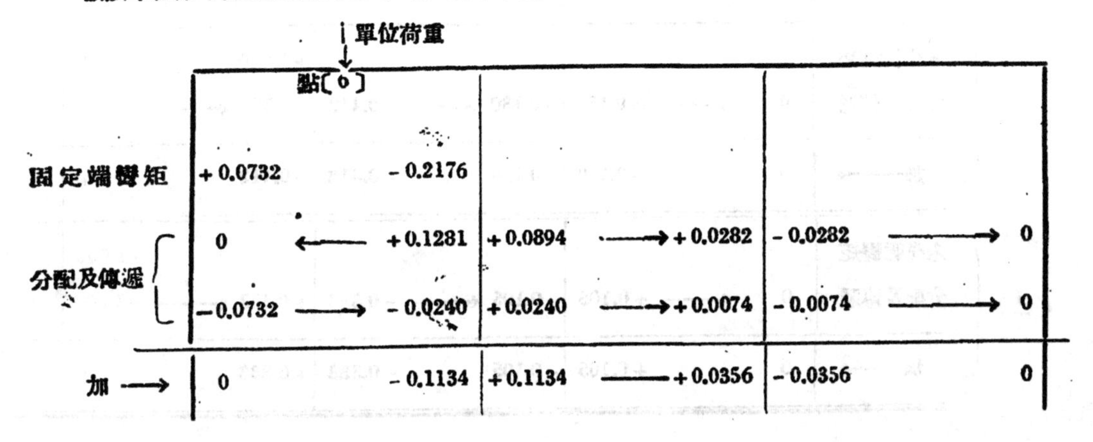

固定端彎矩	+0.0732	−0.2176				
分配及傳遞	0 ←	+0.1281	+0.0894 →	+0.0282	−0.0282 →	0
	−0.0732 →	−0.0240	+0.0240 →	+0.0074	−0.0074 →	0
加 →	0	−0.1134	+0.1134 →	+0.0356	−0.0356	0

25557

使鄰節順時針方向彎矩爲正(＋)

使鄰節反時針方向彎矩爲負(－)

單位荷重移動於橋梁上各點位置時，如用(乙)所述之方法求其彎矩分配，則尤覺迅速與簡便，其法如下：

	支點	1		2			3			4
(子)	未平衡彎矩	+1.000								
	分配及傳遞	-1.000	→	-0.333	+0.333	→	+0.105	-0.105	→	0
	加→	0		-0.333	+0.333		+0.105	-0.105		0
(丑)	未平衡彎矩			-1.000						
	分配及傳遞	0	←	+0.588	+0.412	→	+0.130	-0.130	→	0
	加→	0		-0.412	+0.412		+0.130	-0.130		0
(寅)	未平衡彎矩				+1.000					
	分配及傳遞	0	←	-0.588	-0.412	→	-0.130	+0.130	→	0
	加→	0		-0.588	+0.583		-0.130	+0.130		0
(卯)	未平衡彎矩						-1.000			
	分配及傳遞	0	←	-0.130	+0.130	←	+0.412	+0.588	→	0
	加→	0		-0.130	+0.130		-0.588	+0.588		0
(辰)	未平衡彎矩							+1.000		
	分配及傳遞	0	←	+0.130	-0.130	←	0.412	-0.588	←	0
	加→	0		+0.130	-0.130		-0.412	+0.412		0
(巳)	未平衡彎矩									-1.000
	分配及傳遞	0	←	+0.105	-0.105	←	-0.333	+0.333	←	+1.000
	加→	0		+0.105	-0.105		-0.333	+0.333		0

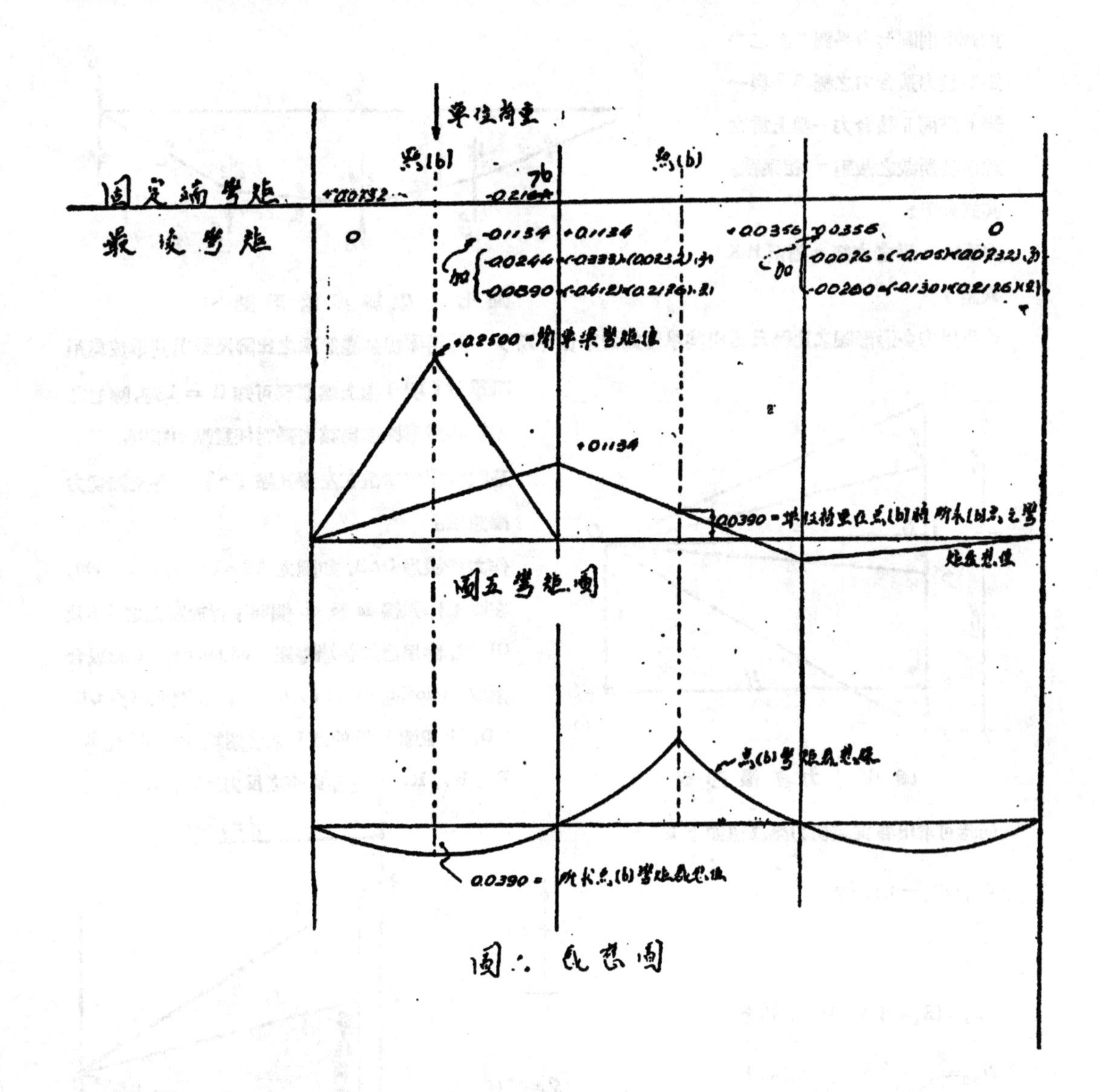

圖五　彎矩圖

圖六　感應圖

其他各點之彎矩感應線，見大圖(2/4)

（二）　各支點切力感應線求法

應用原理說明

設欲用索線多邊形求圖七力系對於 P 點之力矩，先作力多邊形如圖八再作索線多邊形如圖七則圖七力系對P點之力矩等該力系合力之極距H與一經P點而∥該合力一線上爲索線多邊所截之截距y相乘積。其式如下：

力系對P點之力矩＝極距H×截距y

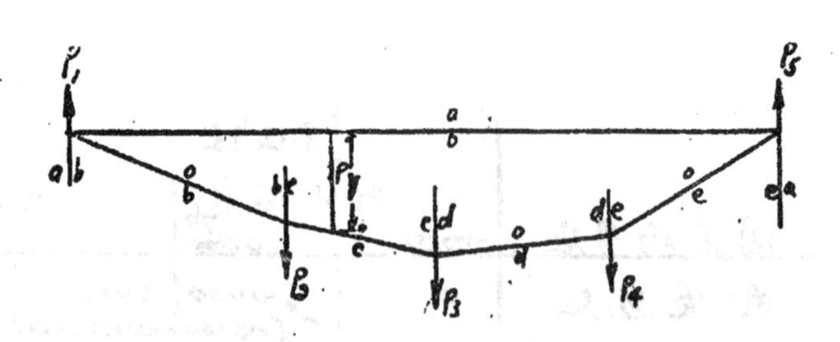

圖七　索線多邊形圖

H乃用力多邊形圖之比例尺量出其單位爲力單位（磅）y乃用索線多邊形圖之比例尺量出其單位爲距離單位（呎）由上述原理可知H＝1時，圖七之索線多邊形圖即爲該力系對任意點力矩圖。

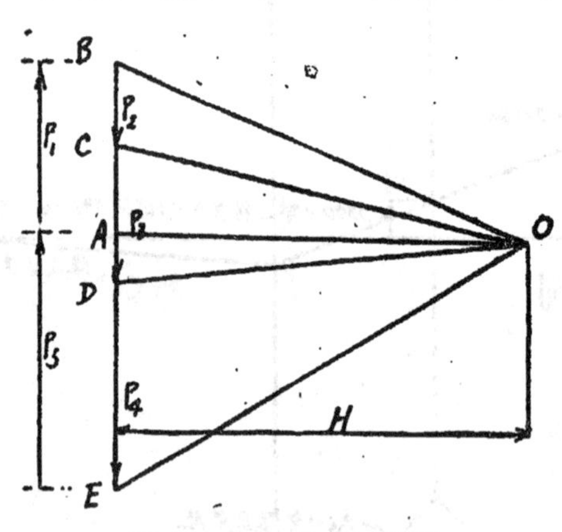

圖八　力多邊形圖

茲試求單位荷重在左旁孔點〔6〕時，各支點切力感應值。

作力多邊形OAB，如圖九AB=1.000，H=1 000，在圖十作索線aa及ob個別平行於圖九之OA及OB線。應用已知各端彎矩(end moments)繪成合閉線 (closing lines) oc, od, oe。在圖九再作OC, OD, OE個別平行於圖十之合閉線oc, od oe。R_1, R_2, R_3, R_4 即爲各支座之反力感應值。

同時可求出各支点切力感應值如下：

$V_{12}=R_1=0.390\uparrow$

$V_{21}=(R_1-1)=0.610\downarrow$

$V_{23}=(R_1+R_2)-1=0.145\uparrow$

$N_{32}=\quad 0.145\uparrow$

$V_{34}=(R_1+R_2)-(1+R_3)=0.040\downarrow$

$V_{43}=-R_4\quad 0.040\downarrow$

其他各切力感應線見大圖 4/4

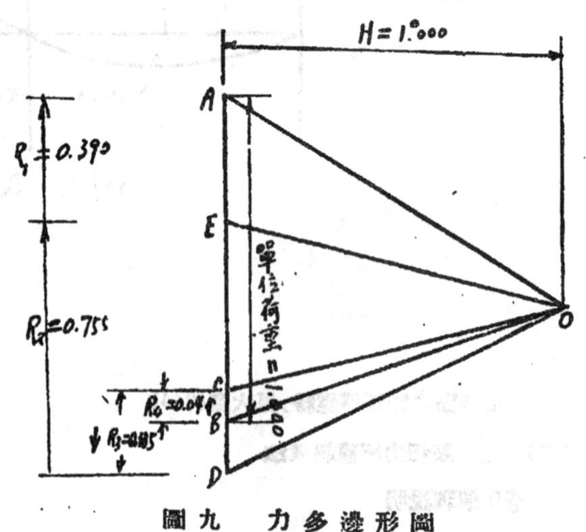

圖九　力多邊形圖

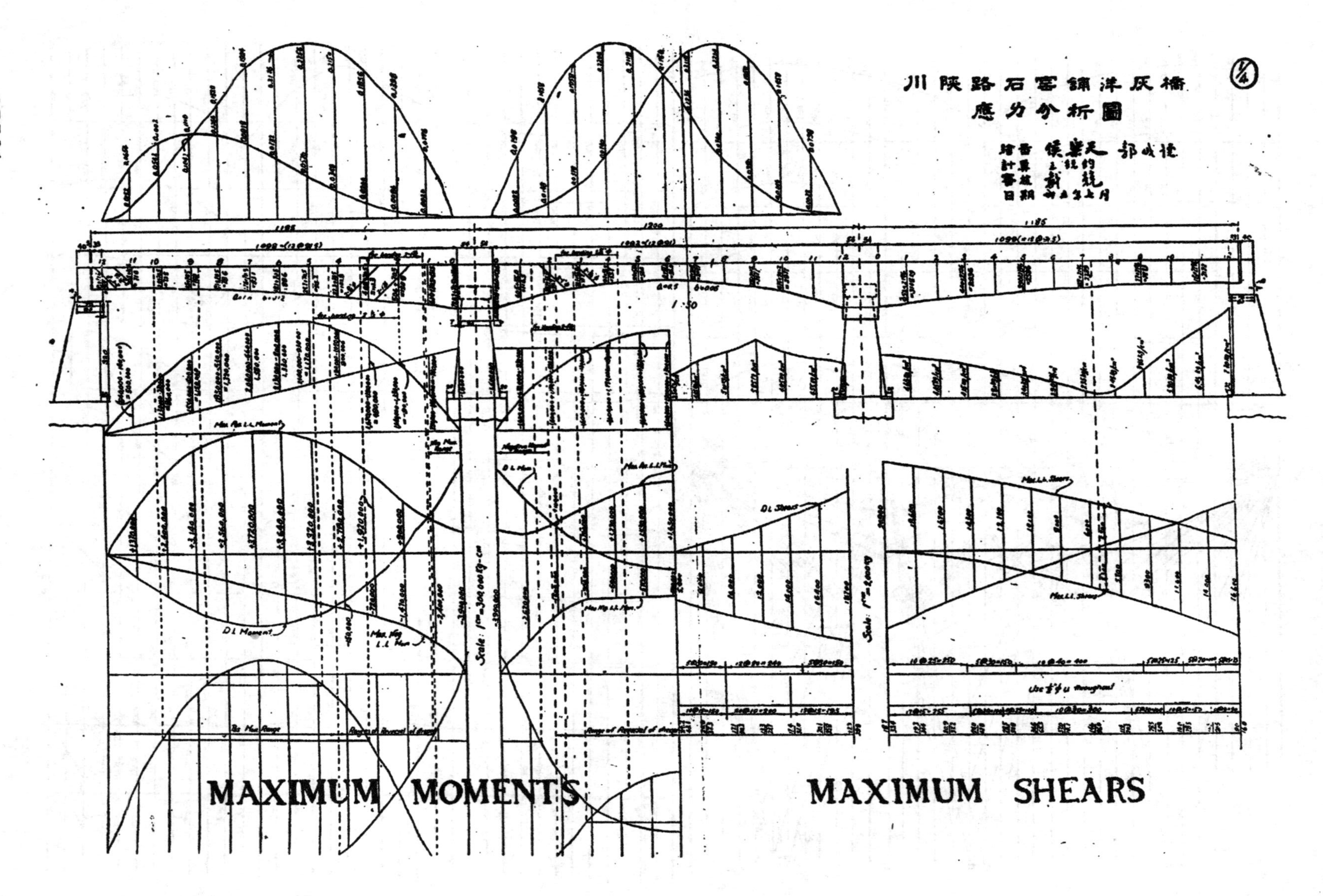
川陜路石窖鋪洋灰橋
應力分析圖
1/4
1185
1200
1:50
Max. Pos. L.L. Moment
DL Moment
DL Shears
Max. L.L. Shears
MAXIMUM MOMENTS
MAXIMUM SHEARS

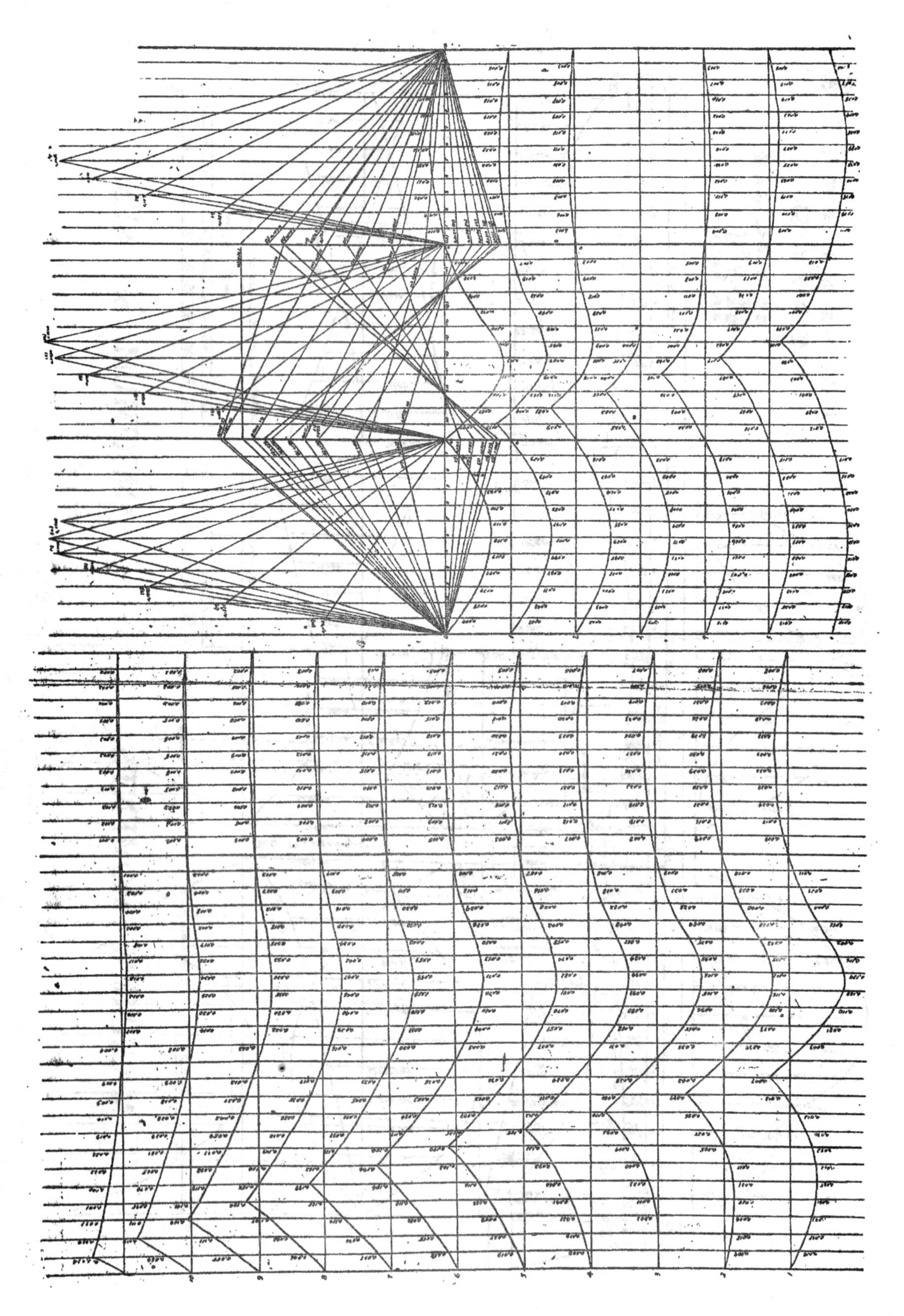

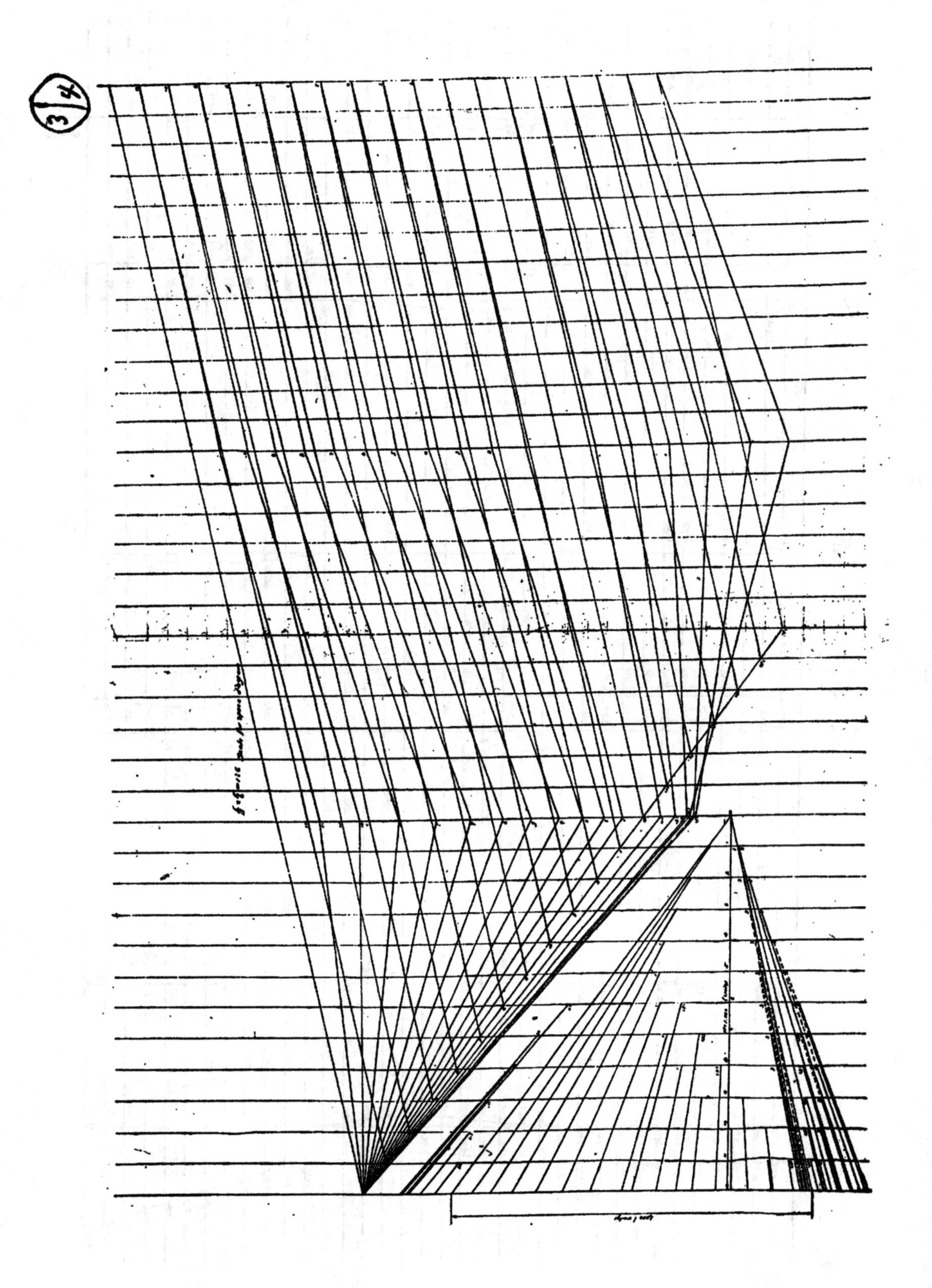

25563

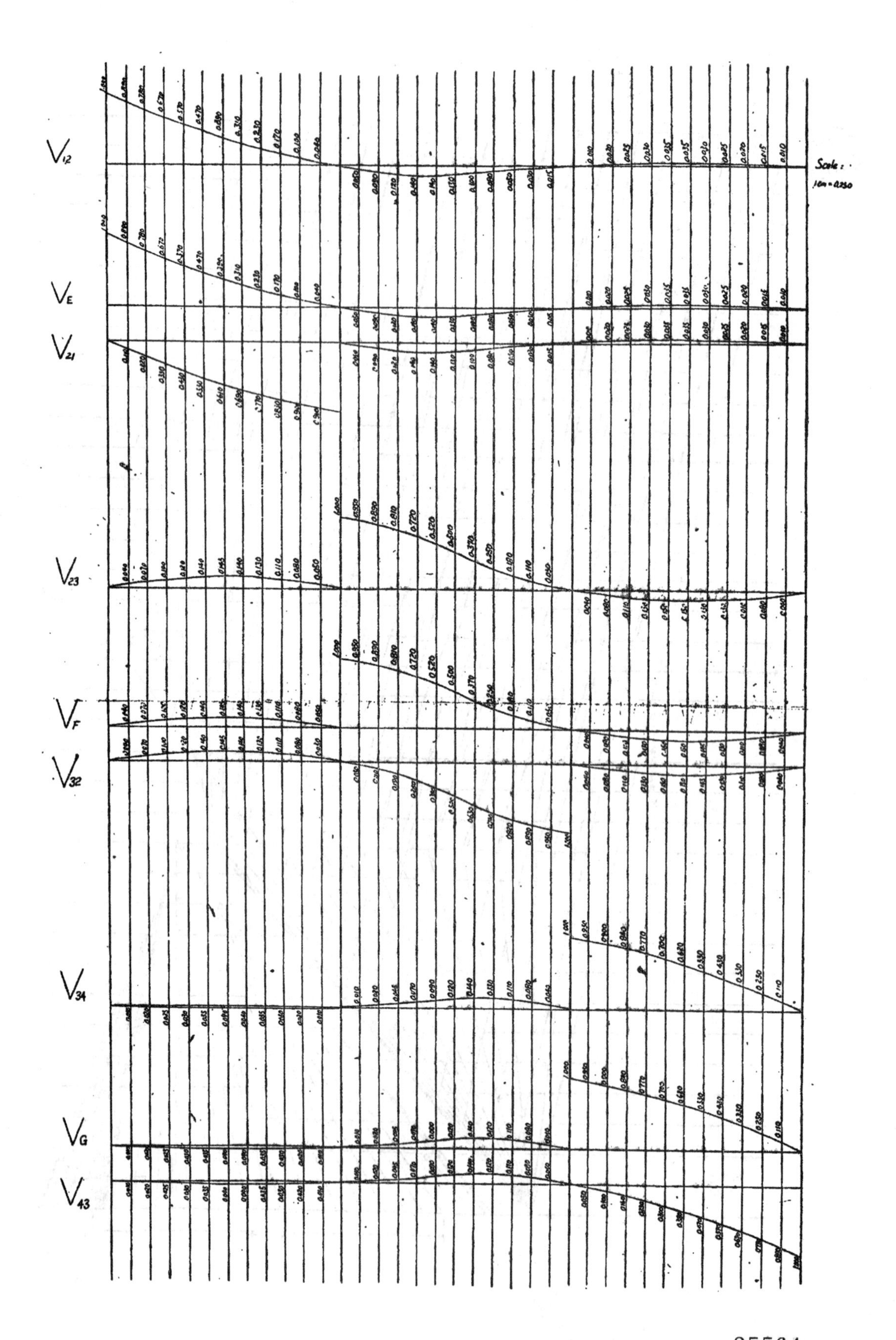

V_{12}
V_E
V_{21}
V_{23}
V_F
V_{32}
V_{34}
V_G
V_{43}
Scale:

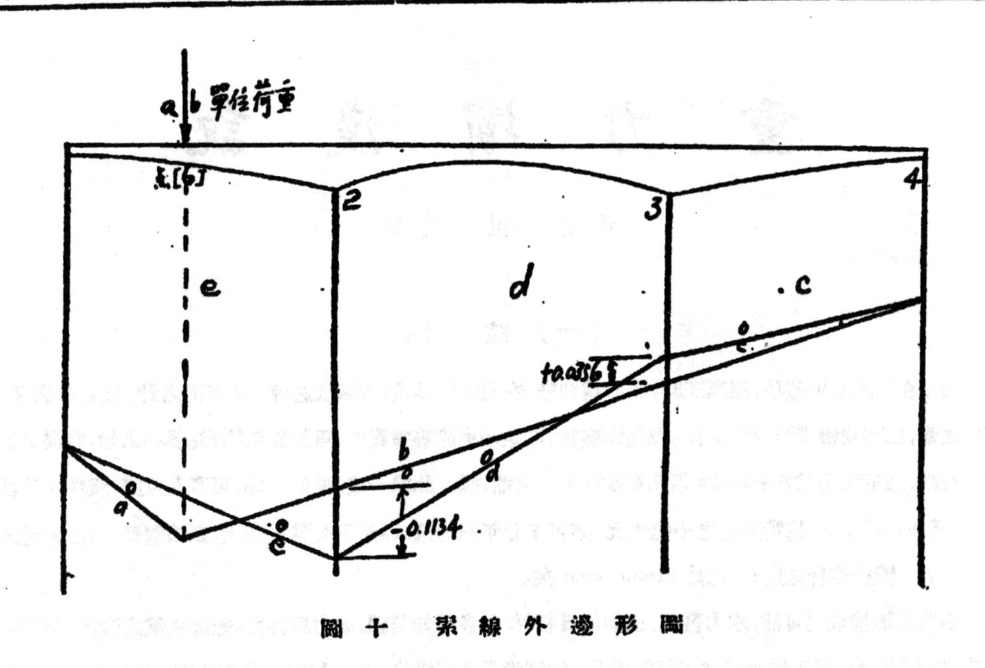

圖十　架線外邊形圖

（三）附言

本文例題乃者爲川陝公路石碚舖洋灰橋設計之計算稿部份。應力分析圖當時爲侯樂天與郭成德兩君幫助繪成，并經戴競先生校核。今引爲文，順此致謝。

主要參攷書：

Cross & Morgan: Gontinuous Frames of Reinforced Concrete.

Russll :Analysis of Condtinuous Frames by The Method of Restraining Stiffnesses.

文中錯漏，誠恐難免，敬希讀者不吝賜教，以匡不逮，幸甚。

重力壩淺說

●潘 世 英●

（一）緒 論

壩工在我國應用甚早，秦惠王時，李冰爲蜀守，在灌縣築離堆，撤岷江之流，灌成都之野，使荒蕪之區，成天府之國，已開我國堵流灌漑之源。惟築壩發電，則除小丰滿等電廠外，可說鳳毛麟角。勝利之始，當局以爲可安心建設，乃斥資美元五十萬，請薩凡奇設計長江大壩，據云此壩一成，萬噸海輪，可直入重慶，漑地一千萬英畝，發電1056萬千瓦，建築所耗之十億美元，亦可于廿年內淸還云。然不久戰火彌天，哀鴻遍野，長江大堤遂如1947年九月號通俗什誌所云，已成 dream dam 矣。

自電力遠輸成爲可能，水力發電，遂如旭日初昇。就多煤如美國，亦力爲啓發。蓋因水電之發展，不獨使人民有廉價之享受，且可得平價之工業製成品，控制世界之市場也。水力發電之基本原理，乃藉水頭以轉動能故堵水壩之建築，實爲最主要之一部工程。

壩之種類甚多，在其排水機能而言，分固定壩與活動壩。在其所用之建築材料而言，分土壩混凝土壩與鋼壩。在其力之分佈而言，分重力壩與拱壩等。土壩之建築，經驗重于學理，且只能用于防潦而不能用于堵流，鋼壩雖然堅固，但易于損壞。拱壩雖較省材料，但設計困難，（須有足以承受大部壓力之岸壁始能應用）施工時所需之技巧亦多。因篇幅所限，今且不論。本文所及，只限于重力式混凝土壩。

重力壩者，乃利用壩身之重量，以抗水壓等外力者也，其所受力爲壩底平均担負，設計與施工時之技巧均較簡單，所用材料之量雖多，但價錢較平，故堵水工程用之者頗多。在中國小丰滿電廠鏡泊湖電廠等所用均屬此種。

（二）設計之基礎

壩之工程甚鉅，且對民生之影响甚大，故在設計之先，壩址之選擇，水量之大小。水頭之提高等，均應有愼密之研討，今分述于後。

1. 壩址之選擇——爲使壩儘量減短，與乎壩成後河道不致有意外之變遷，壩最好建于兩山夾持之河面上。在決定壩應在某一段河面上後，該段之河底，應經詳細之探鑽，岩層之深度，岩石之性質，土壤之情況，可能招至地下水之石縫，甚至水之化學性等等，均需詳細記錄之。爲使壩堅固而經濟，以選岩層淺而密實之處爲好。在選址時，不但應顧及壩之持久與經濟，流沙之宣洩，引水道與電廠建築之困難程度等，亦應計及。爲便于冲沙計，在灣曲之河上，以建于凹岸爲宜。

在一切探鑽與調查之工作完畢後，周圍之地形，應詳細繪于圖上，以作設計時之根據。

2. 水文記錄——水文對壩之影响甚大，對民生之影响更鉅，爲萬全計，在設計之先，應搜集十年以上之水文紀錄，使不致閉門造車，惹無窮之後患。

水文中之最主要部份爲流量與水頭，因能量與流量水頭之乘積成正變也。

流量之紀錄恒用曲線表示之，圖一所示，爲流量曲線，乃以一年之各日爲横坐標，以流量爲縱坐標而繪成。如水頭爲常數，圖上縱坐標之比例尺若1-in＝g馬力，1-in＝t小時，則1-in^2＝gt馬力小時，故圖上之面積爲功甚明矣。

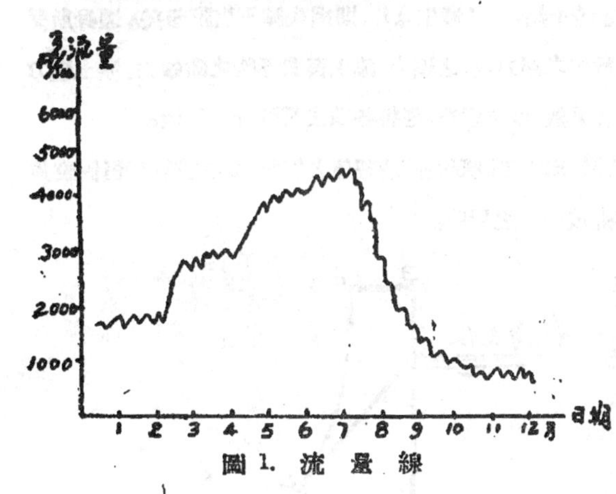

圖 1. 流量線

但一水力之設計，其流量并非全部皆可盡用者，故在得多年之水文紀錄，求得最大最小流量與平均流量後，開發水量之决定，實屬重要。開發水量之大小，恆視電力之用途，水塘之容量，水力之種類，需否熱力之輔助而定。爲使水量不致供不應求，在以前常用最小之流量爲開發水量。但大部之水能，遂不能利用，實屬可惜，且引用較大之水量，工費增加不大，而管理費用并無增加，故近來多採用較大者，寧築塘蓄水，以資調節，再不足時，寧用熱力輔之。

二圖所示，稱流量期間線，乃依水位流量之大小繪成。與流量曲線依時間先後繪成不同，其最大用處爲决定開發水量。普通以一年中二百日常有之水量爲開發水量。

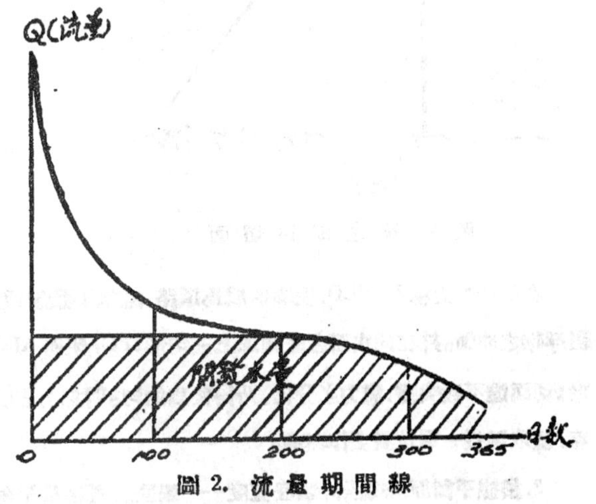

圖 2. 流量期間線

除此而外，水頭亦爲影响電力之主要因素。水量天然之决定性大，人力所可能改變者不多，而水頭則可依自然之環境與需要，由人工加以改變。故在壩工之設計上，佔首要之位置。然在决定提高水位時，除在工程觀點着眼外，國民經濟因之而致之影响，更屬重要。若上游之地域，有因之而成澤國者，則問題之解決，實應切實籌議。

水頭之紀錄，最好亦用曲線表示之。得悉流量之最大值最小值與平均值後，閘門之設計等有所根據矣，最高水頭與最底水頭決定後，則壩之平衡設計有所歸依矣。

設P爲功率，F爲水因水頭提高而得之壓力，V爲水流之速度，則$P=Fv=PA\frac{Q}{A}=h\omega Q$。P爲單位水壓，h爲水頭，$\omega$爲水之密度，Q爲開發流量，由此而觀，因水頭提高所得之功率可計得矣。

（三）設　計

壩爲一鉅大而持久之工程，故設計時應極周密。若不然，一旦發生意外，則損失誠不堪設想矣。壩身所受之力，普通有水之壓力，壩身重量，壩身或壩底可能發生之浮力，水之壓力，流水與飄浮物之衝擊力，淤土壓力與地震等。在設計時，不能有所忽畧。並應剩以一安存系數，以防意外，茲將各項之設計分列于後。

1. 壩之切面——以最高水位爲高度之直角三角形，爲壩之最經濟而最理想之切面，故在設計時恒假設其形狀如此。但爲適合實用計，在設計完畢後，恒使其變成圖4之形狀。

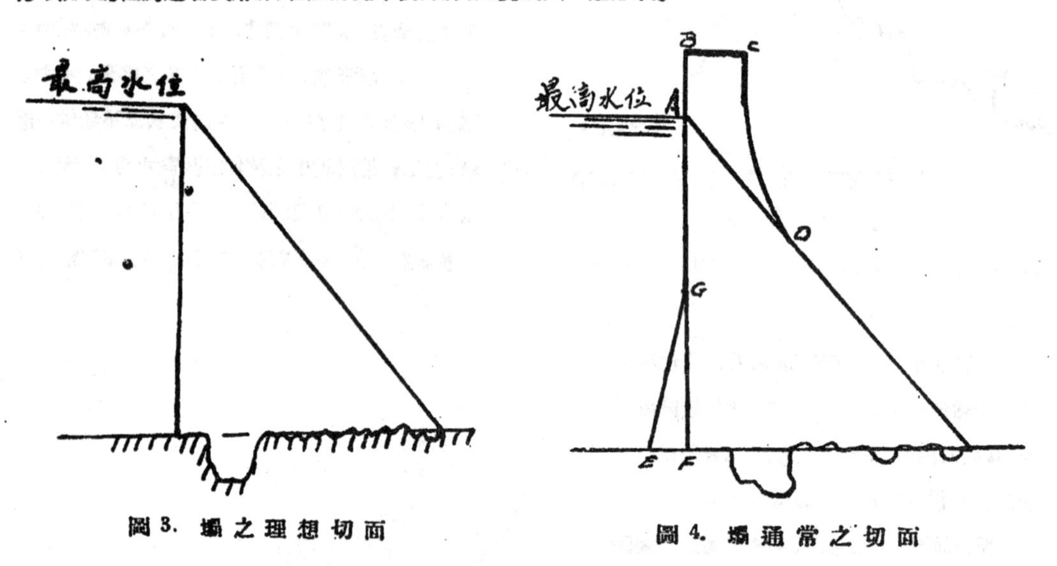

圖3. 壩之理想切面　　圖4. 壩通常之切面

在壩頂加上梯形ABCD，使壩頂成爲道路，流水不致越頂流過，並得足够之抗剪力，(Shearing stress) 抵抗飄浮物之冲衝。并在向水面之底側加上三角形EFG，使與ABCD平衡，令全壩之重心不致改變。CD面更使成弧形；切面遂不致突變，應力之分佈，乃得較爲均勻。所加上之□ABCD，與△EFG大小如何，容後述之。以下之計算，恒取壩長一單位長度作標準。

2. 根據平衡原理據設計壩底寬度——壩底之寬度根平衡原理而設計爲好。因依此原理設計得者當可防，止傾覆，即某一面之抗剪力不足，或底部之摩擦力不足以抵抗水平之壓力時，問題之解決亦被簡單也。在利用此原理設計之先，吾人應研討壩基承受壩重之情形如何。

如圖 5，P_H 爲水平合力，W爲壩重，P爲P_H與W之合力，V爲P之垂直分力，F爲P之水平分力，〔因混凝土與岩石之附着力，只作壩之安全因素(factor of safety)，常畧去不計，故F卽爲壩底之摩擦力。〕則 $P_H=F$，$V=W$，當水乾時，V與W在一直線上，但若水頭增高，則V向右移。今設壩爲一短柱以求V在壩底之分佈。設C爲壩底之單位壓力，則其極大值

$$C_{max}=\frac{V}{b}+\frac{Mc}{I}$$

其極小值

$$C_{min}=\frac{V}{b}-\frac{Mc}{I}$$

圖 5

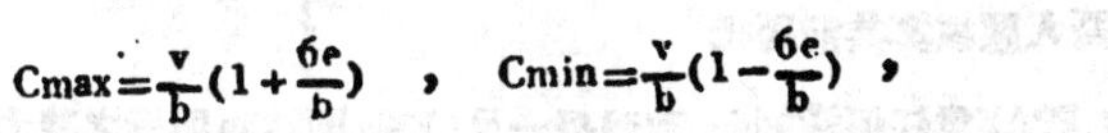

$\because\ c=\frac{b}{2}$，M(bending moment)$=Ve$，$I=\frac{1}{12}b^3$，

$C_{max}=\frac{V}{b}(1+\frac{6e}{b})$，$C_{min}=\frac{V}{b}(1-\frac{6e}{b})$，

設 $e=\frac{b}{6}$時

$C_{max}=\frac{2V}{b}$　　$C_{min}=0$

根據Navier's氏之原理，與上節所述V之位置因P而變之關係，在此情形下之圖，遂如圖6圖7所示。圖6表示水乾時情形，圖7表示水頭最大時之情形。

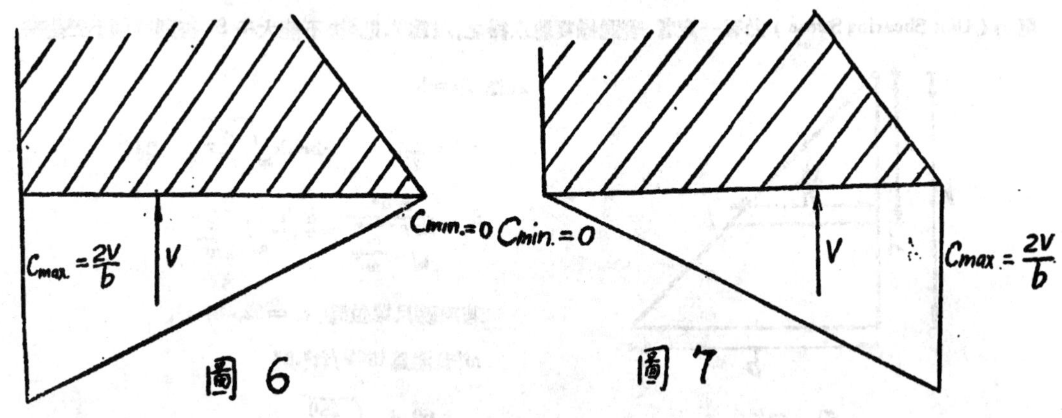

圖 6　　　圖 7

若$e>\frac{b}{6}$時，則 C_{min} 爲負值，卽發生拉力矣。水泥對拉力之承受能力極弱，且拉力發生後，承受壓力V之

25569

面積即行減小。若一有不幸生成裂縫時，與水頭高度水重相等之向上壓力，有使裂縫變大而將壩破壞之趨向，故在設計時，不論如何均不能使壩身有拉力發生，即V之位置，不能離開壩底中點$\frac{b}{6}$。

爲節省材料計，在最高水頭時，恆使V在離開水面$\frac{2}{3}$b處。(離心率$\frac{b}{6}$處)如圖8所示。

因 $P_H=F$ $V=W$ 則

$$W\frac{b}{3}=P_H d$$

$$b=\frac{3P_H d}{W}$$

若水頭甚高，其他之水平壓力相對而言爲甚小，則

$P_H=$水壓$=\frac{h^2\omega}{2}$ $d=\frac{h}{3}$ 則

$$b=\frac{\frac{3h^2\omega}{2}\frac{h}{3}}{\frac{1}{2}b\ h\ \omega'}=h\sqrt{\frac{\omega}{\omega'}}$$ ω'爲混凝土之密度。

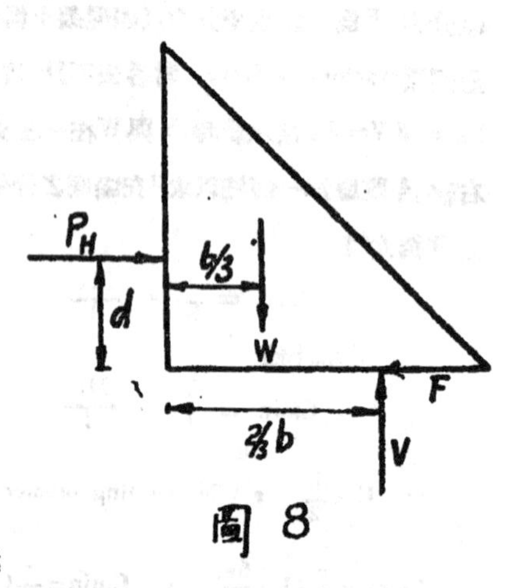

圖 8

故在水頭知道後，b之值可由上二式中之任一式計得之。

3，剪力之核算——壩底之寬度求得後，吾人應核算其能否抵抗因向水方之水平力而生成之剪力。在圖9中，設AB爲任何橫剖面，取壩長一尺計算，則該面所受之剪力當爲$\frac{\omega x^2}{2}$。設S爲作用于該面之單位剪力（Unit Shear），則

$$S=\frac{\frac{\omega x^2}{2}}{\frac{x}{h}b}=\frac{x\omega h}{2b},\quad \frac{\omega h}{2b}\text{爲常數。}$$

由式觀之，即x越大S越大，即單位剪力最大之處，爲壩之底部。故只核算底部即足。同一混凝土之單位抗剪力（Unit Shearing Stress）乃有一定者，普通做實驗求得之，設爲S'，則S決不能大于S'，否則壩即不安全矣。

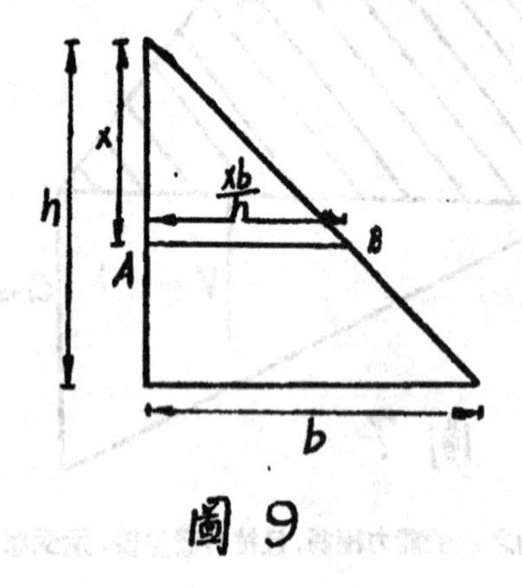

圖 9

在壩底x=h

$$\therefore S=\frac{\omega h^2}{2b}\qquad \because b=h\sqrt{\frac{\omega}{\omega'}}\quad \text{則}$$

$$S=\frac{\omega h^2}{2h\sqrt{\frac{\omega}{\omega'}}}=\frac{\omega h}{2}\sqrt{\frac{\omega'}{\omega}}$$

若用磅尺單位時，$\omega=62.4$#/ft³，

ω'普通爲150#/ft³，則

$$S=\frac{62.4}{2}\sqrt{\frac{150}{624}}h=48.5h$$

依如此之設計，即壩高500尺，水之壓力爲50,000#/ft時，若混凝土混合適當，而質地良好，亦屬安全因

$$總剪力=48.5\times500+\frac{50,000}{500\sqrt{\frac{62.4}{150}}}=24250+155=24400\ /ft^2=169\#/in^2$$

169#/in²之數，良好之混凝土尙勉能抵受也。但實際工作不能用如此大之値者，普通應乘以一安全係數。S'之値通常爲70#/in²，卽壩高在200英尺下，始保安全耳。若壩更高，則底應加大，或用鋼根以助之。

以上之設計，乃假設縱剖面爲一直角三角形者，若壩頂有冰壓力等，則頂點當不能頂受矣，故必須在頂上加寬$\frac{Pi}{S'}$。Pi爲冰壓力等水面浮物施于壩頂之最大力。若Pi爲50,000，而S'=70 /in²，則加寬應爲

$$\frac{50,000}{70\times144}=4.95'。$$

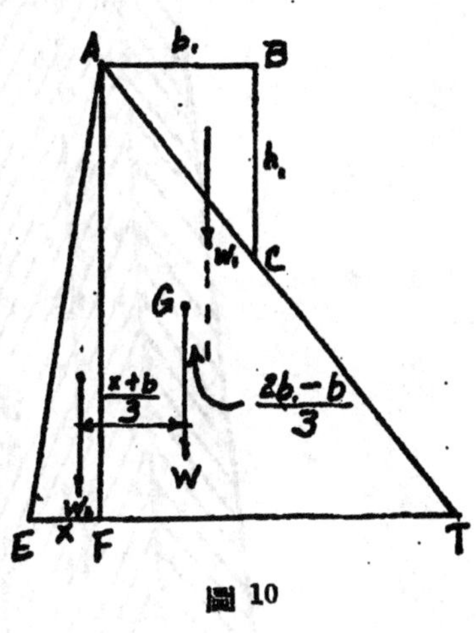

圖 10

但頂部加寬後，原來之重心改變矣，欲使其重心回原位，則AEF必須加上，$EF=x=\frac{-hb\pm\sqrt{(bh)^2+4h(2b_1-b)b_1h_1}}{2h}$若x相對于b甚小，則斜面AE上之水壓力可以不計。如用$b_1=\frac{Pi}{S'}$計出之値不足踣用時，可用三合土板(Slab)加闊之，或將ABC加大再求x。

4. 滑動之核算——作用于壩底以抗滑動之力，有混凝土附着于岩石之力(bond stress)與摩擦力。但在設計上，混凝土附着于岩石上之力，通常并不計算在內，故此步工作，卽核算壩底之摩擦力是否大于所受水平力而已。摩擦力之大小爲fv，f爲摩擦係數。若壩底爲平者，f決不能大于0.42。若fv小于水平壓力時，應將壩底岩石掘成牙形與溝狀，使其摩擦力加大。但爲使岩石與壩底固結一體計，不論摩擦力如何，底部岩層亦以掘成牙狀與溝狀爲好。

5. 若依據壩身不能受拉力之原則設計，則壩身各部均不能發生裂縫，亦不能讓水侵入。設計得當，施工良好之壩，壩身內之浮力，當可避免。但爲愼重計，仍應于牆內安置垂直之混凝土管或石管，以排萬一侵入之水，并限制浮力至埋管處爲止。此外基底之岩石，亦應研究。若有縫隙時，應壓入洋灰漿以塡補之。并在向水面建截流牆，深入岩層，同時在壩脚鋪一層粘性土，以防水入侵壩底。

6. 熱漲冷縮之預防與壩之形狀——冷縮熱漲之定性，混凝土亦不能例外。故在設計時，應使壩身微具伸縮性，以免因拉力或壓擠力而生裂紋。在以前往往使壩身成弧形，使有彈性。但此種形狀，是否合用，現在懷疑之者甚多，爲省料計，故多用直線形矣。但必須在離50或80英尺處設一垂直不透水之縫，使微具活動性。經此幾步設計，壩之縱剖面，約如圖11所示。

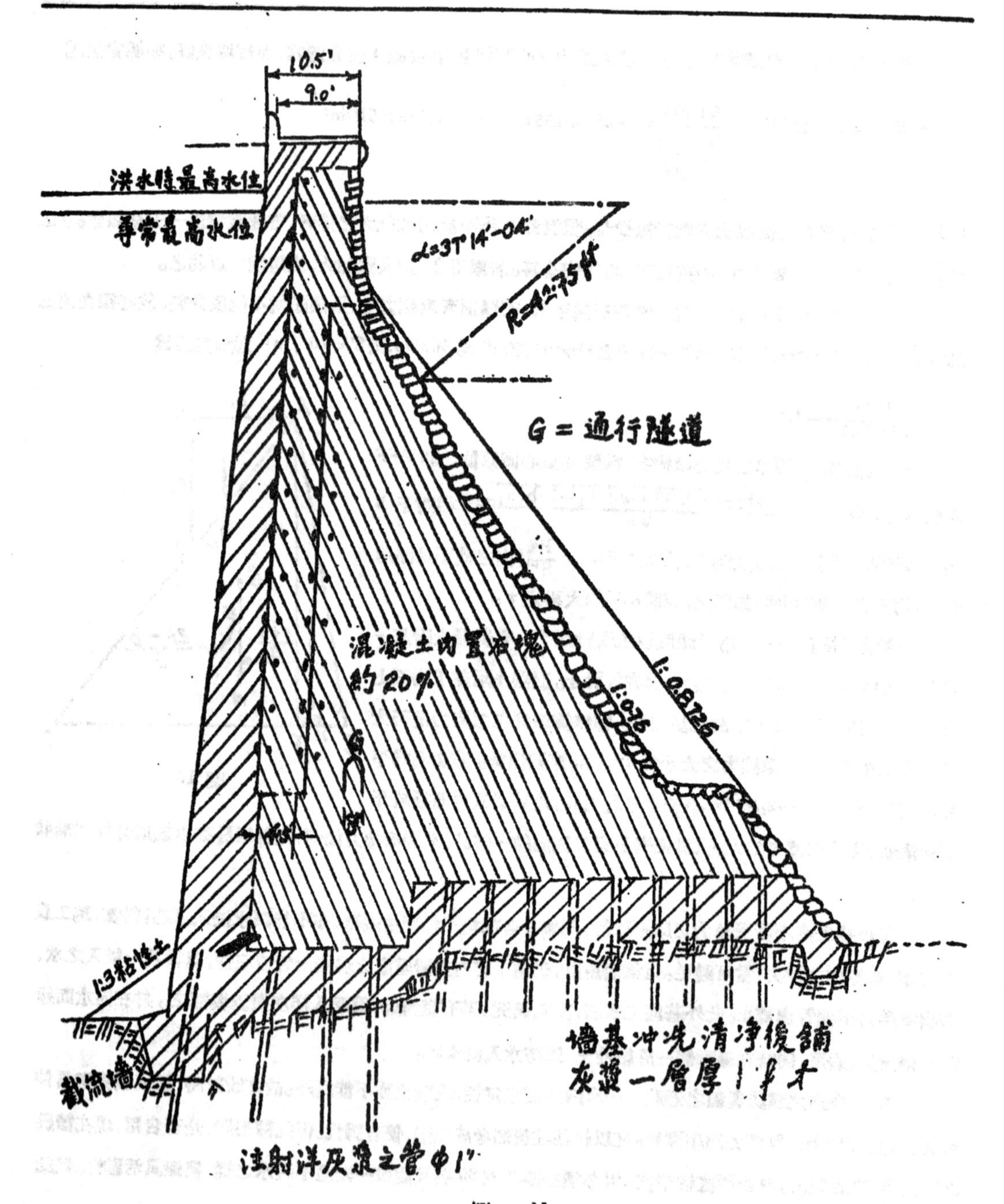
10.5'
9.0'
洪水時最高水位
尋常最高水位
α=37°14'-04"
R=42.75ft
G=通行隧道
混凝土內置石塊
約20%
1:0.75
1:0.8725
1:3粘性土
截流牆
牆基冲洗清淨後鋪
灰漿一層厚1f寸
注射洋灰漿之管Φ1"

圖 11

（四）施　工

1. 混凝土之配合——在以前，重力壩恒用石頭築成但因砌石需要大批之熟練工人，且工作進行甚緩，故自水泥工業發達後，已改用混凝土矣。壩身中灰漿，約佔全體積十分之三四。在壩之中心，常充以十分之三之石頭，以省材料，并使壩身加重。但石頭與石頭之間，應用洋灰固結之。若不然，壩之堅固性減少矣。壩之底部與向水面之防透層，所用之混凝土應含灰漿較多，因灰漿較多則較實，不易爲水侵入也。壩面往往砌石一層，以抵氣溫。灰漿內并常雜有火山噴出岩灰漿，使凝結不致太快，使不致因裂縫而受損。

2. 地基工程——重力壩之重量甚大，且乃持久性者，故必須有良好之地基。不論如何，壩以建于岩層之上爲好。岩層之種類甚多，在施工時必須切實研究。若爲密實而近乎水平，則稱上品。若已被侵蝕或易被侵蝕者，應將已被侵者剷去，以直達完整岩層，將未被侵蝕者加以保護。若岩層非整體而多罅，應將其空隙塡滿。若岩下因地下水之冲刷而空虛，則必須在地下水之上流築牆，將水流截斷，引之他往，并將空處塡滿。水之化學性亦應研究之，察其有否危險之化學物存在。基底之工作，多在沈箱中進行之。若注灰漿時用降底水位法，則可于抽乾水後進行。

3. 混凝土之灌注——灌注混凝土之方法甚多，因篇幅關係，只述二法。

一爲唧機壓灌法，乃1929年德人發明。乃保留原水面諸法中最新而最好者。應用此法，不須抽水，只須圍堰築成，工作即可進行。但圍堰必須密實而不透水，否則初成之混凝土。會受水冲刷變壞。唧灌灰漿時之壓力甚大，故圍堰有時用鋼板築成。圍堰築好，地基檫淨後。可將一尺直徑左右之漏斗形管直通石面，然後在管在水面之一端用唧機將混凝土壓入。管之下端，應隨時保持于混凝土表面三尺之下，使與水接觸之灰漿，只表面一層，新灌下者受此層遮蓋。此法最大送達遠度爲600尺，最大送達高度爲135尺，每小時工作能力爲380立方尺。若工程大者，可同時用多管，但管之距離需在其最大送達距離之內。工作最好一氣完成，以使新舊接口減少至最少限度。若曾停工，則于重新開工之先，最好將接口一層之混凝土鏟去，并鑿成牙形。

但在水下工作，監工困難。施用此法時，上至工程師，下至工人，均須有相當之經驗。且新凝之混凝土，對侵蝕力之抵抗甚弱，故如情形許可，以用降低水位法爲佳。降低水位法最常用者，乃用板樁築圍堰抽水一法。如水非甚深，則可打木板樁。但若水深超過75尺時，則必須用特製之鋼板樁。鋼板樁之形狀，歐美各國不一，但均大同小異，在二次大戰前，以德國出品之 Larseen, Hoesch, 與 Krupp, 三種爲著名。打板樁時應以兩塊爲一組。鋼板樁能擠去碎石，打穿混凝土或樹幹，但過石頭而仍繼續打下，則可能脫鎖或彎曲，而致漏水，在打下應刻刻注意之。板樁打好，水亦抽乾後。則基底之掘泥工作等與灌注灰漿，均可在無水時進行矣，

（五）結　論

1. 重力壩在中國之應用——水力發電在中國之歷史甚短，壩已不多，壩中之重力壩，更不待言矣。但以重

力壩之數目，與全國所有用于堵水發電之壩數相比，則又似不算少，因就東北一地而言，已有小豐滿，鏡泊湖，水豐等三處也。

小豐滿大壩離長吉十九英里，位于猴嶺南喇姑塔嶺之間，截斷松花江，蓄水甚多，形成世界第四位人造大湖。小豐滿電所最高發電量為26億Kwh，即利用蓄水之位能而得。

鏡泊湖大壩乃由火山岩漿天然造成，橫于牡丹江上，水位差達170尺，水塘容量為小豐滿之六分一。據說若電所能將牡丹江水全部運用，可發電卅萬瓩。今之發電量，不過為偽滿五年計劃之初步而矣。

水豐大壩建于中韓分界之鴨綠江上，亦為重力壩之一種。由偽滿朝鮮合資建成，水力發電計劃是以朝鮮新義州上流9.7英里為起點，至建新義州上流274英里止。在中間有水閘七處，流域面積1480平方里，總水量6500億立方公尺，所得電量由中韓兩國平均應用，但經蘇軍佔領後，所生電力，已不復我國所有矣。

上述三處之電力，實足供東北工業與照明之用。但年來不斷破壞，現已支離破碎，不復舊觀。就小豐滿一處而言，勝利之後，八台水力發電機，只剩兩台，目前戰況更緊，二架之發電機能否保存，更成疑問。某巨公曾云，復興中國，必須發展東北，復興東北，必須重興此數電廠，今也如斯，我欲無言。

除東北三廠外，聞說台灣海南島亦有水力發電廠，但材料搜集困難，今畧之。

2. 尾語——由上所述，知中國之水力資源幾未經開發，僅有之幾所，亦多被破壞，此實萬分可惜也。

水力成本低廉，（在美國一啓羅瓦特成本在0.005至0.0063美元之間）如易于發展，實值得啓發。故美國1889年水力發電只佔全電量1.5%，而1927年已進至7.5%，1937年更躍到30%矣。

中國若不願尾隨于人，而欲迎頭趕上，則必須注意水力發電焉。

本人學識淺漏，是文之作，錯誤難免，敢書此文者，惟欲得先進諸先生之惠教耳。

論混凝土中之危害物質

(Injurious Substances)

●吳秉俠●

前言

遠在二千年前，混凝土即被採用於建築上，直迄今日，其受應用範圍之廣，實非吾人可能逐一加以指出者。此無他，乃因混凝土可以製成各式模樣之單位物質。蓋只需將水泥（Portland cement）與適量之砂石及水混和後而鋳置於各模型中，經凝固後即成一强於支持壓力（Compression）而合於設計者心目中足以利用之式樣材料，如在混凝土中放入鋼筋，便成爲建築上之主要材料，是爲鋼筋混凝土（Reinforced Concrete.）。

雖則，混凝土之製法甚爲簡便，然同一尺度大小之混凝土，因其採用混凝組成物比例之不同，則其功效亦異。而同混和比例及同尺度大小之混凝土，其功效亦非一致，此基於內中所含之雜質（Miscellaneous Substance.）及危害物多少而定。普通對於混凝土起效應之因素如下：

A.水泥之性質。

B.水泥與水之混和比率。

C.組成物與水混和後所起之化學反應之完成程度。

D.危害物質之存在。

上述四點中之水泥之性質，則需考及該項水泥之化學成分，提煉方法，鍛燒過程，研磨工作等步驟。至水泥與水之混和比例，經有 Abram 氏之報告及各項實驗示知，至如組成物間所起化學反應之程度方面，則需顧及之點甚多，此如天氣之溫度，濕度，壓力，混合物之均匀分佈等問題皆有關係。然上述三項，皆非本文所欲討論者，現只就內中之危害物質一項加以析述。其有謬誤之處，尚希讀者賜知。

危害物質之探討

所謂混凝土之危害物質，乃指凡物質混和於混凝土中凝固後，足以產生促使混凝土之抗剪力（Shear）抗壓力（Compression）抗張力（Tension）及其耐久性（Durability）等減低之作用者，概稱之謂『危害物』在早期之羅馬建築中，對於沙石之選擇，即有如下之指定：『所用之沙石必需免除塵污，而易於與水泥起黏合作用者』。（註一）至日後粗骨材中（Coarse Aggregate），於各種報告中吾人總得如下之指示誡語如：『所用之粗粒必須

25575

清凈，堅實，耐强，無膠合物存在，無各種損害及腐蝕物。如酸鹼與鹽類。及其他易於粉碎，軟弱之微粒和各種有害雜質之存在』。至於細骨材(Fine Aggregate)，亦有如上述不出左右之指述。由此可知混凝土之製成須簡單，唯如欲得一標準理想之混凝土，則非注意及此等危害雜質不可。

如欲防範此種危害物所生之劣果，則事前必需探討此種物質之存在方式，危害程度，進行危害之步驟如何等，方知其防預之法。今將各種有害物質概括分存四類而加以論述：

(一)促使混凝土不健全之組合物質(Unsound Aggregate Particles)

(二)易於溶解與凝固之物質(Freezing and Thawing)物質

(三)酸鹼與鹽類(Acid, Akaline and Salt)

(四)平寬之長形及薄片結構物質與表面膠結質(Elongated, Laminated Particles and Surface Coating)

上述之分類，乃就各危害物之進行危害步驟類似而定之，作者未敢說爲絕對合理，蓋有多種物質可視爲(一)類而又有(二)類之作用者，比如黑砂石(Chert)。及葉岩(Shale)。遇此情形，則分在各節中解釋，雖有混沌之嫌，然亦非作者本意所在。

(一)促使混凝土不健全之組合物質

混凝土之組合若不健全，則此種組合便難於發揮其本身之黏性。如是，則此種凝固極易趨於崩解(Disintegrte)，隨之，混凝土遂趨於破裂。是以其影响甚大。

此種促使混凝土不健全之組合物，其所產生之害點，最普通者多爲『洞罅』(Spit)與『裂縫』(Spalling)之形成。其中尤以『洞罅』最易產生。此無非由於大部份之『易於粉碎物質』(Friable substance)或易於風化之微粒存在故也。如『葉岩』(shale)黑砂石(註二)(Chert)及黏性物質與軟弱之砂石等是。此等物質混存於混凝土中凝固後，每因曝露於自然間而受雨水之侵蝕及起風化作用，本身體積乃驟然轉變，此局部之體積收縮，遂成小『洞罅』。由於此種洞罅之形成，各種危害物質乃易於浸入其內部，而進行更大之損害，其中尤以此等洞罅如成於混凝土之表面時，則更易爲他種物質所假道。至侵入這種小罅之水份，每因溫度下降而在內凝固，遂產生內力，此亦足以促使混凝土逐漸離解。

如遇有此種物質過多，則形成之洞罅遂多而彼此吻合相連，於是形成顯著之裂痕。尤其在混凝土舖成之道路上，此種裂痕一成，則路面之不潔水及雜質極易進分其侵蝕作用，由是裂縫之分佈更多。

至於『裂罅』(Spalling)之形成，每因混凝土內含有過多之層形組織(Laminated Structure)而易於粉碎之物質，比如黏土(Argillaceous Sandstone)與岩土(註三)及多種之水成岩石粒(Sedimentary Rock)。此等物質每因天氣關係而進行體積轉變，層與層間乃趨於分離(在平寬長形及薄片組織物質一類中另爲詳述)。裂罅遂因之而起。

由於上述，吾等得知在混凝土中如有此種促使混凝土不健全之組合物之存在，則造成洞罅與裂痕，結果此混凝土之耐久性必然減低。故其危害實不能忽視之。

(二)易於溶解與凝固(Freezing and Thawing)之物質

在溫度起伏無常之狀況下，混凝土常隨之而起膨脹或收縮之作用，誠然，此種作用非吾等肉眼所能覺察。因爲在相差非劇烈之溫度變動階段中，其膨脹係數約爲0.0000055每度華氏每單位長度。然而，正因此種連續的作用，混凝土乃起裂縫。此無非因爲混凝土本身缺少張电力 (Tension) 此種裂縫之生成，足以損害其耐用性，故不能不尋求其減少程度之研究。

足以促成混凝土產生裂縫之因素固多，如水分之多寡，水泥之黏性程度、雜質之存在，酸鹼與鹽類之侵蝕(另述於段三)等，然較爲普通之成因乃在於混凝土中含有大量之易於感受溫度轉變與水氣稀濃而起體積變動之微粒物質。此等物質每因天氣溫度之高低，及水氣多寡而起膨脹與收縮之反應。蓋極易於吸收水份而膨脹，反之，即縮凝也。

普通如層形組織之葉岩(由黏土與泥沙結合而成者)微粒，其吸收水份之性能極强，即體積變動亦易。因此種微粒爲一種層形薄片之組織，故每因體積稍起轉變，則本身層與層間乃互相脫離而趨於解散，因之失去本身之存在。換言之，對混凝土之黏合作用全失，而形成局部之崩解。及至其中組成之黏土 (Clay) 放出水份時，因局部之收縮而形成洞孔。抑所含之水分因溫度低降而凝固，遂產生促使混凝土劈裂之內應力。

而有種易於粉碎而本身非完整有小孔 (Porous) 存在之黑礎石，對於混凝土亦甚同樣之損害作用。F.V. Reagal 氏在其著述之『Chert Unfit for Coarse Aggregate in Concrete』中有如下之括語：『當黑礎石處在易於起溶解與凝固作用之混凝土表面位置時，因其膨脹之關係，結果生成促使混凝土遜弱之裂罅』。(註四)

又水泥中如含有大量之鎂化合物，石灰與石膏，則其膨脹敏感度較大。而由砂石，火成岩微粒與砂礫所組成之組合物，其體積膨脹每較由石英與石灰岩微粒所組成者大半倍至一倍之多。此種現象在採用混凝土組合材料時不能不加以注意者。

若混凝土組成物質之分配組合極均勻時，則其因體積變動所受之損害必少，概如均勻之組合，則混凝土無形成爲一單位物質，故苟有體積變動時，則各部之變動率相等，故可免除因局部膨脹與收縮而致分解之弊；其損害因之乃較少矣。

括言之，混凝土體積之轉變(不論組合之全部與局部)越大，則生成之內力亦大，轉變之次數頻頻，則因連續之內力之壓迫，遂造成混凝土產生裂罅。

(三) 酸鹼與鹽類(Acid, Akaline and Salt)

混凝土混製時所用之水如含有過多之酸類，則此酸類每與其組成物起化學反應而減少其效能，最普通之現象爲水中所含之硫酸鹽類，常水解而生成硫酸，乃與混凝土中之石灰起反應而生成石膏（$CaSO_4 \cdot 2H_2O$）致成局部之體積膨脹遂使混凝土破裂，而在其鈣鹽是可溶性之酸類如鹽酸，硝酸及有機酸者，則其爲害更甚。此乃因石灰之耐酸性能不强，以術受極弱之酸侵蝕，亦不能持久。

在鋼筋混凝土中，則酸類之侵入，乃銹蝕內中之鋼筋，而使此鋼筋之抗电力減少。同時鋼筋因受銹蝕而生鐵銹，因之體積乃增大，此種膨脹，惟有向鋼筋表面之混凝土擠壓，於是促成混凝土劈裂。

如混製之用水含有多量之溶解氣體如二氧化碳（CO_2）及硫化氫（H_2S）對混凝土亦起腐蝕作用，尤需注意。

普通每在混凝土中加入適量之矽酸鈉（$NaSiO_2$）其功用足以防腐。蓋因其在混凝土內之空隙間，生出矽酸，而阻止液體流入混凝土內部以進行侵蝕之故。

鹼之對於混凝土，與酸類起同樣之損害作用。混凝土本身因其侵蝕而致瓦解破裂。普通如鈉及鎂之鹽類皆爲主要之腐蝕物。而硫化物之鹽類更易於銹蝕鋼筋，應爲防範。

用以混製混凝土之水，每爲就近河谷井內之天然水，故內中所含之雜質甚多。其中較普通者乃鎂及鈣之鹽類。此乃氯化鎂（$MgCl_2$），氯化鈣（$CaCl_2$），碳酸鎂（$MgCO_3$）硫酸鎂（$MgSO_4$）及硝酸鈣〔$Ca(NO_3)_2$〕等。此等鹽類每易在混凝土中水解而生酸性作用，因而產生酸類對混凝土所起之損害。

於前兩節中作者曾指出水份在混凝土內凝固所起之損害情况，如欲避免此種弊端，每每於混和時加入普通之鹽類以減低混凝土組成物所含之水份之凝結點，以使其不致過於敏銳，至其所加之限量與其效能，則爲「如加入鹽類百分之一於其用以混和之水中，則減低水份之凝固點華氏一度」，又「若混凝土在凝結溫度下進行凝固，則以加入百分之十二之鹽類爲最有效」。（註五）然加入超量之鹽類，則反生不良之效應。蓋多餘之鹽類每侵蝕混凝之組成物也。

在 Wisconsin 大學曾有一實驗記錄爲如加入百分之二之氯化鈣鹽與百分之九之普通鹽類，其所生之效應最强。（註六）

於都市之混凝土建設物中，其受酸鹼及鹽類之侵蝕而遭破壞之例甚多，如化學工廠中之溝渠，每每因倒棄之酸鹽所侵蝕而起大裂縫。普通之混凝土水渠及路面，每因常受不潔水中雜質之侵蝕亦起破痕。蓋此等不潔水多有酸鹽類之存在也，是以在易於與酸鹽類接觸之混凝土建築物中，以水渠而言，則每設法在其表面加上一層隔離物質，以隔離其侵蝕之效能。

（四）平寬之長形及薄片結構物質與表面膠結物

(Elongated Laminated Particles and Surface coating)

在 A.S.T.M (American Society for Testing Materials) 之報告材料中，對於粗骨材，每有指示如『必

需避免包含過多之平濶長形微粒，及各種危害雜質」。因此等小微粒之存在，每影响混凝土之耐力及其稠度(Workability)。如有過多量之此種物質存在鋼筋混凝土中之鋼筋表面時，則此鋼筋混凝土可能支持之抗滑應力(Bond stress)必降低。此乃因該等物質之平面光滑，鋼筋遂易於脫移(Shift)也。

於此種平濶之長形小粒混結後，若從此混凝土之局部剖視之，則其結構無疑爲薄片之組合(Laminated mixture)，此種組合，有如上述，乃最易於相互脫離者，蓋只需一旦受天氣溫度之影响而至內中組成之黏土起風化現象時，層與層間遂起分離徵象。苟同時此混凝土受有壓力或电力，則薄片之間卽起滑溜作用，隨之乃起裂罅。另方面，若多量之此等小粒相聚一起，則必易於產生小洞孔，此乃因彼此間難於吻合也。反之，在多形之粗粒方面，其吻合之或然率尙高。此種小洞之形成，遂足以爲水份及危害物侵入之假道及凝固之空間，隨之而生之劣果實非稚弱也。

然而，究竟何者方稱之謂平濶與長方形之小粒，其濶長與厚度之比應如何方在限制之列？迄今尙無準確之規定。然而，只需吾等感覺其長濶厚之比爲長方形，則卽可產生上述之弊端。是以不能不注意及之。

另外一種危害混凝土之物質。爲表面膠結物(Surface coating)此種膠結物對於混凝土之抗滑力影响至大。此等膠結物多爲極易於粉碎及吸收水份而溶解之物質，故一旦混結於混凝土中後，稍因風化作用，乃立卽消失其膠合作用。普通之黏土膠結物，只需用水冲洗卽去。另有一種膠結物乃因石粒表面附着過多之塵埃而成者，其對混凝土所生之劣果，在Goldbeck氏之實驗此種塵埃膠結物之報告中，可概括而言之爲每百分之一之塵埃存在，則混凝土之破裂係數(Modulus of Rupture)減少百分之一至一又二分一。其壓力之支持則減少自百分之零至二左近。(註八)

如要防範此種膠結物之存在，亦可用水事前洗淨之。蓋此種塵埃之附着力非大，極易於潔淨也。

結論

由上列各節所述，可概括指出者，乃混和在混凝土中之任何危害物質，雖則其進行損害之方式與步驟之先後不同，然其最後之效應，皆在促成混凝土裂縫(Cracks)之形成，而使混凝土之耐用性及持久性減低。在目前之淨混凝土建築中，裂罅之產生似無由避免。蓋只就天氣之驟變，卽可促成混凝土裂痕之產生矣。然若各種危害物與雜質之存在率減少時，則混凝土對天氣之敏感度乃低，是故裂縫之促成率亦減少矣。故對此等物質實有探討之必要。至於其潔除之方法，則非本文所包涵也。

註(一)：轉錄自1935年A.S.T.M.(American society for Testing Materials)之「Repart on Significance of Tests of Concrete and Concrete Aggregates」Committee C.9.第九十六頁。

註(二)：或稱凝石。此處所指者爲由沙泥與易於粉碎之黏土組合而成之一類。

註(三)：苦土者，卽氧化鎂。

註(四):轉錄自1935年A.S.T.M.之『Report on Significance of Tests of Concrete and Concrete Aggregates』Committee C-9 第九十八頁

註(五):錄自1935年之A.S.T.M, Committee C-9 第四十五頁之R.E.Davis及 J. W Kelly兩氏所作之 Volume change of Concrete) 一文中。

註(六)(七):錄自Urquhart及O'rourke兩氏之『Design of Concrete Structures』一書中,第一章第廿七頁。

註(八):錄自1935之A.S.T.M. Committee C-9 第一百頁中。

本文參考書籍

A.S.T.M.:『Report on Significiance of Tests of Concrete and Concrete Aggregates』Committee C-9 1935

McMillan: Basic Principles of Concrete Making

Stillman: Engineering Chemistry

報導

漫談近年美國土木工程系之輪廓及工程雜誌

●區　東●

當近年物理與化學飛躍進步期間，土木工程一門究有何新發展？此在文中有概括之描述。作者區東先生前爲本系講師，對同學極備友善。年前始赴美再求深造。彼於功課煩忙與催稿期限匆迫間，尚能爲本刊撰稿，殊感慶幸也。　——編者誌——

郁開嶺南不覺快半年了，可是不時還從同學的來信中得到一些學校的消息，并且知道你們行將出版南大工程。我本來不該在這裏佔了你們寶貴的篇幅，但是經不起同學們疊次的來信，囑我報導美國的見聞。其實我來此後，整日都忙在圖書館的書堆裏，沒有甚麽可告。我倒願藉着你們的刋物，把我懷念你們的心情，寄給每一位同學，至於文中所云，無非東拉西扯，原是不足一道的。

區東寄於三十六年聖誕節後兩日

在土木工程各門學問當中，鐵路和公路工程已經有點過時，如果要找一些這類的材料，常常就要翻到二十多年前的雜誌。雖然目前的雜誌也有這類的文章，但多半是關於改善和管理方面的。這一類課程的開設，差不多全是爲外國學生，水利和衛生工程朝着堤壩方面走，都市設計這一類課程的課本大多都是很久不曾重訂或修改了。結構工程曾經一度走紅運。但是除了對鋼筋混凝土工程之外，高等結構原理的應用範圍還是少。不過如果小心一點追尋近幾年來鋼筋混凝土設計規範的更改，便知道它所循的途徑還是脫不了高等結構學。土壤力學是土木工程方面近年的最大進展。讀過污工結構（Earth and Masonry structures）的同學知道，其中對於土壤應力是很多假設，以往也很少人作有系統的研究。一直到現在才被人發掘了出來。但是一切還在試驗的階段，談不到大量應用。

以上大概是這裏土木系的一個輪廓。不過範圍太廣了，而我對於許多方面不過是一個門外漢，說錯了也不一定，但大概還不致離得太遠。可是中國的需要是不同，所以在這裏的中國學生，還是選擇自己認爲需要或感興趣的課程。

學術雜誌是研究學術所不可缺少的一部份。由於集中研究和互相交換智識可能得到較大的效果，所以學術雜誌就在不斷的增加。目前在美國，各式各樣的工程雜誌可謂五花八門。有些是工程學術團體出版的會刋，內容比較嚴謹而專門。有些是工程施工的實際紀錄，或是工廠工作的檢討，其中不少是別人的實貴經驗。也有些雜誌是專以廣告和營利爲目的的。我在這裏祇將個人所及而與土木工程有關的會刊（Proceedings）雜誌，畧爲一談。

關於土木工程方面內容比較綜合的當推Proceedings-American Society of Civil Engineers。其中包括土木工程中各方面的論著。讀者（這是指學會的會員而言）如果有甚麼批評，可以儘量在規定期間內提出，作者也可以答辯。那麼重印時便將這些討論附諸文末，以供參攷。重印的版本主要目的是供給各大學或學術研究機關一些資料，和這會刊性質相同而分類較偏的會刊也很多，下面所列舉的不過是其中的少數。Proceedings-American Concrete Institute, Proceedings - American Railway Bridge and Building Association, Proceedings-American Railway Engineering Association, Proceedings-American Road Builders' Association, Proceedings-American Society of Sanitary Engineering 和 Proceedings American Society for Testing Materials 都是讀土木工程的人常常會碰到的刊物。

其他綜合性的工程雜誌也有好幾份。其中最被推崇的 Engineering News Record 是一份綜合性的工程刊物。還有兩份性質相同的是Civil Engineering 和 Engineering Contract Record。內容都不錯，其中附有許多說明圖片。

分門別類的雜誌相當多，都是以圖片爲主。專論很少，廣告却很多。關於鐵路方面的有 Modern Railroads Railway Engineering and Maintenance, Railway Gazette, Railway Signaling, Railway Magazine 等。公路方面的有 Roads and Streets, Roads and Road Construction, Roads and Bridges 等。鐵路和公路在這個國度已達到成熟階段，所以比較通俗的雜誌也就多一點。關於給水和下水道工程的雜誌有 Water and Sewage, Sewage Works Engineering, Sewage Works Journal 等。後一本是 Federation of Sewage Works Association 出版的，內容比較充實。市政工程方面的有 Public Works Magazine 和 The American City 等，不過內容多牛涉到建築工程，這就離開我們的範範了。

關於工程材料的雜誌有如雨後春筍。Materials and Methods 的內容比較廣泛。此外如 Iron and Steel Engineer, Metal Progress, Modern plastics, Wood 等等，多少帶點替出品廠商宣傳的色彩，并不大受人重視。

很倉卒而概括的把我個人較熟悉的一些雜誌都提到了。其中一定還遺漏了不少。工程雜誌是現代研究工程學術的中心，將來的發展是未可限量的。

参攷

重複代入法對於工程上方程式之應用及其表解法

● 潘演强 ●

解决聯梁 Continuous Beam 或其他工程上之問題時，常需借助於聯立一次方程式。若方程式之含有多個未知數者，欲求解答，每緣繁長之苦。玆特介紹一較簡單之法，名爲重複代入法 (Solution by Repetition)。此法所求得之值，雖爲近似值，但可達到任意之準確程度。今先述重複代入法，其要點如下：

(1) 在各方程式中擇一未知數，其係數 (coefficient) 爲最大者，(指絕對值而言)，命其他未知數爲零，而求此未知數之第一近似值 (First approximation)

(2) 再擇另一未知數，其係數爲第二大者，除先前所求得之未知數外，命其他未知數爲零，而求此未知數之第一近似值。

(3) 如此類推，求得各未知數之第一近似值。

(4) 以此法重複計算，至求得適意之結果而止。

此法對於聯立一次方程式之具有相當收歛性者，尤有特效焉

試舉例以說明之：玆有聯立方程式

$$27.75A+6B = 55 \cdots\cdots\cdots\cdots(1)$$

$$6A+28.8B+8.4C = 54.38 \cdots\cdots\cdots(2)$$

$$8.4B+44.8C+14D = 14.38 \cdots\cdots\cdots(3)$$

$$14C+38.5D = 3.75 \cdots\cdots\cdots(4)$$

在方程式(1)中，令B=0，求A之第一近似值，以A_1表之。

則　$27.75\ A_1=55$

以此值代入方程式(2)中，令C=0而求B_1(卽B之第一近似值)。

$$\therefore 28.8B_1=-54.38-\left(\frac{6}{27.75}\right)\times 55$$

$$=-54.38-11.90=-66.28$$

以此值代入方程式(3)中，而令$D=o$，求C_1

$$\therefore 44.8C_1=14.38-\frac{8.4}{28.8}(-66.28)$$

$$=14.33+19.33=33.71$$

以此值代入方程式(4)中而求D_1，

$$\therefore 38.5D_1=3.75-\frac{14}{44.8}(33.71)$$

$$=3.75-(10.52)=-6.77$$

以上所求得者，爲各未知數之第一近似值。

A之第二近似値A_2，可由方程式(1)求得，此時B之值不等於零而爲B_1，

故 $27.75A_1=55$

$27.75A_2+6B_1=55$

$$\therefore 27.75A_1-27.75A_2=6B_1$$

$$=\frac{6}{28.8}(-66.28)=-13.80$$

$$\therefore 27.75A_2=27.75A_1+13.80$$

$$=55+13.80$$

B_2亦可以同法在方程式(2)中求得，但此時$A=A_2$，$C=C_1$

$$\therefore 6A_1+28.8B_1=-54.38$$

$$6A_2+28.8B_2+8.4C_1=-54.38$$

$$\therefore 28.8B_1-28.8B_2=8.4C_1+6(A_2-A_1)$$

$$=\frac{8.4}{44.8}(33.71)+\frac{6}{27.75}(13.80)$$

$$=6.31+2.99=9.30$$

$$\therefore 28.8B_2=28.8B_1-9.30$$

$$=-66.28-9.30$$

C_2及D_2之值，亦可按此法而求。

$$\therefore 44.8C_2=44.8C_1-\left(\frac{14}{38.5}\right)(-6.77)-\left(\frac{8.4}{28.8}\right)(-9.30)$$

25584

$=44.8C_1-(-2.46)-(-2.71)$

$=44.8C_1+5.17=33.71+5.17$

又　$38.5D_2=38.5D_1-\left(\frac{14}{44.8}\right)(5.17)$

$=6.77-1.61$

以同法續求A_3, B_3, C_3及D_3

由(1)　$27.75A_2+6B_1=55$

$27.75A_3+6B_2=55$

$\therefore 27.75A_2-27.75A_3=6(B_2-B_1)$

$=\frac{6}{28.8}(-9.30)=-1.94$

$\therefore 27.75A_3=27.75A_2+1.94=55+13.8+1.94$

由(2)可得　$28.8B_3=28.8B_2-\frac{8.4}{44.8}(5.17)-\left(\frac{6}{27.75}\right)(1.94)$

$=-66.28-9.30-1.39$

由(3)可得　$44.8C_3=33.7+5.17-\left(\frac{14}{28.5}\right)(-1.61)-\frac{8.4}{28.8}(-1.39)$

$=33.7+5.17+0.58+0.41$

$=33.7+5.17+0.99$

又由(4)　$38.5D_3=-6.77-1.61-\frac{14}{44.8}(0.99)$

$=-6.77-1.61-0.31$

計算至此,A,B,C及D之第三近似值,皆已求得矣。

若再以上法計算,可求各未知數之第四近似值。

由(1)　$27.75A_3-27.75A_4=6(B_3-B_2)$

$=\frac{6}{28.8}(-1.39)$

25585

此數值，相對而言，則甚小。故A_3之值，可視爲A之眞值。

又　$28.8B_3-28.8_4 \doteq \frac{8.4}{44.8}(0.99) \doteq 0.19$　(可以略去)

$44.8C_3-44.8C_4 \doteq \frac{14}{38.5}(-0.31)$

$\doteq -0.11$　(可以略去)

同理D_4與D_3相差亦甚微。故A_3，B_3，C_3，及D_3之值可視爲A,B,C及D之眞値

故　$A=\frac{55.00+13.80+1.94}{27.75}=\frac{70.74}{27.75}=2.55$

$B=\frac{-66.28-9.30-1.39}{28.8}=\frac{-76.97}{28.8}=-2.67$

$C=\frac{33.71+5.17+0.99}{44.8}=\frac{39.87}{44.8}=0.891$

$D=\frac{-6.77-1.61-0.31}{38.5}=\frac{-8.69}{38.5}=-0.226$

應用上述之解法之原理，將解法中之各步驟，列而成表，尤顯此法之爲用

將各方程式之係數，排列如表(I)。方程式(1)之係數排列在A列，方程式(2)之係數排列在B列如此類推。)

在A列之第一行，B列之第二行，C列之第三行，D列之第四行之數字，各附一橫線於其下。

在表(II)第一行中，A列之數值由上行移下，而附以括號，此即27.75A之第一近似值也。

將此值照表(I)中第一行各列之比例而分配。

$\frac{6}{27.75}(55)=11.90$是爲表(II)中B列第一行之值，以表(II)中B列末行之値減去此11.90得-66.28，是爲表(II)中B列第二行之值，附以括號，此爲28.8B之第一近似値也。以此值按表(I)中第二行各列之比例分配，塡於表(II)第二行之各相當列中，如$\frac{6}{28.8}(-66.28)=-13.80$

$\frac{8.4}{28.8}(-66.28)=-19.33$

25586

		A	B	C	D
(I)	1	27.75	6	…………	…………
	2	6	28.8	8.4	…………
	3	…………	8.4	44.8	14
	4	…………	…………	14	38.5
		55.00	-54.38	14.38	3.75
(II)	1	(55.00)	11.90	…………	…………
	2	-13.80	(-66.28)	-19.33	…………
	3	…………	6.31	(33.71)	10.52
	4	…………	…………	-2.46	(-6.77)
(III)	1	(13.80)	2.99	…………	…………
	2	-1.94	(-9.30)	-2.71	…………
	3	…………	0.97	(5.17)	1.61
	4	…………	…………	-0.58	(-1.61)
(IV)	1	(1.94)	0.42	…………	…………
	2	-0.29	(-1.39)	-0.41	…………
	3	…………	0.19	(0.99)	0.31
	4	…………	…………	-0.11	(-0.31)
總和		70.74	-76.97	39.87	-8.69
除數		27.75	28.8	44.8	38.5
未知數值		2.55	-2.67	0.891	-0.226

以表（Ⅰ）C列末行減去表(Ⅱ) C列，第二行之值得C列第三行之值，附以括號，此卽44.8C之第一近似值也同法可求得表(Ⅱ)中各行各列之值。

以表(Ⅱ)中第二行A列之值改號後置於表(Ⅲ)中A列第一行中。表(Ⅲ)中各行之值，可做表（Ⅱ）中各值之求法，分別列下，并附以括號。

以上法，可續求得表Ⅳ，表Ⅴ……等，至所得之值相對甚小，可以畧去。然後將每列之有括號者，求其代數和置於該列之下。

以表（Ⅰ）中，附以橫線之數值，分別置於"除數"各列中。以總和與除數相除所得之商，卽各未知數之值也。

舉一例以明之設有聯梁及其負荷之情形如下：圖中AB＝20尺，BC＝35尺，CD＝25尺，DE＝15尺，EF＝30尺

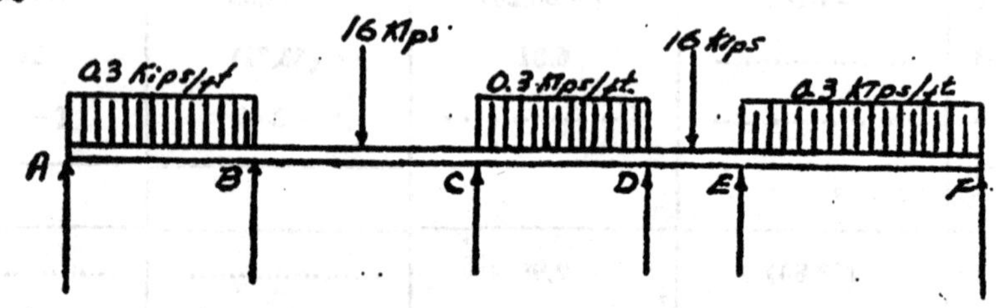

A, B, C, D, E及F爲各支點 M_A, M_B, M_C M_D, M_E 及M_F 爲A, B, C, D, E及F各支點之力矩則按三力矩定理(Theory of three Moments)可得下列各方程式

$$110M_B + 35M_C = -7950 \cdots\cdots\cdots (1)$$

$$35M_B + 120M_C + 25M_D = -8522 \cdots\cdots\cdots (2)$$

$$25M_C + 80M_D + 15M_E = -2522 \cdots\cdots\cdots (3)$$

$$15M_D + 90M_E = -3375 \cdots\cdots\cdots (4)$$

(Ⅰ)		M_B	M_C	M_D	M_E
	1	110	35	…………	…………
	2	35	120	25	…………
	3	…………	25	80	15
	4	…………	…………	15	90
		-7950	-8522	-2522	-3375

25588

(II)	1	(-7950)	-2530	…………	…………
	2	-1750	(-5992)	-1250	…………
	3	…………	-398	(-1272)	-249
	4	…………	…………	-520	(-3126)
(III)	1	(1750)	557	…………	…………
	2	-46.4	(-159)	-33.1	…………
	3	…………	173	(553.1)	103.6
	4	…………	…………	-17.25	(-103.6)
(IV)	1	(46.4)	14.8	…………	…………
	2	-54.8	(-187.8)	-39.1	…………
	3	…………	17.30	(56.35)	10.4
	4	…………	…………	-1.73	(-10.4)
(V)	1	(54.8)	17.4	…………	…………
	2	-10.1	(-34.7)	-7.24	…………
	3	…………	2.8	(+8.97)	1.68
	4	…………	…………	-0.28	(-1.68)
	總和	-6099	-6373	-653.6	-3242
	除數	110	120	80	90
	未知數值	-55.5	-53.1	-8.17	-36.0

故 $M_B = -55.5$ ft-kips

$M_C = -53.1$ ft-kips

$M_D = -8.17$ ft-kips

$M_E = -36$ ft-kips

25589

已知三直線方程式所成之三角形面積之捷徑求法

● 姚 保 照 ●

在普通解析幾何中，若已知三直線方程式（Linear equation），則其所成之三角形之面積，必需聯解得其交点後方可計算。然若能將各方程式之係數代入如下之求証公式中

$$\triangle = \frac{1}{2}\frac{D^2}{C_1' C_2' C_3'}$$

C_1', C_2', C_3', 爲 C_1 C_2 C_3 之子行列式

則其面積可直接求得而省掉解各方程式之煩長手續矣！

若三角形各邊之方程式爲

$$A_1 x + B_1 y + C_1 = 0 \quad\cdots\cdots(1)$$

$$A_2 x + B_2 y + C_2 = 0 \quad\cdots\cdots(2)$$

$$A_3 x + B_3 y + C_3 = 0 \quad\cdots\cdots(3)$$

則三角形之面積 $\triangle = \frac{1}{2}\frac{D^2}{C_1' C_2' C_3'}$

$$但\quad D = \begin{vmatrix} A_1 & B_1 & C_1 \\ A_2 & B_2 & C_2 \\ A_3 & B_3 & C_3 \end{vmatrix}$$

C_1', C_2', C_3', 爲 C_1, C_2, C_3, 之子行列式

證： 設 (2) 與 (3), (1) 與 (3), 及 (1) 與 (2) 之交點之坐標爲 (x_1, y_1); (x_2, y_2); (x_3, y_3)

$$則\quad x_1 = \frac{B_2C_3 - B_3C_2}{A_2B_3 - A_3B_2} \;;\quad y_1 = \frac{C_2A_3 - C_3A_2}{A_2B_3 - A_3B_2}$$

$$x_2 = \frac{B_1C_3 - B_3C_1}{A_1B_3 - A_3B_1} \;;\quad y_2 = \frac{C_1A_3 - C_3A_1}{A_1B_3 - A_3B_1}$$

$$x_3 = \frac{B_1C_2 - B_2C_1}{A_1B_2 - A_2B_1} \;;\quad y_3 = \frac{C_1A_2 - C_2A_1}{A_1B_2 - A_2B_1}$$

$$故\quad \triangle = \frac{1}{2}\begin{vmatrix} x_1 & y_1 & 1 \\ x_2 & y_2 & 1 \\ x_3 & y_3 & 1 \end{vmatrix} = \frac{1}{2}\begin{vmatrix} \dfrac{B_2C_3 - B_3C_2}{A_2B_3 - A_3B_2} & \dfrac{C_2A_3 - C_3A_2}{A_2B_3 - A_3B_2} & 1 \\ \dfrac{B_1C_3 - B_3C_1}{A_1B_3 - A_3B_1} & \dfrac{C_1A_3 - C_3A_1}{A_1B_3 - A_3B_1} & 1 \\ \dfrac{B_1C_2 - B_2C_1}{A_1B_2 - A_2B_1} & \dfrac{C_1A_2 - C_2A_1}{A_1B_2 - A_2B_1} & 1 \end{vmatrix}$$

$$= \frac{1}{2} \cdot \frac{1}{\gamma_1 \gamma_2 \gamma_3} \begin{vmatrix} B_2C_3 - B_3C_2 & C_2A_3 - C_3A_2 & A_2B_3 - A_3B_2 \\ B_1C_3 - B_3C_1 & C_1A_3 - C_3A_1 & A_1B_3 - A_3B_1 \\ B_1C_2 - B_2C_1 & C_1A_2 - C_2A_1 & A_1B_2 - A_2B_1 \end{vmatrix}$$

但 $\gamma_1 = A_2B_3 - A_3B_2 = \begin{vmatrix} A_2 & B_2 \\ A_3 & B_3 \end{vmatrix}$

$\gamma_2 = A_1B_3 - A_3B_1 = \begin{vmatrix} A_1 & B_1 \\ A_3 & B_3 \end{vmatrix}$

$\gamma_2 = A_1B_2 - A_2B_1 = \begin{vmatrix} A_1 & B_1 \\ A_2 & B_2 \end{vmatrix}$

故 $\Delta = \frac{1}{2} \frac{1}{\gamma_1 \gamma_2 \gamma_3} \cdot \frac{1}{A_1B_1C_1} \begin{vmatrix} A_1B_2C_3 - A_1B_3C_2 & A_3B_1C_2 - A_2B_1C_3 & A_2B_3C_1 - A_3B_2C_1 \\ A_1B_1C_3 - A_1B_3C_1 & A_3B_1C_1 - A_1B_1C_3 & A_1B_3C_1 - A_3B_1C_1 \\ A_1B_1C_2 - A_1B_2C_1 & A_2B_1C_1 - A_1B_1C_2 & A_1B_2C_1 - A_2B_1C_1 \end{vmatrix}$

$$= \frac{1}{2} \cdot \frac{1}{\gamma_1 \gamma_2 \gamma_3} \cdot \frac{1}{A_1B_1C_1} \begin{vmatrix} D & A_3B_1C_2 - A_2B_1C_3 & A_2B_3C_1 - A_3B_2C_1 \\ O & B_1(A_3C_1 - A_1C_3) & C_1(A_1B_3 - A_3B_1) \\ O & B_1(A_2C_1 - A_1C_2) & C_1(A_1B_2 - A_2B_1) \end{vmatrix}$$

$$= \frac{1}{2} \cdot \frac{D}{\gamma_1 \gamma_2 \gamma_3} \cdot \frac{1}{A_1} \begin{vmatrix} A_3C_1 - A_1C_3 & A_1B_3 - A_3B_1 \\ A_2C_1 - A_1C_2 & A_1B_2 - A_2B_1 \end{vmatrix}$$

$$= \frac{1}{2} \cdot \frac{D}{\gamma_1 \gamma_2 \gamma_3} \cdot \frac{1}{A_1} (A_1B_3B_2C_1 - A_2A_3C_1B_1 - A_1{}^2B_2C_3 + A_1A_2B_1C_3 - A_1A_2B_3C_1 + A_2A_3B_1C_1 + A_1{}^2B_3C_2 - A_1A_3B_1C_2)$$

$$= \frac{1}{2} \frac{D}{\gamma_1 \gamma_2 \gamma_3} (A_3B_2C_1 - A_1B_2C_3 + A_2B_1C_3 - A_2B_3C_1 + A_1B_3C_2 - A_3B_1C_2)$$

$$= \frac{1}{2} \frac{D}{\gamma_1 \gamma_2 \gamma_3} (-D) = -\frac{1}{2} \cdot \frac{D^2}{\gamma_1 \gamma_2 \gamma_3}$$

若 C_1', C_2', C_3' 爲冠以適當符號之 C_1, C_2, C_3, 之子行列式

則 $C_1' = \gamma_1$; $C_2' = -\gamma_2$; $C_3' = \gamma_3$

故 $\Delta = \frac{1}{2} \cdot \frac{D^2}{C_1' C_2' C_3'}$

二端固定樑力矩之簡易計算法

(Fixed-end Moments)

● 潘應標 ●

譯自 June, 1943, Civil Engineering

在各種不同負荷情形之下，二端固定樑力矩之計算法可以圖一簡化之，如以垂直指示器左右在圖一上移動則可成爲一計算尺矣，此尺可用於集中負荷(Concentrated load)，均匀負荷(uniformly dis tributed load)及均勻增量負荷(uniformly increasing load)上。

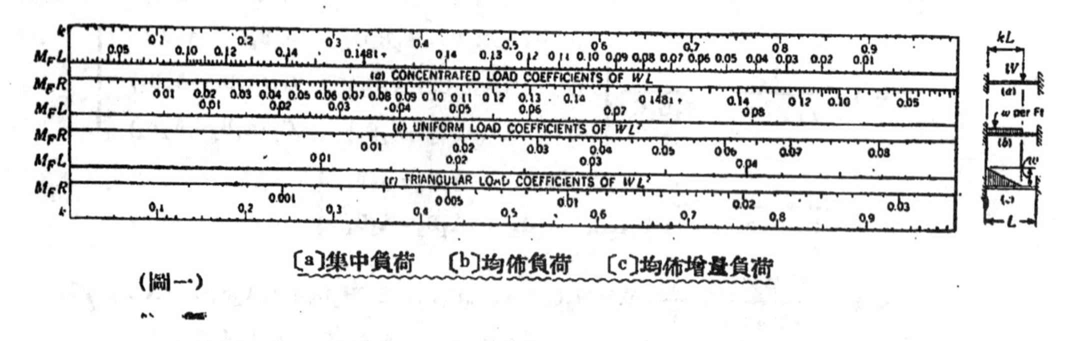

〔a〕集中負荷 〔b〕均佈負荷 〔c〕均佈增量負荷

(圖一)

欲求集中負荷在固定樑左支柱之力矩(M_F L)可用指示器或呎垂直放在圖一之 k 值上，(k 爲負荷在樑之位置)，在左力矩(M_F L)表上可查得一係數，將此係數乘 WL (W=load負荷 L=beam length樑長) 則可得其答案，請注意此係數其數值增至某點然後遞減，因此係數爲感應值 (influence value) 故左支柱力矩之最大 k 值在圖一上爲 $^1/_3$，在右支柱 k 爲 $^2/_3$，此係數實際上爲重複小數 $(0.14814+=\frac{4}{27})$ 如刻在圖一上。

求均佈負荷在固定樑之力矩時，k L 爲均佈負荷之限度由左方起計，以指示器置於 k 值上，在適當負荷表上有 M_F L 及 M_F R 各係數值，將此係數乘 ωL^2 得各支柱之力矩 (ω=uniform load or uniformly varying load 均匀負荷或均匀增量負荷)。

以下爲計算例題，圖一不獨爲利便計算，且爲一感應線圖表 (influence graph)，可藉以劃出感應線 (influence lines) 及最大力矩(max. moment)，爲簡便起見以下之樑長以10呎計，以下數字下有劃者皆由圖一上查出。

25592

例題一：試求下圖固定樑之左右支柱力矩？

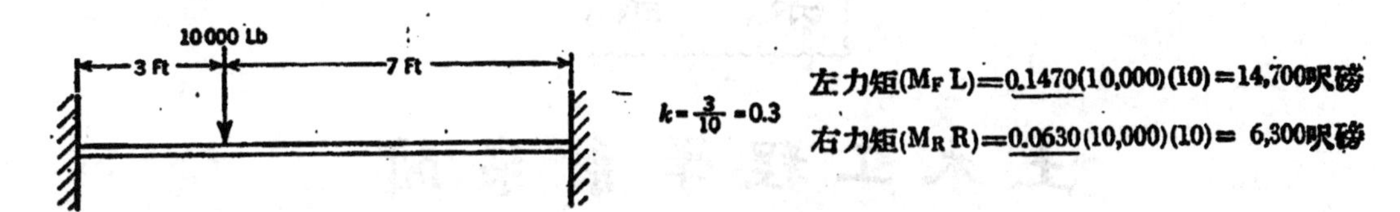

$k=\frac{3}{10}=0.3$

左力矩(M_F L)＝0.1470(10,000)(10)＝14,700呎磅

右力矩(M_R R)＝0.0630(10,000)(10)＝ 6,300呎磅

例題二：試求下圖固定樑之左右支柱力矩？

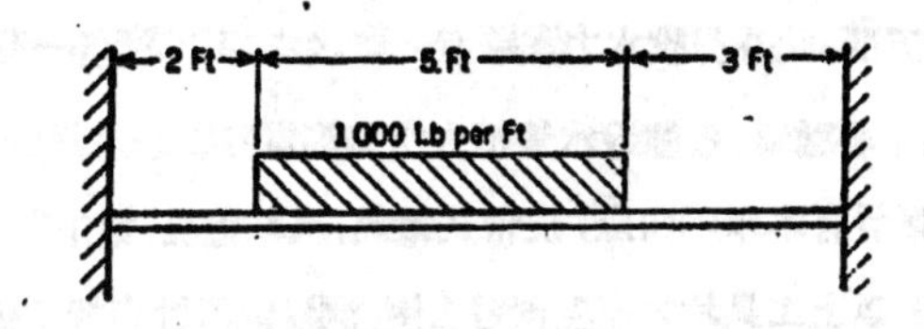

注意在此例題中，在7呎負荷之力矩雖減去在2尺負荷之力矩。

左力矩(M_F L)＝(0.0763－0.0150)(1,000)$(10)^2$＝6,130呎磅

右力矩(M_F R)＝(0.0543－0.0023)(1,000)$(10)^2$＝5,200呎磅

例題三：試求下圖固定樑之左右柱力矩？

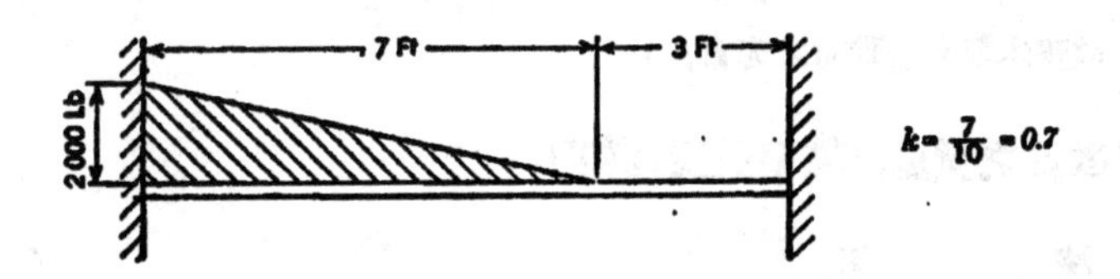

$k=\frac{7}{10}=0.7$

左力矩(M_F L)＝(0.0365)(2,000)$(10)^2$＝7,300呎磅

右力矩(M_F R)＝(0.0166)(2,000)$(10)^2$＝3,320呎磅

系訊

土木工程學系系聞

圖書設備：本系佔有哲生堂之一二兩層，內設繪圖室二間，材料試驗室二間，儀器室一間，電機工程實驗室一間，辦公室六間，及課室五間。此外另設水力實驗室一座，金木工工廠各一間，各實驗室俱裝有水電設備。水力實驗室設有孔口，呑口，各種堰，各種送水管損失水頭等實驗設備及拋物線形與圓形之溝渠衝動水車，離心唧筒等試驗儀器。木工廠置有木車床四座，皮帶式電鋸，風車電鋸，電刨床等各一座。金工廠置有車床三座，刨床，電鑽打磨機各一座，以上工具均爲學生實習之用。測量儀器計有經緯儀六架，水平儀六架，另有精密水平儀一架及六分儀兩架，河流測水速儀二部，上述三種爲南中國僅有之儀器，故時有政府機關及其他大學向本系短期借用材料試驗室設廿五噸萬有應力試驗儀一座，材料堅度儀一座，另有試驗砂石水泥等各種小型儀器。電機實驗室設直流電機五具，交流電機及電動機四具，直流交流伏特計，安培計，瓦特計及電阻箱多具，配速器三具可供二十餘個，實驗之用。本系前身，原爲獨立之工學院，今縮爲土木一系，故種種設備，雖經淪陷期間之損失，仍可應付教學方面之需要。至本系之參考書籍，戰前共有五千餘本，而雜誌亦達十二種之多，惟因戰事搬遷影响，故現未能有若戰前之充實矣。

教員名稱：本系教員之資歷及其專長表列如下：

姓名	職務	資歷	
桂銘敬	教授兼系主任（在假）	交通大學工學士，美國康奈爾大學土木工程科碩士，歷任廣東大學講師，中山大學教授，廣東省建設廳公路處技正，兼工務課課長，粵漢鐵路株韶段正工程師，湘桂路天成路副總工程師，本校教授兼系主任。	鐵路公路
馮秉銓	教授兼本年度代系主任	清華大學理學士，燕京大學理碩士，美國哈佛大學科學博士，歷任本校助教，講師，副教授，哈佛大學講師，特約研究員，本校教授。	電工
黃郁文	教授	美國亞廟工程專門學校工學士，歷任咪吔洋行，富新機器製造廠，禮信洋行等工程師，廣州工業專門學校教授，本校副教授，教授。	機械工程

25594

姓名	職別	學歷及經歷	專長
梁健卿	副教授	嶺南大學工學士，美國麻省理工大學工程碩士，歷任美國公路總局實習工程師，交通部公路總管理處幫工程師，軍委會滇緬公路監理委員會專員，軍委會緬運輸局駐緬副工程師，軍委會運統局公路總處副工程師，美軍部正工程師。	水力工程
韓約瑟 J.A. Hahn	副教授	美國麻省理工大學工學士，曾任美國瑪麗諾大學教授。	航空工程
王鋭鈞	講師	嶺南大學工學士，歷任軍委會技術研究室研究員，技佐，技士，交通部西北公路局幫工程師，廣州市府工務局技士。	
凌鐵錚	助教	湖南大學電機工程學士，交通大學電信工程碩士，曾任上海廣播電台工程師。	電信工程
林炳華	助教	交通大學工學士。	
馮啓德	助教	交通大學工學士。	
江開礦		美國加州大學理學士。	
鄺正文	兼任教授	菲律賓大學工學士，碩士，歷任廣東公路處技正，主任，工程股主任，國民大學教授，本校副教授。	
黃發瑞	兼任副教授	復旦大學工學士，美州愛歐華大學碩士，歷任交通部公路總管理處幫工程師，西南公路局副工程師，交通部公路標準委員會技正，廣東省建設廳技正。	

開設科目：本系開設科目如下：

科目	學期	學分	年級	每週實習時數	教員	
機械畫	一	二	一	六小時	黃郁文	馮啓德
投影幾何	一	二	一	六小時	黃郁文	馮啓德
工廠實習	二	二	一	三	黃郁文	
應用力學	一	四	二	○	韓約瑟	
材料力學	一	四	二	○	韓約瑟	
工程材料學	一	二	二	○	江開礦	
機動學	一	二	二	○	黃郁文	
平面測量	二	十	二	六	桂銘敬	林炳華
材料試驗	一	一	二	三	黃郁文	江開礦

25595

*地質學	一	三	三	○	桂銘敬	
熱力學	一	三	三	○	黃郁文	
*大地測量	一	三	三	三	桂銘敬	
初級結構學	二	六	三	○	王銳鈞	
鋼筋混凝土學	一	三	三	○	王銳鈞	
鋼筋混凝設計	一	二	三	六	王銳鈞	
結構設計	一	二	三	六	韓約瑟	
鐵路測量	一	三	三	三	本系各教授	
土石及基礎學	一	三	三	○	梁健卿	
水力學	一	四	三	○	梁健卿	
水力實驗	一	一	三	三	梁健卿	馮啓德
直流電機工程	一	四	三	三	馮秉銓	凌鐵錚
交流電機工程	一	四	三	三	馮秉銓	凌鐵錚
水紋學	一	二	四	○	黃發瑤	
鐵路工程	一	三	四	○	鄺正文	
道路工程	一	三	四	○	鄺正文	
合約與規範	一	一	四	○	梁健卿	
高級結構學	二	四	四	○	韓約瑟	
鋼筋混凝土橋樑設計	一	二	四	六	梁健卿	
河港工程	一	三	四	○	黃發瑤	
灌溉工程	一	二	四	○	黃發瑤	
水力設計	一	二	四	六	梁健卿	
道路工程材料試驗	一	一	四	三	本系各教授	

上列課程凡有*符號者示本年度未有開設。

研究工作：　數年前本系曾致力於研究用竹竿代替鋼鐵包含於三合土之內，并研究以三合土凝結而成之磚柱作爲房屋中之棟樑其結果雖相當圓滿，但牽涉之技術問題尚多，此當有俟於將來之繼續研究，最近爲研究排洪道之流量將廣東開平縣梁金山水壩造成模形一座，備高年級生作研究之用，本系學生人數爲全院各系之冠，四年級生論文工作甚爲注重。

附　錄

廣州市建築規則

三合土及鋼筋所受之內應力，不能超過下列數目：

受力種類	符號	三合土成份 1:2:4 kg/cm²	1:1.5:3 kg/cm²	1:1:2 kg/cm²
(1) 短柱三合土壓 ($h<40.r$)	fc	30	40	50
(2) 長柱三合土壓 $fc=0.2f'c\left(1.33-\frac{h}{120R}\right)$	fc	30	40	50
(3) 三合土樓面樑桁之壓力	fc	55	60	65
(4) 三合土引力		0	0	0
(5) 三合土剪力(無鋼筋)	V	3	5	6
(6) 三合土剪力(有鋼筋)	V	9	12	15
(7) 三合土與鋼筋黏力(竹節鋼筋)	U	8	9	10
(8) 三合土與鋼筋黏力(光身鋼筋)	U	6	7	8
(9) 鋼筋壓力		十五倍三合土之壓力		
(10) 鋼筋引力	fs	每平方公分 1260公斤 (即1260kg/cm²)		

鋼鐵材料能勝任之力量，不得超過下表之規定：

力別 ＼ 鋼鐵種類	生鐵	熟鐵	鋼
引力 (每平方公分)	200公斤	750公斤	1200公斤
壓力 (每平方公分)	1200公斤	750公斤	1200公斤
剪力 (每平方公分)	200公斤	500公斤	750公斤

鋼鐵柱之載重規定如下：

柱高與最小寬度之比	10	20	30	40	50	60
熟鐵每平方公分	720公斤	680公斤	650公斤	620公斤	580公斤	550公斤
生鐵每平方公分	600公斤	580公斤	550公斤	520公斤	490公斤	460公斤
鋼每平方公分	1080公斤	1030公斤	970公斤	930公斤	880公斤	830公斤

建築所用之木料，其應力不得超過下式之規定：

木之種類 ＼ 受力之種類	壓力(每平方公分)	引力(每平方公分)	剪力(每平方公分)	彎力(每平方分公)
杉木，松木，櫻木，樟木，雜木	70公斤	70公斤	10公斤	70公斤
柚木，抄木	100公斤	100公斤	15公斤	100公斤
坤甸，鐵抄	120公斤	120公斤	20公斤	120公斤

木柱之定限應力，須照下式計算 $F_1=F(1-\frac{L}{10D})$，L代表柱高度，D代表柱寬度：

木之種類 \ L/D	0	10	15	20	25	30	35	40
杉木，松木，樱木，樟木，雜木（每平方公分）	F=70公斤	63公斤	60公斤	56公斤	53公斤	49公斤	45公斤	42公斤
柚木，抄木（每平方公分）	F=100公斤	90公斤	85公斤	80公斤	75公斤	70公斤	65公斤	60公斤
坤甸，鐵抄（每平方公分）	F=128公斤	120公斤	120公斤	96公斤	90公斤	84公斤	78公斤	72公斤

計算建築物各部之本身重，應以下表所列建築材料之單位重量爲最低之依據：

建築材料	白麻石	沙石	磚墻	士敏三合土	煤屑三合土	磚屑三合土	鋼筋三合土	鬆土	實土	杉木松木	雜木樟木樱木	坤甸柚木抄木	生鐵	鋼	空心磚
每立方公尺重	2600公斤	2400公斤	1800公斤	2200公斤	1500公斤	1800公斤	2400公斤	1600公斤	2000公斤	700公斤	800公斤	900公斤	7200公斤	8000公斤	1400公斤
建築部份	單層瓦面	雙層瓦面	木樓面單層階磚	木樓面雙層階磚	天花批盪	木板天花	窗門	門	大門	單隅磚（無批盪）	雙隅磚（無批盪）	磚墻批盪每面			
每平方公尺重	100公斤	150公斤	120公斤	180公斤	50公斤	30公斤	15公斤	30公斤	50公斤	200公斤	400公斤	20公斤			

廣東各地逐月平均雨量表

單位 MM

區域	站名＼月份	1	2	3	4	5	6	7	8	9	10	11	12	全年	紀錄年份
西江	蒼梧（梧州）	34.5	55.0	191.8	155.9	204.5	198.5	154.9	177.4	96.0	43.2	39.4	40.6	1396.7	1900—1938
	新興	36.2	51.6	82.2	101.8	184.9	145.4	200.8	202.2	154.4	58.5	36.6	26.2	1230.7	1927.8—1938
	高要白土	31.8	53.7	72.8	121.3	265.3	136.3	280.0	256.0	212.9	53.0	24.2	42.8	1550.1	1928.4—1938
	高要宋隆	40.9	53.9	74.5	143.7	273.6	240.2	293.8	232.0	174.8	62.9	29.7	28.0	1707.8	1923.5—1938
	高明三洲	45.5	43.0	87.7	99.4	250.1	195.1	205.1	204.6	270.4	98.3	16.0	25.8	1546.3	1934.6—1938
北江	連縣（連州）	45.4	90.3	144.4	199.8	238.9	206.2	165.4	178.5	75.0	58.2	49.3	47.2	1543.6	1921—1938
	南雄	49.1	100.6	166.3	230.9	274.8	249.0	134.1	187.7	107.3	55.0	36.4	51.9	1643.1	1918.3—1938
	樂昌	32.2	70.3	131.7	187.3	246.0	256.2	168.6	184.3	93.8	64.6	41.3	45.9	1522.2	1920.2—1938
	曲江	49.9	94.5	133.0	210.0	266.8	245.8	117.9	180.6	81.6	47.4	35.2	41.7	1504.4	1918.3—1938.9
	英德	40.8	97.4	130.1	225.9	320.7	328.3	173.2	238.6	101.1	50.0	38.5	32.0	1776.6	1918.5—1938.9
	翁源	48.9	139.2	175.8	262.6	379.8	33.4	261.9	325.4	143.7	45.0	11.5	43.8	1871.0	1935—1938.9
	清遠	49.4	79.2	118.1	196.9	338.6	481.1	356.3	340.4	170.3	85.9	37.1	47.8	2351.1	1932.6—1938.8
東江	龍川	47.3	101.6	147.4	146.3	246.2	218.3	171.8	199.6	120.8	39.2	35.8	38.9	1513.2	1920.2—1926.6 1927.6—1933
	河源	42.9	98.9	161.0	229.0	294.9	293.3	253.0	242.2	131.5	43.7	33.7	38.4	1867.5	1920—1938
	石龍	39.6	84.0	113.8	178.1	222.1	265.9	268.3	268.2	143.5	33.0	42.5	36.5	1695.5	1920—1938.9
三角洲	三水	38.3	63.5	117.1	176.1	279.2	249.8	235.5	251.7	157.4	73.7	41.6	28.0	1711.9	1900—1938.9
	廣州	44.7	71.0	92.9	149.3	250.6	268.9	251.7	243.4	138.2	57.6	39.6	34.6	1642.5	1908—1938.9

25599

稻作需水量

1. 中大農科15—17年測定水稻平均每週需水量(mm)(試驗地廣州)

早造

週別	1	2	3	4	5	6	7	8	9	10	11	合計
需水(mm)	23.11	21.31	38.87	48.81	34.73	34.73	63.39	54.95	58.10	55.13	59.30	496.94

晚造

週別	1	2	3	4	5	6	7	8	9	10	11	合計
需水(mm)	32.12	49.58	39.43	42.53	45.00	53.19	48.70	31.27	51.33	58.98	69.05	521.18

2. 中大農科民國15—17年三年間平均每週人工灌溉量(試驗地廣州)

早造

週別	1	2	3	4	5	6	7	8	9	10	11	合計
(需水)灌溉水量(mm)	12.11	0.77	2.97	15.31	5.94	——	——	9.97	12.58	——	——	59.65

晚造

週別	1	2	3	4	5	6	7	8	9	10	11	合計
灌溉水量(mm)	0.42	13.08	8.45	16.77	22.27	21.67	36.65	13.03	29.60	32.67	40.83	245.39

3. 整田期之患旱情形：春雨不足延誤植期者多於晚造(試驗地廣州)

民國十至十八年整田期之雨量及蒸發滲漏量(mm)

(蒸發、滲漏及水道上之損失1%—2%)

早造

年份(民國)	10	11	12	13	14	15	16	17	18
雨量	80.2	94.3	131.8	14.5	81.7	24.2	43.9	223.9	126.0
蒸發滲漏量	48.1+20.7	35.6+20.7	36.8+20.7	46.9+20.7	42.0+20.7	77.4+20.7	60.6+20.7	44.1	49.9+20.7
較差	+11.4	+38.0	+74.3	−71.1	+19.0	−73.9	−37.4	+159.1	+55.4

晚造

年份(民國)	10	11	12	13	14	15	16	17	18
雨量	20.2	—	346.11	69.8	30.4	135.3	248.0	69.8	210.7
蒸發滲漏量	66.7+20.7	—	52.5+20.7	71.8+20.7	81.8+20.7	58.3+20.7	48.9+20.7	108.9	48.7+20.7
較差	−67.2	—	+273.0	−22.7	−72.1	+59	+131.1	−66.7	+131.3

4. 設計灌溉需水量之標準(其他各地須酌量各地雨量氣候土質變更之)依照中大試驗觀察其結論如下：

a. 早造在插秧期前整田期間，晚造在水稻孕穗期以後，患旱最甚，供給灌溉水量，應以適應此兩個期間需要爲合宜。

b. 早造插秧期無雨，田土雖極乾旱，灌溉 120 mm. 已足。

c. 晚造以民國十七年爲標準，旱期23日共需256 mm(八寸)即每萬畝所需灌溉之水流量爲 0.705 m^3/sec. 抽水機每秒抽水 1 ft./sec. 可灌田 10)市畝，蓄水庫所需要之蓄水量須另計蒸發及滲漏損失，渠道輸水須另計渠道滲漏損失。

編餘語

事前，該說明一下我們的編輯態度。

爲了目前專門學科名詞上迻譯的不統一，因而對於原文稿件(指英文而言)，我們無從拒絕。於是在編排上不能不稍作考慮.本來，把英文與中文稿件各編在一起，對觀瞻上是妥善的。然而，對於讀者的翻閱方面，不無麻煩之嫌。經過酌量後，我們決定以稿件的性質來打破中英文的界限，而概括地編彙在一起，相信讀者也同意吧。

本期內容共分轉載，專著，論著，評述，報導，參考，系訊，與附錄等八欄.稿件未算充實，這故然是編者失咎之點，然而，多少該牽連到費用方面.爲了經濟上拮据的羈束，使我們不能不縮狹了六分一的編幅。因之三數篇大作，皆在割愛之列，這對於各讀者與作者無疑是一個大損失；然而，除此以外，別無他法。我們特此致歉。

對於這一期，我們不敢作任何大抱負。因爲這期只是本刊今後出版的初次嘗試。我們願意在這期中面臨困難，而找出其預防的根源，遇到漏洞，而尋求其補救的良方，這樣，自後本刊的出版，必然進步。編者並非値此以推卸其失責無能之點，只要讀者對於這一期不至過於失望，我們已感到萬分慶幸了。

最後，該補充一點的就是爲了教授與同學們教務和學業上的煩重，以至徵稿的時日雖久，而着手編輯的時間實少。因而編排上難免差錯，尙希各讀者原諒。

李耀記

·潔淨·工程專家·

專營

衛生設備

冷熱水喉

鑽井地渠

工程

兼售

洋瓷潔具

五金器皿

鋼窗鋼門

建築材料

香港德輔道中三十七號 電話二三〇三三

補行鳴謝

本刊經費籌集，荷蒙校内外人士熱烈贊助暨各商號惠顧廣告費於前，並得嶺南大學，及各仝學特別捐助於後，本刊同人，不勝雀躍，謹此補行鳴謝，兹將其芳名列後。

嶺南大學捐助國幣六百萬元

姚保照　鍾福華　林思進　潘演強　曾文達　牟基球　鄧錦榮　陳子浩

李克勤　周公海　林壽庚　吳乘俠　以上同學各捐助港幣壹拾元

南大工程

康樂再刊第一期

嶺南大學工程學會出版

每本暫售國幣拾萬圓

主編者： 吳秉俠

發行者： 嶺南大學工程學會

經售處： 嶺南大學工程學會會所

出版日期： 中華民國卅七年二月二十日

南大工程

桂銘敬

工學會日紀念特刊

THE JOURNAL OF THE LINGNAN ENGINEERING ASSOCIATION

康樂再版　　　　Vol. 2

第二期　　　　May 16th. 1948

嶺南大學工學會刊印

民國三十七年五月十六日出版

廣州蔚興印刷場承印

南大工程

•第二期•

工學會日紀念特刊

出版委員會

顧問：梁健卿教授　劉載和教授　黃郁文教授

主席：趙浩然

編輯：趙浩然　黃漢基　陳潤初

廣告：楊民安　李克勤

校對：何思源　歐陽讓

總務：羅明享

電油發動：Concrete Mixer, Aircompressor,
Piling Machine, Stone Breaker,
Rock Driller;

Electric Motor, Centrifugal Pump, Plastic Cables

建築工程材料，英國漆油，化學工業原料

總代理

新志利洋行

香港　法國銀行大厦四樓　電話22765

廣州　沙面敦睦路四號　電話13024

香港油蔴地輪船有限公司

中環德輔道中一四四至一四八號　電話三一三五一　三一三五二

THE HONGKONG & YAUMATI FERRY CO., LTD.

Phone 31351 - 31352. 144-148, Des Voeux Road, Central, Hongkong.

香港開長洲綫

	a.m.	a.m	p.m	p.m	p.m
香港開	7.15	9.30	★10.30	1.00	5.15
坪洲開	8.25	10.40	——	2.10	6.25
梅窩開	8.50	11.05	——	2.35	6.50
到長洲	9.30	11.45	11.50	3.15	7.30

長洲開

	a.m.	a.m.	p.m.	p.m.	p.m.
長洲開	6.30	9.45	1.00	5.00	5.30
梅窩開	7.10	——	1.40	5.40	——
坪洲開	7.35	——	2.05	6.05	——
到香港	8.45	11.05	3.15	7.15	6.50

★星期日及公衆假期加開

南頭綫

香港開上午七時卅分　南頭開上午十二時

汽車船

香港開 上午六時二十分至 下午十時四十分止

佐頓道開上午六時四十分至下午十一時止

旺角綫

旺角開頭渡上午六時二十分
尾渡下午十一時

香港開頭渡上午六時三十分
尾渡下午十時廿分

深水埔綫

深水埔頭渡開上午六時二十分
尾渡下午十一時

香港開頭渡上午六時卅分
尾渡下午十一時廿分

佐頓道客船

佐頓道開頭渡上午七時四十分
尾渡下午十一時

香港開頭渡上午六時二十分
尾渡下午十一時

大澳青山綫

香港開	2.00p.m.	大澳開	7.30a.m.
水門開	3.00 „	屯門開	9.15 „
屯門開	3.45 „	水門開	9.45 „
到大澳	5.45 „	到香港	11.00 „
每逢單日順泊東涌		每逢雙日順泊東涌	

南 大 工 程

康樂再版●第二期

工學會日紀念特刊

—目 錄—

鳴謝

本刊此次出版，經費龐大，幸得留穗劇人于四月廿八日，廿九，五月一，二，三一連五晚假長堤青年會禮堂舉行演出曹禺名作三幕白話劇,「日出」爲我們出版基金籌欵，乃克完成，高誼隆情，至以爲感。其次并得林文贊先生踴躍介紹廣告，陳潤初，呂惠炎，黃文添三位同學不辭勞苦親身赴港接洽徵求廣告，更蒙各界人士愛護帮忙，并得陳樹彬先生捐助港幣50元，姚德霖先生捐助港幣50元，本刊同人，至深紉感，特此一併鳴謝。

25612

25613

25614

專載

南大工學會日成立的經過和意義

——劉載和——

距今二十年前，本校受鐵道部的委託，開設工程學院，來訓練土木工程人才。到民國廿三年，第一屆工科學生畢業，當時人數很少，後來漸漸增加。民國廿七年，因爲適應當時環境的需要，把原有的文理學院分而爲二，工學院則和理學院合併而成爲理工學院，僅設土木工程一系。民國廿七年秋，廣州淪陷。本校遷往香港復課，仍開設工科。至卅年冬，香港又告失陷，本校乃內遷粵北，因爲儀器設備無法內運，乃將理工各科暫行停辦。當時在本校修習工程學科的學生，多轉學別校。卅四年秋，日寇投降，本校搬回廣州原址，再恢復理工各科，至今年夏始有工科學生畢業，是爲抗戰後的第一屆，和抗戰前最後一屆相距足足有八年之久。因爲中途經過長期間的停辦，而且前後的組織又不相同，所以抗戰前畢業的同學和抗戰後的畢業同學，儼同別校，非常隔膜。在校的同學有見及此，去年曾計劃成立工學會日，請各舊同學返校聚首，藉資聯絡。後來因爲功課太忙，卒未能實現。本學期開始的時候，舊事重提，在校的同學皆覺到有此需要，因是遂積極進行。至於舊同學方面，不特感到新舊同學間甚多連繫，而已畢業的舊同學亦缺之聯絡。且認爲母校的工科學生與社會常常接觸，所以有成立工科同學總會的提議，並推舉同學與在校同學密切合作，積極籌備。並定成立之日爲「嶺南大學工科同學總會成立日」簡稱工學會日。這一天除請各舊同學返校團聚之外，還請本市各大學的工學會及各工程團體機關派代表參加。至於這天的秩序，則分爲學術和聯誼兩方面。學術方面包括出版特刊、成績展覽、開放實驗室、公開試驗建築材料各項；聯誼方面則開紀念會、聚餐、球賽、和晚會項。這是工各學會日成立的大概經過！

這次工學會日的成立，是本校工科自開辦以來的第一次。因爲時間的傯促，所以沒有充份的準備，成績或有未能盡如人意的地方，但這一日的意義，則非常重要。現在把牠對於社會、學校、和學生三方面的意義，分述之如下：

對於社會方面

（1）發揮嶺南的真精神：凡爲嶺南人，都知道「爲神爲國爲嶺南」的校訓。這就是說，在校的時候要爲嶺南，離開了學校的時候要爲國家，不論在校或離校都要爲神服務，換句話說，就是要造福社會。這是嶺南對於每一個學生的期望。在工學會日那天，新舊同學聚首一堂，在校的新同學，把各實驗室整理好，開放參觀；把各種成績搜集齊，公衆展覽，這種爲嶺南的精神，藉着這個成立日傳播到校外去。畢業的舊同學，曾不避艱苦的服務社會，曾竭誠盡意的効忠國家，這種爲國的精神，亦在這一天帶返校內來，在融和歡洽的空氣中，交流着爲國爲嶺南的精神，提高每個同學高潔的意志，發揮嶺南的真精神。

（2）改革嶺南的舊作風：以往的嶺南學生，多養尊處優，一切衣食住行固然和當時一般人不同，而對於社會一切的事，都抱着超然的態度，和社會格格不相入。所以嶺南的畢業生，在任何的機關裡，在工作上都陷於孤獨無援的境地。我們這個工學會日成立的目的，就是想改革以往的舊作風，除去惟我獨尊的超然態度，而盡力和社會打成一片，爲他日服務社會的階石。

（3）利用嶺南的好環境：嶺南有豐富的圖書，有充足的儀器，有美麗的校園，有幽靜的環境。我們成立工學會日就是想社會知道嶺南的好環境而能充份利用它。我們希望工商界的領袖能利用我們工業上的人才和設備，來幫助他們解决工業上的難題，利用我們良好讀書的環境，爲他們培植優秀的幹部。

對於學校方面

（1）使國家信任學校：政府對於幾間國立大學，特別注重，所以在國立大學畢業的工科學生，都由政府派往各機關實習，而在私立大學畢業的工科學生，則甚少留意。這因爲政府對於私立大學缺乏認識所至，我們在工學會日那天，可以使政府知道嶺南有優良的讀書環境，有充足的儀器設備。嶺南的工科學生，都是力學不倦的。嶺南的畢業同學，在每一個機關裡都能表現出優異的工作，使國家從此信任學校。

（2）使社會了解學校：在工學會日那天，我們所展覽的，是學生在烈日之下用汗和墨所測繪而成的圖表，是學生在實驗室裡用汗和泥所造成的模型。漆黑的面孔，使社會人士知道我們並非官仔少爺的貴族學生，樸實的裝束使社會人士知道我們已無洋氣十足的派頭。

（3）使家長明瞭學校：做家長的，用他們辛辛苦苦所得到的金錢，來栽培他們的子弟讀書，都想他們的子弟將來學能致用。所以對於他們的子弟的學識和技能，常常發生疑問。在工學會日裡，做家長的可以見到他們的子弟的成績，知道學校是怎樣的在教導他們的子弟。

對於學生方面

（1）檢討自己的工作：他們可以藉着工學會日的成績展覽，來檢討他們一年來的工作。

（2）表露自己的能力：他們可以利用工學會日的各項工作來表露自己的辦事能力。

（3）開闢自己的前途：在工學會日那天，有已畢業的舊同學，有各團體的代表，有各機關的首長，他們可以利用這個機會，交接工程界的先進，來開闢自己的前途。

由以上幾點看來，足見這一個工學會日是含有很重大的意義。希望從此以後，在校的同學和畢業的同學都能珍重這一日的意義，盡力來光大它！

工學會日節目表

上午10:00－11:00　各實驗室開放展覽

11:00－11:00　舊同學報到

地點：工學院

12:00－1:00　紀念會

地點：工學院101課室

下午 1:00－2:00　叙　餐

地點：大學膳堂

2:00－3:00　討論會

地點：工學院101課室

2:30－4:30　各實驗室開放展覽

3:00－4:30　球類比賽

地點：小學球場

4:30－6:30　遊　覽

7:30－10:00　晚　會

地點：工學院201課室

工學會日籌備委員會

顧　問：馮秉銓教授，黃郁文教授，劉載和教授，鄺正文教授，
梁健卿教授，黃發瑤教授，王銳鈞教授，江開礦教授，
林炳華教授，馮啓德教授，麥鉄錚教授，韓約瑟教授，
J.A.Hahn

主　席：吳乘俠

副主席：趙浩然，楊民安

總　務：曹文達

(一)交通：呂惠炎　(二)招待：歐陽讓

(三)傳宣：關學海　(四)佈置：梁耀顯

財　政：潘演強

文書，聯絡：尹行賢

學術，展覽：黃漢基

(一)材料試驗室：黃文添，黃康道，余文傑，趙浩然。

(二)電機試驗室：陳潤初，潘應標，鄔境厚。

(三)水力試驗室：鄺國良，劉益信，潘世英。

(四)金 木 工 廠：黃漢基，徐良佐，李偉文，區元侃。

(五)測量實驗室：譚傑靈，謝國光，鄔堅栢

活　動：梁謹紹

攝　影：韋基球　巢永棠

廣州私立嶺南大學工科同學總會會章

第一章　定名

第一條　本會定名爲「嶺南大學工科同學總會」。

第二章　宗旨

第二條　本會以聯絡感情交換智識以達到互助互勵爲宗旨。

第三章　會址

第三條　本會會址暫設在廣州康樂嶺南大學。

第四章　會員

第四條　凡在嶺南大學工科畢業或肄業之同學及曾在校任教職之教員皆爲本會會員，

第五條　凡在本校附屬各校畢業或肄業之同學而從事工程業務或研究者經由會員二人以上之介紹及得理事會通過者皆可爲本會會員。

第六條　凡本會會員皆得享受本會一切權利並有遵守會章服務本會之義務。

第五章　會費

第七條　本會每年常費交由理事會決定徵收之如遇有特別需欵時得由理事會提出交監事會通過再行徵收之。

第六章　組織

第八條　本會組織採用理事會及監事會制理事會及監事會由大會選舉理事及監事組織之。

第九條　理事會之職權如下：

(一)有代表本會對外之資格。

(二)處理本會日常之事務。

(三)執行大會議決事項。

第十條　理事會設常務理事十一人皆由上屆理事會提名交由全體會員票選之並由理事中互選出理事長一人副理事長一人文書二人事務二人福利二人會計研究交際等幹事各一人。

第十一條　監事會之職權如下：

(一)稽核本會財政之收支。

(二)監察本會一切對外對內會務。

第十二條　監事會設常務監事五人均由全體會員提名票選之並由監事中互選一人爲監事長。

第十三條　本會理監事任期皆定爲一年於每年工學會日爲新舊交接日期但連選者得連任。

第十四條　本會得聘請國內有名之工程界先進爲顧問。

第七章　會員大會

第十五條　全體會員大會爲本會最高之機關於每年工學會日舉行之如遇特別事故得由理事會或經三分二會員之連署召開之。

第八章　附則

第十六條　本會會章如有未盡事宜經會員五人之提議三分二以上之會員決議得修改之。

潭岡測量實習歸途前留攝

黄埔港碼頭前留影

水力試驗室之一部

「新廣東」船前得意的一群

繪圖忙，在潭岡

材料試驗時之一瞥(竹之拉力試驗)

新自英國送來之刨牀

竹之拉力試驗

「新廣東」船上得勝者

材料試驗室機器之一部(拉力試驗機)

金工廠機器之一部

水力實驗室儀器之一部(流速儀)

論　著

黃埔港工程上的幾個問題

——陶　述　曾——

在工程觀點上看，黃埔港是問題最少的港。有幾個問題，解決都不困難。現在提出來簡單地談一談：

第一，風浪問題　黃埔是內河港，水流方向確定，流速不大。四週遠近都有山，颶風不常經過。江面之寬足够停充分數量的船隻，但最寬處不過二公里，因此風浪不大。無須建築防波隄。江岸有天然的小岔港，小船可以避風，不須專做避風塘。

第二，潮汐問題　黃埔一帶山洪的漲落與潮汐合起來最大的水面高差不過四·三四公尺，每日潮汐之差只一·六公尺。碼頭地面高出尋常低水面只須四·三公尺。大小輪船停泊都不發生上下不便的困難。換句話說，就是碼頭建築不必因水面漲落而作特別的計劃或設備。

第三，地基承托力問題　這一帶江岸江底的土質都是粘土和粗沙相間的冲積層。承托力很大，地面十五六公尺以下就是沙石層。碼頭和高大建築物的地基都不須打很長的樁，普通房屋地基可以不打樁。

第四，給水問題　海潮可以上溯到黃埔。江水在冬春水小的季節含鹽質百萬分之一千二百五十。不合飲用和鍋爐用。這一帶的山溪來源都很短。附近山澗也沒有看到可築大規模蓄水庫的地方。所以戰前黃埔開埠督辦公署和前年的行政院工程計劃團都注意到淡水供給問題。最近研究的結果，知道這一帶鑿井深不到十公尺就有可飲的淡水。但水量不大。山溪可以利用。但源頭不遠，中下游低原農村人口稠密，水量小而不潔淨。當地農村近三十年來以械鬥著名，原因只爲爭水。利用附近溪水可能引起人事糾紛。所以這一問題不能在黃埔附近求得滿意的解決。

好在黃埔離廣州市不遠，可以和廣州市水廠共一水源。增步在廣州的上游，水的質和量都够標準。離黃埔港最遠一端也還不到四十公里。中間沒有高山，裝置水管工程並不大。開港初期，水管沒安好以前港內人數不多，可用井水。

第五，排水問題　這一帶平地高度都在高潮水面以下。每三五年伏秋大汛淹沒一次。爲時五六天，水深一兩公尺。碼頭堤岸的高度根據四十年來的水文記錄，可填到高潮水面以上。但將來發展的市區地面寬闊，便不能都填到相同的高度。只有把沿江隄岸連續築成基圍。圍內做成排水系統。排水溝口建築水閘。市內多挖大面積的蓄水塘。濕水季節滿塘水面盡量放低以備儲蓄大汛時的雨水。水塘四週和排水幹溝兩岸作成園林住宅區。因此這一個都市必須設計成為一個花園都市。

因為珠江的洪水位並不太高，洪水持續時間也不長，所以排水問題並沒有漢口和天津那樣嚴重。

第六，航道水深問題　黃埔港區以內錦鼓沙附近航道有一段淺灘叫做第一沙。港區以南十五公里，蓮花山附近有一段淺灘叫做第二沙，這兩處水深在低水位以下只有五公尺。虎門以外，珠江口有一段長二十公里的淺灘叫做零丁沙。水深六公尺。除這三段淺灘外航道水深都在九公尺以上。萬噸洋船要進到黃埔港，這三段淺灘必須濬挖。估計挖泥量有二百八十萬公方　挖成以後，保持航道深度每年得挖十萬公方。

用濬挖來保持航道深度自然是最笨的辦法。最好是建築潛水壩，束水攻沙，使航道自然刷深。

總理　在建國方略中即曾提出建築潛水壩的方案。但虎門以外水面寬泛，壩工不容易做。又珠江下游支流繁密如蛛網。在各支流的水文沒有弄清楚以前，壩的設計也不好冒然着手。而弄清這種繁複的水文，就需要相當長的時間和不少的經費。

保持航道水深是黃埔港唯一的大問題。但比起上海，塘沽，天津來，困難却小得多。黃埔江經過五十年的整理，到現在每年還得挖泥四百萬公方。海河經三十多年的努力，現在只有五公尺深，比目前的珠江下游還淺一公尺。塘沽新港想挖到黃埔目前的深度相同，不知道要費多少年的力量。在技術上似乎還沒有十分把握。三處比起來，黃埔港在這一方面可說是得天獨厚的。

一九三四年航道測量圖記載第一沙最淺之點水深四、八公尺，第二沙五・--五公尺，零丁沙六・四〇公尺，前年測量的結果，第一沙最淺之點水深五・〇〇公尺，第二沙五・〇八公尺，零丁沙五・七八公尺。第一沙在一九三六年浚挖過，所以現在比一九三四年還深。其他兩處十二年間，淤澱最多之點不過〇・六二公尺。大部分情形並無顯著變化。航道濬深以後，水流比較集中，淤澱可能減少，所以航道淤淺的情形並不怎樣嚴重。

減少河水含沙量最主要的辦法是上游水土保持工作。珠江流域的氣候雨量最適於種草造林。人口較密水土保持工作容易普遍推行，見效也快。這也是海河流域所不及的。國家如果有十年二十年的安定。珠江上游的童山必可普遍變綠。珠江的水必能逐漸澄清。源頭既清，下游航道的淤塞問題就更容易澈底解決了。

其他的問題都是一切港工所共有的，這裡不多說了。就以上的幾個問題說，其中以保持航道水深工程為最難做，而航道淤淺的情形並不嚴重。我所以說黃埔港有幾個問題解決都不困難。

華南國道設施之要点

——王節堯——

我國之有汽車公路，迨始於民國初年。其時所謂公路，大都因陋就簡，無技術可言。而且各地各自爲政，支離散漫，尤無系統可尋。及民國十六年以後，經中央交通當局之努力，擬定全國公路網與公路工程標準，對於新路之修築，舊路之改善，汽車之聯運，建制漸增完備。至抗戰前夕，不特公路路線延長甚多，而公路業務之發展，亦有顯著之進步。中日戰事發生之後，國土大部淪陷，原有公路被破壞者達百分之六十以上。迨卅四年八月，日寇投降，國土重光，中央爲統籌全國公路，重行設立公路總局，歸交通部管轄。並將全國公路劃爲九區，每區設局管理。區局之工作爲(一)國道之興建改善及養護，(二)區內之公路交通管理，(三)督導區內省道之興建改善及養護。華南國道劃爲第三區，由第三區公路工程管理局管轄。

本區所轄及督導區域，跨越粵桂閩台四省，劃爲國道之路線，計長七八九七公里，分爲幹線十八條及支線一條。區域旣廣，路線又長，橋涵設備多屬臨時性質，其鋪有路面之路線，不及十之二三。故爲整飭路務，其施政原則如下：

(一)改善及修建公路採取「三不」主義

華南粵桂閩台四省公路路線，全長達二八·二七二公里，已算不少。然計劃新建或改建舊路使標準提高之路線，仍然甚多。該項新建路線，其施工標準，須依照下列三原則：

(1)坡度不可太陡　　公路運輸，速度與安全爲最要。若坡度太陡，則行車速度降低，運輸量減少，平時妨碍貨運，戰時影響交通，對於經濟國防，皆非上策。故新建路線，坡度不宜太陡，而原有公路之坡度太陡者，使其改緩。

(2)灣道不宜太急　　灣道太急，其影響所及，與坡度太陡並無異致。對於行車速度與安全，皆有妨碍，故宜避免。

(3)路基不應太狹　　一公路之運輸，每與年遞增，苟路基太狹，則不特目前行車危險，將來發展，亦感困難。故在建築之時，應預留地步，體察實際需要，以及未來之發展，盡量放寬。

(二)公路交通應本「三先」原則

華南國道路線，在抗戰期間，曾經嚴重之破壞。勝利之後，雖曾加以搶復，後因大水爲災，再被冲壞。近因匪禍，隨修隨毀，故公路運輸，大受影響。以後爲加强公路運輸，以配合勘亂工作，故提倡「三先」原則。

(1)養護先於修築　　中國公路之最大毛病，乃標榜新線之增築，而忽視舊路之養護。其實養應先於築。

築路而不養，有路之名，無路之實，徒耗工款，無補於交通。苟能養先於築，則築路之時工程雖然草率，若保養得宜，運輸力量亦可因是得以加强，而成爲一良好公路。

（2）通車先於求備　目前華南各省公路因先天不足，後天又復限於經費，多半失調，致橋毀路破，時有發生，影響運輸，至爲巨大。爲加强運輸，溝通有無，故路線之修通，尤急於路線之改善。

（3）守路先於修路　公路之目的，乃爲發展交通。然目前路線所經，多爲匪徒嘯聚之所。路基橋涵，固時被破壞，而搶車劫貨，亦時有所聞。至於派往沿線工作之員工，亦被擄去數次。行車安全，既不能保，員工生命，亦朝不保夕，安能發展交通，故目前最重要急切者，乃爲路線之守衛，而非路線之增修也。

（三）路政設施實行「三公」政策

華南爲我國革命之搖籃，人民思想甚爲前進，而公路交通之需要又非常迫切，故對於路政設施，非實行「三公」政策，不足以充分發展華南路政：

（1）公家築路　爲使公路技術得以提高，公路標準得以劃一，公路行政得以簡化，故國道之建築皆須由國家統籌辦理。

（2）公開營運　公路運輸應採開放主義，獎勵民營，以促進行者有其路。

（3）公共守護　際此匪患日趨嚴重之時，公路路線之保衛，若靠政府兵力，實防不勝防。公路路基之養護，若全靠職工，實應付不暇。故應與沿線人民合作，共同保養，方能有濟。

（四）施政目的在完成「三路」任務

公路交通，對於國計民生之重要，已爲吾人所習知。故一切路政設施，皆須配合國家政治經濟國防上之需要，以期完成「三路」任務。

（1）政治路線　修築公路以連絡各政治中心，發展荒僻邊地，便利人民之往來，加速文電之傳遞，使國家施政之效率提高。

（2）經濟路線　修築公路以聯絡各商業中心，工業重點，運輸樞紐，使商業得以繁榮，工業得以振興，交通得以發達，使國家日趨於富强之康莊大道。

（3）國防路線　修築公路以聯絡各港口要塞以及其他軍事要地，使軍運便利，邊疆安穩，以鞏固我國之國防。

以上所述，爲針對現時局勢，適應目前環境所定之國道設施要點數則，用爲華南區內國道設施之南針者也。

湛江建港計劃

——桂銘敬——

（一）總論

（1）沿革

湛江原名廣州灣，位於廣東省雷州半島東岸，於一八九九年十一月十六日中法條約租與法國，爲期九十九年，其範圍包括赤坎、西營、東營、及孟島、硇䑸島、東頭山島、特呈島、烏冠島、東海島、𡉕州島等地。赤坎在昔日原已爲商業區，至一八九九年法人租借後，始將其南十二公里之西營闢爲港口，並劃爲行政區，而商業則仍集中赤坎。法人初擬在特呈島烏冠河之間（又名巴蒙港）闢港，設有機械修理廠及煤站等，旋因故放棄。祗在西營築堤岸千餘公尺，帆船避風塘一處，及長約三百餘公尺碼頭一座，以資木駁船起卸之用。此外並在𡉕州島、烏冠島、東海島等地，設置航道標誌燈塔等以指示航行。迨抗戰軍興，京滬廣州相繼陷，此地遂成爲我國唯一港口，亦爲沿海各地轉運孔道，故漸趨繁盛。尤以香港失陷之後，更爲繁榮，爲法人治下之全盛期。此時法人曾擬有計劃發展該港，將原有碼頭伸長以利便船隻靠泊，展築堤岸約一千公尺，建築船塢二處，並開闢東營港。惜日軍於卅二年二月六日登陸佔領，至計劃未獲實施。在日僞佔領期間，僅沿用舊設備，並無新建設。迨卅四年秋日軍投降，同年九月廿一日我軍進入接收，國土重光，乃改名湛江市，設市政府，隸廣東省。市區範圍，仍包括赤坎、西營、東營、及各島嶼。惟中央見及此次抵戰，西南各省，貢獻特大，而西北對於建國，亦非常重要，乃計劃完成西南鐵路網，以發展西南，開發西北。湘桂黔鐵路爲西南鐵路網之幹線，而此幹線之主要出口則爲湛江，故交通部除將湘桂黔鐵路展築至湛江之外，復在西營建築港口，以爲我國西部海陸聯運之總樞紐。此乃湛江建港之大概經過也。

（2）位置

湛江位於東經一一〇度廿五分北緯廿一度十二分，東距香港約二四〇海里，西距越南之海防約三二〇海里。西界雷州半島，東界吳川縣，北以寸金橋爲界，而與遂溪縣之麻章毗連，南有𡉕州島、東海島、烏冠島、東頭山島、及特呈島爲屏藩，面積約九〇〇平方公里。就地理方面而言，可分爲三區，（甲）麻斜河岸右區（卽西營、赤坎，一帶）爲邱陵起伏地帶，地質多沙土，僅一部份黏土質。（乙）麻斜河左岸區，（卽東營一帶）地勢較爲平坦，土質亦較肥沃，甚宜墾殖。（丙）群島區，包括東南部各島嶼，多爲石山荒嶺，童山濯濯，並無樹木，表土爲雨水所冲刷，變成磽土，沙石流入海中，淤塞水道，影响航行。故此區應植樹種草，以保護土質。

關於港址之選擇，法人前曾擬在特呈島、烏冠河之間築港，惟以水淺港狹陸地區域不多，與各地交通不便，卒而放棄。亦有議在硇州島築港者，以該處水深港闊，巨輪進出便利，惟該島兀立海中，四無蔭蔽，常受颶風侵襲，且與內陸交通不便，殊非築港之地。港址之選擇仍以西營東營，麻斜河一帶爲最宜。南自馬其尖角起，北至北度尖角止，長約八公里，寬約一五八公里，聊廣水深，東南有諸島爲之屏藩，以蔽風浪，實爲商港及軍港之良好處所，而西營毗連赤坎，爲法人經營五十年之城市，可資利用，故決定以東營西營一帶爲港址。

(3) 形勢

本港北連大陸，南通深流。橫過其中之麻斜河，鳥冠河，皆爲天然之優良深水港，港灣深，水深均在一〇公尺以外，可容巨舶。雖進口水道在硇州島及東海島附近各有淺沙一處，然在低潮之時，深度亦有六・四及五・八公尺，數千噸輪船可以暢航無碍。萬噸以上之巨輪，亦可候潮進出，至於聯接大陸，有公路可通，將來鐵路完，成交通更爲便利，必成爲我國西北西南通海惟一孔道，故本港在經濟上之地位，甚爲重要。

本港形勢，不特在經濟上佔重要地位，而在軍事上亦非常重要。因其港灣深廣，東西北三面皆連接大陸，南面有諸島環繞，屏障不虞風浪，可發展爲優良軍港且位於太平洋之北岸，爲歐洲至遠東航線所必經之地。其南爲海南島，楡林港在焉，日俄戰事之時，俄國波羅的海艦隊曾寄碇於此，淪陷期間，日人經營甚久，計劃建爲軍港。其西南之法屬安南，爲法國在遠東最大之殖民地。渡南海而南，則有南洋羣島，英屬之新加坡，英人曾在新加坡建築一堅固優良之軍港。其東南則爲菲律濱羣島，爲美國在太平洋之前哨。故本港實爲我國南部邊陲要地，內可併緊國內各軍事要點，外可控制歐亞及南洋羣島海上之交通。在此次中日戰爭其中，日本不惜破壞中立，强行佔領湛江，因爲對我國作戰之據點，不爲無因。其後美軍亦曾選定此處爲登陸中國作戰之地。其在軍事上之重要性，可想而知也。在昔日鐵路未築，交通不便，已爲各國軍事家所注意，將來鐵路完成，交通便利，則其地位形勢之重要，更爲國防建設所不可忽視者矣！

(4) 港灣

本港之港灣，皆自天成，共分爲內港，外港，進口水道三部份。內港包括西東營市鎮，即麻斜河一帶，南北長約八，〇〇〇公尺，東西寬度以東西營間爲最狹，約一，四〇〇公尺，其他各處寬度多在二，三〇〇公尺左右。低潮水深在十公尺以下者，水面面積約六・三〇〇・〇〇〇平方公尺。西營沿岸較爲平坦，沙灘伸出岸邊約三四百公尺。水深在八公尺以外者多距岸約四五百公尺，幸航道中部水深多在一〇至二五公尺間，亦有超過三〇公尺者，足供巨輪停泊。東營沿岸地勢較陡，水位亦較深，宜於巨輪灣泊，爲船舶之安全區，巴蒙港寬約六〇〇公尺，長約二五〇〇公尺，低潮時水深由四・五〇公尺至七公尺。外港則包括自馬其尖角以南至東頭山一帶，水深港闊，亦即原來所稱之廣州灣也。

進口水道分歐司多水道及古流水道兩處。歐司多水道乃自特呈島馬其尖角西南行，沿東頭山島西邊及東海島西北，迂迴西南行出海，道狹水淺，沙灘横亘，僅爲帆船出入便道而矣！故曾擬堵塞此水道使水流集中古流水道，以增加古流水道之流量及深度，然仍須待詳細測量研究之後，方能決定。古流水道爲現時船輪進出航道，法人曾沿各島設置標誌，指示航行。船隻自馬其尖角起航，正對東頭山島標誌，南行長約三·四公里，折東南行，經東頭山島北部，後對伯打利尖角附近標誌前行，長約五公里，水深均在十二公尺至三十公尺之間；稍再微偏對東海島標誌航行，長約四·八公里，本段航道較淺，且沙洲數處侵迫，低潮時，最淺處水深僅有六·四公尺，萬噸以上之巨輪，須候潮進出，過此折緇東北，對烏冠島標誌航行，長約二·六五公里，水深均在十一公尺以外，又折東南，沿烏冠島及東海島之中航行，長約七·四公里，水深均在24公尺以外，此即爲湛江入口處，位於東海島東北角及硇州島之北。過此再折東南行，對烏冠島及南山群島標誌，長約六·六公里，最淺處低潮時水深在七·七公尺。再轉向東，沿所謂孟觀番道航線出海。惟沿海口均有沙洲橫亙，最淺處深度爲低潮下五·六公尺，幸面積不太，過此以後，水深海濶，分往世界各大港口，毫無障碍矣。總計進口水道全長約三三·三公里，寬約在三〇〇公尺至一·〇〇〇之間。深度以硇州島北出海處之五·六公尺爲最淺，次爲東海島北之六·四公尺。平均潮差約二·五公尺，最大潮差約四公尺，數千噸之輪船可以自由航行無阻，萬噸以上輪船則須候潮始能進出。將來加以濬深，使數萬噸之巨輪亦可以進出無阻，則必成爲頭等港，而與世界各大名港相埒矣。

（5）氣象

湛江港之氣象及水文資料，法人曾有長期之詳細紀錄，惜在日人佔領期間，卷帙多已散失，現所能搜集者，僅得一九三九年至一九四四年之西營氣象紀錄，一九三六年全年之湛江硇州潮水紀錄，及一九四六年七月一日至一九四七年三月卅一日之西營潮水紀錄，以供參考矣。

氣候：本港緯度甚低，位於亞熱帶，故熱季時間甚長而寒季時間甚短，幸有海風調節，寒熱尚不甚懸殊。全年平均溫度爲二三·五度。最熱月份爲六月份平均溫度爲二八·六度。最冷月份爲十二月份，平均溫度爲一七·一度，相差僅一一·五度。其最低溫度爲三·二度。可知本港夏無盛暑，冬不結冰。

雨量：本港雨量充沛，每年雨量平均爲一，七五四·九〇公厘。每年降雨日數平均達一二四·三〇日，雨量多集中於五，六，七，八，月．此數月之降雨量爲一，〇六二·三〇公厘，約佔全年雨量百分之六十。冬季雨量稀少，十二月及一二月之降雨最小爲八六·四公厘，僅佔全年雨量百分之五而已。又本港多暴風雨一日間之雨量達達一〇〇公厘以上之紀錄甚多。在二十四小時內最大之降雨量，有達二四五·二〇公厘者。雨雷多發生於五．六，七，八，等月，平均每年雷雨日數達三十五日之多。

蒸發量：本港因位於亞熱帶，終年多風，故蒸發量甚大，每年平均達九三六·四〇公厘，一日之蒸發量亦有達一一·八〇公厘者。冬季空氣乾燥，風力亦較大，故溫度雖低，而蒸發量比之夏季反高。二，三，四，月間因溫度甚大，蒸發量反形減少。

濕度：本港因地近海洋，且地勢低窪，故濕度甚大，全年平均相對濕度爲百分之八三·六。

風：本港終年有風，最多風向爲東風，冬季多東北風，夏季多東南風，夏秋間常有風暴侵襲。

霧：本港冬末春初，常有薄霧，惟濃度甚微，且時間甚短，多出現於破曉至午前九時之間，十時即散。有霧日數每年平均約爲二十二日，多集中於三，二，四月。此數月平均有霧日數達十七日，佔全年總目數幾達百分之八十。

潮汐：過去潮汐紀錄皆已遺失，無從查考。現所得者僅法屬殖民地航海小冊內所載湛江硇州一九三六年全年潮汐水紀錄及雷州關西營支所載卅五年七月一日至卅六年三月卅一日之西營潮水紀錄兩種而已，茲分別列下，以資參考。

一九三六年湛江　州潮水位公尺		卅五年七月至卅六年三月江西營潮水位公尺
最高潮水位	四・三〇	五・三三四
最底潮水位	〇・四〇	〇・〇五一
最大潮差	三・九〇	四・五二〇
大潮平均高水位	三・八三五	三・九六六
大潮平均底水位	〇・九〇	一・〇八一
大潮平均潮差	二・九三五	二・八八五
小潮平均高水位	三・四五一	三・四二一
小潮平均低水位	一・三五四	一・三二三
小潮平均潮差	二・〇〇七	二・〇〇八
平均潮水高水位	三・六四三	三・六九四
平均潮水低水位	一・一二七	一・二〇二
平均潮差	二・五一六	二・四九二

潮速：關於潮水流速，以往並無紀錄。本年四月二十一日至五月三十一日，本處曾在內港西岸之西營測量潮水平時之最大速度，施測結果，盛潮時可達每秒一公尺，低潮時僅每秒〇・四公尺。潮水速度受風向及風力影响甚大，需長期施測，始能準確。至全潮流速變化情形，則須待詳細測驗之後，方可明悉。

（6）經　濟

過去貿易概況　湛江過去之出入口貿易數量，法人及雷州海關皆有記載，惜所記錄者皆不完整，只可見其大概，而不能詳悉其實在數量。現所知者，由一九二二年至一九二五年每年出入口貿易總值，約爲越幣一千萬元。一九三九年至一九四〇年每年出入口貿易總值，約爲一千萬美元。一九四三年至一九四四年每年出入口貿易總値約爲國幣三億餘元。一九四五至一九四六年每年出入口貿易總値約七十三億餘元。

光復後，本港貿易比戰前銳減，據一般估計光復後一年內，本港對外出入口貿易總值約爲三百卅萬美元。

腹地面積及人口　本港腹地，如將來西南西北兩鐵路網皆已完成，則範圍甚廣，可包括西北西南之全部。惟就目前而論，現交通部積極籌築由本港起經柳州貴陽，重慶，成都，天水以達蘭州之鐵路，苟此線完成，則本港腹地，亦可伸至甘肅，其範圍約包括廣東三分之一廣西大部分雲南小部分，貴州四川全部以及甘肅大部分。其腹地面積及人口數約計如后：

省別	面積(平方公里)	人口	備考
廣東	七三，〇〇〇	一〇，九五〇，〇〇〇	佔三分之一
廣西	一，五七〇，〇〇〇	一五，五〇〇，〇〇〇	佔大部份
雲南	七六，〇〇〇	二，三三八，〇〇〇	佔大部份
貴州	一七六，〇〇〇	六，九九〇，〇〇〇	全省
四川	二五四，〇〇〇	三七，五〇〇，〇〇〇	全省
甘肅	一九〇，〇〇〇	二，八一三，〇〇〇	大部份
合計	二，三三九，九〇〇	七六，一四一，〇〇〇	

交　通　本港交通，水陸兼備。水上交通，因港闊水深，可以容納巨舶，近可以行駛本國沿海各埠，遠則可以與歐美各埠通航，故將來水運發展，實未可限量。

至陸上交通，則分公路與鐵路兩方面。公路方面，現所有者，北行有西南公路交通網之柳湛線，由本港起經鬱林，貴縣，直達柳州。東行有西南公路交通網之廣湛線，由本港東營起，經梅菉，電白，陽江，台山，以達廣州。西行則有省道以達雷州半島之徐聞，海康，安遠等地。鐵路方面，在湘桂黔鐵路之來湛段完成以後，則由本港起，北行經貴縣，來賓，以至柳州，再分二線，一東至衡陽而與粵漢路相接，一北行入黔而達貴陽。如將來繼續展築，經重慶，成都，廣元，天水而至蘭州，則交通更爲便捷矣。

空運方面，法人曾在西營西北約七公里之西廳鄉，築有飛機場，倘重加修建擴充，可作航空站，以發展國內外之空運也。

物　產　經本港輸出之物產可分爲本港附近所出產者及本港腹地所出產者，本港附近所出產之物產，最大宗者爲鹽，沿水東，梅菉，雷州半島，及海南島一帶，均爲產鹽區。產量甚豐，以東海，硇州，烏石，流沙爲最著，年產量約一〇萬公噸。大戰前本區鹽斤一部份運銷日本及越南，大部份運銷粵，桂湘，贛，黔，諸省，惜因交通不便，輾轉裝運，費事失時，銷量不廣，影响產量低降。將來鐵路線完成，所有鹽斤，可逕由鐵路運柳轉銷西南各缺鹽省份，預計年產量可增至二十餘萬公噸。

農業方面，以米爲大宗，其次爲甘蔗，山芋，小麥，蕎麥，花生，荳類，水菓，瓜菜及糖，生油，草蓆，藥，豬，牛羊，鷄，鴨，及蛋類，年產量頗巨，除供應本港附近各城市之外，大都運銷廣州香港澳門各地。

本港附近港灣有天然之蔭蔽，魚類甚爲繁殖。而硇州島附近，爲魚羣常經過往，沿海居民，多以捕魚爲業，收獲頗豐，多醃製以運銷內陸，至於腹地物產，廣東西南部及廣西省有穀米，桐油，青麻，菓品，蔬菜，牲口，豬鬃，八角，桂皮，牛皮，藥材，錫，鎢，鐵，煤，等爲最大宗。其餘貴州，雲南，湖南，四川，各省，則以藥材，桐油，棉花，木材，牛皮，錫，鎢，鐵，煤，及雜糧等爲大宗。過去因運輸不便，雖向外運銷，將來鐵路通車，各省物產，由鐵路線運至本港。轉外輪運，便利迅速，當能大量增加出口貿易數額也。

（7）國防

本港不特在經濟上甚有價值，而對於國防亦甚爲重要玆分述如下：

保衛華南海岸　我國海岸線甚長，倘不設軍港，造戰艦，練海軍，則處處可被敵人侵入，而尤以南方海岸線爲最甚。蓋南方海岸線距首都較遠，且海中小島星羅棋佈，易爲敵人所佔領，以爲進攻中國之基地。又因接近英法美等强國之屬地，易被利用爲侵犯中國之跳板。故欲鞏固海疆，必須先自華南。本港在華南海岸線之中點，與海南島之榆林港及廣州之黃埔港互成倚角之勢，倘建立軍港，駐以艦隊，則可保衛南疆國土，永不被敵人侵入之虞。

鞏固西北邊疆　我國之西北及西南，皆深入內地，遠洋海，而以本港爲最快捷之通海出口。將來邊疆有事，係敵入侵，如須仰賴國際友人之援助，則前方糧食之給養，彈械之補充，皆可由本港起卸分運前線，以鞏固邊疆。

控制南洋海上交通　由遠東至南洋各島嶼之航線，固經本港海外，而由遠東至歐洲各國之航線，亦經本港出海外。故當對外發生戰事時，倘駐艦隊於本港，旣可遏制敵艦之活動，又可阻斷敵國之給養，使遠東與南洋以及歐洲各地之航運，完全斷絕！

（二）建港計劃

（A）商港方面

（1）整理本港灣　本港港灣係南北向，自馬其尖角起，迄北渡尖角止，計長約八，〇〇〇公尺。港口寬度在馬其尖角約爲一，五〇〇公尺，入港口後漸漸寬展，約達二，五〇〇公尺，至東營西營間爲最狹，約一，四〇〇公尺，越此又復展寬，約在二，二〇〇公尺至二，八〇〇公尺之間，至北渡尖角，寬約二，三〇〇公尺。港內水面，水深在低潮水位二公尺以下者約一〇，五〇〇，〇〇〇平方公尺，水深在低潮水位一〇公尺以下者約六，三〇〇，〇〇〇平方公尺，足以容納多量巨輪，不虞擁塞。港灣中部水深，均在低潮位一〇公尺至二五公尺之間，亦有超過三〇公尺者，實足與世界各名港水深相埒。

西營東營適居港灣中部，東南區遙相對峙，爲本港中心，應先行開闢，將來再向南北伸展。西營沿岸臨水線長約七公里，擬築岸堤碼頭，其淤淺處加以濬挖，塡築堤內，濬挖深度爲低潮水位下五，五公尺至九公尺，可泊三千噸級船隻約二十餘艘。東營自烏冠河口起，至村環止，沿岸臨水線長約四，五公里，擬築堤岸式碼頭，東營水位較深，風浪較少，宜於巨輪靠泊，堤外擬濬深至低潮下一〇・五公尺，可泊二萬噸級至五萬噸級船六艘。大頭嶺至三桃村一帶，擬浚深至低潮下八公尺至九・五公尺，可泊公千噸至一萬五千噸級船六艘。新村環一帶浚深至低潮下五・五至六公尺，可泊三千至五千噸級船四艘。所浚深之沙泥，將之塡築堤內，可得新塡地約三，七七〇，〇〇〇平方公尺。除利用建築碼頭，鐵路，軌道，倉庫，堆棧外，餘可標售，補助工費。

烏冠河東端井頭村至竹頭村及巴蒙港南端東特呈村附近，均築橫壩堵塞，以減流沙淤積，並可增加本港航道流速，維持水深。烏冠河灣內，擬加以浚深，爲船隻避風區域。

環繞港灣附近之光禿嶺，一律植樹種草，以防冲刷流淤港內。

（2）浚深水道　進口水道以硇州島附近之蒙硯番蓮進口航道爲最淺，水深在低潮水位下僅五・六公尺。東海島附近水道雖較深，但在低潮水位下水深亦僅有六・四公尺，祇數千噸輪船可隨時暢航無阻，萬噸巨輪，則須候潮進去，數萬噸大輪。則無法移動矣，爲適應將來擴展及巨輪進出之需要，擬將硇州島附近蒙硯番蓮航道及東海島附近航道，暫行浚深，至低潮水位下八・五公尺，寬二〇〇公尺。使萬噸巨輪，隨時航行無阻，數萬噸巨輪亦可候潮進出。如照上述寬度浚深，則硇州蒙硯番蓮航線應浚挖面積約爲一，三八〇，〇〇〇平方公尺，浚挖數量爲一，九三二，〇〇〇立方公尺。東海島附近航道應浚挖約面積爲四八〇，〇〇〇平方公尺，浚挖數量約爲六六，〇〇〇〇立公方，共計浚挖數量約二，五九二，〇〇〇立公方。

（3）修築碼頭　本港港灣中部，水深多在一〇公尺以外，惟西營沿岸甚淺，低潮下六公尺水深處，多距岸約三百餘公尺。東營畧較水深，惟低潮下六公尺水深處，亦距岸百餘公尺。如採用突出式碼頭，其長度須達四〇〇公尺以外，始可達相當水深，末端仍須築丁形臨河臂，以便船隻靠泊，計長共約五〇〇公尺，僅堪泊船一艘，昂費效微，殊非經濟。若造船澳式，則堤勘總長度過長，工款浩大，且凹入船澳，水流緩慢，常有沉澱淤積之虞。查本港堤似臨水線長度足用，故以堤岸式碼頭爲宜。擬沿西營東營兩處岸邊最低潮水位線建築堤岸式碼頭，並將堤外浚挖至適當深度，以資船隻靠泊。所浚挖淤泥，塡築堤內。並在碼頭上建築驗貨廠，貨倉，舖設鐵路線，以便利貨物起卸轉運。將來兩岸堤成，河床收束，流速增加，港內淤積自可減免。

堤岸高度，須視潮水高度而定，現在堤岸高度，合標高五八・五〇至五八・八九公尺，而本港潮水位達標高五八・五〇公尺者，年有數次。去年最大潮水，曾達標高五八・八九公尺，而至沿海岸街道商店，多被淹沒。故新築堤面高度，須高出最高潮水位一公尺。茲擬堤面高度爲標高六〇・〇〇公尺，向內漸漸傾斜，至與原地面標高相接合。

碼頭建築在西營方面，爲配合鐵路器材運輸起卸急需，首先將原日碼頭伸展至水深六公尺處，使三，〇〇〇噸船隻，能直接靠泊起卸。在原有碼頭填土部份長一七九·四〇公尺，寬一六·六〇公尺。棧橋部份爲鋼筋混凝土建築，長一五二·四〇公尺，寬七·一〇公尺。棧橋末端水深僅三公尺，活載重每平方公尺一，七五〇公斤（每平方呎三五〇磅）計需伸長四十四公尺，可達六公尺水深。其中三十二公尺，寬七·一公尺，與原有棧橋相接。其餘十二公尺，寬一〇〇公尺，成丁字形臨河臂，以供船隻靠泊起卸貨物之用，亦係用鋼筋混凝土建築。設計標準棧橋部份，活載重與原有棧橋同，仍用每平方公尺一七五〇公斤。丁字形部份，活載重每平方公尺二，五〇〇公斤（每平方呎五〇〇磅）以備堆積起卸貨物之用。臨海一方，每隔二十公尺有繫船設備，以牢繫船隻，并設木墊櫬以減低泊船時之衝擊力。臨岸一邊，設斜面式梯道，以利普通客貨上落。梯道寬二·二公尺，坡降一比八，載重每平方公尺一七五〇公斤，原有棧橋路面已磨擦損壞，擬加鋪五公分洋灰沙漿路面，以保護棧橋平地。並在原日避風塘之南端，新建碼頭頭一座，末端，以利車輛接運。在此新舊兩突出碼頭近末端處，建築堤岸碼頭長約七百餘公尺堤邊浚挖至低潮下八公尺，約可泊八千噸級船四艘。而採用吸揚式浚挖機，將挖出沙泥填築堤內，約得新填地二一二，四〇〇平方公尺。堤岸碼頭之上，加鋪路軌，以便利水陸接運。舊碼頭之北平樂頭止，築堤岸式碼頭長約二，六〇〇公尺，其中七五〇公尺堤邊浚挖至低潮位以下七公尺，計可泊六千噸至七千噸級船隻三艘，其餘一，八五〇公尺，堤邊則浚挖至五公尺及六公尺，約可泊四千噸及三千噸級船隻各四艘。新碼頭之南至伯打利尖角，計尖角計築堤岸碼頭長約三，七〇〇公尺，其中一，一〇〇公尺堤邊浚挖至七公尺及七，五公尺深，可泊六千噸級船四艘。其餘二，七〇〇公尺則分別浚深至七，六，五，公尺，可泊三千噸級四千噸級六千噸級船隻各一艘。

東營方面海水較深，自烏冠河口至新厝兒止，築堤岸式碼頭長約一，八〇〇公尺，堤邊浚深至低潮下一〇·五公尺，可泊二萬噸至萬噸級船隻六艘。由新厝兒至新村環則築堤岸式碼頭長約二，七〇〇公尺浚挖深度分九·八，六，五，公尺，可泊一萬五千噸，一萬噸，八千噸，四千噸，三千噸等級船隻各二艘。

（4）闢關市區　西營方面，沿海岸及新塡築地帶，應闢爲碼頭倉庫區。原日法人經營之行政區，將仍爲本港之行政區。原有商業區及北部地帶，則劃爲商業區。平樂頭一帶，宜闢爲住宅區，南邊尤其坑至海濱一帶密邇碼頭及鐵路線，宜闢爲工業區。原有西廳機場，則闢作航空站。

東營方面沿海濱新築堤岸一帶，深水碼頭，爲巨輪停泊之所，鐵路支線，亦止於此，故定爲碼頭區。新厝兒至麻斜一帶，毗連碼頭車站，宜劃爲商業區。西山村至麻斜埠頭一帶近海處闢爲碼頭區。其北爲鐵路線，南爲烏冠河，宜作爲輕工業區及造船區。大頭嶺三桃村煙樓村一帶宜闢爲重工業區。烏冠河擬加浚深爲船艇避風區。其近井頭村至竹頭村處。擬築橫壩堵塞減少淤積，壩面鋪築鐵路，以通巴蒙港。此外並擬沿特呈島南岸巴蒙港東部及烏冠河西北岸，闢植樹帶，寬約二〇〇公尺，藉以減低風勢並增園林之勝。

（5）充實設備　設備分港內設備及市內設備兩種，港內設備有航行標誌，浮標，碼頭，倉棧，廠塢等項。

市內設備則有電力，電話，給水交通等項，茲分述如下：

航行標誌：港口及進口水道，原設有航行標誌，惟多已殘舊，且無燈光，不便夜航。擬沿進口水道及港口增建航行標誌及浮標，並修理舊標誌一律加裝燈光，以便夜航。

碼頭：沿堤岸碼頭線設軌道起重機，以便移動裝卸之用，鋪築堤邊鐵路軌道，以便車輛靠近起卸之處。

倉棧：建驗貨廠以供貨物檢驗，建貨倉貯存貨物，留空地以爲露天堆棧及貯煤場。

廠塢：設船塢及修船廠以修理輪船，闢避風塘以資船艇避風之需。

電力：應建一中央發電廠以供給全港市電力，本港附近無大規模之水力發電可資利用，故應採用火力發電而尤以蒸汽渦輪機爲最宜。蓋此種機式具有優點甚多，即開辦費低，容量巨大，維持費廉，轉運安定。適應負載變化，佔地面積較少，故適合大動力廠之採用。本港用電趨勢將以工廠負荷爲主體，倘有廉價之電力，則輕重工業之興盛，必可預卜也。

電話：應建置新式之自動電話機，並須設立完備之無線電台，俾與世界各主要城市作迅速有效之通訊。

給水：西營東營及赤坎各地，均面臨海灣爲邱陵起伏地帶，地面水源缺乏，現唯靠鑿井取水。將來鐵路通車，船隻停泊增多市面擴展，用水數量，必大爲增加，故食水將成爲嚴重問題。西營西南約十五公里有湖光岩，爲火山噴口湖，面積廣濶，水深達二十餘公尺，水量充足，水質清潔，可用管輸至西營赤坎供給。惟東營水源困難，亦唯有用水管由西營經海底輸送而已。

(B) 軍港方面

(1) 基地建設　海軍基地必需深廣港灣，足以避風浪者，以便艦隻停泊修理，及補充抑資。並須有適宜陸地面積，以建倉庫，船塢，辦公處，及指揮部等。港灣位置更須外能控制海洋航線內與本土交通便利，戰時不虞斷絕聯絡。港外及港口，尤須有陸地防衛區，不易受敵艦威脅，進可以出擊，退可以堅守，方稱良好之軍港而本港水深灣庋廣東西北之面接連大陸，南有諸島環繞屏障，不虞風浪，故優良軍港之條件，皆已具備，擬將巴蒙港上北涯及特呈島一帶，劃爲軍港區，而以巴蒙港爲軍港中心，設船塢機械廠修理廠以資船艦修理，沿上北涯海岸，建築堤岸碼頭。堤外加以浚深，以備船艦靠泊補充煤水彈糧。沿岸設倉庫，存儲軍需糧彈，鋪築鐵路與東營鐵路支線聯接，以利轉運，建海軍辦公處所，以利指揮。巴蒙港內並加浚深，以便船艦避風，特呈島建營房，爲海軍仕官及陸戰隊訓練基地。

(2) 外圍構築　本港外圍，有碙州，東海，烏冠，南山諸島爲屏藩，形勢險要，苟能在此數島上構築堅强工業，則本島防務，必堅如磐石。碙州位於本港進口水道之外緣，擬在其最高處，設瞭望台，以爲本港耳目，並建堅固砲壘，以資防衛。其餘東海，烏冠，南山諸島亦依其地形，據其險要建築砲壘及工業，使敵艦無法侵入。

(3) 海空聯絜　現在戰爭乃爲立體戰爭，故空軍地位，非常重要。雖有强大艦隊，苟無空軍掩護，亦不能發揮其最大戰鬥能力。雖有堅固之要塞，苟無空軍保衛，亦易爲敵人所攻陷。故海軍基地必須建築機場，以爲空軍活動之根據地。在西營西北之西廳鄉,法人曾建有機場,應將之擴充,使成爲完備之軍用機場,以利海空配合作戰。

(三) 實施方案

(1)建港機構 湛江建港工程,規模宏大,宜設建港工程機構,專司其事。第一期工程,除工程外,並無別種業務,擬設一純粹工程機構之築港工程間以主其事。進入第二期後,除繼續未完工程外,並兼辦港務管理,及征收碼頭,倉庫,堆棧等使用費用與土地劃分標售等,業務繁多,故擬將工程局擴大爲港務局,兼辦工程及港務兩方面業務。

(2)征收土地 本港地價,必隨港市之發展而增加漲,私有土地自應從新評定地價,征收地稅,並酌征港市建設費,專用以建設市內公共事業,如排水道路,公園衛生等。至公有土地,及港岸新塡地亦應早爲計劃,除新塡地由築港機構自行支配外,公地爲私有土地則擬由築港機構會同市政府及地政局等合組地籍整理委員會處理之此項公地或規定稅額,將出租或評定價值,將之標售。所得價款,則移作公用事業給水,電廠電話等建設費。至於新塡地面積,共約三,七七〇,〇〇〇平方公尺,除碼頭,倉庫,堆棧,鐵路,道路等。所須用者外,其餘新塡地面積,可由築港機構劃分地段標售,將價款補助築港工費。

(3)儲備器材 築港所需用之器材,分築港器械及築港材料兩方面。築港器械有以下各項:

浚渫機械(浚渫機械之選擇,每隨土質,土量,水深潮汐,風浪,氣候,搬運距離,卸下處理,竣工期限等入定。就本港情形而論,以選用吸揚式浚渫船爲宜,蓋其浚渫能量較大,具有輸送廢土以塡地之優良點。其吸而管末端,並附有鋼齒輪,雖遇礁石亦可緊碎吸入。在一千公尺輸土距離之內如用六五〇馬力少啣筒與六〇四馬力鋼齒輪之吸揚式挖泥機,每小時之挖泥量爲一五〇公方,首期工程需用此種挖泥機三艘,在第二期三期工程中,因浚挖岩石,除增加此種挖泥機六艘外,尙應備 揚挖泥式機,二艘,以其齒力强大機障阻較小故也。至於硇州口外航道浚渫,因距岸頗遠,風浪較大,宜採用自航吸揚式挖泥機三艘,以配合工作。

起重機械:蒸汽與柴油兩式均應備用,其能量以五噸至三十噸爲宜。計首期工程需用71-B雙座摩擦鼓形蒸汽起重機三十噸柴油起重機三部,另浮船式起重機兩部。第二三期工程則須大量之電動式起重機矣。

建築器械:建築器械,計汽錘打樁機六部,壓錘打樁機六部,附有電動機成套之離心式及往復式抽水機共十五副,速自立式之量桶之混凝合機捌副,附同柴油原動機或電動機之打風機六部,土質試鑽機二套,連同蒸汽式柴油原動機之碎石機四部,潛水器捌套,附有配件設備之沉箱六套。

修理器械:爲修造本港各種機械及電機設備,應就實際需要,設立木工,金工,電工,鑄鐵等廠,各廠皆配備充足良好修理器械。

築港材料種類頗多,數量亦大,計首期工程約需洋灰十五萬桶,鋼筋捌仟公噸。鋼板樁五千萬噸,洋木料八千公噸。至第二三期工程,規模加大,所須材料更多。在築港工程中,石料最爲重要,硇州島及東海島均有石礦，可資採用。石灰須用數量亦大，宜擇石礦就近設窰燃灰，其他如鋼梁，木料，海砂等材料，需用亦多。應事先預爲籌備對於工業進行亦無妨碍。

(4)施工程序　施工程序，共分三期第一期工程計劃，爲完成港市區域及進口水道測量，舉辦水文氣象觀測，展築西營原有碼頭，在原日避風塘南端另築突出新碼頭一座，貨倉四座，並於兩突出碼頭間築堤岸式碼頭七百餘公尺。浚深堤外至低潮下八公尺，在堤岸碼頭上鋪築綫路，設軌道起重機，築驗廠貨倉等。在巴蒙港內特呈東村附近由烏冠河井頭村至竹頭村，築塡堵塞，修理及增建航道標誌及浮標。並配燈光。闢西營爲市區，建設給水工程。並在環繞港灣之荒山禿嶺種植樹草。本期工程擬自卅六年至卅八年，爲期三年。

第二期工程計劃，西營方面舊碼頭，向北展築堤岸碼頭七五〇公尺自避風塘新碼頭南口展築堤岸碼頭一一〇〇公尺，將堤外浚深至七公尺與七·五公尺之間，堤內則塡築爲平地，堤上則敷設軌道建築倉庫堆棧等。並將西廳鄉之機場重修，擴充爲本港航空站。在東營方面，興建廉江東營鐵路支線，沿烏冠河口至新厚兒築堤岸碼頭，堤外浚深至一〇·五公尺，堤內塡爲平地。堤上鋪路軌，建驗貨廠貨倉堆棧等。并開闢東營市區工業區及住宅區，安置底輸水管，由西營輸送淡水往東營 浚深烏冠河，在其北岸設造船塢。在進口水道方面，將硇州口及東海島附近航道浚深至低潮下八·五公尺，寬二〇〇公尺。本期工程計劃，擬由卅九年至四十二年，計期四年。

第三期工程計劃，在西營方面，完成沿海堤岸碼頭。計北至平樂頭止，長約一·九〇〇公尺，南至伯打利尖角止，長約二·六〇〇公尺，堤岸北段浚深六公尺，南段浚深爲七，六，五，公尺三級，堤內塡築爲新塡地，並在堤岸碼頭上鋪設軌道，增建貨倉廠棧等，在東營方面自新厚兒向北展築堤岸碼頭以至新村環，長約二·七〇〇公尺，堤外浚深分爲九·八六五公尺四級，堤內塡築爲平地堤上鋪築軌道，建貨倉，闢露天堆棧等。此外並擴充烏冠河造船塢，浚渫巴蒙港，並建海軍船塢及海軍基地倉庫。在進口水道方面將硇州口及東海島附近航道再浚深至低潮下一〇公尺。本期工程，擬自四十三年始至四十六年止，爲期四年。

（四）結　論

湛江港水深灣闊，形勢天成，爲一優秀良之港口，已如上述，將來湘桂黔鐵路之來湛段通車及西北南兩大鐵路網完成，則其在商業以及軍事上之地位其爲重要。故本港之興築，對於民生經濟以及國防建設，皆急不容緩。本處奉命修築來湛段粵境鐵路線，兼辦湛江建港工程茲已積極進行，一方面加緊各種測量，一面搜集各項資料。一俟完竣，卽着手作詳細之研究與設計，上所述者，不過初步計劃而矣。

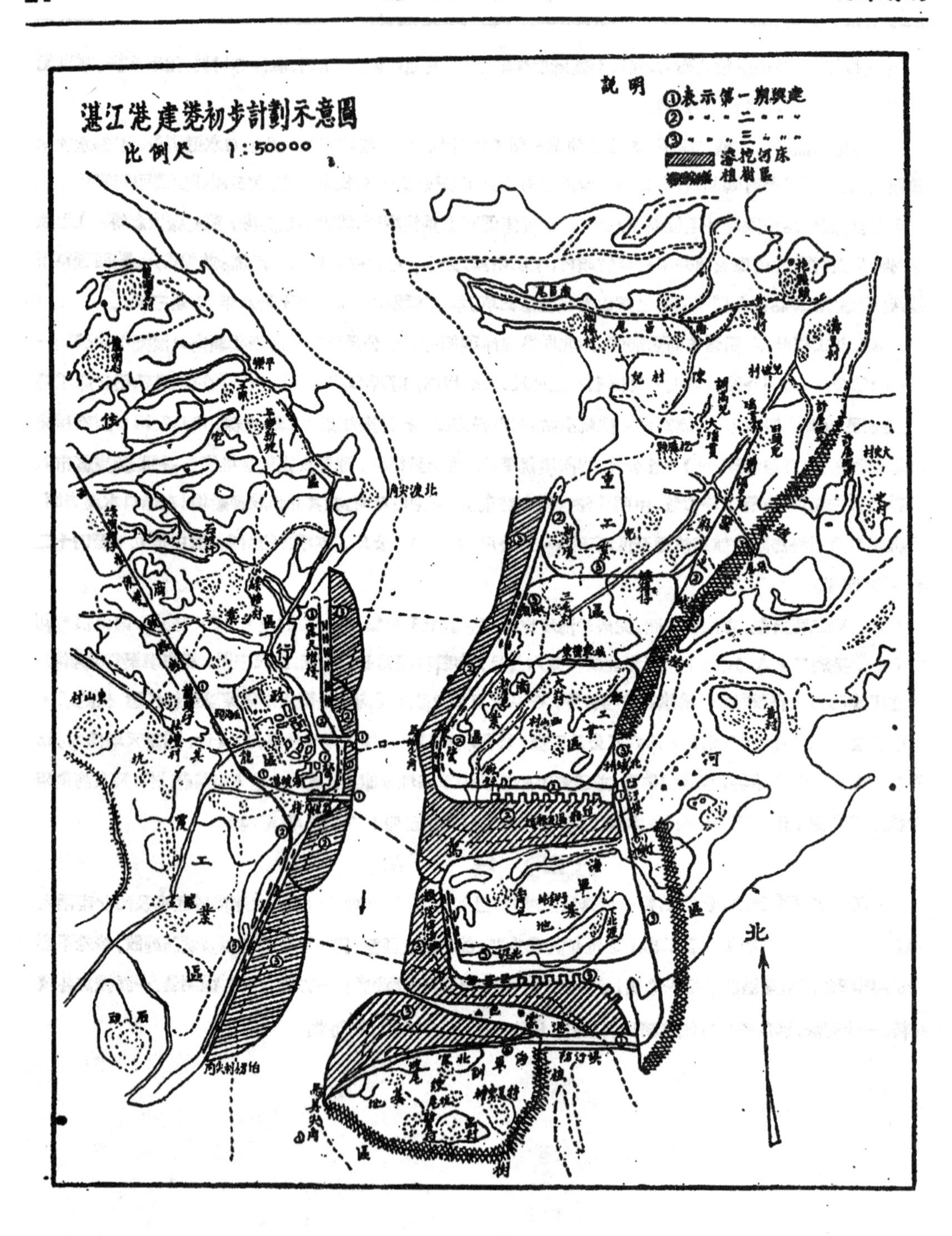

25640

原子炸彈與未來都市建設

麥蘊瑜先生主講　　歐陽讓記錄

一談到原子炸彈，凡是在第二次世界大戰身歷其境的人，都知道它的威力。它的破壞力不但等于兩萬噸的T. N. T. 炸藥，而且間接的毀滅力，確使以前認為最大破壞力的 T. N. T. 也退讓幾分。

原子彈的威力是這樣的偉大，許多人都認為是沒有方法防避，還談甚麼都市的建設？不錯，原子彈投下來確是無法躲避的。不過，它正如其他的普通炸彈一樣；據戰時的統計，直接被炸死的僅佔百分之三十，而被炸彈的破片及旁壓力致死的却達百分之七十，將這死亡的數字一比，我們就很顯明地知到一個原子彈的投下，直接致死的僅及旁威力影响而死的七分三。所以正如防避普通炸彈一樣。都市的建設，未嘗不可以減少原子彈的壓力的。

都市建設是防空工程的一部門。防空工程是一個最近代的科學，即使在第一次世界大戰後仍未受人注意。直至一九三一年至一九三三年間，德、意、法才有科學家研究這問題。在我國，則在民國廿五年間，麥蘊瑜先生倡議在大學課程中增加防空工程，可是都不得到教部的同意；而設此課程者，只有勷勤大學而已。

在第二次世界大戰未爆發前，泰西諸國對都市建設主張大概有三派：其一為法國派，主張高空都市；就是將全個城市的建築物築高至六十層以上，而最頂的一層，用四公尺厚的鋼筋三合士為頂。這種的建築，係用以防止炸彈投正建築物時，只能毀去一層三合士，或更毀去在頂的數層，但對整個建築物而言，却無多大影响（因為在原子炸彈未發明以前，一噸的炸彈僅能破壞深度四十公尺的泥土，所以四公尺厚的三合士，是相當于四十公尺的泥土的防禦力的。）

其二是英國派，此派則駁斥高空都市主的張；因為炸彈的投下，非垂直的落下，往往因氣流或飛機俯衝的關係，成四十五度以上的角度落下，那末，僅能防止由正頂投下炸彈的高空都市就被毀無餘了。所以這一派的主張，是平面發展；即在都市中的建築物採用平房式。這種作用，即使有炸彈投下，其破壞的建築物甚少。

其三是俄國派，就是將圓形都市化為帶形都市，這幾派的學說，爭持了許多時候。但在原子時代的今日，最合理的都市，應該將帶形都市再改良成串珠式都市。

串珠都市的建設，就是將每一小個型都市，利用交通的工具，來連絡一串都市；而每個都市的人口，不可多過十萬。（因為一方面政府的管理容易，別方面人民的負担稅項減少）而這許多小型都市，靠水陸交通的發達來連繫，每一小型都市內，不可劃分住宅區，工業區，商業區等。因為這樣可減少轟炸的集中目標。正如在抗戰時，在粵北的韶關到仙人廟，其間有很多小村鎮，無形中代表許多小珠，而粵漢鐵路，便是一條聯繫小珠的縋了了；當時敵機雖頻有襲擊，但被破壞的很少。假如將來的都市能照這個方式來建設，那麽原子彈的威力雖大，但其所能收到的效果，不能與其價值相抵的。

不過，這些都是基本的原則，至於技術上的問題，還要待我們人類去尋求與努力。不過，這些都是用來應付戰爭的建設，我們很竭望世界和平，使這種建築都市方法失其效用，但是，在國際風雲紛擾的今日，我們能保証以後的和平嗎？要防止別人侵畧，防止原子彈的大屠殺，我們應該在未來都市建設裡去探討！

鋼筋混凝土圖解法

——劉耀鈿——

近世工程問題之發達，大多數重視於數量之分解；但此種之推求，往往引用複雜公式之計算，并且會得學者感覺困難。于是本人因爲此一問題，從事撰圖解法一部份來討論。

圖解根本之原理，包括靜力體規則及幾何圖畫之定例；若欲明瞭圖解之用處；學者先要能通達該兩種規則及定例爲佳。

用圖解法來分解工程問種，其優點極顯易，其方法甚淺，手續簡單及工作迅速，并可避免公式之煩惱；本人所討論之問題，係鋼筋混凝土之一主要部份。

對于鋼筋混凝土理論之重要定義，本人姑信學者明瞭矣，故在此再示提出研究：

I. 普通矩形樑：在此一種樑，其內容可分三種來研究．．

(一)假定已知梁之大小及其安全應力而求梁之安全彎性率。

(二)假定已知梁之大小及其彎性率而求其應力。

(三)假定已知其彎性率及其應力而求梁之大小。

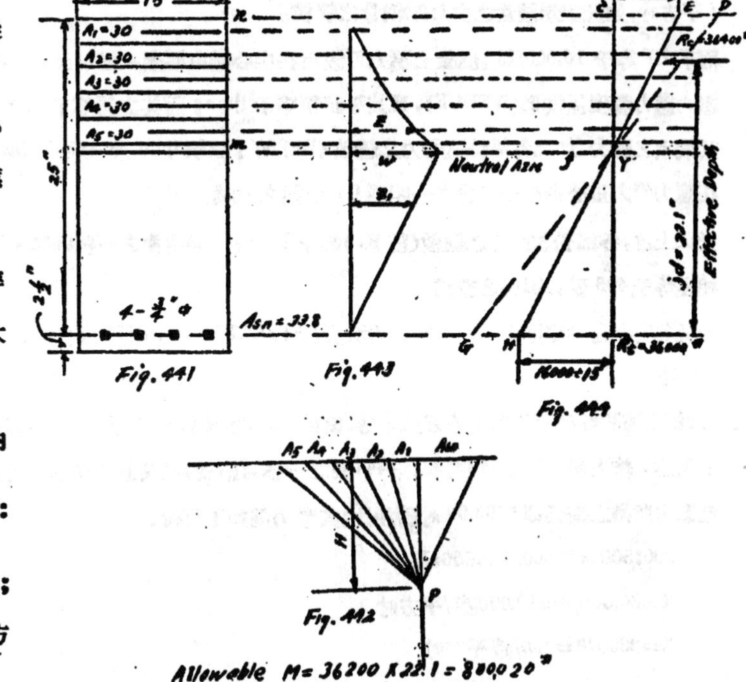

(一)：試取梁爲441圖內用$4-\frac{3"}{4}\phi$鋼筋，假定：

$f_c=650$磅/平方吋；

$f_s=16,000$磅/平方吋。

將梁受壓部分分作數相等小部份，每部份與彎性軸平衡及分至中線軸而止，分部之面積分別等于 A_1，A_2……等等；而由各部份之重心繪上施力線與彎性軸平行，并由鋼骨中點亦繪上同樣之平行線。鋼骨之面積假設等于 A.；而若用 n（假定n=15）乘此面積，其係數等于 A. n。

由 x 點在第 442 圖用一適宜之比例尺，向右邊繪一有方向力線等于 A，n；并向左邊依着次序亦繪上A_1，A_2……等等，然後擇一極點 p，x－p 線最適宜係垂直，但此非係必需要之辦法。

由第 442 圖，依着次序，在第 443 圖繪一連鎖多邊形，其交出點 O 決定中心軸之位置。由第 443 圖；在 O 點之上隨取一距離Z，與由面積重心之力線平行，此距離 Z 乘極距 H 等于此梁在 Z 距離以上該部份之第一力距（FIRST MOMENT）；而且週環該距離 Z。若仍以前法令取一距離 Z，在 O 點之下，此距離乘極距 H 則等于此梁在 Z，以下所受壓力部份之第一距（FIRST MOMENT），而且週環該距離 Z_1。

假定將 Z 距移下，Z_1 距移上，而至距離 W，如此Z 與 Z_1 相等；于是此種情形表示週環該 W距上之第一力距等于 W 距下之受應力第一力距；但其符號相反，此爲 YOJ 線，就係梁受應力部份之重心而同時係中心軸。

玆在中心軸，取一適宜點 Y，及繪上基線 CF（第 444 圖）；用一適宜比例尺 CD 等于 CY，繪上 DY，而伸長至 G；以比例尺量 GF；若將此係數乘 15，即等于鋼筋之應力，以上述試題 GF 乘 15 已超過 16,000 磅／平方吋，如此表示該量之力量歸鋼骨之管核。

而若 HF 等于 16,000/15；混凝土壓力可假定在中心軸由零漸次增加至 560 磅/平方吋在梁上端；如此該壓力應力體積係楔形（WEDGE），及其平面積等于 ijmn；而其立體等于 CEY，于是其平均應力等于一半其最大之應力。如此各部份壓應力之總數（Rc）等于面積 ijmn 乘一半 CE 則等于 36,400 磅，其施力點位在應力體之重心則等于三份之一距離 CY 由梁之上端。

仍如上述，各部份拉應力之總數（即 Rt）等于 FH 乘15乘鋼骨之面積則等于36,000磅，如此可見兩應力之相差等于 400 磅，約1%之差錯。

如此其安全彎性率則等于梁之有效力深度（EFFECTIVE DEPTH）乘 Rc 或 Rt（或其平均）則等于 800,000 吋磅。

（二）：第二種問題亦可依照上述方法分解；由第 441 圖，彎性率等于 500,000 吋磅；n=15；試求梁之應力。如上所述，繪上第442，第443兩圖；第444圖之一亦可繪就；惟其應力務須先行假定，求得一彎性率；然後假定應力所得之彎性率 500,000 吋磅相比；其應力則如下可得：

800：500＝16000：X＝560：X

X＝80,000/8＝10,000磅／平方吋

X＝2800/8＝350/磅平方吋

（三）：第三種問題，情形較爲複雜，其繪法以前不同。

假定需設計一梁，其彎性率等于 1,500,000 吋磅，fc=650 磅/平方吋，fs=16.000 磅/平方吋，n=15，試求梁之大小及鋼骨面積之份量；在受此彎性率中要能達到梁之兩種材料安全應力爲目的。

設計次序如下：

假定梁之寬爲15吋，厚度至鋼骨爲$27\frac{1}{2}$吋（如第445圖）

附　圖

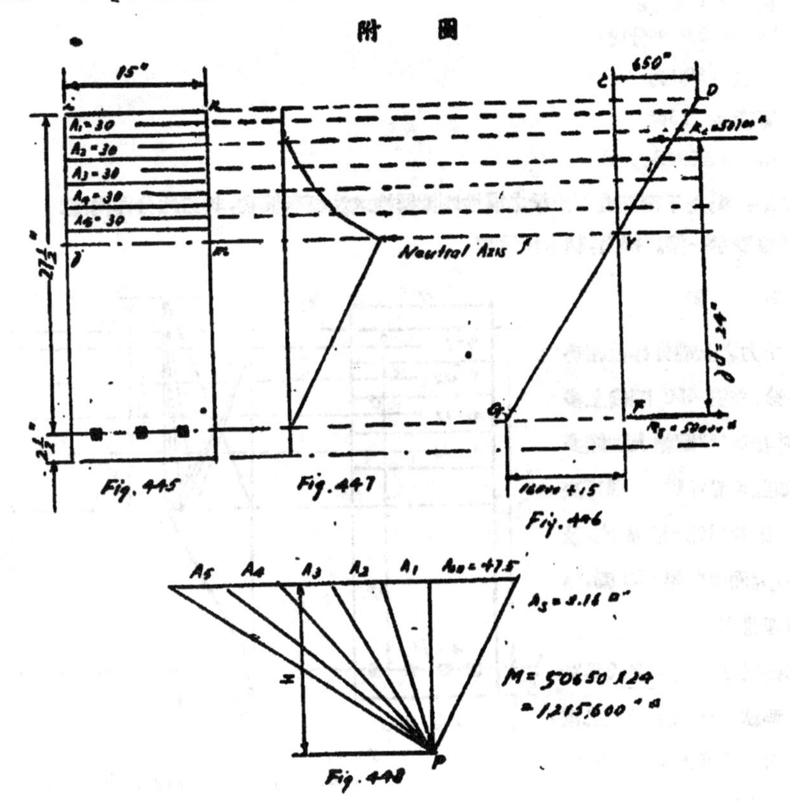

在第 446 圖用一適宜比例尺繪 CD=Cy,GF=fs / 15。如此 CF 與 DG 交於點 Y 則決定中心軸之位置，惟所用鋼骨之多少，亦能影响中心軸之位置，但此位置之變更與 fc 及 fs 之比全無關係。

決定中心軸位置之後，即將該梁受壓力之上端分作數相等小部份，及由各部份之重心繪上其施力線，由鋼骨之中点亦繪其施力線（如第 441 圖）。

在 449 圖，繪上 A_1，A_2……等等成力線之多邊形，及在第 447 圖繪上連鎖多叠形；由此 O－R 線則成該多邊形之聯結線；及在第 448 圖，由P 点繪一線與O－E²平行，及伸長與戗重線交於于 Z点，如此 ZX 則等于 AsN；并 ZX/15=As；若以比例尺：As=3.16 平方吋如前第 444 圖；合力Rc 及 Rx 亦可決定，如此以假定梁；M＝50,650×24＝1,215,000 吋磅于是假定梁之大小不足以應規定之彎性率，但以理論上：

a. 梁之抵抗彎性率與梁之寬度成正比。

(b)梁之抵抗彎性率與梁之厚度平方成正比，于是若梁之 M＝1,500,000 吋磅

則：

(a) 厚度＝27.5吋，

寬度＝$18\frac{1}{2}$吋。

As ＝3.9 平分吋。

(b) 厚度＝30.6吋

寬度＝ 15吋

As ＝3.52平方吋

II.矩形梁之慣性率：對于鋼筋混凝土梁之慣性率圖解推求法極有興趣，其方法分作兩種：

(一)第449圖表示一梁之剖面，試求慣性率．

附 圖

先將其受力之上端分作數相等之小部份，在第 450 圖繪上多邊形，并在第452圖繪上連鎖多邊形；如此中點軸點 O 則可決定；并在第421 圖定極點 p'，及其多邊形；而由此第 452 圖，繪上連鎖多邊形。

上述之繪構法，是可之氏署圖解慣性率法，如此矩形梁之慣性率可用下公式而求．如上圖：

$I＝150×10×10.4＝15,600in^4$

若欲得AsN，可用As乘15，如此可表示慣性率是在混凝土範圍之內。

得到慣性率之後，梁之彎性率鋼筋之應力及混凝土之應力可用下列公式而求：

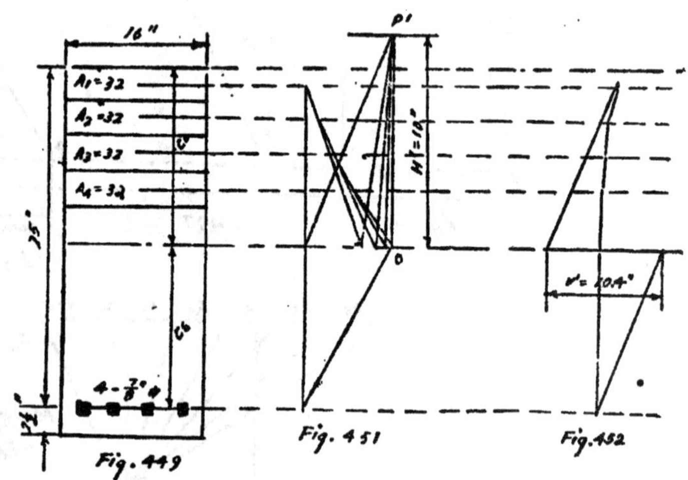

I = (H)(H')(V')
= 150 × 10 × 10.4 = 15600 in.⁴

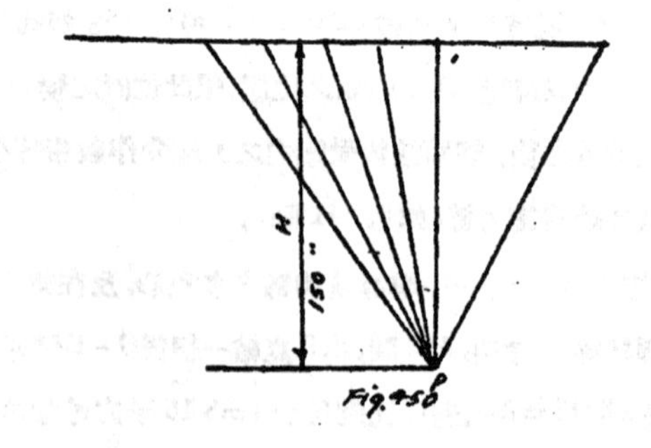

M/S=I/C

fs=MCc/I

fs=MCcN/I

(二).第453 454及455圖,表示第二種方法而求慣性率,此係摩氏圖解慣性率法(Mesh method)其公式如下:

I=(H)(2)　(第455圖之連鎖多邊形內之面積)。

附　圖

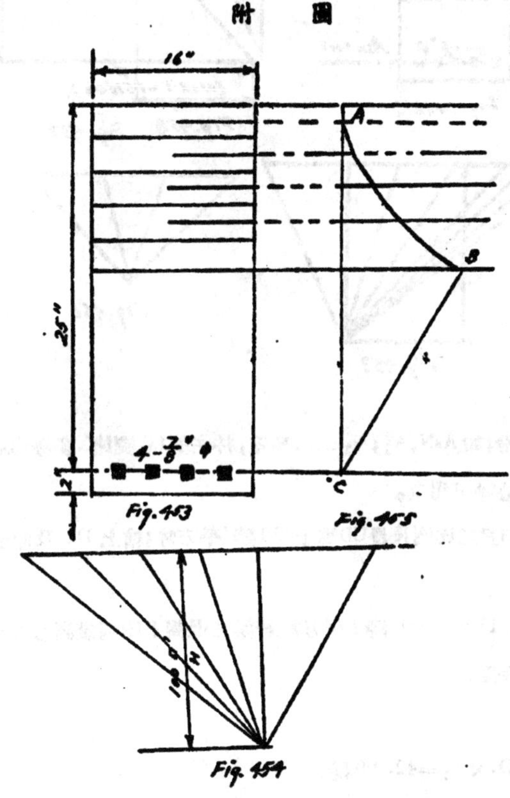

III. 丁字梁:在設計丁字梁中,所用之公式不顧及梁腹所受之壓力,此種畧定引出多少差錯,假若腹部係寬闊共差錯則反之。

惟圖解設計法有顧及着梁腹部之壓應力,于是數量設計之差錯,可以避免.

如矩形梁之設計,丁字梁亦分三種:

(一)假取一丁字梁如第456圖,試求其彎性率:

fc=650磅/平方吋　　fs=16,000磅/平方吋, u=15.

25647

附　圖

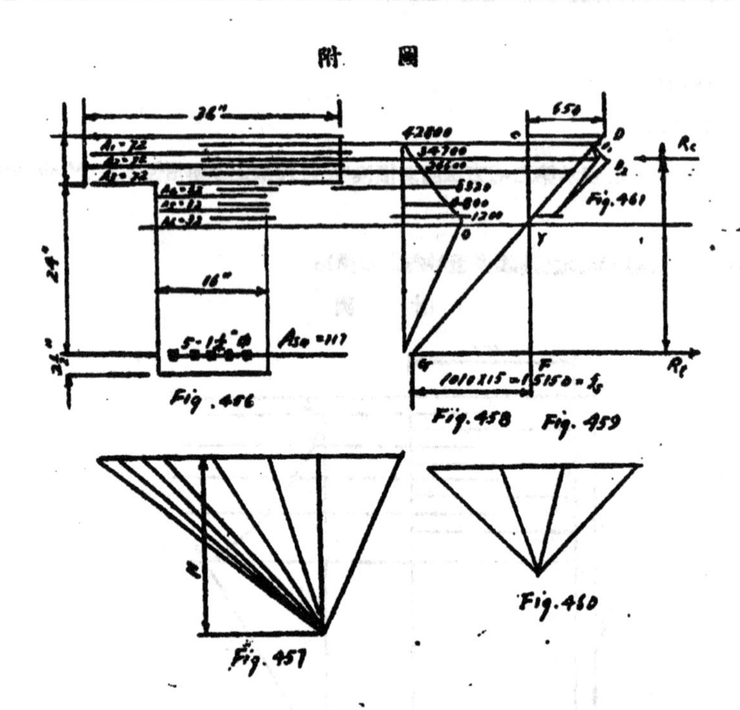

將梁之壓力端分作數小部份;如 AsN, A_1, A_2 ......等差,并在第457圖繪X多邊形。而由該圖繪上連鎖多邊形在第458圖;如此梁之中心軸可得矣。

在第459圖繪上 CYF 及用適宜之比例尺量CD等于650磅/平方吋;繪上 DY 及伸至 G 點,如此得 LG 等于10100。

由此:鋼骨之應力= 1010 × 15 = 15,150 磅 / 平方吋若梁之混凝土達到最高應力——650磅/平方吋 A_1之平均應力則等如 C_1D_1 之長度.

由此:

上部份所受之壓力$=C_1D_1\times A_1=42,800$磅.

第二部份所受之壓應力$=C_2D_2\times A_2=34,900$磅.以上所述之辦法其他部份之壓應力亦可得矣,得此合力之後,繪上第460圖,而且由第460圖,繪上第461圖,由此 Rc 之施力線位置及方向可得矣.如此;

Rc= 118,620 磅

Rt= 118,170 磅

$M=(118,620+118,170)\times\frac{1}{2}\times26.8=3,173,000$ 吋磅

(二)假定梁之大小及其彎性率而求其應力:

如上述繪上第456，457，458，459及460圖，然後將CD作一假定適宜應力，由此應力求其彎性率，得此彎性率後，如在矩形梁內所述之辦法，用比列尺則可得其所需之應力矣。

（三）對于第三種問題之引用，試設計一丁字梁，其彎性率等于3,300,000吋磅，fc=650磅/平方吋，fs=16,000磅/平方吋，n=15

附圖

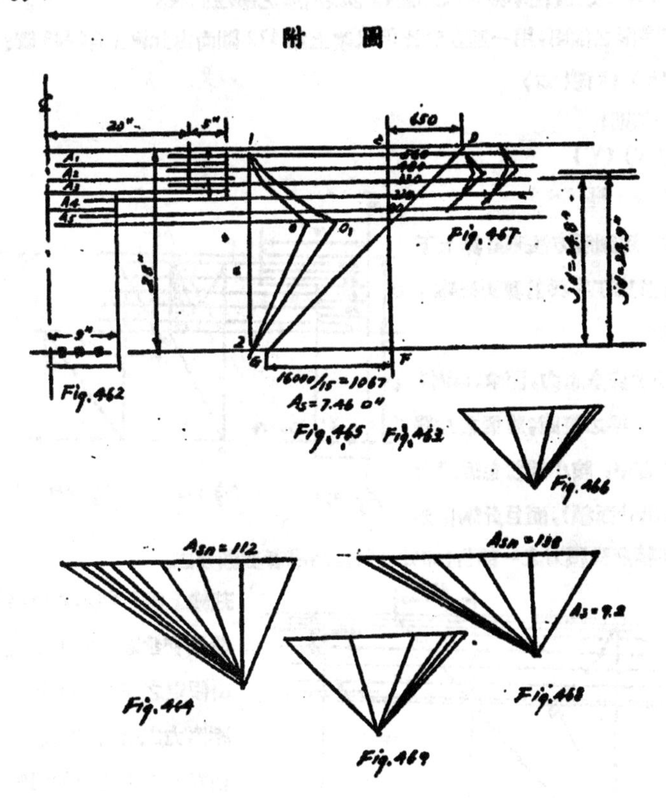

第462圖表示假定之梁，并由第463圖決定中心軸之交岔點Y，然後由第464及第465兩圖用交岔點O求鋼筋之面積，則等于約7.46平方吋，再由第466及第467兩圖，決定壓應力之合力位置及其有効力之嗥度可得矣。

假定所得之：M=2,976,000吋磅

則所得之彎性率比梁實說之彎性率淡小，于是該假定梁應增大，而其增大之辦法有下，略述如下：

(a)將梁上端及腹部之寬度同時增加，由此辦法相當寬度及鋼筋之面積可用比例而求。

(b)單將梁上端之寬度增加：

假設用第二方法而求；梁上端增大10吋，則每邊增5吋，如此由第408圖及其交岔點O，則得相當鋼筋面積等于9.2平方吋，幷由第469及470兩圖，若壓力之合力施力線位置可得矣；其有効力之厚度亦由此而得，則用此新厚度第二彎性率可得矣.

由此兩假定空梁之彎性率；梁之上端每吋之增加與彎性率之增加之比例則可得矣；而由此梁實載之彎性率應用梁之大小可用比例而求。

IV.丁字梁之慣性率：丁字梁之慣性率亦可以如前矩形梁所述之邊法而求。

第471圖表示一丁字梁之剖面，用一適宜定比例尺繪上第472圖而由此繪上第433圖：

如此：$I=(2)(H)$（面積）gr）

或用極點p' 及第474圖則：

$$I=(H)\ (H')\ (V')$$

附　圖

Fig.471　Fig.473　Fig.474　Fig.472

V.上下鋼骨混凝土梁：用圖解方法來計算上下鋼骨混凝土梁之推解比丁字梁計算更複雜，如前題可分作三種：

（一）假定梁之大小及安全應力，而求其彎性率，第475圖表示一梁之腹面；將梁之上端分作數小部份之當中，該小部份包混凝土之面積及（15）（鋼骨面積），而且此鋼骨亦減小該混凝土面積所受壓力之一部份；如此在計算時學者要顧及之。

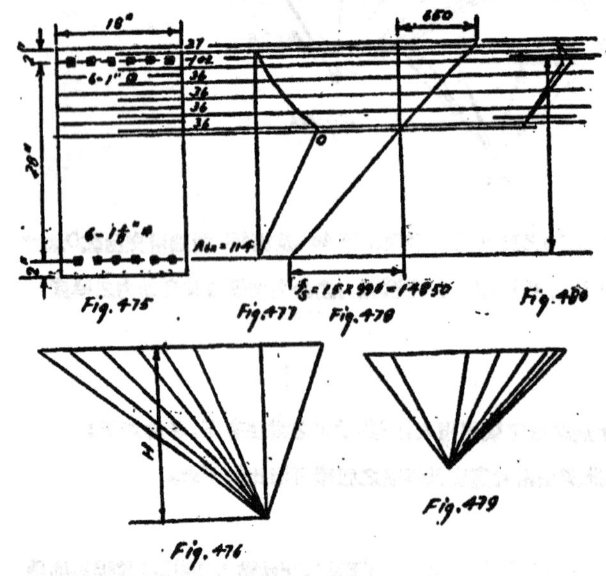

茲繪上第476圖；幷由第477圖，中心軸之位置及交岔點O則可決定。

由假定之安全應力繪上478圖，及由此各壓應力之合可定矣。

由第479圖及480兩圖則可決定合行之施力線位置及方向；而有效之厚度亦可得算等如26.7吋，如是M＝3,042,500吋磅

（二）假定梁之大小及其彎性率，而求其安全應力；此一種可用比例方法，如前矩形及T字形兩題所述法而求：

（三）試設一上下鋼骨混凝土梁：

彎性率＝3,200,000吋磅

$f_c=650$ 磅/平方吋。

$f_s = 16,000$ 磅/平方吋。

$n = 15$

假定一梁寬度等于20吋及FGT$=\left(\frac{1}{15}\right)(f_s)$，如此交岔点Y則決定中心軸之位置矣。由第483及484兩圖，則得Asn。

如此：

拉力鋼骨之份量$=As''n/n = As''$

如用As''；混凝土之應力fc務須要達到650磅/平方吋，井鋼骨之應力fs亦達到16,000磅/平方吋，

于是：

$M=(As')(jd)(fs)=(4.62)(26.23)(16,000)$

$M=1,939,000$ 吋磅

但此彎性率比實較之彎性率小，相差1,261,000吋磅，此相差之彎性率務須壓力鋼骨來荷之。

此鋼骨之距離=28吋.

于是：

載重$=1,261,000/28=45,000$磅

而鋼骨之面積$=45,000/16,000 = 2.81$平方吋。

如此： $As= 2.81 + Asn'' = 2.81 + 4.62 = 7.43$ 平方吋

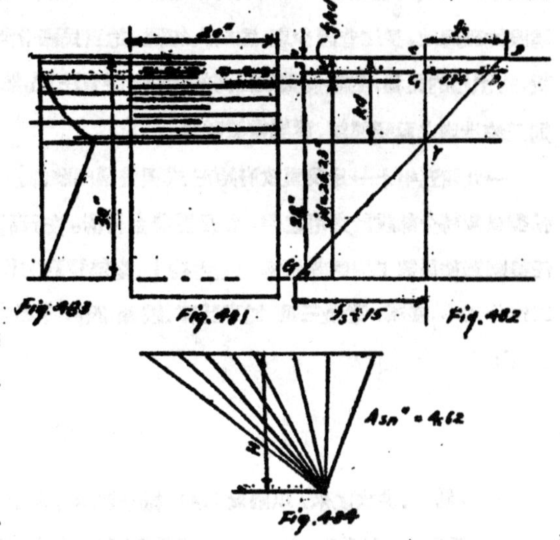

但混凝土之應力係在壓力鋼骨之中，及等如C,D（第482圖）；始此受壓力鋼骨之應力等于(15)(C,D)等于(15)(534)

于是：

需要鋼骨面積$= 450,000/15\times534 = 5.62$平方

25651

三峽水利工程建設

——林炳華——

現在來談籌劃未來水利建設，實在是一種未雨綢繆，而且是極重要的基本工作。我國是一個水利建築落後的國家，過去因為水的災害，動力的缺乏，航行的阻隔，影響我國經濟的發展。經過幾年堅苦卓絕的抗戰，確立了國家民族獨立自由的基礎，我們今後的任務，應該是如何使它發揚光大，以垂久遠，所以必須要確切認明年來積弱的癥結所在，而將戰後水利建設列為首要，然後建國大業纔能循着正軌，作有計劃的發展。但是按照目前國內人口，資源分佈的情形，和戰後工業區域劃分的需要看來，長江三峽的水壩建設，應該是我國水利建築中的一個最切要的問題。

長江在青海發源地拔海達五六〇〇公尺，在川滇境內，山嶺重叠，水行谷中，河床坡度極大，迨出三峽，坡度轉緩。本人曾兩度航行川江峽谷，沿途所見，山如劍排，水如湯沸，流勢喘急。宜昌正當三峽門戶，乃一水利建設的理想地，長江昔因少患，遂為人忽視，迨自民國廿年大水後，國人始稍加注意，然其目標乃在消極的防災，對積極的興利，對整個國家經濟的榮枯，鮮有注意。然以長江流域面積之廣，人口之衆，工礦之尚待開發，則三峽造壩之亟須興辦，實屬當然者也。

一九四五年十一月我國政府始正式與美國內政部簽訂揚子江水閘建設計劃之合同，同年十二月即由資源委員會將全部設計費用之半，廿五萬美金付清。在設計美國大苦力電廠及波多壩電廠之薩凡奇博士，與現任美國墾務局總工程師華格揚先生主持下，全部設計工作於一九四六年七月在丹佛(Denver)城開始矣。全部工作分土木，機械，電機三組，從事計算，製圖，設計等事宜，謹將造壩之目的，及其各部工作建造計劃等簡述如下：

造壩之目的

一、防洪：洪水之來，以雨量為主，揚子江流域雨量，自北至南，逐漸遞增，依據歷年雨量統計，贛粵邊境，暴雨最早，鄱陽湖四月即漲，六月反形低落；洞庭湖常穩鄂錫湖水而來；川省雨量以八月為甚，且上游流域面積最廣，又乏湖泊之調節，以其承下游滿漲之後，故為勢最猛，是為下游洪水災害主因也。

防洪原則有四：(1)改進河道，以增加河流本身之水容量。(2)加寬一部分河道使減少流速。(3)利用岸、牆、堤、渠以攔阻洪水。(4)利用水庫以貯洪水，此四法必須先行研究，比較各法利弊，經濟因素，及工作之難易而決定取捨。

今對支流之整治，實無影響宜昌至漢口沿江一帶之最大洪水位，同時欲開挖引渠以導溢流，該處又無天然之地位。堤岸旣爲攔阻洪水期最高水位之用，若建築過高，超過安全限度，甚不合用。故以當地自然環境視之，以三峽水壩形成之水庫貯存川江一部份洪水，乃爲最合經濟原則之方法。三峽水壩有調整流量之功，更能阻遏下游洪水泛濫，則影響所及，爲拯救無數無辜生命財產，居民得能安居樂業！

二、發電：水庫之建造，于防洪方面可有極大之貯水量，同時可產生鉅大數量之水力，此等水力，配合透平及發電機，成爲廉價之電能，更可利用而使農業與工業互相配合，保持平衡。減輕家庭與農業工作，增加家庭收入。鄉村電化更有助於公共衞生，生活標準，人口分佈，水土保持，以及在此流域中普遍之經濟及社會之改進。

三、灌溉：工程完成後，水壩上游爲一長二五〇英里之蓄水池直達重慶，上游之水位普遍被提高一六〇公尺，可以容水五千萬畝一一呎，故低窪之地，可闢渠引水，就其坡降，須流交分以灌田畝，高地則藉閘壩剩餘電力，引水灌溉，壩築成後，吾人可預見果實纍纍溝渠縱橫，深得灌溉便利之繁榮農村。

四、航運：長江上游水流湍急，灘險密佈，宜渝段可記錄之險灘，約二百八十餘處之多，險況則因水位漲落而異。流速之大小與地點及水位之漲落有關，通常在中高水位時，每秒約二公尺至四公尺之間，故航行上游之輪船，其馬力須較航行下游者大三四倍。壩之築成，因能集中低水時期之流量，整個上游河道化爲一大靜水湖，加大河道寬度，消滅淺灘威脅，計劃使六五〇〇噸汽船能直航重慶，減低航行費用，影響貨物交派匪淺。

工　程

一、壩：壩可分高壩與低壩兩種，設計所需壩之數量與其性質有關，低壩所需之數量恆較高壩多；雖則建造低壩較高壩經濟，然低壩提高水位不大，狹曲之河道便航行遲緩而困難，況且在洪水期間，更不能阻遏船閘附近水近的急劇變化；採用低壩旣不能防洪，亦無補於長江上游航行深度之整治，更少有利用水力發電之價值；另一方面，三峽一帶之地形，地質亦頗有利於高壩之建造；且現三峽兩岸因無連續之鐵路與公路系統，三峽中建造高壩並無關乎經濟上之限制，故此一偉大水力工程應根據建造高壩原則而設計。三峽可能之閘址共有五個，其中第三第四閘址較有希望，上述二閘址均在宜昌南津關附近，第三址江面窄，水較淺，築閘費用較廉，但不能防禦萬年一次之大洪水；根據薩凡奇計劃，壩高出最低基礎爲二百五十公尺，可提高上游水位一百六十公尺，壩長達七百六十公尺，爲一重心式大壩，全部建壩工程須用水泥一千五百萬立方碼，約爲哥崙布河之大苦力水壩之一倍半。

二、洩水道：洩水道之爲用，乃在洪水時期部份洪水由此排出，使水位不致超越不溢水部分，整個結構之穩因乃基于洩水道作用成功與否，故安全係數必須大也；洩水道之高度大致與所設計水庫最大容量之高度相等；三峽水壩爲混凝土壩，洩水道可爲壩中之一部分。洩水道之形式及其建造地點爲經濟條件，地形及安全所決定；洩水之道具有頂門者，高于頂部之控制積水可加利用也，三峽水壩由九扇一三五呎×五六呎的鋼製數形大門控制，另有五二根一〇二吋的洩水管，能容揚子江一萬年一次之全部洪水，爲避免形成水壩上游二五〇英里蓄水池附近之城市及農田受水災之危險，則水庫中之水位，除極大之洪水外，務使其維持不變，則洩水道之設計，須使其排出最大水 ，而不危及壩之安全爲原則，長江在南津關附近成一大彎形流速，計劃通蓮群山，岸旁開鑿隧道共二十五條，每條直徑爲五十呎，以改大水流速度，施工時謂之池水隧道，便於在築壩時將水引入支流，以利建壩；工成後稱爲引水隧道，此一組隧道可宣洩揚子江頻率一百年之洪水，

三、航行設備：普通壩均有船閘設備，以利航行，船閘設計基于具有下列各水力設備之 1.各水道需使閘室儲水在一適當之時間內充滿或放出，且對船隻擾動最小，2.控制水流之活門及其動作之機械，3.閘門及其動作之機械，4.檢驗閘門，閘室或閘座時將水放出所用之臨時壩。

設計中之三峽大水壩，東西兩面水位差達五三〇呎，則需建造一連串之船閘，逐個遞升，工程浩大；故在三峽水壩設計中，改用一水流隧道；若船由下游溯江而上，先經過四分之一英里長之隧道，由八架起重機 (Lifting Machine) 分別將之舉起五百三十呎，而達西面之大蓄水湖。

四、其他設備：每一引水隧道裝置發電機四部，預計每部發電一〇五，〇〇〇基羅瓦特，總共發電達一〇，五〇〇，〇〇〇基羅瓦特，較田納西河發電量大五倍，能供給直徑一千哩區域中之二萬萬居民使用，各種輕重工業因三峽電力之成功而勃興矣，預計其電費每度備合美金一厘，折合戰前國幣值三厘三。

造壩前之準備工作

目前水壩尚在籌劃階段，吾人應儘量收集有關資料，如作水庫附近之地形測量，河床測量，流量測量；製就之地量作爲決定閘壩與水庫位置之根據；計算迴水 (Backwater) 長度，估計水庫面積對土地價值與開發之影響，以及水庫附近公路與鐵路之重行定後。地質探鑽工作實屬重要，在選擇壩之可能位置，鑽探至于基礎石層，査驗每一規劃之可否實行，最後乃得一最經濟最合理之位置；水文觀察站之設立亦屬必要，以從事記錄雨量，逕流，河流挾沙量等數字使設計者便於決定各種構造之位置及其主要特性。

此一偉大工程預計要動員經驗豐富工程師四千名之多，工作十年後始能完成，全部工程費用估計須十三億美元，其中九・五億美元屬造壩部分，其餘三・五億美元用以建造隧道及其他設備，據云十三億美元之建築費在廿年內就可償清，至于工業勃興而引起之繁榮尚未計算在內。

目前國內政局激蕩，使三峽水力工程籌劃工作無形停頓，但三峽水力工程關係整個國計民生，現在的任務，應該是如何斟酌建設的時間和國家經濟條件，來決定實施之步驟，我們是政治，國防，經濟建設一切落後的國家，以往因爲缺乏近代立國條件，幾使我們淪於萬劫不復之地位，今後必須抓住問題的焦點，儘最短時間，完成三峽水利工程！

堤壩概説

——趙浩然——

（一）引言

築堤防水，人皆知其爲消極的防水患，不是積極的興水利。在科學落後的我國，既不能大事疏河，築壩，去淤，又不能廣事植林及種植苦導，調節山洪，以興水利免水患，惟有依襲古例築堤以防之。

築堤之要訣，第一要選擇土質，第二要施工精細，換句話說：卽要找有黏性之泥土，分層鋪平，擊碎之而加以水潤，拉牛練實，自可以收防水之効。但隄之成矣，外水可防，而圍內積水，無法宣洩，淹沒農作物，雖不似缺隄災害之大，然失收之數，尙屬驚人，此爲晚近之提倡築隄防方所當注意也。

（二）堤之種類

隄之種類很多，依其所選用之材料而分類，然其設計之原則在使浸潤線不與隄身相切，使浸潤線之水不得由隄身透出，以免滑倒而危及隄身之安全。防水壓滲透可用混凝土牆，黏土或打板樁，至計算滲壓力之大小在學理上幷無可靠之法，因其所荷之重不得確切求得也，其一方受水及濕土壓力，另一方受空氣及土壓力，堤身受浮力，使滲透水壓不能固定而時常變更，其種類大致因所用材料不同可分下列三種：

（一）石：用極透水之粗石堆于隄脚，使浸潤綫銳降，使水由該部分滲出，不致危害隄坡。滲出之水，可于隄脚掘溝容之導往適當之處，或以抽水機抽出隄外。

（二）黏土：防滲層須達不滲水地層或築成截流牆式或于其下打板樁。黏性土層厚度不得小過一公尺，其厚度自上向下漸增，蓋水壓力愈深愈大也。

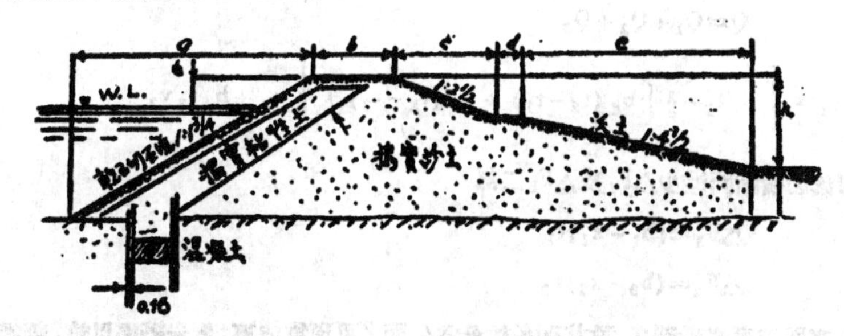

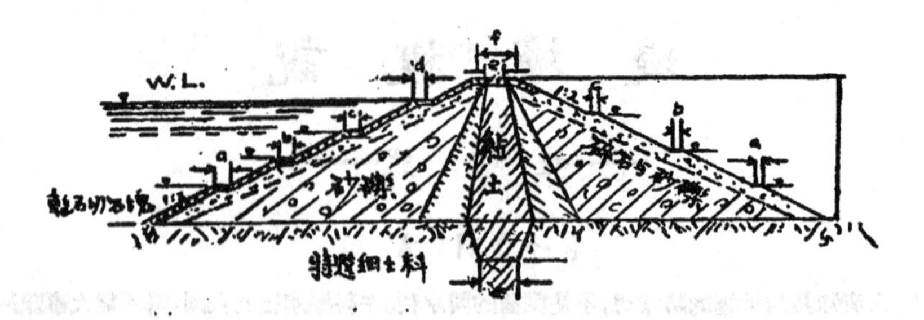

（三）三合土：混凝土牆之築設應在隄土之前不遠處爲之，其深度應達不透水地層，如不透水層過深，則須打板樁以代之，以免發生消縮縫而漏水，每隔二十公尺留一不漏水之伸縮縫。

（三）堤線之規劃

堤線之規劃，須求二堤之間，能暢瀉多量之水，並使堤身之位置十分安穩。而規定堤距，必以下列各項爲標準：

（一）最大洪水位之高度與水面之比降爲丁。

（二）未造堤以前，洪水期內之河流橫斷面。

如圖所示爲所測得之洪水位橫斷面，用 LL 及 RR 二縱線分爲三部，各部之平均流速爲 V_1, V_2, V_3（自左至右計之），故全斷面之流量爲：

$$Q=Q_1+Q_2+Q_3$$

$$Q=b_1 t_1 v_1+\left\{b_2 t_2-\frac{1}{2}\left[b_4(t_2-t_1)+b_5(t_2-t_3)\right]\right\} v_2+b_3 t_3 v_3$$

堤成後，河流之橫斷面減少 $\triangle F_1$ 及 $\triangle F_3$ 二部

$$\triangle F_1=(b_1-x_1)t_1$$

$$\triangle F_2=(b_3-x_3)t_3$$

橫斷面既減少，水面被束必致漲高，設其漲高數量爲Z，而Z可預爲估算；蓋未築堤以前，經過 $\triangle F_1$ 及 $\triangle F_3$ 二斷

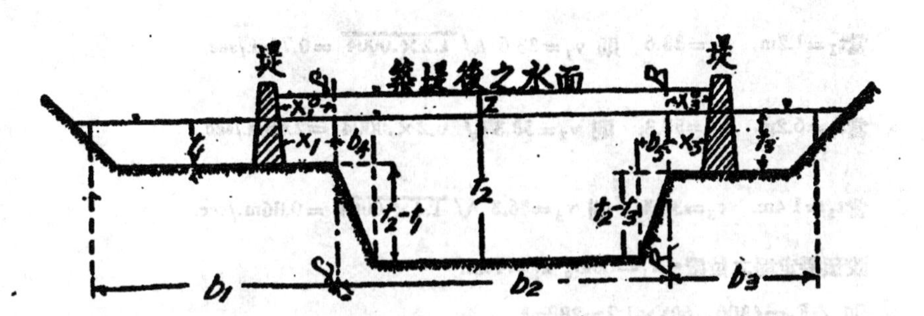

面流出之水量，于堤成之後，須與從（$x_1+b_2+x_3$）Z 面積流出之水量相等，假設 $x_1^0\cong x_1$，$x_3^0\cong x_3$，并估計漲高水面 Z 之平均速率較舊槽之水約大 5/4 倍，于是得：

$$\triangle F_1 V_1+\triangle F_3 V_3=\frac{5}{4}Z(x_1 v_1+b_2 v_2+x_3 v_3)$$

此式中之 x_1 及 x_3 可預爲擬定，作爲已知之數，而 V_1，V_2，V_3 可以從流速公式算出：

$$V_1=C_1\sqrt{t_1 J}$$

$$V_2=C_2\sqrt{t_2 J}$$

$$V_3=C_3\sqrt{t_3 J}$$

C，C_2，C_3 可從赫馬克（Hermanek）氏計算流速係數表得之，由此可直接求出

$$Z=\frac{4(\triangle F_1 V_1+\triangle F_3 V_3)}{5(x_1 V_1+b_2 V_2+x_3 V_3)}$$

既得 Z 值，須考驗堤成以後，其河流上段之水面漲高爲幾何，又因水之深度增大，其流速加大幾何均須妥洽而後決定 x 值爲合作，否則須增加 x_1 及 x_3 之值再行計算：

例題：

如上圖：設

$b_1=300$ 公尺（M.）； $t_1=1.2$m.

$b_2=300$m.； $t_2=6.2$m.

$b_3=100$m.； $t_3=1.4$m.

$b_4=30$m.； $J=0.0004$

$b_5=29$m.；

依赫馬克氏係數表：

25657

當$t_1=1.2m$，$c_1=33.6$　則 $v_1=33.6\sqrt{1.2\times.0004}=0.74m./sec.$

當$t_2=6.2m$．$c_2=53.3$，則 $v_2=53.3\sqrt{6.2\times.0004}=2.65m./sec.$

當$t_3=1.4m$．$c_3=36.3$，則 $v_3=36.3\sqrt{1.4\times.0004}=0.86m./sec.$

設預擬定堤之位置：$x_1=60m$；$x_3=40m$.

則 $\triangle F_1=(300-60)\times1.2=288m^2$

$\triangle F_3=(100-40)\times1.4=84m^2$

$$\therefore\quad Z=\frac{4(288\times0.74+84\times0.86)}{5(60\times0.74+300\times2.65+40\times0.86)}=\frac{1141.44}{4369}=0.26m.$$

Z之值既定後，乃考驗堤前後之每秒鐘流量而比較之，如相差無多，則前所假定之值亦均合用矣。

築堤以前之流量爲：

$Q=300\times1.2\times0.74+[300\times6.2-\frac{1}{2}(30\times5.0+29\times4.8)]\times2.65+100\times1.4\times0.86$

$=4933m^3/sec.$

堤成以後之流速：

$V_1=37\sqrt{1.46\times0.0004}=0.89m./sec.$

$V_2=53.5\sqrt{6.46\times0.0004}=2.72m./sec.$

$V_3=38.5\sqrt{1.66\times0.0004}=0.99m./sec.$

故築堤以後之流量應爲：

$Q=60\times1.46\times0.89+[300\times6.46-\frac{1}{2}(30\times5.0+29\times4.8)]\times2.72+40\times1.66\times0.99$

$=5022m^3/sec.$

前後流量相差不大而中泓洪水流速V_2由2.65m./sec.增至2.72m/sec.倘屬無礙。

如洪水之流量增加，則堤距亦須放寬。設遇河灣，則堤距較之直河更應加大，因堤線最忌灣曲，不可專循河灣之岸築堤，宜依據河灣處之洪水斷面，使凹岸之堤，離近其岸，凸岸之堤，遠離其岸，成一平緩之堤線，而介于凹岸頂點及堤之間，尤宜留存外灘，藉免水力之直接侵堤，而有潰坍之虞也，且外灘亦須善爲掩護，莫使被水冲刷。總之兩岸之堤宜力求平行，蓋洪水流之斷面，倘驟寬驟狹，則氷凌易于壅積騰高，而越堤外溢，冰經堤頂，則損及堤身矣。

河岸之地質，若爲堅壤，則易于築堤。設地質疏鬆或當卑溼之地，則堤身自有陷落之虞，將來建築與修養等費，必甚昂貴，施土之前，須預先踏勘，估計始可精確，凡遇大村落城市，或有價值之設備，均宜圍入堤內，卽堤線稍有彎曲，亦所勿顧；而卑溼之區與深塘，須完全留在堤外，否則足以妨礙堤身之安全，不可不愼也。

規定堤坊橫斷面時，最宜顧慮者爲水之壓力，水流之冲刷力，波濤之冲擊力，冰澌凝塞時勿使水越堤頂，貛鼠與地羊之爲患，堤上植樹最爲危險，蓋樹根入土，四面通出，水從隙入，積而外洩，堤土乃坍，故掩護堤面，以草爲適宜。

（四）築堤之材料，季節及建竇位置

築堤材料，百分九十幾在坭，坭以黏性不透水爲宜，有時以築堤地方爲冲積岸，土質多沙，找黏土不易，宜先去表土，除去有機腐質，掘至深層始可採用。然慮在我國每每因爲採土而毀壞民田，容易引起反感，馬虎從事，不選擇而築之，後患堪虞，終無底止。

築堤在我國現情下，談不到機械化，蓋以車運坭，成本高于人力，用機壓土，其費廉不過牛蹄，不過用人有用人之難處，愚而自私爲我國人之通病，坭工尤甚甚謁。築堤者恐堤之不固，坭工祇恐收入之不高，故一般坭工，每乘監管人員監管不到，或以沙作坭，或鋪土不平，或厚度超額，硪練失効，致造成後患。

築堤，人多以爲最適宜的季節爲秋冬季節，以天氣晴和，採運方便。但以晴天日久，坭土乾躁，取水困難，土無水潤，壓練不貼，空隙必多，堤身弱點，就由于此，春夏之交，雨水多而氣候熱，工作效率低，自然之理也，但如能在春夏季節築成之，其坭土之貼實，較勝秋冬季節所築成之堤也。

任何地方築堤，不能不建竇洞以宣洩內水，竇洞位置，多騎正原日之流水涌，以本人兩次見敝邑兩次建竇之經驗，深覺其竇洞位置，應宜在坑涌之側，不宜騎正涌坑之中，蓋以坑涌之中，流水日久，積而冲，冲而積，不知已若干次，積回之土，浮蕩如漿，清之而後打樁，糊狀難免，施工困難，若從側爲之，施工容易，底土實而不浮，至竇成之日，將流水道畧事修改便可，是則建築時工作上之艱困可減，竇洞永置于實土上之功可收；若騎正涌心，縱使杉樁够長而又够密，將仍不免有些微之變態。

（五）築堤經過

本人生處高要新江下游之東，時遭水患，童年時代目覩長輩提倡築圍已成白熱化，不幸爲鄙人阻撓，結果功敗垂成，及至民國廿五年吾粵歸政中央，肇慶有三區行政督察專員之設，李公磊夫其始受命焉，鄉人以李公施政方針，首重水利，故向之求助，頗許之，代向省行貸款，乃得興工，一切計劃均依照近代科學原理，并得許寶漢工程師督導一切，堤心厚度爲八英呎，外坡坡度爲一：三，內坡坡度爲一：二，然當時鄉人以鑑于前清築圍失敗及連年水患失收，急不暇擇趕快快而築之，其施工之草率，選土之馬虎，祇求其速成而不計其後患，當是時表面觀之，實似小丘，大而宏偉；十七個月草告完成，卽大錯之鑄成也。六年四崩，考其原由，卽在于此。

本圍自民國廿六年春興築，廿七年夏告其所謂完成，完成至今，計十有一年矣，中間經過民廿七，廿八，廿九，卅，連年水位不高故得免于決。至民卅一年七月下旬西江水位突過一一五·六米，本圍之下察竇側，突然潰決，竇側潰決，竇洞上之護土牆亦爲冲崩，缺口深入數丈，竇下木樁粗似一呎直徑者亦有被冲折，細察致決原因，其爲建竇時，竇之位置騎正坑涌，工程進行時，爲宣洩流水而在竇側另開水道，至竇將成之日，西潦驟至，忙個不了，憑水堵坭而塡築之，旣不能椿又不能練，坭成糊狀求其阻水而已；幾經冬季，水份揮發，土質收縮，罅隙與焉，一交潦漲，乘隙乃決；此爲本圍第一次潰決之主因。

民三三年七月中，西江水位超過一一六・二，本圍烏郊段地帶冲積，土質多沙，興築施工時不合理以此爲最，且乾燥季節，運築之土，大塊而乾，旣無水潤，尤欠牛練，故于崩決之前，發覺滲漏不過半小時，便巨响發出，泛濫遂成，潰決之由，旣如上述，無待探索。

民三四年八月中，慶日敵投首之餘，不圖立秋節過，而潦竟還超過一一五・五，本圍烏郊段搶險多次，以爲可免崩決，豈知事隔一天，潦水稍退而小塘段竟發覺內坡基脚滲漏，搶救半日，結果潰決二十八丈餘，旬餘水退，基底畢現，見横形小涌由外而內長約十餘丈，闊八九尺，深五尺餘，涌之底，涌之側土質黏固，以其浮土被冲淨盡也。小涌之來，無從考据，然致決之由，人多謂涌滲漏而成之。

民三六年本圍因鑒連年崩決原在堤身土質滲透，有見及此，即在外坡加築一厚一英呎之三合土牆直落基底，牆下更打長八呎木樁以加强其抵抗力。及六月下浣，西潦來得特早，水位超過一一六・一，本圍以新造之三合土隔水牆，完工未幾，西潦突至，掘出之土，不能從容塡回，塡回惟恐不及，水位過隔水牆上二三呎，水由塡回之土乘鬆而入，搶救逾一晝夜，卒之泛濫全圍。

考以上四次崩決致禍之由，仍在于興築時缺乏學理上研究探討，選土不愼，操練不良，施工草率。謹寫出以證上述學理之不謬與築堤者之參考。

GENERAL DISCUSSIONS OF SEDIMENT TRANSPORTATION

By

Kam Kwong Yee

M. Sc. in C.E., University of California

Sediment consists of small stones, gravel, sand, silt and other insoluble materials carried by water into a flowing stream and its tributaries. As a large amount of sediments transported by streams, either as suspended matters or as bed-loads, are accumulating or growing in water, difficult problem arise in many projects for flood control, soil conservation, irrigation, navigation, and water power development. Costly maintenance, loss of efficiency, and, in many cases, complete destruction of important engineering works have been experienced due to the filling of reservoirs by sediments, filling or scouring in navigation and irrigation channels, and erosion and gullying on arable lands. It may be expected that the accumulation and deposition of the sedimentary materials carried by rivers will become a problem even more serious in the future.

In rivers, canals, and their tributaries. it is necessary to minimize the excessive deposit of silt. In storage reservoirs, the accumulation of sediments continually decreases the storage capacity and in some cases limits the value of a storage reservoir to a comparatively short period. As the years go by, the process of sedimentation continues indefinitely until one of the two things can happen: all the feasible dam sites in the stream will be used up and only a series of sediment-covered structures will remain; or the highly developed communities and farm areas will be abandoned. Therefore, it is desirable to have a better knowledge to control sediment movement and to have more information regarding the nature and amount of sediment carried by the streams.

Silt begins with erosion, which frequently takes away the best top soil of the farm lands, or form ugly, damaging gullies. The soil, carried into a stream by rain run-off, can take up valuable space in reservoirs intended for storage of irrigation water. It can clog irrigation canals and cover good crop-producing lands with layers of useless soil.

Man cannot eliminate forces of nature which create sedimentation. However, he can devise ways and means of reducing the effects of sedimentation and decreasing its useless effects. From the future point of view, the quicker we can obtain a sure means of controlling sedimentation, the better future we will have.

Erosion may cause great damage to farm lands. Erosion and siltation usually occur during times of floods or severe storms. Heavy damage is done when the soil is washed into the rivers. Clearance of the channels in streams which are overloaded with sediment and dredging along the bed of all navigable streams are required. Because sediments cause river beds to rise and therefore stronger and higher levees are needed to hold back the water. When the stream bed rises as high as the surrounding farm lands, drainage of the farms becomes difficult or impossible. Thus, causing ground water levels to rise as high as to waterlog the farm land, making it unproductive. Moreover, during storm seasons, heavy floods may occur and cause great destruction to properties and lives. Erosion control can be worked through maintenance of forest and other vegetal cover, by land management, the use of good farming methods, and installation of small erosion control structures.

There are different methods to prevent the excessive deposite of silt:

1. Design the diversion works on the stream and canal headgates to decrease the amount of silt carried by the water diverted in the canal.
2. Provide sand boxes or sand traps on the canal so that some of the silt and sand will be deposited and removed from the canal.

In certain areas where it has been recognized that sediment carried in canals would cause undue maintenance expenditures, elaborate works have been devised to remove most of the sediment before it enters the canal. The cost of removal of sediment from reservoirs has been found to be far in excess of the benefits derived.

Until the present, some proposals have been made to construct dams on tributary streams to enable it to store the sediment, thereby reducing the amount carried in the main stream.

It is important to know the character and amount of sediment as well as its source and manner in a stream channel.

Two streams may have the same hydraulic characteristics; namely, slope, rate of discharge, etc. but they may differ greatly in their sedimentary characteristics, magnitude of sedimentary load and mechanical composition of the sediment.

Sediment or solid matter which is transported by running water may be classified as follows:

1. Traction, whereby solid materials moved by rolling, sliding, or siltation along the bed of a watercourse, motion being produced by the action of water flowing along the inclined plane of the bed.
2. Suspension, whereby solid material of very small, fine particles carried in the water does not come in contact with the bed of the stream for an appreciable time, motion being produced by upward velocity components of the current.

G. K. Gilbert attempted to establish a basic relationship between traction and velocity of the current. European investigators, following the theories of Du Buot and Du Boys, have developed a simple relation between traction and the slope and depth. In May, 1932, Leighly presented an American treatise based on the theory of tractive force. His contribution confirms the results and conclusions presented in many ways and helpfully broadens the application of the theory of tractive force to channels of finite width and variable depth.

In most rivers, large particles are moving along in contact with the bed. Comparatively coarse material is transported just above the bottom in suspension or saltation or both. Fine particles are carried in suspension.

The bed-load passing any section of a natural stream varies to a greater or lesser extent with the discharge, slope, cross-sectional area or hydraulic roughness, which in turn affect the stream velocity, hydraulic roughness, the availability, characteristics of sediment, and possible others. Measurements of bed-load transportation in a river will indicate the net effect of all the variables as they interact upon each other.

The fine material in suspension is very uniformly distributed in the water from the surface to the bottom, but the greater part of the coarse material is found nearer the bottom.

Because the different parts of the load cannot be measured with the same instrument, it is difficult to determine the total quantity of sediment discharged in a unit time.

To determine the amount of material being carried in suspension, it is necessary to use a suspended sediment sampler to find its concentration by taking a sample of the flowing water to obtain its weight and measure the weight of sediment in it in proportion to the weight of the sample. It is also necessary to determine the water discharge. From these data, the weight of sediment discharge per unit time can be computed.

To determine the amount of the heavy material moving on the bed, it is necessary to measure the quantity of sediment moving per unit time for a known width of the stream, so that the total quantity can be computed. This may be done by using a bed-load sampler or a suitable trap that permits water to pass through but retains the solids by screens or by reducing the water velocity inside the trap.

It is also important to observe the rate of movement at a sufficient number of points across a stream and see the variation between the sampling points accurately. Plot on a graph with the rate of sediment discharge per unit width of stream as ordinate and the position of the observing points across the stream as abscissa. The total sediment discharge is determined by the area under the curve thus computed.

Du Boy's equation is the most common type of formula used for finding the rate of transportation. In the equation, the rate of movement is plotted against a function of tractive force. The Schoklitsch-Gilbert type is also very common in which slope, discharge, and grain size are considered of primary importance. In some investigations, bed-load movements have been related to bottom velocity, and in others to mean velocity. More recently, Einstein has developed a dimensionless bed-load function from theoretical reasoning and experimental observations.

CONCRETE PREFABRICATION

By

Kong H. Go

A short survey of Concrete construction trends

Prefabrication has become one of the latest innovations in the field of concrete construction. Though not a new method in the field of engineering, concrete prefabrication has been brought to the fore mainly due to the present dire housing situation and the great shortage of steel.

One form of prefabrication, the concrete blocks, have been used for many decades, though its use has not been too popular. The main use of concrete blocks was for the purpose of erecting partition and walls. By the use of light weight aggregates, the total weight of large structures are tremendously reduced. Insofar as strength and durability is concerned the concrete blocks are comparable to the clay brick that has been used so widely throughout the world. For natural beauty, the clay brick generally have the upper hand but due to the ever increasing factor of economy, concrete blocks have come into greater use especially in the construction of commercial and industrial structures. The latest uses of cinder concrete blocks, at present, is the large scale housing project of the New York Housing Authority where apartment houses are built to hold from 1100 to 1900 families each. Each housing unit are constructed of concrete frames ranging from 11 to 14 stories in height. The exterior and interior walls are to be of light weight concrete blocks.

The newest form is the on-the-job prefabrication of structural members. With the forms to be constructed on the ground, pouring could be carried out with the least amount of work. On top of this, the ground could serve as part of the form or support thereof, which eliminates one side of the form as practiced in upright pouring; thus, the greatest economy is afforded.

By the use of high early strength cement, prefabs have been ready for use within 20 hours. These units are raised from their forms by cranes or other equipment. A unique method was designed for the lifting of concrete prefabs from the forms by the use of compressed air and another by the use of suction.

This new method has been unprecedented for the construction of concrete; and simulates the factory production of many of our present day construction materials. Though new in method, the basic principle of construction is still dependent upon the long well known trades of masonry and carpentry. By the use of this method, the most uniform quality of concrete is produced with the minimum of labor waste and greater all around economy. The main disadvantage lies in the fact that for any construction, a large working area would be required.

For small housing units or large industrial units, some distance from the center of population, this method has proven profitable and practical. In the case of small housing units, the exterior walls were poured flat on the ground next to the site of construction. For large industrial buildings where beams and columns are of uniform size throughout a plant, there is a decided advantage in the re-use of the molds and forms.

Most of the precast concrete beams and columns are produced by use of bolted reinforced concrete sections that are similar to steel channels. By the radical use of these hollow structural members, the engineers have eliminated much of the dead weight, and combined it with rigidity. As stated before precast concrete is approaching more and more the process of other construction materials.

As in all concrete, beams and girders require steel reinforcement to take up the tension stresses. Regular steel have been used with wire meshing for the reinforcing of all members, including the roof slabs, which were only 1¼ inch in thickness. Connection were cut to a minimum and made at the points of least stress. Dowels were used at column bases; and for the connecting beams between columns, splicing was made by use of welding the reinforcing steel. This type of construction was employed by the United States Navy Department's Bureau of Yards and Docks in the erection of warehouses that required large floor area at Mechanicsburg, Pennsylvannia.

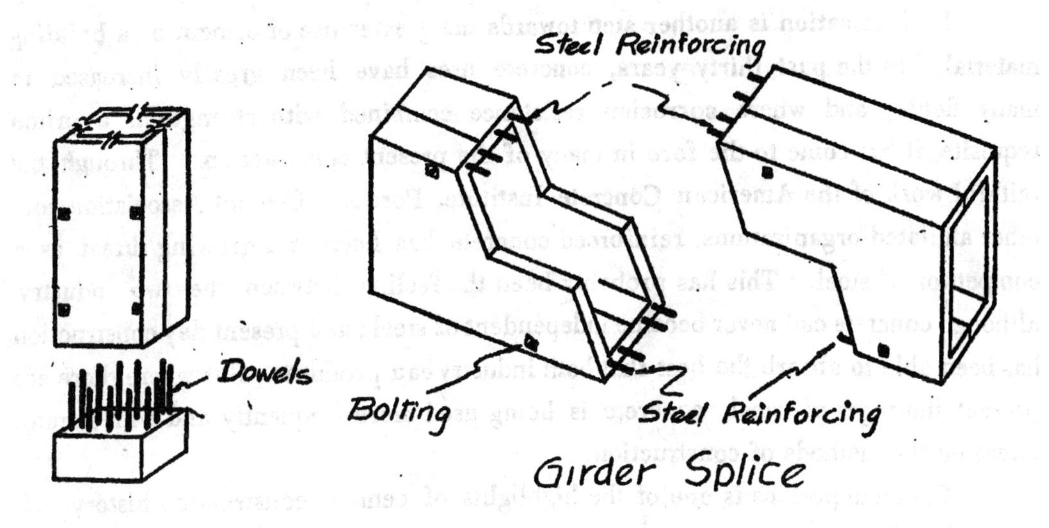

Column Detail

In areas where there has been a shortage of steel the concrete prefabs have come into prominent use. Europe has gone for this type of building material in the mass production of housing units as well as pioneers in the use of concrete as bridge members.

In England, the critical housing situation after the war was an immediate problem. Tnis problem led to the use of the prefabrication of concrete slabs for construction of compact bungalow type houses for the middle class people. Their quick action have all but solved the housing problem.

France has of late launched a program of reconstructing their war torn bridges with plans for a number of reinforced concrete girder bridges way beyond the dreams of the conservative engineers. They too have gone into the use of hollow reinforced concrete units. The bridge spanning the Marne at Luzanzy, France, now under construction, are composed of prestressed concrete chords over 180 feet in length. The bold use of piano wires as reinforcement is another feature of its radical design. With a total stoppage of engineering research during the war, with the great loss of manpower and equipment, and with only two years after the close of the war, this one feature in the engineering field is a measure of the fast reconversion of France and a return to normalcy.

Prefabrication is another step towards the greater use of cement as a building material. In the past thirty years, concrete uses have been greatly increased in many fields; and where corrosion resistance combined with strength is a prime requisite, it has come to the fore in many of the present construction. Through the faithful work of the American Concrete Institute, Portland Cement Association and other affiliated organizations, reinforced concrete has become a growing threat as a competitor of steel. This has probably been the feeling between the two industry, although concrete can never become independent of steel; and present day construction has been able to absorb the best that both industry can produce. But where there are present shortages of steel, concrete is being used more frequently and with greater stress on the methods of construction.

Concrete prefabs is one of the highlights of cement construction history. It shows that through greater research there are innumerable methods of solutions to the same problem, which only time and human effort will bring out. To the engineer it is just one of the normal routine which they have played in the building industry.

Concrete itself is an ideal building material. Its greatest asset lies in its property of being able to increase its strength indefinitely when it is properly handled. Noticed the term properly handled. Properly handled meaning "control"; that is, control of the making of the concrete from the beginning to the end, from the selection of the aggregates of the curing of the finished structure.

There are five steps to the process and all require the proper control and supervision during its operation. First, there is the choice of aggregates; second, proportioning of the mix, choice of the water cement ratio, choice of the yield; third, mixing and transportation of the mixture to the site; fourth, placing of the mixture; fifth, curing of the completed members.

Though all the steps are important routine work, the second step is the most important in considering the economy of the structure. It has been found by experimentation that the strength of concrete is solely dependent upon the water cement ratio, the strength increasing logarithmically with a decrease in the water cement ratio. Thus, we can choose a water cement ratio of the proper strength for various part of a structure. This, of course, can only be done when all factors are under strict control.

For instance, where a water cement ratio of 7.5 gallons per sack is required to obtain a standard 3000 pound concrete; or in parts of building where strength is not a prime factor, a 2000 pound concrete could be used by the use of a water cement ration of 9.0 gallons. Or if a 5000 pound concrete is required for beams or girders, a water cement ratio of 5.0 gallons could be used. For maintaining a minimum slump, the mix will have to be varied according to the workability. For a given water cement ratio, we can calculate the yield of any mix by the absolute volume method using the known values of specific gravity and unit weights of the materials. Typical values of the yield for a 3 to 4 inch slump are approximately as follows:

No.	mix	water cement ratio (gallons)	yield (cu. ft. of concrete per sack cement)	cement factor (sack cement per cu. yd. concrete)
1.	1:1.7:3.0	5	3.88	6.96
2.	1:2.8:4.8	7½	5.90	4.60
3.	1:3.5:5.7	9	7.00	3.80

From these figures one can draw his own conclusions that economy is afforded in the use of controlled concrete. By using a water cement ratio of 9 instead of 5, there is a saving of 2.1 sacks of cement to the cubic yard of concrete, which represents a 43% saving on the cement. Even with the use of 9.0 gallons instead of 7½ gallons, there is a saving of 0.8 sack of cement per yard of concrete which is 17½ % saving. Since cement is the greatest cost in the use of concrete, any economy in the use of cement is a large factor in the cost analysis of concrete structures.

This control then is the prime requisite of economy. From these well known principles, old ideas are like new and through the applications of these facts, greater efficiency, greater economy, and greater use can be had with concrete.

China has been using cement for many years and she has used it quite extensively in building construction mainly as floor slabs and columns. For many years, it was limited in its use due to the high cost, and even up to the present with local cement manufacture, the extensive use is prohibitive. Though the use of cement is limited, that concrete that is put into use seems to be under poor control resulting in poor efficiency. In order to bring out the best there is in concrete, strict measures must be faithfully carried out in the field.

In all engineering, there is no better proof than that of old and tried methods of design, but there comes time when the facts of the past must be used to forward progressive ideas of efficiency and economy. Especially is it true in the materials available here where the engineer wonders where the next steel bar is coming from when one is taken from the storeroom and put to use.

In order to conserve that which we have, investigations must be made of the available local materials as well as imported materials, of available data and information of new up to date processes, designs, and methods of construction. Combined with this must be a coordinated program whereby the investigations could now be applied and to further the participation of all those interested. Of course, original investigation should be kept in mind at all ttmes and encouraged. Though research and application must go hand in hand, the matter of application must be stressed by the engineer. It is only through this type of work will the university be able to make progress and to fulfill the purpose and aims of a true university.

Bibliography

Engineering News Record
Civil Engineering

AN OUTLINE OF THE DESIGN OF A DAM

By

Chan Yun Cho

One of the most important factors in the development of all industries is the acquirement of cheap power in great quantities. This, together with the continually growing uses of power in the home and the high cost of coal have, in the past few years, focused public attention uqon the need of comprehensive water power development.

In China a total of 22 mill, potential horsepower of water power is estimated as available, with less than 4000 hp. developed. The Yangtze River, one of the two main rivers of China and a great artery of transportation, furnishes many opportunities for power development on its course, as well as on its tributaries, with a total potential power of upwards of 10 mill. hp. in an area which is densely populated, but with practically no water power development at present.

In carrying out of projects for water power, dams often involved. The purpose of the dam in a water power development is primarily to afford a head of water, but its function in creating pondage or storage is also often of great importance. When designing a dam, an investigation must be made upon the drainage area and a long term mean precipitation of the entire area must be obtained.

Mean Precipitation on a Drainage Area.—It frequently is desirable to determine the mean precipitation for a given drainage area over some period of time, such as for a particular storm, month, year, or long period, etc. This may be done by plotting "isohyetals," or lines of equal precipitation, properly interpolated from available records and obtaining a weighted average by planimeter.

A simple graphical method often used, known as Theissen's method, is based upon the assumption that the record at any one station should be used for the portion of the drainage area nearest that station. As will be seen from Fig. 1 where A, B, C, D are precipitation stations to be used in determining the mean amount upon an adjacent drainage area, by erecting perpendiculars halfway between adjacent stations the portions of the area nearest any particular station may readily be defined and in the figure are marked a, b, c, d, the percentage of the total area corresponding to a, b, c, d, determined by planimeter, is then applied to the respective station records A, B, C, D, the sum of these amounts being the proper weighted precipitation for the drainage area.

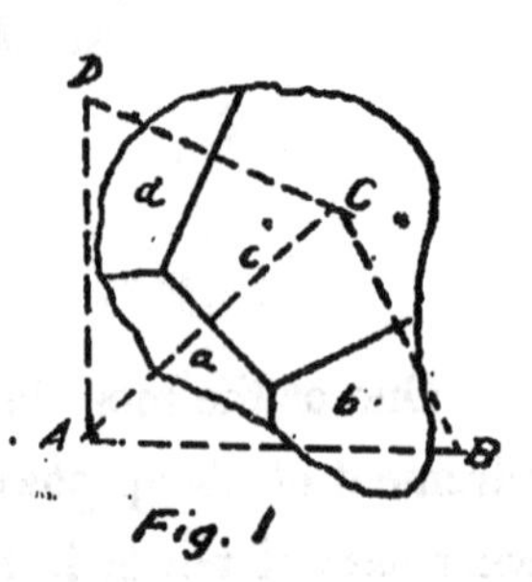

Fig. 1

In determining the mean precipitation for a given area, the records at individual stations should either include the same years or time of record or, when determining a long-time mean, preferably be modified by comparison with some long-time or index record near by, assuming the relation that is found between the short-and long-time record at the index station to hold for the other stations.

Illustrative Problem.—Given a map of drainage area, the annual precipitation at stations near drainage area, and the area of quadrilaterals following:

Middle latitude	Sq. miles
44° 22′ 30″	213. 72
44° 37′ 30″	212. 82
44° 52′ 30″	211. 91

LAMOILLE RIVER, MILTON, VT—CLARKS FALLS PROJECT
YEARLY PRECIPITATION—INCHES

Year	Burling-Ton Elev. 398
1820	
21	
22	
23	
24	
25	
26	
27	
28	43.30
29	
1830	
31	
32	39.59
33	49.44
34	
35	
36	
37	
38	30.83
39	27.99
1340	37.28
41	32.71
42	33.85
43	26.75
44	31.21
45	36.04
46	29.66
47	38.55
48	31.33
49	26.35
1850	37.51
51	31.83
52	28.82
53	33.05
54	25.45
55	38.30
56	36.82
57	37.25
58	30.21
59	35.30
1860	33.88
61	42.56
62	37.74
63	41.18
64	36.04
65	32.43
66	39.26
67	26.59
68	29.01
69	46.07
1870	31.66
71	29.91
72	33.25
73	25.92
74	31.94
75	26.94
76	27.53
77	33.11
78	41.45
79	24.27

Year	Burling-Ton Elev. 398	St. Albans Elev. 140	North-Field Elev. 842	Enosburg Falls Elev. 422	St. Johns Bury Elev. 711	Bloom-Field Elev. 930	Newport Elev. 700
1880	25.26						
81	26.99						
82	25.64						
83	29.34						
84	33.37						
85	33.64						
86	28.47						
87	31.13						
88	35.12		45 89				
89	38.21		36.66				
1890	36.92		38.17				
91	29.12		31.11				
92	42.24		32.67	49.99			
93	29.04		31.36	37.24			
94	22.96		28 92	38.97	27.15		
95	28.69		35.20	35.70	34.89		
96	28.38		33.82	38.13	30.67		
97	23.44		39.14	46.94	42.64		
98	31.78		30.52	40.35	40.85		
99	37.25		27.36	37.37	33.07		
1900	34.24		34.11	48.29	38.70		
01	33.88		31.42	52.30	35.15		
02	38.36		38.33	48.65	40.85		
03	32.86		29.09	36.73	31.24		
04	29.71		27.66	37.24	29.67		
05	34 73		32.31	39.28	33.12		
06	29.87		34.75	35.67	33 74	42.47	
07	29.67		37.77	38.61	42.35	31.84	
08	23.49		29.07	31.90	33.23	36.48	
09	35.76		32.34	38.65	29.78	36.02	
1910	31.63		31.71	37.83	36.46	33.77	
11	26.32		27.92	32.66	35.53	37.93	
12	34.13		37.00	46.84	40.10	35.45	
13	25.75		31.35	40.99	38.14	26.50	
14	22.62		30.08	30.45	31.06	40.94	
15	25.68		28.95	32.88	33.10	38.71	
16	28.45		31.36	36.97	34.14	38.96	
17	31.71		32.13	47.92	37.43	45.10	
13	42.18		37.16	42.34	39.06	40.17	
19	30.10		29.97	42.47	31.47	44.96	
1920	41.08		37.61	40.71	43.92	32.61	
21	27.17		30.85	31.90	27.32	40.51	
22	34.26		34.15	38.87	33.33	40.96	
23	29.25		29.81	35.85	31.71	39.59	
24	28.12		29.94	38.88	30.85	39.80	
25	37.17		34.23	43.18	35.25	37.29	
26	31 72		34.90	36.52	32.95	41.69	
27	35.65		40.38	46.35	32.53	37.07	
28	26.82		30.67	44.77	32.71	36.48	
29	33.87		31.43	45.78	32.39	32.69	
1930	29.49	32.03	28.95	38.27	31.32	39.84	
31	32.04	35.11	30.58	46.19	31.72	41.59	34.50
32	37.05	36.48	32.34	44.31	31.71	36.90	—
33	29.90	—	27.23	37.32	34.34	41.64	—
34	27.33	—	24.83	36.70	30.42	—	—
35	33.80	—	30.95	40.61	37.00	—	—
36	36.24	—	41.32	47.11	37.93	44.93	44.72
37	32.05	—	37.50	—	39.27	—	—
38	—	—	—	—	—	—	—
39	—	—	—	—	—	—	—
Years of Record	103		50		44	29	
Mean	32.54		32.86		34.69	38.39	

25673

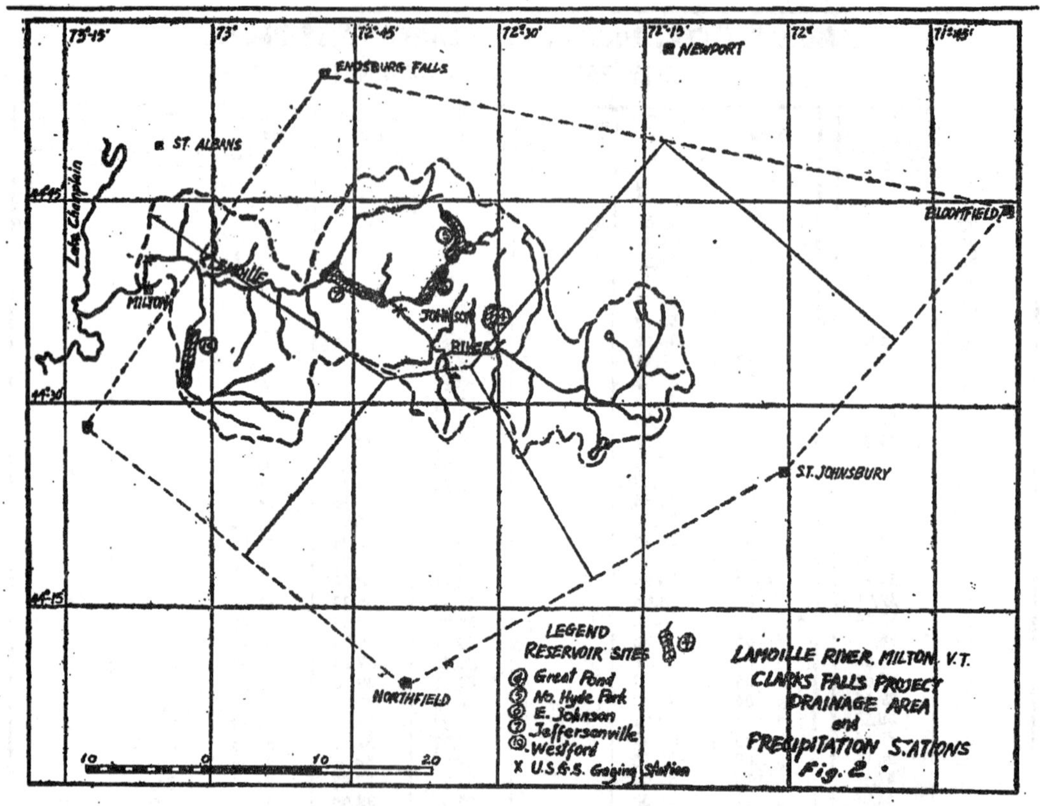

1. Determine the scale of the map accurately from the area of a quadilateaal. Graphically divide the drainage area into the areas governed by each rainfall station shown on the map and determine the area in sq. miles for each station, the total drainage area and the percentage weight for each station.

2. Find the ratio of the mean annual precipitation at Burlington for the period of the North field record 1888-1937 to the long term mean at Burlington and apply the ratio to Northfield mean to obtain a long term mean at Northfield.

3. In similar manner, adjust the mean annual precipitation at the other three stations to the long term mean by comparison with Burlington Record.

4. Using the long term means as determined in (2) and (3) (including that of Burlington and the percentage weights in (1). Compute the probable long term mean annunal precipitation for the entire drainage area.

Solutions.—

1. Drainage area governed by station Burlington .... 173 sq. mi.
   Drainage area governed by station Enosburg Fall .... 312 sq. mi.

Drainage area governed by station St. Johnsbury .... 186 sq. mi.
Drainage area governed by station Northfield .... 27 sq. mi.

Total drainage area .... 698 sq. mi.

Therefore, % wt. for D. A.$_1$ $= \frac{173}{698} \times 100 = 24.8$

% wt. for D. A.$_2$ $= \frac{312}{698} \times 100 = 44.7$

% wt. for D. A.$_3$ $= \frac{186}{698} \times 100 = 26.7$

% wt. for D. A.$_4$ $= \frac{27}{698} \times 100 = 3.8$

2. Mean annual precipitation at Burlington from 1888 to 1937 ....31.75 in.

Long term mean to Burlington = 32.54 in.

Therefore ratio of mean from 1888-1937 to long term mean at Burlington = 31.75/32.54
= 0.975

Northfield mean = 32.86 in.

Hence, long term mean at Northfield = 32.86/0.975
= 33.70 in.

3. Mean annual precipitation at Burlington from 1892 to 1936 ....31.46 in.

Therefore ratio of mean from 1892-1936 to long term mean at Burlington = 31.46/32.54
= 0.967

Enosburg Falls mean = 40.15 in.

Therefore long term mean at Enosburg = 40.15/0.967
= 41.52 in.

Mean annual precipitation at Burlington from 1894 to 1937 ....31.29 in.

Therefore ratio of mean from 1894-1937 to long term mean at Burlington = 31.29/32.54
= 0.962

St. Johnsbury mean = 34.69 in.

Hence, long term mean at St. Johnsbury = 34.69/0.962

= 36.06 in.

Mean annual precipitation at Burlington from 1907 to 1936 (lacking the year 1935) ....32.30 in.

Therefore ratio of mean from 1907-1936 to long term mean at Burlington = 32.30/32.54

= 0.993

Bloomfield mean = 38.39 in.

Hence, long term mean at Bloomfield = 38.39/0.993

= 38.66 in.

4. Probable long term mean annual precipitation for the entire area = (24.8 × 32.54 + 44.7 × 41.52 + 26.7 × 36.06 + 3.8 × 33.7) ÷ 100

= 37.58 in.

Maximum Stream Flow or Flood Flow.—The second thing which must be studied and investigated upon in the design of a dam is the maximum stream flow at the dam site. This maximum flow or flood flow can always be calculated by the following formula:

Fanning's formula:

$$Q = 200\ M^{\frac{5}{6}}$$

Murphy's formula:

$$Q = \left(\frac{46{,}790}{M + 320} + 15\right) M$$

Kuichling's formula:

$$Q = \left(\frac{44{,}000}{M + 170} + 20\right) M$$

where Q = maximum flow in c. f. s.

M = drainage area in sq. mi.

Illustrative Problem.—Based upon the data from the preceding problem, compute the maximum runoff for the basin using different empirical formulas.

By Fanning's formula :

$$Q = 200\ M^{\frac{5}{6}}$$
$$= 200 \times 698^{\frac{5}{6}}$$
$$= 46{,}860 \text{ c. f. s.}$$

By Murphy's formula :

$$Q = \left(\frac{46{,}790}{M + 320} + 15\right) M$$
$$= \left(\frac{46{,}790}{698+320} + 15\right) 698$$
$$= 42{,}600 \text{ c. f. s.}$$

By Kuichling's formula

$$Q = \left(\frac{44{,}000}{M + 170} + 20\right) M$$
$$= \left(\frac{44{,}000}{698+170} + 20\right) 698$$
$$= 49{,}400 \text{ c. f. s.}$$

Hence, the average maximum flood flow = 46,287 c. f. s.

Investifation of Dam Sites.—Thorough investigation to determine the most desirable and economic site for a dam should precede the designing and constructing of the dam. Such an investigation will include surveye, topographic mapping, geologic studies, and subsurface investigations. Also tests will have to be made on the materials in the foundations and on the materials of which the dam may be composed, and the results of the tests will have to be studied and their significance realized.

Types of Dams.—Where a rock foundation is available, the solid masonry dam is the most permanent type of construction. Hollow masonry dams are in quite general use and may be cheaper than solid dams where cement and concrete materials are high in cost. Arched masonry dams, i. e., curved upstream in plan, located in a relatively narrow valley with steep slopes suitable for arch abutments, are of low cost and may be used where conditions permit. The multiple arch dam is also an economical type of construction.

Earth dams may be built either with or without a core depending upon the height and the permeability of the material. The rockfill dam, as the name indicates, is a rough fill of loose rock, leakage through which is minimized usually by a deck or apron or wood, concrete, etc.

Timber dams may be of log cribwork filled with rock, with plank deck or apron—a common type of construction in a new country. or of frame and deck construction for low dams.

Combinations of Types of Dams.—It will frequently be found advantageous to combine different types of dams at a given site. A very common arrangement consists of a concrete spillway or overfall section with earth wings at one or both ends of the dam, either with or without core wall. This is particularly useful in the very common situation at dam sites where ledge rock exists near the surface, perhaps in the river and on one bank, permitting a concrete spillway section here, but is at considerable depth on the other bank.

A hollow spillway section of dam with either concrete or earth abutments is also common.

Forces Acting on Masonry Dams.—The forces acting on masonry dams and stresses due to these forces will be discussed in the following order: (1) water pressure, (2) earth pressure, (3) ice pressure, (4) other forces, (5) weight of dam, and (6) stresses in masonry.

1. Water Pressure. a. Abutment or Bulkhead Section.—Referring to Fig. 3, it will usually be found more convenient to use components of water pressure so that for any height h and water at w lb. per cubic foot, $P_h = \frac{1}{2}wh^2$, acting at a depth of 2/3. h.

Fig. 3

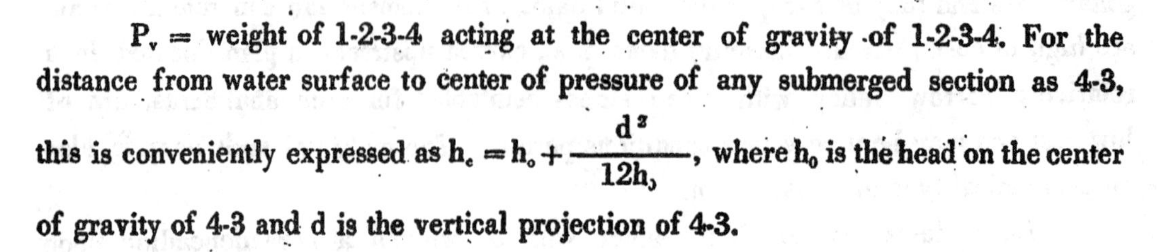

P_v = weight of 1-2-3-4 acting at the center of gravity of 1-2-3-4. For the distance from water surface to center of pressure of any submerged section as 4-3, this is conveniently expressed as $h_c = h_o + \frac{d^2}{12h_o}$, where h_0 is the head on the center of gravity of 4-3 and d is the vertical projection of 4-3.

Forces due to backwater on the toe of the dam with head h_2 would be computed similarly to those at the heel.

b. Spillway Section. — Referring to Fig. 4, vertical and horisontal components of water pressure on 1-2-3, or any portions thereof, will be determined as for the abutment section, keeping in mind the additional head H due to surcharge. It is common practice to neglect the vertical component of water pressure on 3-4-5-6 but to include the horizontal component on 3-4.

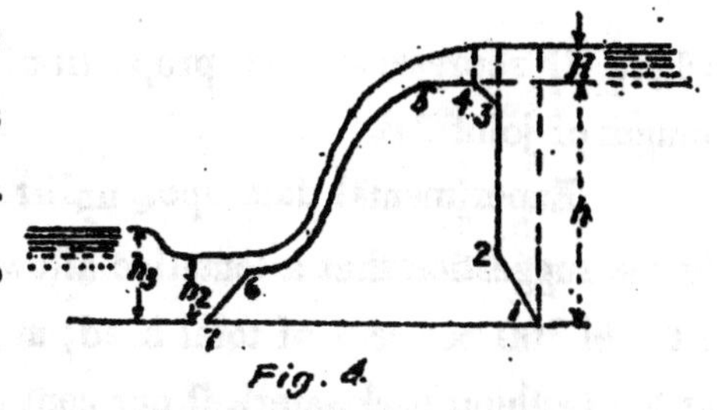

Fig. 4

Backwater at the toe of the dam h_2 may be practically nothing, due to high velocity of flow over the spillway and small depth at h_2, and the formation of an hydraulic jump, shown by h_3. If the depth of water below the spillway is sufficient, the jump will be "drowned out" and backwater due to h_3 will occur.

The force due to impact of flowing water will seldom be large in amount as compared with static pressures. It will seldom be necessary to consider impact as one of the forces acting on the dam.

Upward water pressure or uplift may be a factor of importance, due to pervious masonry or foundations, or both.

Referring to Fig. 5, if AB is a very small opening or passage through a masonry dam with water loads as shown, it is evident that at A an upward pressure on the masonry will exist due to the head h, and at B due to head h_2 (neglecting any velocity head due to flow through AB). Now, considering the entire level of AB longitudinally through the dam, if such openings or passages occurred at frequent intervals, a considerable total upward water pressure might be exerted at this level. If the openings aggregated 50 per cent of the entire length of the dam at level AB, the upward force exerted would correspond to one-half the area of the stress diagram 1-2-3-4; if 30 per cent of openings, the total up-

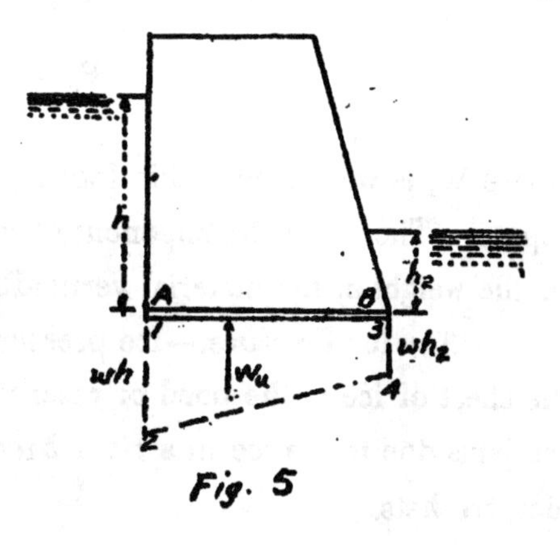

Fig. 5

ward force would be 30 per cent of 1-2-3-4, etc. We may therefore express such uplift per linear foot of dam by the equation

$$Wu = Cwl \frac{(h + h_2)}{2}$$

where C represents the proportion of effective uplift as regards area, and l the length of joint AB.

Experimental data upon uplift pressure at 10 dams studied by Houk resulted in the suggestion that a tentative allowance for lift would normally be as follows: At heel 100 per cent of total head; at 10 per cent toward toe, 50 per cent of head; at toe (without backwater), 0 per cent of head; with uniform variation in pressure between the above points. This would give an average up lift over the entire base of 0.30H, with resultant acting about 0.30 of base from heel of dam.

2. Earth Pressure. A dam may often be subjected to earth pressure due to filling to some level above the foundations. Such a fill on the upstream side of the dam, of course, will be saturated and diminished in weight accordingly. Thus, with earth which in the dry weighs 105 lb. per cubic foot, its weight under water, if one-third voids, will be 105-62.5 (1 - 0.33) = 63.5 lb. per cubic foot, or not greatly different from the weight of water. Under such conditions, it would be assumed to act as a liquid weighing 63.5 lb. per cubic foot.

Where not liquid, if α is the angle of repose of the earth, earth pressure, i.e., the horizontal component, would be computed by the Rankine formula

$$P_e = \frac{W_e h^2}{2}\left(\frac{1 - \sin \alpha}{1 + \sin \alpha}\right)$$

where W_e is weight per cubic foot of earth, h is the depth in feet, and α its angle of repose. The vertical component of earth pressure on sloping faces of the dam will be the weight of the material vertically abave the plane, just as for water pressure.

3. Ice Pressure.—Ice pressure against a dam may be due to expansion of the sheet of ice in the pond or reservoir formed by the dam or to the formation of ice jams due to the ice in a river breaking up and going out quickly in time of sudden freshets.

Suitable allowance for ice pressure will evidently vary greatly with the location of the dam as to latitude, its arrangement with reference to the side shores, of the reservoir, the slopes and condition of those shores, and the "reach," or distance, upstream from the dam to the opposite shore.

4. Other Forces.—These may include air pressure on the spillway section, due to a partial vacuum forming under the nappe of water flowing over the dam due to imperfect aeration at the ends of the sheet of water. This results in unequal air pressure on the heel and toe faces of the dam and, hence, an additional force tending to push the dam downstream. This may not be great enough in amount to be troublesome, but strong vibrations are sometimes produced by periodic making and breaking of this partial vacuum. The remedy is so to arrange the ends of the dam or any piers that the formation of a vacuum under the sheet of falling water is prevented.

Wave pressure of itself is not usually of importance as a force to be withstood but may dangerously increase the head acting on the dam, particularly in the case of earth dams, which may not be safely overtopped with water.

Wind pressure is not of importance in dam design, as the maximum loading due to wind would always be small as compared to water loading, and these loads cannot occur simultaneously to their full extent.

5. Weight of Dam.—The weight of concrete masonry varies principally with the weight of its aggregate and somewhat with the proportions of the mix. It is from about 140 to 160 lb. per cubic foot, depending on circumstances. The weight of stone masonry is usually somewhat greater than that of concrete.

6. Stresses in Masonry.—In Fig. 6, ABCD represents a cross section of a solid masonry dam assumed of unit length, above an assumed joint level AD, subjected to water loading P, with total weight of masonry above joint AD = W.

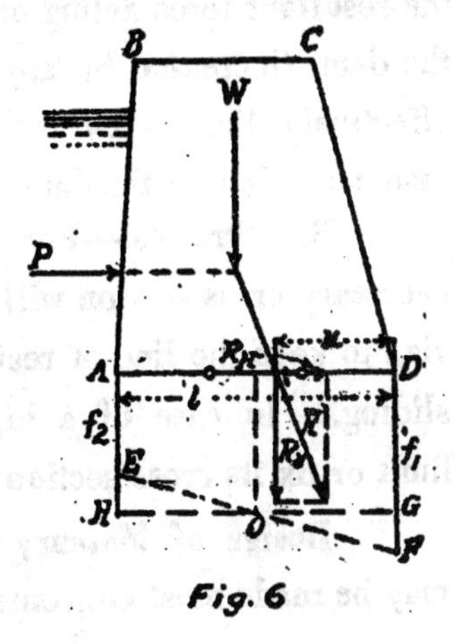

Fig. 6

The resolution of P and W gives a total force R acting on the joint, with vertical component R_v causing compressive stress on the joint (diagrammatically shown as AEFD), and horizontal component R_h, which tends to make the section ABCD slide on AD.

25681

For the conditions shown, the stress diagram AEFD will be a trapezoid made up of (1) direct stress AHGD, and (2) flexural stress EHOGF, and the stress at any point in the joint will be

$$f = \frac{R_v}{L} \pm \frac{R_v\left(\frac{L}{2} - u\right)\frac{L}{2}}{L^3/12}.$$

The above is merely the result of the application of equation

$$f = \frac{P}{A} \pm \frac{M_v}{I}.$$

Requirements for Stability of Masonry Dams.—A masonry dam of gravity section, whether of spillway or abutment type, must be able to withstand with suitable factor of safety: (1) overturning, (2) sliding, (3) stresses, and (4) other possible stresses or conditions requiring special design.

1. Overturning.—Theoretically, overturning would take place about the toe of the dam if the resultant forces acting at any joint level should come outside the joint at the toe. It is not desirable, however, to have any tendency to tension at the heel of the dam, and, hence, the rule commonly adopted of requiring the line of resistance to lie within the outer third point of the dam, which incidentally results in safety as regards overturning.

2. Sliding.—The tendency to slide is caused by the horizontal component of the resultant force acting on any given joint level. It is withstood by the weight of the dam (increased by any effective vertical components of earth or water pressure), effectively limited by the coefficient of friction of the materials forming actual or assumed joints in the dam or its foundations.

3. Stresses.—For dams of ordinary height, say, up to 75 or 100 ft., the necessary cross section will usually be determined by one of the foregoing provisions, viz., to keep the line of resistance inside the middle third or to be adequate as to sliding. In case of a high dam, stresses in the lower part of the dam are likely to limit or fix its cross section at these levels.

Design of Masonry Dams—Abutment section.—The design ef a masonry dam may be made most conveniently by assuming a tentative eross section based on the theoretical considerations which have been discussed, modified to meet other practical considerations, most of which will be obvious.

Considering the three sections (Fig. 7) shown of height h subjected to water loading, by taking moments it will be found that, in order that the resultant of the forces acting on the base b shall just lie at the outer third point A, the necessary width of base b will be as follows (where w and y are unit weights of water and masonry, respectively):

$$(1)\ b = \sqrt{\frac{w}{y}}h \qquad (2)\ b = h \qquad (3)\ b = \sqrt{\frac{w}{y}}h$$

or for ordinary values of w = 62.5 and y = 150 lb. per cubic foot

$$(1)\ b = 0.65h \qquad (2)\ b = h \qquad (3)\ b = 0.65h$$

At the ratio w/y is always < 1, section (1) would be more economic than (2) and have only half the material required for (3).

Considering the base section for the solid dam, with b = 0.65h, in respect to stability against sliding

$$\frac{0.65h^2}{2} \cdot y \cdot f = \frac{1}{2}wh^2$$

or

$$f = \frac{w}{y \times 0.65} = 0.64$$

for ordinary values of w and y. This is a little high as a safe value for f. Taking the latter at 0.55 would require b = 0.75h. On the other hand, modifications in the practical section to give the dam some height above water level and some thickness at and above water level will add to the weight of the section. Hence, a base about 0.7h will ordinarily be sufficient.

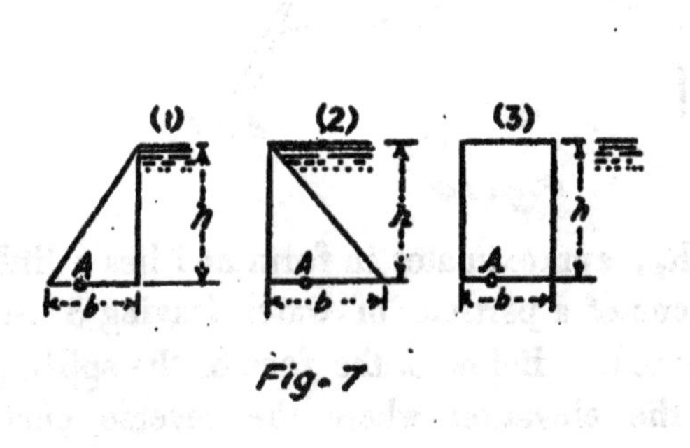

Fig. 7

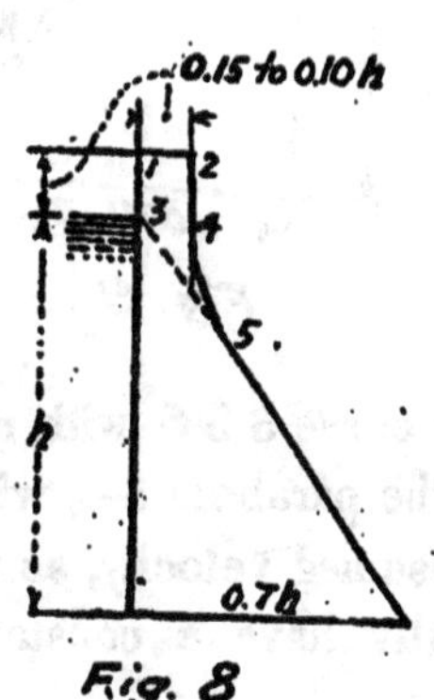

Fig. 8

The top which 1-2 (Fig. 8) of a solid masonry dam is usually from about 0.15h for low dams to 0.1h for high dams. The top of the dam is also about 0.15h to 0.1h above ordinary water (or spillway) level, although this dimension 1-3 will vary somewhat with particular conditions, such as spillway length, flood discharge over spillway, length of reservoir reach as affecting height of waves, etc. The toe face of the dam will be curved somewhat, as shown by 4-5.

Design of Spillway Section.—Design of the spillway section must be based on the maximum water load. An important consideration must be kept in mind in designing the upper portion of the spillway section, viz., to so proportion the curve of the upper part of the spillway that the overfalling sheet of water will not leave it but rather be guided steadily to the lower level and by a reverse curve started horizontally downstream again without undue shock or wearing on the toe of the dam or its foundations.

The theoretical or base section for the spillway will be a trapezoid, as shown in Fig. 9, which is the portion of a triangle of height (h + H) and base about 0.7 (h + H).

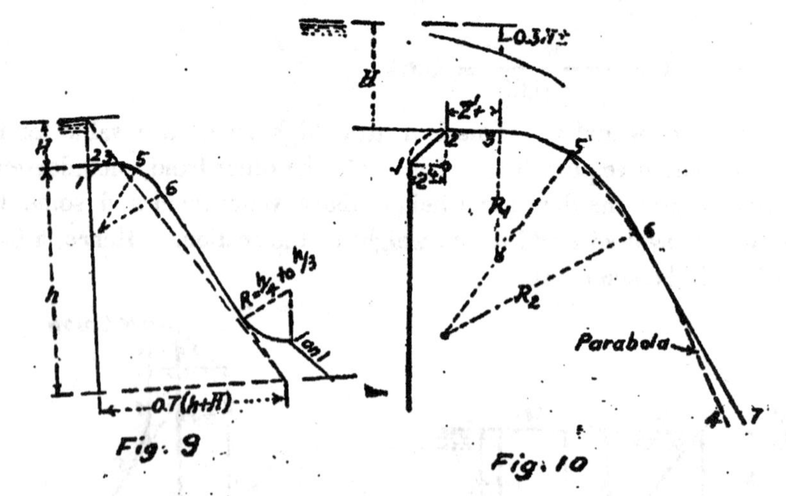

Fig. 9 Fig. 10

The curve 3-5-6, with radio R_1 and R_2, approximates in form and lies a little outside of the parabola 3-4, which is the locus of a particle of water leaving 3 with a certain assumed velocity, as further explained. Below 6, the face of the spillway would usually have a constant slope to the elevation where the reverse curve begins.

With the water area known or assumed at 3, the mean velocity past 3 may be computed. Then, if this is v_m, and x and y are coordinates of the parabola at any point

$$x = v_m t \qquad y = \tfrac{1}{2} g t^2$$

whence $y = \frac{1}{2} g x^2 / v_m^2$, the equation of the parabola.

The velocity of a particle of water at point 3 is less than v_m, the latter lying usually about one-third of the depth upward from point 3. The exact value of the theoretical mean velocity at any point on the parabola may be computed by noting that the velocity v_p at any point on the parabola is $\sqrt{(v_m)^2 + (gt)^2}$. The area of the sheet of falling water W_p at any point will be Q/v_p, and plotting $W_p/3$ normal to the parabola toward the spillway face will give a point on the lower nappe of the falling sheet of water to be used in fixing the curve of masonry of the spillway face. It will be found, however, that substantially the same result with materially less computation will be obtained by assuming v_m to act at point 3 and constructing a parabola beginning at point 3. The curve 3-5-6, with radii R_1 and R_2, approximates in form and lies a little outside of such a parabola.

Earth Dams.—The essentials in the design and construction of an earth dam are:

1. The foundation must be practically impervious, i.e., of such material and formation the seepage under the dam will be slight, or at least there must be no tendency for water to carry along fine material and hence undermine the structure or cause "piping," as it is often called.

2. The dam itself above the foundations must be practically impervious in in the same sense as the foundations.

3. The side slopes of the dam must be stable under all conditions which may occur during or after construction and evidently must be less than the angle of repose of the material composing the dam. The allowable slopes will vary with the material used for the fill.

The uqstream slopes, which will be directly subjected to water action, must usually be somewhat flatter than the downstream slopes at corredponding elevations, because the effect of saturation tends to decrease the angle of repose of the material. A common relation of the two slopes is, to illustrate numerically, 1 on 3 upstream with 1 on 2½ downstream, or a difference of about 1 on ½ between the slopes.

4. Overtopping must never occur; in other words, there must be adequate spillway or gate capacity to handle flood discharge under all conditions.

5. The outlet pipes through the earth dam, if there be any, must be so designed aud constructed that there can be no opportunity for leakage or rupture in a way that would endanger the safety of the dam by washing away of its materials.

6. Settlement of the dam after it is put in use should not occur to any great extent.

Earth Dams with Core Wall.—The most common design is to use a core wall of concrete, usually plain but sometimes reinforced. A typical section of an earth dam with concrete concrete core wall is shown in Fig. , with suggested ordinary dimensions or range in these dimensions.

The top width may be greater than that shown where the dam is used as a roadway. The distance above spillway level of highwater lavel will of course vary with circumstances. The essential is a fraeboard or distsnce between assumed high-water level and top of dam sufficient to allow for wave action and possible uncertainty in the estimated high-water level. For low and unimportant dams, the freeboard may often be less than that suggested.

The core wall is thin and depends upon the adjacent earth fill for its stability. It should extend to impervious material—ledge rock—unless the latter is at too great depth. Its faces commonly slope or batter, as shown, to such a depth as will give

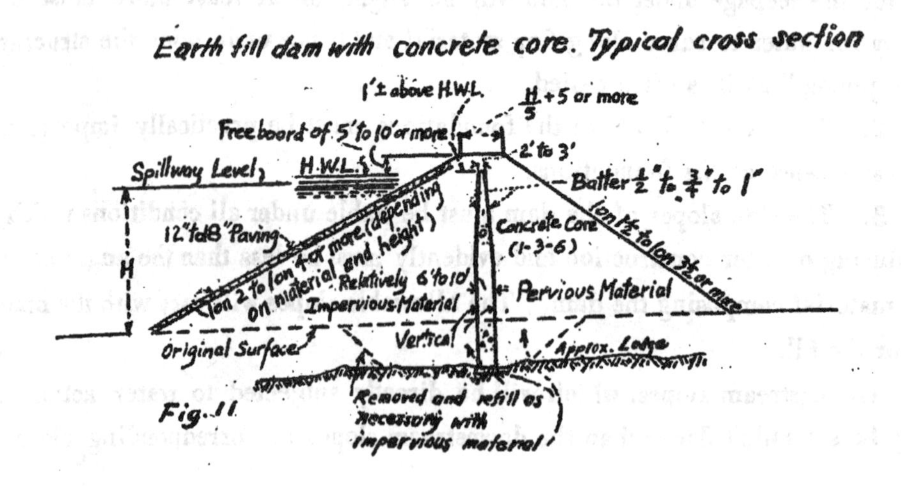

Fig. 11

a width of 6 to 10 ft., depending on the height of the wall, below which depth the sides are vertical. The bedrock at the base of the core wall should be suitably prepared—often by excating a cutoff trench—suitably to bond concrete and rock and prevent seepage at this level.

Arrangement of Spillway.—A spillway of concrete must be used, of course, with an earth dam, and it should be placed to best advantage. This may mean a location in the bed of the stream, at one side, or even in a different valley or hollow in some cases, under which condition the earth dam may be continuous. The presence and location of rock foundation will largely determine where the spillway should be placed. Where the earth dam butts against the spillway section, there must be concrete retaining walls, which will be expensive if very high and will require some care in location.

If the spillway section is located at one side of the river, it will be necessary to provide a channel below the dam to conduct overflow back to the river. This may require some excavation, if in rock; or heavy paving, if in earth.

Illustrative Example.—1. Data: (1) Topographic plan of dam site and vicinity including ledge rock elevations on line of heel of dam. The elevation of the spillway of the dam will be at Elev. 285, (top of flashboards, Elev. 288), and westerly end of spillway approximately at a point where rock is at Elev. 261. The length of the spillway will be 400 ft.

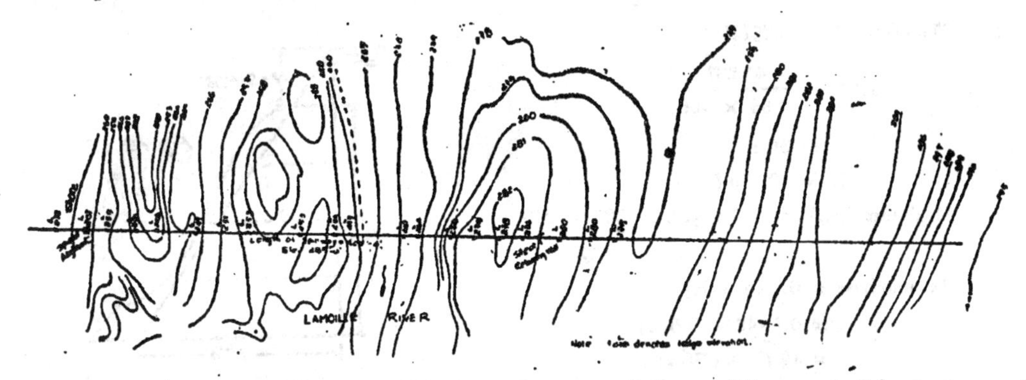

(2) East of spillway section is to be a retaining wall on rock foundation and earth dike with concrete core wall all of suitable dimensions.

(3) West of spillway is to be a concrete abutment.

2. Design of spillway section: (a) Spillway section to be designed under a head of water based on h from the max. runoff of the preceeding problem, on the crest without upward water pressure. Assume tentative section of maximum height,

and investigate for line of resistance, sliding, etc., modifying section as necessary. Line of resistance should be within the middle third and the sliding coefficient should not exceed 0.55. Show conditions for each 10 ft. point level of final section.

(b) Also for the section, as determined in (a), show line of resistance and direction of resultant joint pressure assuming upward water pressure according to Houk's conclusions.

(c) Abutment section on west bank need not be designed in detail and may be based upon dimensions of spillway section.

3. Design of earth secton.

Show proper maximum section of earth dike with concrete core wall running to rock. Also show design of retaining wall between dike and spillway.

4. Appurtenance.

Prepare details of flash-boards, pins, etc., specifying kinds with dimensions. Boards to be of 1" unplaned spruse and are to go out when head of water on the top reaches about 2 ft.

Design of Spillway Section.—(a) Maximum runoff from the preceding problem = 46,287 c. f. s.

Length of spillway = 400′ – 0″.

Use formula for weir with triangular cross-section, C = 3.85,

Therefore $Q = 3.85LH^{3/2}$

$$H^{3/2} = \frac{46,287}{3.85 \times 400}$$

$$= 30.1$$

$$H = 9.676 \text{ ft.}$$

Maximum height of spillway, $h = 285 - 245$

$$= 40 \text{ ft.}$$

Assume base $= 0.7\,(h + H)$

$$= 0.7\,(40 + 9.676)$$

$$= 34.77, \text{ say } 35 \text{ ft.}$$

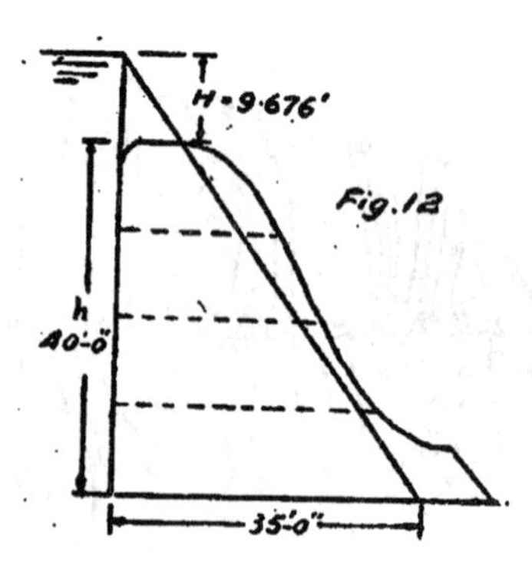

Parabola for determining the curve:

$$v_m = \frac{Q}{400 \times 0.7H}$$

$$= \frac{46,287}{400 \times 0.7 \times 9.676}$$

$$= 17.1 \text{ ft. per sec.}$$

Substitute this into $y = \frac{1}{2} gx^2/v_m^2$, therefore

$$y = \frac{1}{2} \times 32.2 \times \frac{x^2}{(17.1)^2}$$

$$y = 0.0552x^2$$

From this parabola, the upper part of the downstream face of the spillway section can be determined.

Horizontal water pressure upon the upper 10 ft.

$$P_2 = \frac{9.676 + 19.676}{2} \times 62.5 \times 10$$

$$= 9{,}160 \text{ lb.}$$

Horizontal water pressure upon the upper 20 ft.

$$P_4 = \frac{9.676 + 29.676}{2} \times 62.5 \times 20$$

$$= 24{,}600 \text{ lb.}$$

Horizontal water pressure upon the upper 30 ft.

$$P_6 = \frac{9.676 + 39.676}{2} \times 62.5 \times 30$$

$$= 46{,}300 \text{ lb.}$$

Horizontal water pressure upon the whole upstream face:

$$P_8 = \frac{9.676 + 49.676}{2} \times 62.5 \times 40$$

$$= 74{,}200 \text{ lb.}$$

Distance of line of action of P_2 from the 10-ft. plane:

$$\bar{x}_1 = \frac{9.676 \times 10 \times 5 + \frac{1}{2} \times 10 \times \frac{10}{3} \times 10}{9.676 \times 10 + \frac{1}{2} \times 10 \times 10}$$

$$= 4.44 \text{ ft.}$$

Distance of line of action of P_4 from the 20-ft. plane:

$$x_2 = \frac{9.676 \times 20 \times 10 + \frac{1}{2} \times 20 \times 20 \times \frac{20}{3}}{9.676 \times 20 + \frac{1}{2} \times 20 \times 20} = 8.31 \text{ ft.}$$

Distance of line of action of P_6 from the 30-ft. plane:

$$x_3 = \frac{9.676 \times 30 \times 15 + \frac{1}{2} \times 30 \times 30 \times \frac{30}{3}}{9.676 \times 30 + \frac{1}{2} \times 30 \times 30}$$

$$= 12.00 \text{ ft.}$$

Distance of line of action of P_8 from the base:

$$x_4 = \frac{9.676 \times 40 \times 20 + \frac{1}{2} \times 40 \times 40 \times \frac{40}{3}}{9.676 \times 40 + \frac{1}{2} \times 40 \times 40}$$

$= 15.50$ ft.

Wt. of concrete of the upper 10 ft.:

$W_1 = (7 \times 10 + \frac{2}{3} \times 10 \times 13.3)\ 150$
$= 23,900$ lb.

Wt. of concrete of the upper 20 ft.:

$W_3 = 23,900 + (20.3 \times 10 + \frac{1}{2} \times 5.3 \times 10)\ 150$
$= 58,400$ lb.

Wt. of concrete of the upper 30 ft.:

$W_5 = 58,400 + (25.6 \times 10 + \frac{1}{2} \times 5.4 \times 10)\ 150$
$= 101,000$ lb.

Wt. of concrete of the whole section:

$W_7 = 101,000 + (31 \times 10 + \frac{1}{2} \times 13 \times 10)\ 150$
$= 157,000$ lb.

Distance of the center of gravity of the upper 10 ft. from the heel,

$$y_1 = \frac{7 \times 10 \times 3.5 + \frac{2}{3} \times 10 \times 13.3 \times 8.5}{7 \times 10 + \frac{2}{3} \times 10 \times 13.3} = 6.3 \text{ ft.}$$

Distance of the center of gravity of the upper 20 ft. from the heel,

$$y_2 = \frac{1000 + 20.3 \times 10 \times \frac{20.3}{2} + \frac{1}{2} \times 5.3 \times 10 \times 22.1}{159 + 20.3 \times 10 + \frac{1}{2} \times 5.3 \times 10}$$

$= 9.4$ ft.

Distance of the center of gravity of the upper 30 ft. from the heel,

$$y_3 = \frac{3655 + 25.6 \times 10 \times 12.8 + \frac{1}{2} \times 5.4 \times 10 \times 27.4}{389 + 25.6 \times 10 + \frac{1}{2} \times 5.4 \times 10}$$

$= 11.4$ ft.

Distance of the center of gravity of the whole section from the heel,

$$y_4 = \frac{7675 + 31 \times 10 \times 15.5 + \frac{1}{2} \times 13 \times 10 \times 35.33}{672 + 31 \times 10 + \frac{1}{2} \times 13 \times 10}$$

$= 14.1$ ft.

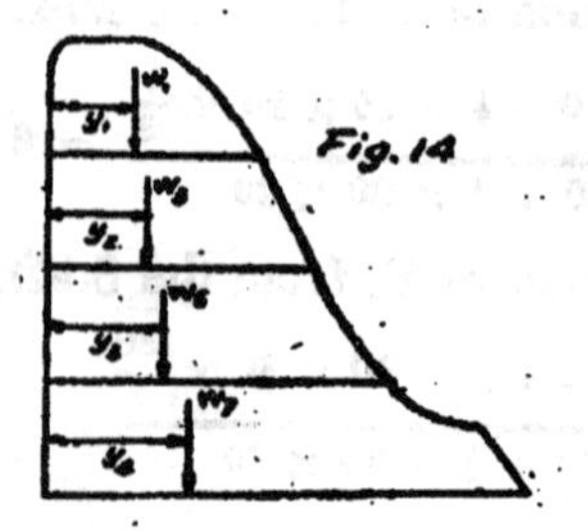

Checking for the line of resistance:

It is found that the line of resistance intersects the base of the spillway section within the middle third, therefore the assumed section is alright. (Fig. 15).

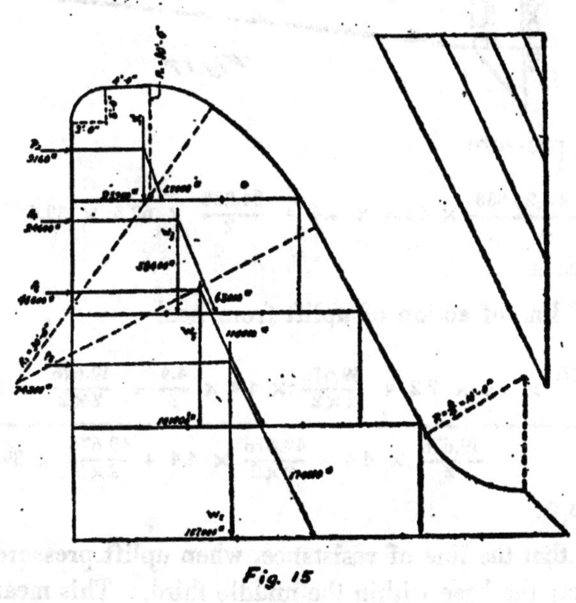

Fig. 15

Check for stability against sliding:

Along the 10-ft. plane: assuming f = 0.55

$$f = \frac{P}{W}$$

P = fW = allowable pressure

$P_2 = 0.55 \times 23{,}900$

$= 13{,}100$ lb. > 9,160 O.K.

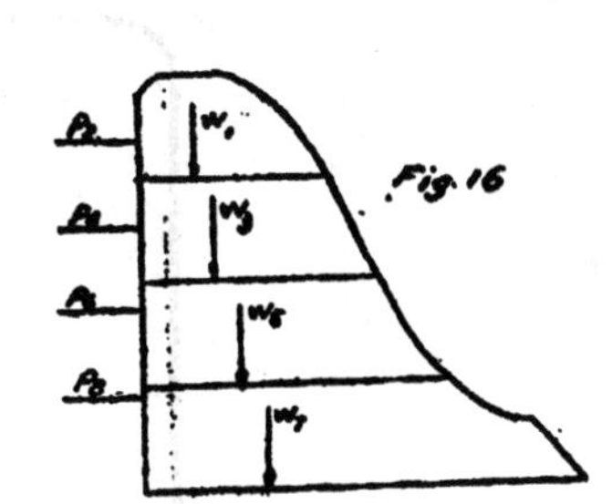

Fig. 16

Along the 20-ft. plane:

$P_4 = 0.55 \times 58{,}400$

$= 55{,}600$ lb. > 24,600 O.K.

Along the 30-ft. plane:

$P_6 = 0.55 \times 101{,}000$

$= 55{,}600$ lb. > 46,300 O.K.

Along the foundation:

$P_8 = 0.55 \times 157{,}000$

$= 86{,}400$ lb. > 74,200 O.K.

Check for stresses:

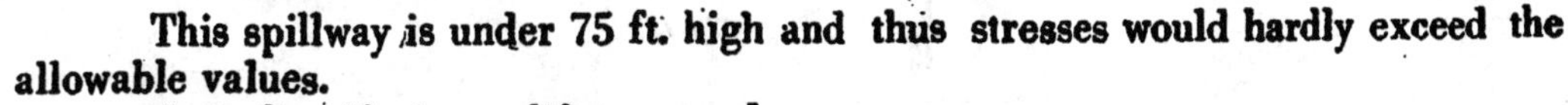

This spillway is under 75 ft. high and thus stresses would hardly exceed the allowable values.

(b) Considering uplift pressure:

25691

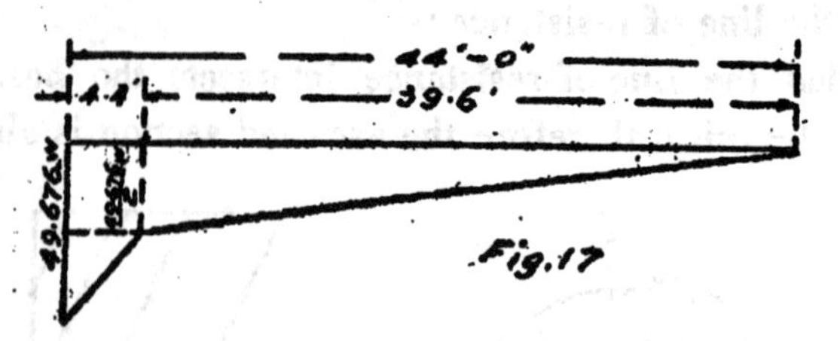

Fig. 17

Total uplift pressure

$$= \frac{49.672 + 24.838}{2} \times 62.5 \times 4.4 + \frac{24.838}{2} \times 62.5 \times 39.6$$

$= 41{,}000$ lb.

Distance of line of action of uplift from heel

$$= \frac{\frac{49.676}{2} \times 4.4 \times 2.2 + \frac{49.676}{2\times 2} \times 4.4 \times \frac{4.4}{3} + \frac{49.676}{2\times 2} \times 39.6\left(4.4 + \frac{39.6}{3}\right)}{\frac{49.676}{2} \times 4.4 + \frac{49.676}{2\times 2} \times 4.4 + \frac{49.676}{2\times 2} \times 39.6}$$

$= 13.68$ ft.

It is shown that the line of resistance, when uplift pressure is taken into account, also intersects the base within the middle third. This means that the section is O.K.

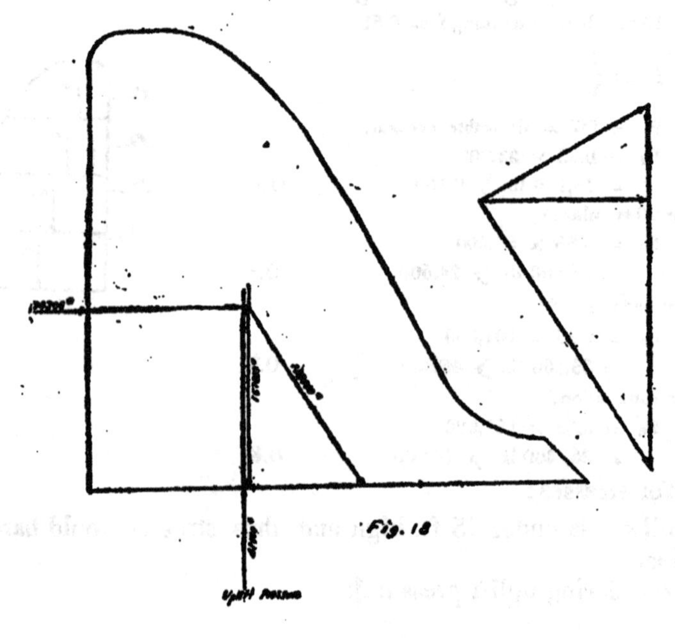
Fig. 18

Check for stability against sliding along the foundation when uplift pressure is taken into account;

$$
\begin{aligned}
P &= fN \\
N &= W - \text{uplift pressure} \\
&= 157{,}000 - 41{,}000 \\
&= 116{,}000 \text{ lb.} \\
P &= 0.55 \times 116{,}000 \\
&= 63{,}900 \text{ lb.} < 74{,}200 \text{ lb by } 10{,}300 \text{ lb.}
\end{aligned}
$$

To secure safety, a trench is excavated along the spillway in the ledge rock near the heel and the spillway section is so constructed to key into the trench that sliding is prevented when the uplift pressure occurs in the manner as above.

(c) Abutment section on west bank :

Elev. of abutment = Elev. of dam
= 285 + 9.676 + Free board
= 294.676 + Free board

Assume free board = 5.324 ft.

Therefore elev. of abutment = 300 ft.

Elev. of rock bed on west bank = 261 ft.

Height of abutment = 300 − 261
= 39 ft.

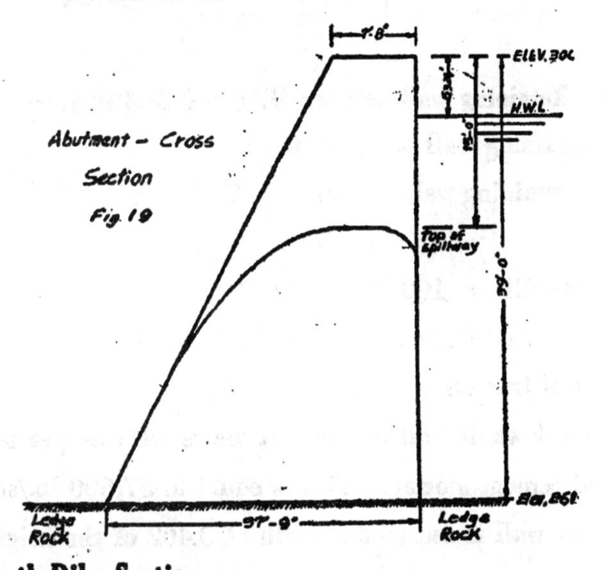

Abutment – Cross Section
Fig. 19

Design of Earth Dike Section.—

Elev. of earth dike = 300 ft.

Maximum height of dike = 300 − 275
= 25 ft.

Top width of dike $= \frac{25}{5} + 5$

$= 10$ ft.

The details of the maximum cross-section of the earth dike is shown in Fig. 20.

CROSS-SECTION OF EARTH DYKE

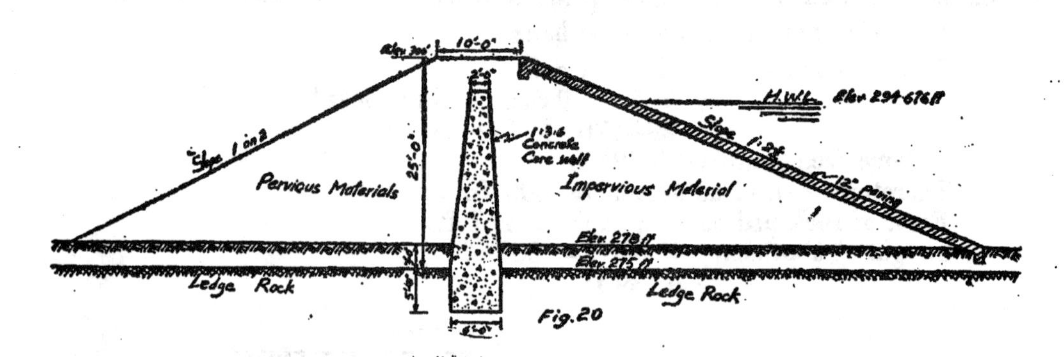

Fig. 20

Design of Retaining wall between Dike and Spillway.—

Elev. of retaining wall = 300 ft.

Height of retaining wall = 300 − 277

= 23 ft.

Wt. of earth fill = 100 lb. per cu. ft.

Angle of internal friction of earth fill $\phi = 40°$.

Coefficient of friction = 0.55

Maximum allowable unit pressure at toe is 400 lb. per sq. in. for limestone ashlar in portland cement mortar. This is equal to 57,600 lb./sq. ft.

The base of wall is assumed a width of 0.462 of the height.

Width of base $= 0.462 \times 23$

$= 10.626$ ft., say 11 ft.

Horizontal earth pressure:

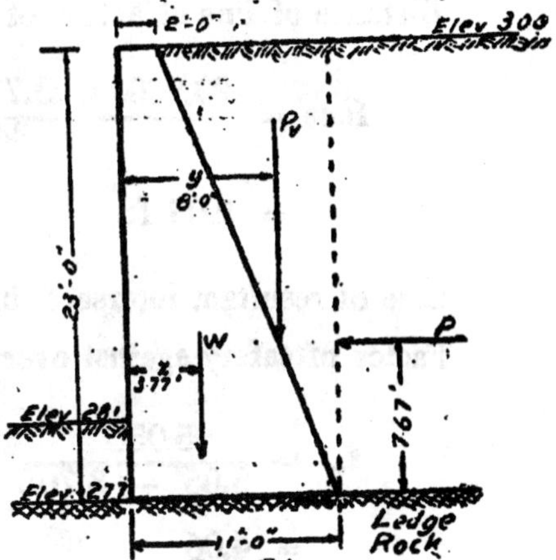

$$P = C\frac{wh^2}{2}$$

$$\theta = 0^\circ \qquad \phi = 40^\circ$$

$$C = \frac{1 - \sin 40^\circ}{1 + \sin 40^\circ}$$

$$= 0.217$$

Therefore $P = 0.217 \times \dfrac{100 \times 23 \times 23}{2}$

$$= 5{,}750 \text{ lb.}$$

Fig. 21

Assume width of top of wall $= t/6$

$$= 11/6 \text{ say } 2 \text{ ft.}$$

Weight of wall $= \frac{1}{2}\,(2 \times 11) \times 23 \times 160$

$$= 23{,}920 \text{ lb.}$$

Center of gravity of wall from the vertical face:

$$x = \frac{2 \times 23 \times 1 + (9 \times 23/2) \times 5}{\dfrac{2 + 11}{2} \times 23}$$

$$= 3.77 \text{ ft.}$$

Vertical component of earth pressure:

$$P_v = 100 \times (11 - 2) \times 23 \times \tfrac{1}{2}$$

$$= 10{,}350 \text{ lb.}$$

Distance of line of action of P_v from the vertical face:

$$y = 2 + \frac{2}{3} \times 9$$

$$= 8 \text{ ft.}$$

Total vertical force acting upon the foundation:

$$= 23{,}920 + 10{,}350$$

$$= 34{,}270 \text{ lb.}$$

Distance of line of action of this resultant from the vertical

$$\text{face} = \frac{23{,}920 \times 3.77 + 10{,}350 \times 8}{34{,}270}$$

$$= 5.05 \text{ ft.}$$

Line of resultant intersects base at 3' – 8¼" from toe, or 3.708 ft.

Factor of safety against overturning :

$$f_{ot} = \frac{5.05}{5.05 - 3.708}$$

$$= 4.06$$

Factor of safety against sliding :

$$f_{s'} = \frac{0.55 \times 34{,}270}{5{,}750}$$

$$= 3.28$$

Check for stress at toe and heel

$$\text{Stress at toc} = \frac{W}{b}\left(1 + \frac{6e}{b}\right)$$

$$= \frac{34{,}270}{11}\left(1 + \frac{6 \times 1.792}{11}\right)$$

$$= 6{,}160 \text{ lb. per sq. ft.}$$

$$\text{Stress at heel} = \frac{W}{b}\left(1 - \frac{6e}{b}\right)$$

$$= \frac{34.270}{11}\left(1 - \frac{6 \times 1.792}{11}\right)$$

$$= 65.5 \text{ lb. per sq. ft.}$$

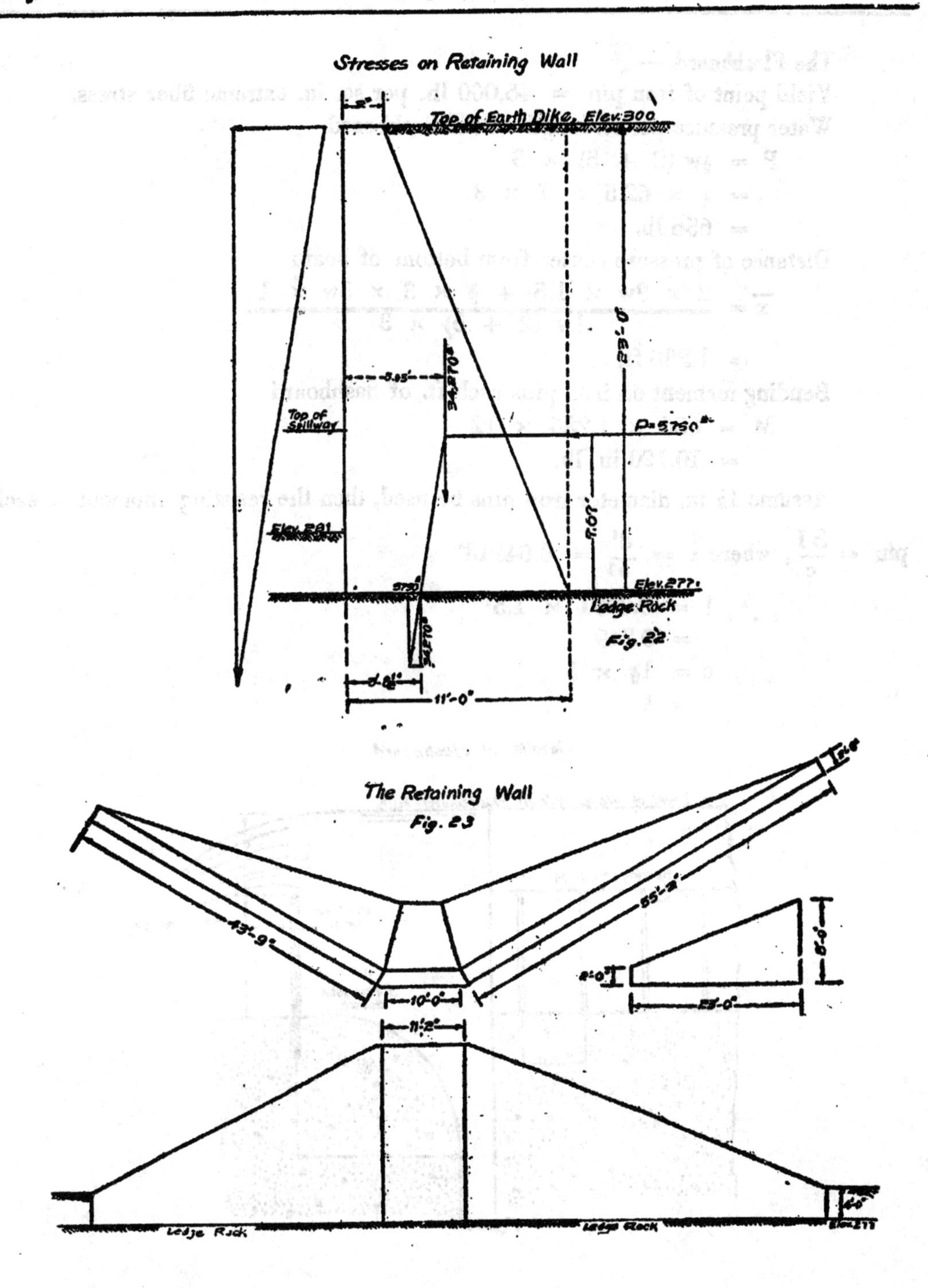
Stresses on Retaining Wall
Top of Earth Dike, Elev. 300
Top of Spillway
Elev. 281
Elev. 277
Ledge Rock
P=5750
11'-0"
The Retaining Wall
10'-0"
11'-2"
Ledge Rock

Fig. 22

Fig. 23

The Flashboard.—

Yield point of iron pin = 45,000 lb. per sq. in. extreme fiber stress.

Water pressure per ft. length of the flashboard

$$P = \tfrac{1}{2}w\,(2 + 5) \times 3$$
$$= \tfrac{1}{2} \times 62.5 \times 7 \times 3$$
$$= 656 \text{ lb.}$$

Distance of pressure center from bottom of board

$$\bar{x} = \frac{2 \times 3w \times 1.5 + \tfrac{1}{2} \times 3 \times 3w \times 1}{\tfrac{1}{2}w\,(2 + 5) \times 3}$$
$$= 1.286 \text{ ft.}$$

Bending mement on iron pins each ft. of flashboard

$$M = 656 \times 1.286 \times 12$$
$$= 10{,}120 \text{ in. lb.}$$

Assume 1½ in. diameter iron pins be used, then the resisting moment of each pin $= \dfrac{SI}{c}$, where $I = \dfrac{d^4}{64} = 0.0491d^4$

$$\therefore I = 0.0491 \times 1.5^4$$
$$= 0.249$$
$$c = 1\tfrac{1}{2} \times \tfrac{1}{2}$$
$$= \tfrac{3}{4}$$

Details of Flashboard

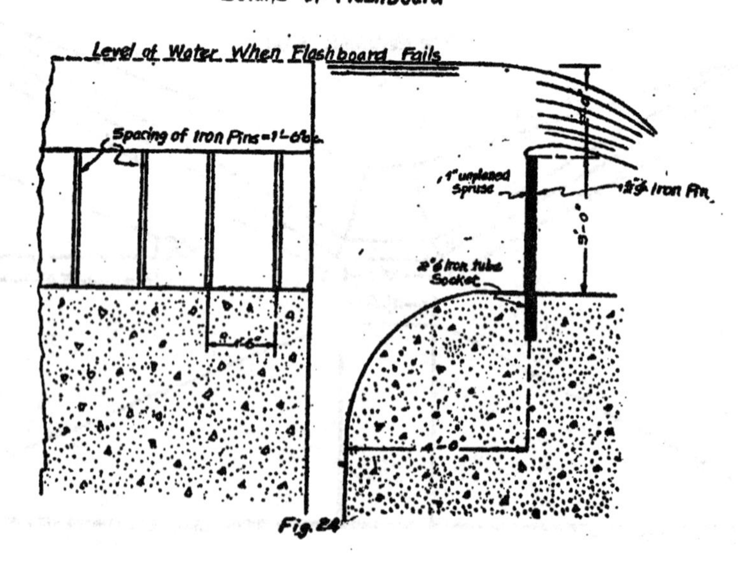

Fig. 24

$$\text{Therefore } M_r = \frac{45{,}000 \times 0.249}{\frac{3}{4}}$$

$$= 14{,}940 \text{ in. lb.}$$

$$\text{The spacing of pins} = \frac{14{,}940}{10{,}120}$$

$$= 1.48 \text{ ft. say } 1 \text{ ft. } 6 \text{ in. c to c.}$$

Eventually, however, the writer has to emphasize that most of the details of this article are introduced from Professor H. K. Barrows' "Water Power Engineering." Also the following references are comprehended :—

Hanna and Kennedy, "Design of Dams"

Greager, Justin and Hinds, "Engineering of Dams," Vols. I, II & III.

The data of the illustrative problems were formerly given as assignments to the studeuts of Massachusetts Institute of Technology by Professor Barrows, but the solutions are worked out here by the writer of this article.

THE END

A SHORT ACCOUNT ON THE PRINCIPLE OF SIMILITUDE.

By

Lui Wai Yaw

Wong Hon Ki

Historical Notes

Experimental studies with models in many an engineering field are based on the principle known as the "Principle of Similitude" and the adoptation of which dates back as far as the Greek philosophers. Aristotle, the great Greek philosophers was the first to seek the help of models for solving problems. The consequent significance was, however, overlooked, and whose importance as a guide to solution of problems was not resumed until the time of Sir Isaac Newton. Newton was probably the first to clearly state the ideas of similarity and it was when this principle was expressed in the form of an equation of conditions which the four scale ratios of length, time, force and mass must satisfy.

In the late sixties of the last century an embodiment of the principle was made by William Froude and whose studies led to the development of a model law which was named after him. In fact, the principle assumed its real form through William Froude. Other Pioneers were Dubeat, Bazin and Fargue. Later Pioneers were Professor Osborne Reynolds, of the University of Manchester, England, Professor Hubert Engels, of the Technical University at Dresden, Germany, Sir William Vernon-Harcourt, in France, Dr. G. de Thierry, of the Technical University in Berlin, Professor John R. Freeman of the United States, etc.

Definitions :—

Similitude: The term is definied as the "Relation between two figures irrespective of magitude."

Model:—A model may be defined as a system by whose operation the characteristics of other system or systems may be predicted.

Prototype:—The existing structure of any kind in nature which is either dimensionally represented by a miniature, or in some cases, by e model of bigger scale, and whose static or operating characteristics are being studied, is called the Prototype.

(a) The principle of geometric similarity.

The studies of problems under this principle requires that the shapes in both prototype and model be similar.

(b) The principle of kinematic similarity.

The criterion requires the similarity of shape, path as well as time.

(c) The principle of dynamic similarity.

The criterion requirés the similarity of shape, paths, time and forces involved.

Any problem, which is to be studied by means of models, may be analysed by any one of the above mentioned principles. (In case the existing conditions warrant such a choice.)

When a model is so constructed that the relative ratio between two points in model is the same as that of two corresponding points in the prototype the model may be said to geometrically similar to the prototype.

Apart from being geometrically similar, if the time involved in the studies is correspondingly similar to that in the prototype the case will fall under the criterion of Kinematic Similarity. This case requires also the similarity of paths raversed by particles in both prototype and model.

In addition to all those stated above, if the force element involved is similar in both systems the problem can be solved by the Principle of Dynamic Similarity.

Nomenclature :—

※ ※ ※ ※ ※ ※ ※ ※ ※ ※ ※ ※

N. B. The following discussions are mainly pertaining to the Science of Hydraulics.

※ ※ ※ ※ ※ ※ ※ ※ ※ ※ ※ ※

In most cases the letters with a subscript "m" is used for the designation of quanties of the model, while ordinary letters without subscripts are adopted for quantities in Nature.

A = area of cross-section

L = length
H = head
S = slope
V = velocity
T = time
Q = discharge
R = hydraulic radius
P = wetted perimeter
G = acceleration of gravity
F = force per unit area.

Physical Forces :

The forces that affect the operation of hydraulic structures and which exist in the universe are

(1) Terrestial gravitation
(2) Universal gravitation
(3) Initial fluid friction
(4) Elastic forces
(5) Capillary forces.

Each of these forces is capable of producing a definite physical result on a particle or a body. In case several forces do exist at the same time and acting on the same particle or body, the problem may be assumed to be actcd upon by only one force if the other forces are so insignificant in comparison to the one in consideration that they may be neglected without resulting any noticeable error. For instance, the internal friction and the viscous forces of fluid are usually neglected; i.e. a perfect fluid is imagined. There are other forces which do not come under consideration in operating models such as the normal forces acting on the external surfaces (as well as in the interior) of rigid bodies, the tangential forces acting internally in rigid bodies, the normal forces acting internally or externally in an incompressible fluid, pure damping resistances.

THEORY

Geometric Similitude

(1) Lengths

If the length in the prototype be represented by L and the corresponding element in the model be designated by L_m, then the ratio of these two quantities is

$$\frac{L}{L_m} = \lambda \ (\text{Lambda})$$

where λ is the scale ratio and is a demensionless number.

(2) Areas

$$\text{Area} = L \times L = L^2$$

$$A_m = L_m \times L_m = L_m{}^2$$

$$\therefore \frac{A}{A_m} = \frac{L \times L}{L_m \times L_m} = \frac{L^2}{L^2{}_m} = \lambda^2$$

(3) Volume

$$\text{Vol.} = L \times L \times L = L^3$$

$$\text{Vol}_m = L_m \times L_m \times L_m = L_m{}^3$$

$$\therefore \frac{\text{Vol}}{\text{Vol}_m} = \frac{L^3}{L_m{}^3} = \lambda^3$$

(4) Kinematic Similarity

$$\text{Time} = \frac{T}{T_m} = \tau \ (\text{Tau})$$

where τ is a ratio of time element in nature to the corresponding element in model.

(5) Velocity

$$V = \text{length / unit time} = \frac{L}{T}$$

$$V_m = \frac{L_m}{T_m}$$

$$\therefore \frac{V}{V_m} = \frac{L}{T} \times \frac{T_m}{L_m} = \frac{\lambda}{\tau} \left(\frac{\text{Lambda}}{\text{Tau}}\right)$$

(6) Acceralation (Gravitational)

The gravitational acceralation is considered the same in both model and prototype as it is only a function of the variation of latitude and altitude its variation is usually negligible.

(6a) Acceralation

$$a = \frac{V}{T}$$

$$a_m = \frac{V_m}{T_m}$$

$$\therefore \frac{a}{a_m} = \frac{V}{T} \times \frac{T_m}{V_m} = \frac{\lambda}{\tau} \times \frac{1}{\tau} = \frac{\lambda}{\tau^2}$$

Dynamic Similarity.

By Newton's Law

$$F = Ma, \text{ then}$$
$$F_m = M_m a_m$$

$$\therefore \frac{F}{F_m} = \frac{M_a}{M_m a_m} = \frac{M}{M_m} \cdot \frac{\lambda}{\tau^2}$$

If ρ (rho) be the density of the mass M and

ρ_m " " " " " " M_m

$$\therefore M = \rho V$$
$$M_m = \rho_m V_m$$

$$\therefore \frac{M}{M_m} = \frac{\rho . V}{\rho_m V_m} = \frac{\rho}{P} \lambda^3$$

Then

$$\frac{F}{F_m} = \frac{M}{M_m} \frac{\lambda}{\tau^2} = \frac{\rho}{\rho_m} \cdot \frac{\lambda^4}{\tau^2} = \frac{\rho}{\rho_m} \left(\frac{\lambda}{\tau}\right)^2 \lambda^2$$

$$\therefore \frac{V}{V_m} = \frac{\lambda}{\tau} \quad \&$$

$$\lambda^2 = \frac{A}{A_m}$$

$$\therefore \frac{F}{F_m} = \frac{\rho}{\rho_m} \left(\frac{V}{V_m}\right)^2 \frac{A}{A_m}$$

or

$$\frac{F}{F_m} = \frac{\rho}{\rho_m} \left(\frac{V}{V_m}\right)^2 \frac{A}{A_m} = \kappa \rho V^2 A \quad \text{or}$$

$$\frac{F}{F_m} = \kappa P_m V_m^2 A_m$$

The constant κ is the same for both equations and should used together.

Knowing the relationship of all the elementary scale ratios, the various quantities of nature can be determined off-handedly. For instance,

Q = discharge in nature

Q_m = " " model

$$\because Q = \text{Vol} / \text{time} = \frac{L^3}{T}$$

$$Q_m = \text{Vol}_m / T_m = \frac{L_m^3}{T_m}$$

$$\therefore \frac{Q_m}{Q} = \frac{L_m^3}{T_m} \times \frac{T}{L^3} = \frac{\tau}{\lambda^3}$$

where λ is the scale ratio of linear dimensions between prototype and model.

If the gravitational force is assume acting, by Newton's Law of Forces

$$\tau = \sqrt{\lambda} \,*$$

Then

$$\frac{Q_m}{Q} = \frac{\sqrt{\lambda}}{\lambda^3} = \frac{1}{\lambda^{2.5}}$$

$$\text{or } Q_m = \frac{Q}{\lambda^{2.5}}$$

i.e., the discharge in model is equal the corresponding discharge in prototype divided by the ratio of the prototype linear dimension to that of model raised to the 2.5 power.

$$* \; \frac{F}{F_m} = \frac{M g}{M_m g_m} = \frac{M}{M_m} \frac{\lambda}{\tau^2}$$

$$\therefore \frac{\lambda}{\tau^2} = \frac{g}{g_m} = 1$$

$$\therefore \tau = \sqrt{\lambda} \qquad \text{Q. E. D.}$$

25705

COMPLETE, INCOMPLETE AND APPROXIMATE SIMILARITY.

Complete dyanamic similarity may be attained for a model when there is only one physical force acting in nature and when the model forces are proportioned in accordance with the model law for such a physical force.

If two or more forces are acting on a model, and if the model studies were to be directed with consideration to only one major force and the effect of the minor one or ones were to be neglected, the case falls under the criterion of "in complete dynamic similarity"

In case the physical forces acting in the prototype can be neglected in the model studies and the model be operated without any regard to the model law, the case is an example of the "Approximate Similarity."

† Poto–

Photo-Elastic Determination of Shrinkage Stress Proc. A.S.C.E. May 1935;

Applications:

The applicaticationsˑ of this principle is in fact numerous. Apart from the simple units enumerated above, various other units in many fields of science can be similarly deduced, and their consequent results be applied to experimental studies. In the studies of stress distribution in hydrauic structures or other structura-l members in bridges etc., a question often arises which is how to similate the structure or whatever that may be. Are we going to build it in full scale? This is absurd. Impossible. Under any circumstance these studies must resort to the constructson of models, and the law governing the operation characteristics of which is the Principle of Similitude.

This principle is widely adopted in laboratory studies in Hydraulics, the structures in close coordination witth it, in Structure, in Resistance of Materials, etc.

Early in France the river Seine was built to study its operation characteristics and improvements were deduced. In Germany the whole course of Huang-Ho was constructed under the sponsorship of Dr. Hubert Engels. Recently, a model of part of the River Youngtze has been built in Denver, U. S. A. and whose performance are being studied. The functional details of the great Y-V-A Dam is being studied in Denver, mostly by Chinese Engineers, under the instructions of Dr. Salvage.

The stress in structural members can also be determined through models. The resistances, strain etc., of various materials can be determined by models with the aid of the principles of Photo-Elastiicity.

There are still many other examples to which the principle has been applied and its adoptation is ever increasing.

* Eng'g News Record Vol. 108. 1932, pp. 828-832

Conclusion:

In the experimental stueies of operational characteristics of problems the model may be either fabricated with the horizontal and vertical scales the same or the model may be built distortedly. Many European experimenters have expressed themselves unmistakably as opposed to the use of distorted models, nevertheless it is impossible in many cases to adhere to true similarity of reproduction on a small scale. In fact the result attained in the model may be so small that it is beyond the possibility of being measured accurately. This is when the distorted modols come into function.

The materials used for the fabrication of models may be either of the following:

(1) Steel
(2) Concrete
(3) Wood
(4) Plaster of Paris
(5) Pyralin, Lucite
(6) Glass
(7) Sand.

In the same model several of the above materials may be used simultaneously. The choice is left to the judgement of the experimenter.

In the preceding paragraphs The simplest case is noted whereas in the case of open-chaunel tests the friction of bed and wall become important. The experimnet requires much more experience and skill and the studies will necessitate much caution. The reader who is interested in the further acount of the latter examples are requested to refer to the literature of models.

The present paper is aimed only to give a conception of the menning of the principle in its simplest form. The main purpose is to cast forth a clue to the further study of the theories and principles.

Bibliography

List of Referrence.

(1) Dimension Ananlysis and Principle of Similarity, Kenneth C. Reynolds.
(2) Handbook of Applied Hydraulics By Davis.
(3) Discussion of Theory of similarity and Models., Transactions A. S. C. E. Vol. 96 1932, pp. 308-325. By B. F. Groat.
(4) Practical Rive Labory Hydraulics, Proceedings, A. S. C. E. Nov. 1933, By Vogel, Lt. Herbert D,.

轉 載

(TRANSFERRED)

AN ABRIGED ACCOUNT ON THE WORLD GREATEST DAM—BOULDER DAM

※ ※ ※ ※ ※ ※ ※ ※ ※ ※ ※ ※ ※ ※

The demand for power, cheap power, is increasing with time and is becoming more urgent daily. Even in such a highly industrialised and well-developed nation as the U.S.A., with (according to statistics) the highest coal annual output, the construction of dams for cheap power is still increasing.

The function of a dam, is in reality many-fold. However, the most significant of all, apart from other benefits derived there-of, are the reclamation of arried lands, the consequent immurity of fear of flood and the generation of cheap power, which can even be considered as a by-product, would alone refund in full the investment in a pretty short time.

The reasons for the present outlines of the Boulder Dam is to cast forth, a fair information of the greatest Dam in existance, to call the public to the attention of water-works, and other fields in conjunction with water so that with the cooperation of the whole nation, the calamities derived from the existing waterways of China may be obliviated.

Should the design of the Y-V-A be realised and the construction made feasible, the Boulder Dam will fall in the second place. Our Youngtze Dam would be 750 feet.

However, this may be, the realisation of the project would be impossible with the civil war going on. Hoping for the best, we may have a chance to see it done.

(The Editor)

※ ※ ※ ※ ※ ※ ※ ※ ※ ※ ※ ※ ※ ※

History:

From its discovery in 1540 until it was harnessed by Boulder Dam almost four centuries later, the Colorado River was America's most dangerous stream,

Friendly rivers in other sections provided safe highways to lead the pioneers into the wilderness, but the surly Colorado, sulking in its canyons, could not be used. On the other hand, it could be crossed only at widely separated places along its 1,700-mile course from the Rocky Mountains to the Gulf of California.

Like other wastern desert streams, this giant fluctuated each year through a cycle which ran from a roaring, flood-swollen torrent when snows were melting, to a sandy-bottomed, sluggish creek during the long, dry summers and autumns.

Man's crying need for water in this thirsty West, however, caused him early to turn calculatingly upon the Colorado in an effect to devise some means to make a servant of this untamed stream.

Before Boulder Dam, whenever he tampered with the river he brought disaster upon himself. Farmers, tempted by dry, fertile desert soil in the Imperial Valley of California and near Yuma, Ariz., tapped the river for irrigation water to create vast and rich gardens. But the unregulated Colorado took its vangeance upon them. Annually it sent destructive floods and annually, by fading to a trickle; it cut off the water supply upon which their crops and their lives depended.

A great cry arose for control and conservation of waters of this river, the waters which were the most valuable natural resource of a vast desert empire.

Agitation for action increased, and in 1922 representatives of the Federal Government and of the seven States in the Colorado River Basin met in Samta Fe, N. Mex., to draft a crp., act for the division of the waters of the Colorado River.

In 1928 the Congress passed the Swing-Johnson bill authorizing the Boulder Canyon Project; by 1930 it had been ratified by the required six of the seven States, an construction was begun by the Bureau of Reclamation in 1931. In 5 years Boulder Dam was completed, and man had won his victory.

The Colorado River now is a useful and reliable friend to the people of the Southwest. Floods cannot pass the dam, which saves the flood waters and uses them by generation of electricity to turn factory wheels 250 miles away, and to provide an unvarying supply of domestie and irrigation water for rural and urban communities from Los Angeles, Calif., to Yuma, Ariz.

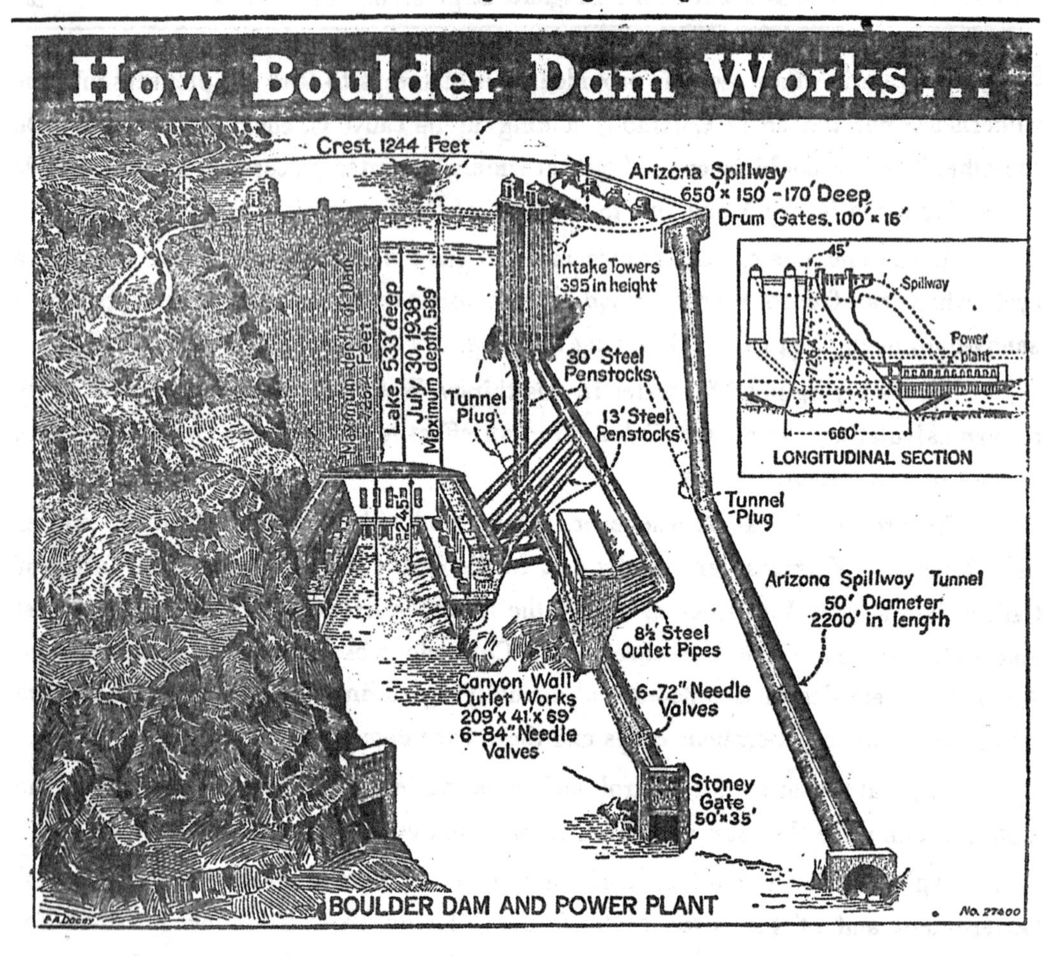

Achievements:

The achievements of Boulder Dam run the scale in the field of water conservation from flood control to provision of a valuable wild waterflow refuge. In regulating the treacherous Colorado River, Boulder Dam has changed its character entirely for 565 miles from the lower end of Grand Canyon to its mouth at the northern tip of the Gulf of California.

The farmers in the Imperial and Yuma Valleys and elsewhere along this stretch of the river now are provided with a steady and trustworthy supply for the irrigation upon which they reply. The domestic water supply of 13 cities, 250 miles west of the Colorado River on the southern Calfornia coastal plain in the vicinity of

Los Angeles, is being augmented through the Metropolitan Water District's aqueduct as a result of construction of Boulder Dam. The floods which once raged through Black Canyon, pouring a destructive force against the communities for downstream, now are halted here by Boulder Dam and the water saved for use. Navigation is possible through the magnificent canyons 115 miles above Boulder Dam, and navigation had been improved on the lower river, A new recreational area of major importance has been created by Boulder Dam. The dam and the lake now are visited by half a million people a year. The lake is stocked with fish, and already it rivals the most popular national park as a tourist and sportsman's attraction. As a will-fowl refuge the lake assumes new importance, since it is on one of the major migration fly ways and offers a haven for waterfowl in a vast area where none previously existed.

Boulder Dam traps the muddy, silt- ; laden waters which pour through Grand Canyon and releases them settled and clear.

And, in addition, Boulder Dam has made possible the generation of 1,835,000 horsepower of electric energy, which is being sold under contracts which will return its entire cost and create a surplus in 50 years.

With the initial installation in the powerhouse complete and six big generators and small one in operation by the fall of 1938, Boulder Dam was producing more than 130,000,000 kilowatt-hours of enery and returning in excess of $290,000 to the U. S. Treasury each month.

While power is a by product of the dam, the fact that Boulder Dam has made available much cheap energy is important to the future welfare of the whole Southwest.

Irrigation:

The West is arid and semiarid. Farming in the West is dependent on irrigation. Below Boulder Dam lie some of the world's most forbidding desert areas, as dry as the Sahara, recipients of 3 inches or less of rain in a year. Without the artificial application of water these areas would be totally useless. With irrigation, their rich soils and warm climates make them gardens of almost unmatcdeh productivity.

There are, below Boulder Dam and capable of being served by it, approximately 1,900,000 acres of such lands. At present about half of this area is irrigated and in production. Large, successful irrigation projects are in operation in the Palo Verde Valley in California, about 200 miles downstream, and in the Imperial Valley in California, and in the Yuma-Gila Valleys in Arizona, 300 miles downstream. Eventually this irrigated area will be expanded. At present, however, there are no homesteads available.

River Flow:

The greatest recorded flow of the Colorado River is 240,000 cubic feet of water per second, but there are evidences that floods as great as 300,000 second-feet have been experienced.

The larger floods come each spring with the melting of snows in the mountains. In the late summer the flow may be reduced to little more than a trickle.

Records disclose that the annual average run-off of the river at Boulder Dam exceeds 15,000,000 acre-feet of water, an acre-foot deep. Lake Mead, therefore, can impound the entire average flow of the river for about 2 years, since its capacity is 30,500,000 acre-feet.

The Plan for Repayment:

Built at a cost of $120,000,000, Boulder Dam is an investment which will be repaid with interest in 50 years to the Federal Government under contracts now in force for the sale of power. When the repayment has been completed, the dam, the power plant, and machinery and all appurtentant works still remain Government property.

The firm power, of which there will be about 4,330,000,000 kilowatt-hours a year, is sold as falling water measured at the voltage delivered at the switchyard for 1.63 Mills per kilowatt-hour. As the power is sold as falling water, the cost of operating and maintaining the generating machinery is placed upon the power purchasers.

Contracts for the sale of all the Boulder Dam power were completed in 1930, before construction of Boulder Dam was began. Principal contractors are the city of Los Angeles, the Metropolitan Water District of Southern California, the California-Nevada Power Co., and the Southern California Edison Co. Arizona and Nevada each have the right to buy and dispose of 18 per cent of the power.

Power generation was begun September 11, 1936, when Presendent Franklin D. Roosevelt in Washington pushed a golden key starting the first generator. The initial installation of four 82,500 kilovolt-ampere generators was completed in March 1937, and all power purchase contracts became operative June 1, 1937, when Secretary of the Interior Harold L. Ickes notified Los Angeles that the Bureau of Reclamation was prepared to furnish firm energy henceforth.

The Government had received $3,751,210 from the sale of Boulder Dam power to January 1, 1939.

The original power purchase contracts not only contemplated complete amortization of the investment in the dam and power plant with interest in 50 years but also the payment of a certain percentage of revenues to Arizona and Nevada in lieu of taxation and the accumulation of a surplus which would be used in further pevelopment of the Colorado River.

You Will Want To Know That:

Boulder Dam is the world's highest dam.

Lake Mead is the world's largest reservoir.

Boulder power plant is the world's largest.

Elevators descend from the dam's crest, 529 feet, equal to a 44-story building.

Max. water pressure on the dam's base is 45,000 pounds per square foot.

If Statistics Interest You

Boulder Dam is	726.4 feet high.
Its crest is	1,244 feet long.
At top it is	45 feet thick.
At bottom it is	660 feet thick.
Concrete content of dam	3,250,000 cubic yards.

Lake Mead is	115 miles long.
Its capacity is	30,500,000 acre-feet.
Flood-control reserve	9,500,000 acre-feet.
Max. depth	589 feet.
Lake Mead covers	146,500 acres.
Power-plant capacity	1,835,000 horsepower.
Large generators	15
Capacity of each	82,500 kv. a.
Small generators	2
Capacity of each	40,000 kv. a.
Large turbins	15
Each of	115,000 horsepower.
Small turbines	2
Each of	55,000 horsep.
Spillways	2
Capacity of each	200,000 cu. ft. a sec.
Drum gates each	100 by 16 feet.
Spillway tunnels	2
Diameter of each	50 feet.
Intake towers are	395 feet high.
Diameter of each	75 feet.
Capacity of outlets	90,000 cu. ft a sec.
Excavation totaled	6,480,000 cu. yards.
Steel and metal used	96,000,000 pounds.
Valves, gates, hoists	33,000,000 pounds.
Steel in penstocks	89,000,000 pounds.
Total concrete	4,360,000 cu. yards.

譯述

INFLUENCE LINES FOR CONTINUOUS STRUCTURES BY GEOMETRICAL COMPUTATION

用幾何計算法求連續結構物之感應線

By

Dean F. Peterson, Jr, Assoc. M. ASCE

——歐陽讓譯——

因爲近年來連續鈑梁橋(Girder Bridges)與剛架橋(Rigid Frame)之跨度比前增長,故應用感應線(Influence Line)來求幾個荷重合成所產生之影响,與求安置荷重所生之極大値漸形重要;雖然吾人可以結構物(Structure)在不同節點上荷重,而利用感應線來求應力與力矩,但用此嘗試法求所需荷重節點是十分麻煩的。吾人也可以利用一結構物在彈性彎度之平等値(Equiality)內荷重時,求其一定點之應力所生之單位變形(Unit deformation)及在此點之應力感應線,而求出鈑架之感應線;此等原理,曾爲George Beggs, M.ASCE, Otto Gottschalk 等應用于所有不同之不定結構物上(Indeterminate Structures)。此等關係很容易伸展到單方向受彎折(Fluxure)之連續結構物感應線分析計算上面,而用模型來做機械式的解析。許多不同的連續結構物之感應線也是依照此種手續和利用靜力分析法來演算出來的,至於決定感應線縱座標之位置及極大値與不變横剖面(Constant Cross section)之桿(Members)與感應線所包括下之面積……均可利用公式推算。

基本關係

(Basic Relationship)

在Müller-Breslau關係所包括之基本原理之發展,係人所共知的,其原理亦以麥氏(Maxwell's)之倒數偏差(Reciprocal deflections)及疊上原理(Principle of Superposition)爲依歸,但結構物所用之材料,係以有彈性能力爲限!

彎曲性之幾何基本方程式

(Fundameutal Equations of Flexure Geometry)

此章包括彎曲性方程式之作用:

(一)求靜力解答；

(二)計算彈性曲線縱座標 (Elastic curve ordinates)

此彎曲性幾何之基本微分方程式

$$\frac{d^2y}{dx^2}=\frac{M}{EI}$$ 直接應力與剪力所生之扭歪 (Distortion)不算在內。

上式爲普通彈性方程式之特殊情形。此方程式有下列之第一次與第二次積分：

$$\left.\frac{dy}{dx}\right|_x=\int_0^x\frac{M_\varepsilon}{E_\varepsilon I_\varepsilon}d\varepsilon+C_1\text{.................................}(1a)$$

及

$$\left.y\right|_x=\int_0^x\frac{M_\varepsilon}{E_\varepsilon I_\varepsilon}(x-\varepsilon)d\varepsilon+C_1x+C_2\text{............}(1b)$$

在上式中，x與y係變數座標，M_ε在ε時之彎性能率(Bending Moment at ε)；E_ε係在ε時之彈性係數；I_ε係在ε時之慣性力矩(Moment of Inertia)；C_1與C_2均爲積分常數．上式有時可用積分算出，但多數用限制積分來算 (finite integration) 但在不定結構物範圍內，因其範圍與情況之複雜而增多，所以必需應用一列同時方程式 (Simultaneous Equations) 或其數學當量 (Mathematrial Fquivalent) 但近世工程師所認爲最捷之方法就是應用以上方式來解答．

求解答之步驟

(Procedure for Solution)

力矩，剪力，衝擊與抵抗力之感應綫係由結構物之彈性曲綫(Elastic curve)而定，但此種結構物受力時要有單位軸旋度(unit axial rotation)，橫軸與縱軸位移或反抗力之偏差方向。其次吾人必需先以靜力學分析結構物所需之位移荷重，用下列任何一種方法均可：力矩分配，斜度偏差，彈性橫曲線…在普通來說，如果所有的桿都沒有荷重(即沒有受本身重力，風力，靜水力或普通之荷重)，其分析法是比包含有其他荷重的桿爲簡單的。

結構物之靜力學解答完成以後，每桿之端旋度(End Rotation)及偏差便能算出，假如用斜度偏差或彈性彎曲導綫來分析，端旋度及偏差可直接算出；若用其他方法則需用彎曲性幾何方程式，或用其他之方程式來求需要之端旋度及偏差，求彈性曲綫可用第二方程式或其他推展式求之。

間接方法

(Indirect Method)

如結構物之荷重爲單位偏差或旋度，其靜力學上之解答可用力矩分配法推演出來；此法以下列三点爲依歸：

25716

(一)設每一個節点都具有分離的不平衡力矩(Unbalanced Moment)。

(二)求其每個情形下所必需的力矩，

(三)將原來分配比例之荷重所產生的幾個定端力矩合併在一起。或用習慣上之荷重來求端力矩，而與在上之位移(Imposed Displacement)成比例或合併來算出。(此法在例題九裡由力矩分配法及用彈性彎曲導錢法詳細指明。

桿之彈性曲線計算

(Calculationg Elastic Curve of Members)

依數學分析吾人應用下列符號：縱座標橫座標一樣。在＋X方向作反時針方向旋轉時爲＋，由＋X方向之切綫，產生一正偏差；彎性向上之力矩爲正號。正力矩沿着＋X方向產生正旋度及正偏差。任何一桿，其軸必爲直綫。

設 y_A 爲原軸A点之偏差，

θ_A 爲A点之切綫到彈性曲線之斜坡，

則在任一点 X_n 之y爲：

$$y_n=y_A+\theta_A x_n+\sum_{\varepsilon=1}^{\varepsilon=n}(x_n-x_\varepsilon)\frac{M_\varepsilon}{E_\varepsilon I_\varepsilon}\Delta\varepsilon \quad\cdots\cdots(2)$$

普通上逐一點一點來計算斜度及偏差比較簡單，其式爲：

$$y_n=y_{n-1}+\theta_{n-1}\frac{\Delta x_{n-1}+\Delta x_n}{2} \quad\cdots\cdots(3a)$$

$$\theta_n=\theta_{n-1}+\frac{M_n}{E I_n}(\Delta x_n) \quad\cdots\cdots(3b)$$

分析解答：在圖1中爲一洋灰托梁(Haunched Beam)

設：$\theta_A=-\frac{1000}{E\ I_d}$ ；　$\theta_B=\frac{300}{EI_d}$ ……(4)

$y_A=\frac{20,000}{EI_d}$ ；　$y_3=O$ ……(5)

I_d ＝慣性力矩(Moment of Inertia) (in⁴)

d ＝中心濶度，

E＝彈性係數，(Kips/sq.foot)

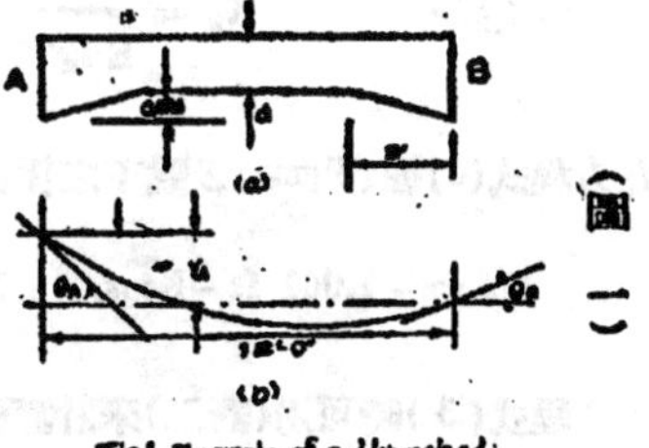

Fig.1 Example of a Haunched Beam

(圖一)

端力矩可用任一種方法求出，通常上可用方程式(1)化簡後把有界限總和(Finite Summatin)代入積分：

$$1{,}300 = M_A \sum_1^n K_n \left(1 - \frac{X_n}{12}\right) \triangle X_n + M_B \sum_1^n K_n \frac{X_n}{12} \triangle X_n \quad \cdots\cdots\cdots\cdots (6a)$$

$$-8{,}000 = M_A \sum_1^n K_n \left(1 - \frac{X_n}{12}\right)\left(12 - X_n\right) \triangle X_n$$

$$+ M_B \sum_1^n K_n \frac{X_n}{12}(12 - X_n) \triangle X_n \quad \cdots\cdots\cdots\cdots (6b)$$

（表 一）

表一　第六方程式中M_A，M_B之解答

Segment	X_n (ft)	K_n (Eq.7)	$\triangle X_n K_n$	$\frac{X_n}{12}$	$\frac{X_n}{12}\triangle X_n K_n$	$1-\frac{X_n}{12}$	$(1-\frac{X_n}{12})\times\triangle X_n K_n$	$12-X_n$	$\frac{X_n}{12}(\triangle X_n K_n)\times(12-X_n)$	$(1-\frac{X_n}{12})(\triangle X_n K_n)\times(12-X_n)$
(1)	(2)	(3)	(4)	(5)	(6)	(7)	(8)	(9)	(10)	(11)
1	0.5	0.353	0.353	0.0416	0.0417	0.9583	0.338	11.5	0.169	3.885
2	1.5	0.510	0.510	0.1250	0.0638	0.8750	0.446	10.5	0.670	4.480
3	2.5	0.737	0.737	0.2083	0.1640	0.7916	0.623	9.5	1.557	5.920
4	3.5	1.000	1.000	0.2916	0.2916	0.7083	0.7083	8.5	2.478	6.015
5	4.5	1.000	1.000	0.3750	0.3750	0.6250	0.6250	7.5	2.815	4.685
6	5.5	1.000	1.000	0.4583	0.4583	0.5416	0.5416	6.5	2.979	3.520
7	6.5	1.000	1.000	0.5416	0.5416	0.4583	0.4583	5.5	2.975	2.520
8	7.5	1.000	1.000	0.6250	0.6250	0.3750	0.3750	4.5	2.810	1.688
9	8.5	1.000	1.000	0.7083	0.7083	0.2916	0.2916	3.5	2.480	1.020
10	9.5	0.787	0.767	0.7916	0.623	0.2083	0.1640	2.5	1.558	0.410
11	10.5	0.510	0.510	0.8750	0.446	0.1250	0.0638	1.5	0.669	0.096
12	11.5	0.353	0.353	0.9583	0.338	0.0416	0.0147	0.5	0.169	0.007
Σ	......	....	9.300	....	4.6493	....	4.6493	....	21.329	34.446

由上式・

$$K_n = \frac{E\,I_d}{E\,I_n} \quad \cdots\cdots\cdots\cdots (7)$$

在方程式(6)裡(Sigma) Σ 號下之係數可由(表一)計算出

$M_A = -1{,}061$ ft - Kips.　　$M_B = 1{,}341$ ft—Kips.

用方程式(3)時可用(表二)來計算彈性曲綫：

（表二）

表二　偏差曲線計算（圖一）

X_n (ft) (1)	$M_A(1-\frac{X_n}{12})$ (2)	$M_n\frac{X_n}{12}$ (3)	M_n (4)	$K_nM_n\Delta X_n$ (5)	θ_n (6)	$\frac{\Delta X_{n-1}+\Delta X_n}{2}$ (7)	Δy (8)	y_n (9)
0.0	-1,061	0	-1,061	....	-1,000	....	....	20,000
0.5	-1,018	56	- 962	-340	-1,340	0.5	- 500	19,500
1.5	- 930	168	- 762	-389	-1,729	1.0	-1,340	18,160
2.5	- 841	279	- 562	-442	-2,171	1.0	-1,729	16,430
3.5	- 752	391	- 361	-361	-2,532	1.0	-2,171	14,260
4.5	- 664	503	- 161	-161	-2,693	1.0	-2,532	11,728
5.5	- 576	614	38	38	-2,655	1.0	-2,693	9,035
6.5	- 487	726	239	239	-2,416	1.0	-2,655	6,380
7.5	- 398	839	440	440	-1,976	1.0	-2,416	3,964
8.5	- 310	950	640	640	-1,336	1.0	-1,976	1,988
9.5	- 221	1,060	839	660	676	1.0	-1,336	652
10.5	- 134	1,173	1,039	529	- 147	1.0	- 676	- 24
11.5	- 44	1,283	1,239	443	300	1.0	- 147	- 171
12.0	- 0	1,341	1,341	....	....	0.5	150	- 21
Total	....	....	....	....	....	....	....	121,882

圖解法：

若不用（表二）時，方程式（3a）可用繩多邊形（Funicular Polygon）圖解法求出。如（圖二）所示，經過樑末点之线其斜度必等于末切綫（End Tangent）之已知斜度。

Fig.2 Construction of Deflection Curve For the Beam in Fig.1

（圖　二）

三稜形樑（Prismatic Members）：

定剖面之樑雖可用分析或圖解法來處理，但假如端旋度及偏差已知時，可應用方程式（2）積分之和來推算另一方程式而求其彈性曲綫。設AB爲一樑異而沒有荷重（圖三）

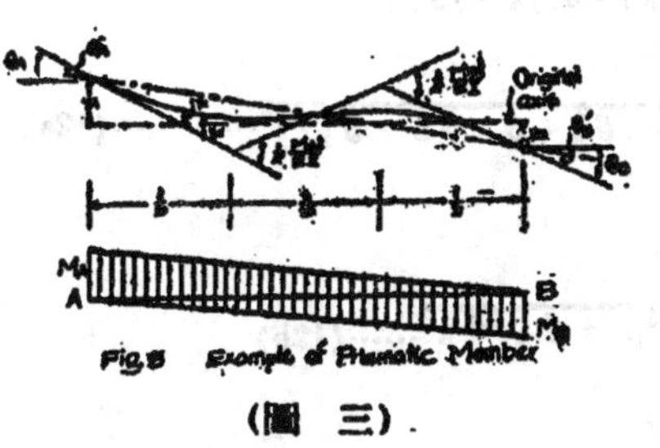

Fig.3 Example of Prismatic Member

（圖　三）

在末端A之旋度爲θ_A，在B端爲θ_B，設y_A與y_B爲A與B至原來位置之垂直位移，則由斜坡偏差方程式或用幾何求出：

25719

$$M_A = -\frac{2EI}{l}\left(2\theta_A + \theta_B - 3\frac{y_B - y_A}{l}\right) = -\frac{2EI}{l}(2\theta'_A + \theta'_B) \cdots\cdots (8a)$$

及：
$$M_B = \frac{2EI}{l}\left(\theta_A + 2\theta_B - 3\frac{y_B - y_A}{l}\right) = \frac{2EI}{l}(\theta'_A + 2\theta'_B) \cdots\cdots (8b)$$

因：$M_x = M_A + (M_B - M_A)\frac{x}{l}$，由方程式(2)積分後得：

$$y_x = \frac{1}{EI}\left[M_A\frac{x^2}{2} + (M_B - M_A)\frac{x^3}{6l}\right] \cdots\cdots (9a)$$

由幾何則得：

$$y_x = y_A + \frac{y_B - y_A}{l}x + \theta'_A x + y_{1x} \cdots\cdots (9b)$$

以方程式(9a)代入(9b)，同時以方程式(8)之M_A及M_B代入，化簡集項後得：

$$y_x = y_A + (y_B - y_A)(k_1)(1 - k_1^2 + k_1 k_2) + l(\theta_A k_1 k_2^2 - \theta_B k_1^2 k_2) \cdots\cdots (10)$$

上式中，$k_1 = \frac{x}{l}$，　　$k_2 = 1 - \frac{x}{l}$

假如 y_A 與 $y_B = 0$ 時此公式可與Ralph W. Stewart M. ASCE 所得之情形相似。

極大荷重點之決定

(Location of Maximum Load Point)

當集中荷重時，吾人應注意能產生極大効果剖面之位置；在方程式(10)可以寫爲：

$$\because k_2 = 1 - k_1$$

$$\therefore y_x = y_A + [l\theta_A + l\theta_B - 2(y_B - y_A)]k_1^3 + [3(y_B - y_A) - 2l\theta_A - l\theta_B]k_1^2 + l\theta_B k_1 \cdots\cdots (11)$$

以k_1爲變數，微分後，令第一次微分等于0，則得到一四次方之方程式：—

$$k_{1m} = \frac{1}{3[l(\theta_A + \theta_B) - 2(y_B - y_A)]}\Big[l(2\theta_A + \theta_B) - 3(y_A - y_B)$$

$$\pm\sqrt{9(y_B - y_A)^2 - 6l(y_B - y_A)(\theta_A + \theta_B) + (\theta_A^2 + \theta_A\theta_B + \theta_B^2)l^2}\Big] \cdots\cdots (12a)$$

假如 $y_B - y_A = 0$

則
$$k_{1m} = \frac{1}{3(\theta_A + \theta_B)}\left[(2\theta_A + \theta_B) \pm \sqrt{\theta_A^2 + \theta_B^2 + \theta_A\theta_B}\right] \cdots\cdots (12b)$$

極大感應線之縱坐標值

(Value of Maximum Influence Ordinate)

當 $y_B - y_A = 0$,吾人可推出一極大感應線坐標之表示式。以方程式(12b)之 K_m 值代入方程式(11)之 k;化簡可求出:

$$y_m = \frac{1}{27(\theta_A + \theta_B)^2}\left(\theta^3_A + 4\theta_A\theta_B + \theta^3_A \mp (\theta_A - \theta_B)\sqrt{\theta^2_A + \theta^2_B + \theta_A\theta_B}\right)$$

$$\times\left[(\theta_A - \theta_B) \mp \sqrt{\theta^2_A + \theta^2_B + \theta_A\theta_B}\right] \quad \cdots\cdots(13)$$

感應線所包括之面積

(Area under the Influence Line)

一稜形桿之感應線所包括之面積,可將方程式(11)來積分,其極限爲 $k_1 = 0$ 及 $k_1 = 1$,則產生另一方程式:

$$\theta = \frac{1}{2}(y_A + y_B) + \frac{1}{12}l(\theta_A - \theta_B) \quad \cdots\cdots(14)$$

設 θ 爲感應綫所包括之面積。

方程式(14)在均佈載重最適合,如偏差曲線爲不連續或跨度中有一斜度爲不連續時,上式不能直接應用。如桿之橫剖面爲變值時,則感應線所包括之面積等于各感應線縱坐標之總和。

例　題

(Examples)

下列例題中之結構物負有變形荷重 (Deformation Load) 可用各種不同的靜力學分析方法解出。以下所列出之結構物雖極簡單,但可應用同一原理到複雜的結構物上。

例題一:求一橫剖面不變兩端固定之梁在一定剖面力矩之感應綫。

可應用 Mr. Stewart's 導綫法,設在一固定剖面之旋度爲 θ,則在圖 4(a) 中 B 點之偏差爲 y_B 可決定;設與 M_B 所成之導綫角爲 4,則在 BC 長度裡之相當角必爲 5,因其亦與 M_B 所合成也。應用在 B 點之偏差相等,設△(圖 4b)爲端力矩之相關曲率 (Relative Curve):

則： $\Delta_1=\frac{5}{4}\Delta_2+\frac{9}{8}$ ……………………………(15)

再由力矩圖表示可求出：$M_B=M_A+\frac{4}{9}(M_C-M_A)$ …(16)

但 $M_A=(1)\Delta=\Delta_1$ ；

$M_B=(1)\ 4=4$ ；

$M_C=0.8\Delta$ ，

以上三式之值在入方程式(16)，同時加方程式(15)解之得：-

$\Delta_1=\frac{36}{7}$ ， $\Delta_2=-\frac{45}{14}$ ，

再由A到C點，用旋度之總和法求出 $\theta=-\frac{243}{4}$，如 $\theta=1$，則導精角必依次等于 $\frac{72}{243}$，$\frac{56}{243}$ 及 $\frac{45}{243}$，由幾何學觀之：

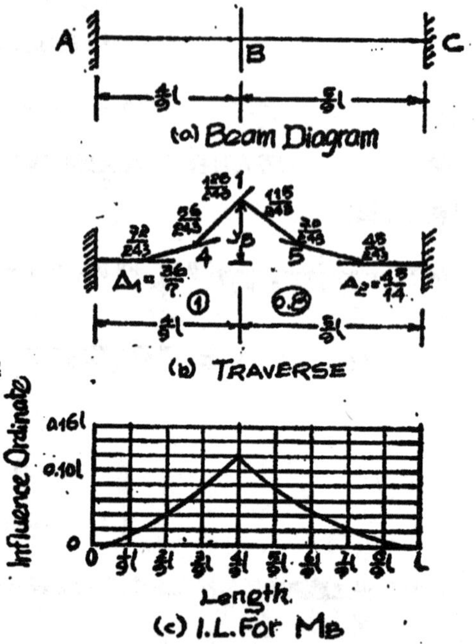

Fig.4 I.L. for M. at a Given Section of a fixed-ended beam with constant cross section.

(圖 四)

$$Y_B=\left(\frac{72}{243}\right)\times\left(\frac{8}{27}l\right)+\left(\frac{56}{243}\right)\left(\frac{4}{27}l\right)=0.122\ l$$

在圖 4C 之其他縱座標，可用方程式 10 來算出：

例題二：求一固定洋灰托梁固定端力矩之感應綫：

參照(表三)，利用梁A端之單位旋度，用柱之類似法求出梁之靜力狀況。將其所得結果列在（表三）內，設在A端有一單位角荷重(unit angle lood)則末端之力矩當依次爲

$$M_A=\frac{1}{26.15}+\frac{11.21\times11.21}{697.7}=+0.2189\ ;$$

$$M_E=\frac{1}{26.15}-\frac{11.21\times8.79}{697.7}=-0.1031\ ,$$

最後計算感應綫之縱坐標與(表2)之編排相同。

（表三）

表三　端力矩之計算

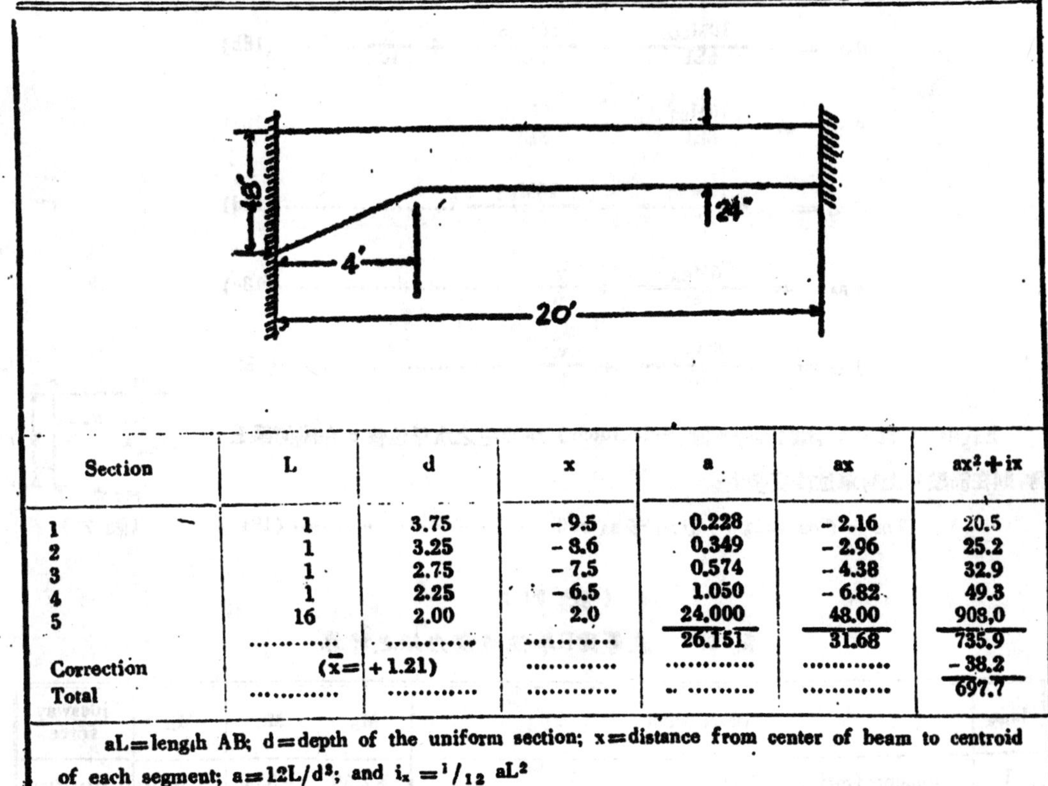

Section	L	d	x	a	ax	ax^2+ix
1	1	3.75	-9.5	0.228	-2.16	20.5
2	1	3.25	-8.6	0.349	-2.96	25.2
3	1	2.75	-7.5	0.574	-4.38	32.9
4	1	2.25	-6.5	1.050	-6.82	49.3
5	16	2.00	2.0	24.000	48.00	908.0
	…………	…………	…………	26.151	31.68	735.9
Correction	($\bar{x}$=	+1.21)	…………	…………	…………	-38.2
Total	…………	…………	…………	…………	…………	697.7

aL=length AB; d=depth of the uniform section; x=distance from center of beam to centroid of each segment; $a=12L/d^3$; and $i_x={}^1/_{12}\ aL^2$

例題三：求一兩端固定梁在其梁托點（Support）之剪力。

設一均勻橫剖面樑，如（圖5）所示，在A點有一單位旋度，在此例題中，其屈折微分方程式之配度爲：　$y_A=1$；$y_B=0$；$\theta_A=0$；$\theta_B=0$

其解答最普通爲應用（方程10）化簡爲

$$y_x=1-k_1(1-k_2^2+k_1k_2)\cdots\cdots\cdots(17)$$

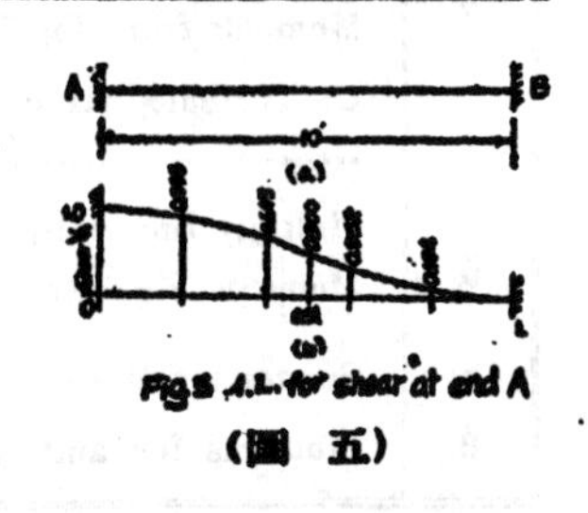

Fig.5 A.L. for shear at end A

（圖五）

方程式（17）即可計算出感應綫。（如圖5b）

例題四：如圖6所示，求一均勻剖面載活重10 kips之架（Frame）之M_C最大值。（同時求節點C負有1.5 kips/Ft平均分佈荷重之力矩）

25723

由斜度偏差方程式得。

$$\theta_{DC}=0=\frac{10M_{DC}}{3EI}+\frac{10M_{CD}}{6EI}+\frac{y}{10}\quad\cdots\cdots\cdots\cdots(18a)$$

$$\theta_{CD}=-\frac{10M_{DC}}{6EI}-\frac{10M_{CD}}{3EI}+\frac{y}{10}\quad\cdots\cdots\cdots\cdots(18b)$$

$$\theta_{CB}=\frac{10M_{BC}}{6EI}+\frac{10M_{CB}}{3EI}\quad\cdots\cdots\cdots\cdots(18c)$$

$$\theta_{BC}=-\frac{10M_{BC}}{3EI}-\frac{10M_{CB}}{6EI}\quad\cdots\cdots\cdots\cdots(18d)$$

$$\theta_{BA}=\frac{6M_{BA}}{3EI}+\frac{y}{6}\quad\cdots\cdots\cdots\cdots(18e)$$

$$\theta_{AB}=-\frac{6M_{BA}}{6EI}+\frac{y}{6}\quad\cdots\cdots\cdots\cdots(18f)$$

上式中 y 爲 B.C. 兩點因旁傾側(Side lutch) 而發生之水平位移。在幾何學上言,則在節點 C 上有單位旋度發生:

$$\theta_{CD}-\theta_{CB}=1;\qquad \theta_{BC}-\theta_{BA}=0\quad\cdots\cdots\cdots\cdots(19)$$

Fig. 6

(圖 六)

(表 四)

表四　在節點B單位旋度力矩之計算

Line	Descriprion	M_B	M_C	M_D	Sidesway force
1.	Moments from Fi;g. 7(c)	+2.50	-8.00	-4.50	....
2.	Moments from Fig 7(d)multiplied by 1.389	+9.72	-2.78	+1.39	....
3.	Undistributed fixed-ended moment at poin t D	....	....	+17.00	....
4.	....	+12.22	+10.78	131.9	4.504
5.	Multiply line 4 by 1.467/4.504	+ 3.98	+ 3.51	+4.53	1.467
6.	Momeuts Fig. 7(a)	- 7.90	- 2.00	-1.00	-1.467
7.	Summation lines 5 and 6	- 3.02	+ 1.51	+3.53	0
8.	Moments for unit rotation(divide line 7b 6.8)	0.442	0.224	0.520	....

由靜力學上言:

$$M_{CD}=M_{CB};\quad M_{BC}=M_{BA}\quad\cdots\cdots\cdots\cdots(20)$$

假設H及V爲在A點反抗力之水平及垂直分力,則:

$M_{LC}=4H+10V$；$M_{CD}=10V-6H$；$M_{BC}=6H$；$M_{CD}-M_{BC}=10V$………(21)

由方程式(18)至(20)同時解出：

$M_{BA}=M_{BC}=-0.0444\,EI$；

$M_{CD}=M_{CB}=0.0972\,EI$；

$\theta_{CD}=0.603$；　$\theta_{CB}=-0.398$；　$\theta_{BC}=0.310$；

$\theta_{BA}=0.311$；　$\theta_{AB}=0.444$

由方程式(12b)得：

$$k_{1m}=\frac{1}{3(0.310-0.398)}(2\times0.31-0.398\pm\sqrt{0.31^2+0.398^2-0.398\times0.31}$$

$$=0.531$$

由方程式(10)得：

$$y_m=10\left[0.310\times0.531\times0.469^2+0.398\times0.531^2\times0.469\right]=0.888$$

由方程式 (14) 可求出感應曲線所包括之面積

$$=\frac{1}{12}(10^2)(0.31+0.398)=5.90$$

當活重 10 kips 放在離 B 點(如圖5) 5.31ft 時，在節點 C 上因負活重而產生之最大力矩爲 8.88 ft-kips，如受均佈載重時則在節點 C 之力矩 5.9×1.5=8.85 ft-kips.

例題五：參看圖六，用彈性彎曲導線來做撓力分析，求在節點 B 之最大力矩值．

設在節點 B 之旋度爲 θ，在圖 7 中，因 $M_{BA}=M_{BA}$，吾人可由 C D 柱之較低彈性角起，設其值爲 I，由圖 (7a)，$\theta=2.8+4.0=6.8$，因旁力尙未計算，故此等值均需更正．未平行之水平力爲 1.467EI；從圖 (7b)中，假設此架不依照節點 B及 C 來旋轉而向旁屈折，而指定△爲其導線角，M 爲 CD 末端之力矩，因在垂直桿中其旁傾側相等，故AB柱之導線角一定等于0.833△ 及 $M_{BA}=\frac{5}{6}\times\frac{5}{3}\times M=1.389M$(參看表 4 第二行)。以圖 7c及7d 可求出節點 B 及 C 之不平衡力矩，在每個節點之不平衡力矩爲 17 。假如設在圖 7b 之M等于 17，則圖7b之力矩可算出。同時旁改正(Sideway Correction)可求出。最後之力矩必需以6.8 除之來減低節點之單位偏差．則由D點起，可依此等力矩變到相當角單位 (Equivalent angle unit)，正確之導綫構成如圖 7e 所示．

由方程式12b得：

$k_{1m}=0.474$

由方程式 10 得：

$y_m=0.838.$

例題六：如圖六所示，在BC中心之垂直偏差感應綫求法可以以一單位垂直荷重放在中心跨度上，先求出其最後之偏差曲綫而求出之。

由第四第五例題中要得之 θ_{BC} 及 θ_{CB} 之值在入方程 (10) 可求出 M_B，M_C 對此荷重之值：

$$M_B=10(0.369\times0.5^3+0.295\times0.5^3)=0.83$$

$$M_C=10(0.310\times0.5^3+0.398\times0.5^3)=0.885$$

由靜力分析，$V_B=0.4945$，$V_C=0.5055$，則圖 8a 之力矩圖表可以繪出。

在 B 與 C 二點間，應用方程式 (2) 求出：

$$\theta_{BC}=-\frac{2.02}{EI}\qquad\theta_{CB}=\frac{1.92}{EI}$$

則彈性曲綫（所求之感應綫）可應用方程式 3a 算出（參閱圖 8b）。

Fig 7 Analysis of an Unsymmetrical Bent

（圖 七）

例題七：求在一對稱剛架受垂直荷重之 M_{BC} 感應綫。

（設有一水平均佈載重在 AB 桿上時，其中節點B之力矩為何？參閱圖 9）。

在此例題中，以應用力矩分配法較為適宜，設在 B 點旋之度為一單位，則在桿 BC 之固定端力矩當依次為 $M_{BC}=\frac{4EI}{l}$ 及 $M_{CB}\frac{2EI}{l}$

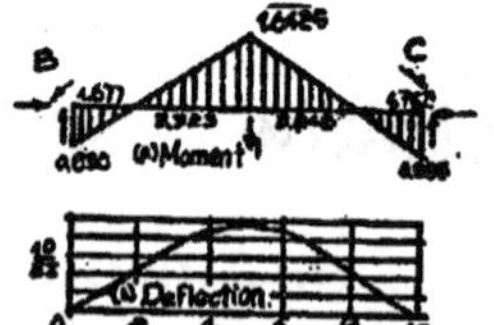

Fig.8 I.L. for Deflection at the Center of the Span

（圖 八）

圖9a之解答為：$M_B=-1.615\frac{EI}{l}$ ；

$$M_C=0.461\frac{EI}{l}\;;$$

$$M_D=-0.230\frac{EI}{l}\;;$$

未平衡之剪力為 $0.231\frac{EI}{l}$，由圖9b求出旁傾側之改正數及由圖9a求出旁傾側准許更正力矩，由端旋度及端力矩之關係可得：

$$\theta_{BC}=\frac{1.09}{3}+\frac{0.273}{6}=0.409$$

$$\theta_{CB}=-\frac{1.09}{4}-\frac{0.273}{3}=-0.273$$

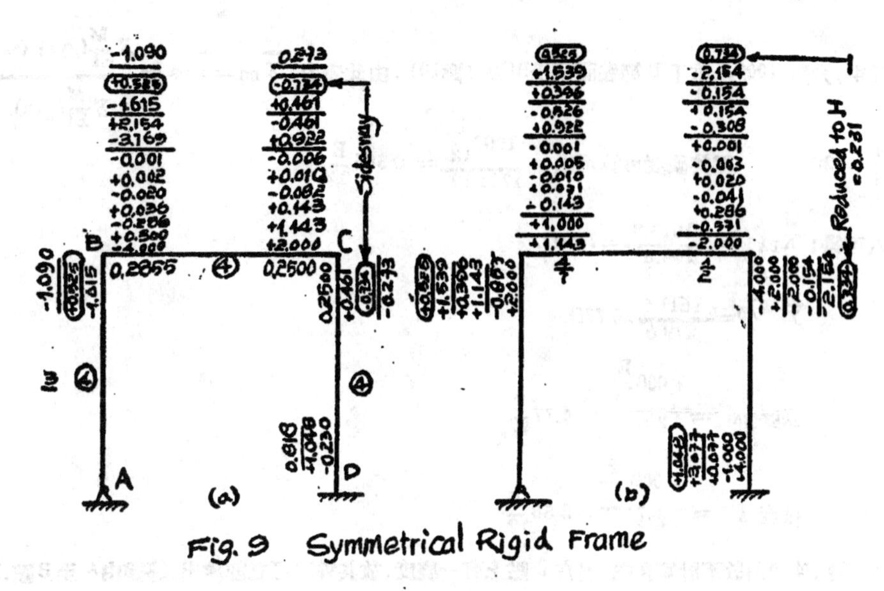

Fig. 9 Symmetrical Rigid Frame

(圖　九)

因 $\theta_{AB}=\theta_{BC}-1=-0.591$

由方程式8可求出AB桿之M_{AB}：

$$M_{AL}=0=-2\left(2\theta_{AB}-0.591-\frac{3y}{l}\right)\cdots\cdots(22a)$$

$$及\ M_{BA}=-1.090=2\left[\theta_{AB}-2(0.591)-\frac{3y}{l}\right]\cdots\cdots(22b)$$

在上式中y爲由旁側所生之位移，同時可解出 $\theta_{AB}=-0.046$ ； $y=-2.277$ ；

由方程式12b中則可解得BC桿之$k_{im}=0.456$；

由方程式10得：$y_m=0.862$；

由方程式14得：$Q=\frac{10}{2}(-2.277)+\frac{1}{12}10^2(-0.046+0.591)=6.848$.

桿AB受一500磅/ft之平均分佈活重時，則在B點之力矩爲$6.848\times500=3424$ft－lbs.

例題八：洋灰托之剛架——在圖10之架中，BC梁受垂直荷重，用導綫法求出M_B之感應綫，設有1.5 Kips/ft之平均分佈荷重，求M_B之值。若桿之一端受荷重時則其導綫角必在M/EI表之重心之中間。今離重心點之距離爲$\overline{x}$，則從表五及表六中，可找出各桿之硬度(Stiffnesses of Members)及各節點之解答手續。

在表五中，力矩 $1000\frac{E}{12}$施于B節點而至于BC桿（圖10），由此表中，$\overline{X}_{BC}=\overline{x}_{CB}=\frac{\Sigma\frac{M}{EI}(\triangle x)(x_n)}{\Sigma\frac{M}{EI}(\triangle x)}=$

$\frac{35,075.4}{1,711.18}=20.44ft$，至其硬度可算可出$=\frac{1000\frac{E}{12}}{1711.19}=0.585\frac{E}{12}$

由表六可得：$\overline{X}_{BA}=\frac{1613.8}{209.8}=7.71ft.$

$\overline{X}_{AB}=\frac{1611.6}{279.6}=5.77ft.$

硬度$_{BA}=\frac{1,000\frac{E}{12}}{209.8}=4.77\frac{E}{12}$,

硬度$_{A}=\frac{1,000\frac{E}{12}}{279.6}=3.58\frac{E}{12}$

由圖（10C），旁傾側數不計算在內，因在B點上有一旋度，故其導綫可立即繪出。導綫由A至B點，D點對導綫無影响，因$M_{BA}=M_{BC}$，如右邊旋度乘以$\frac{15.040}{23.830}$則左右相連，其正確之旋度均在圖上指出。

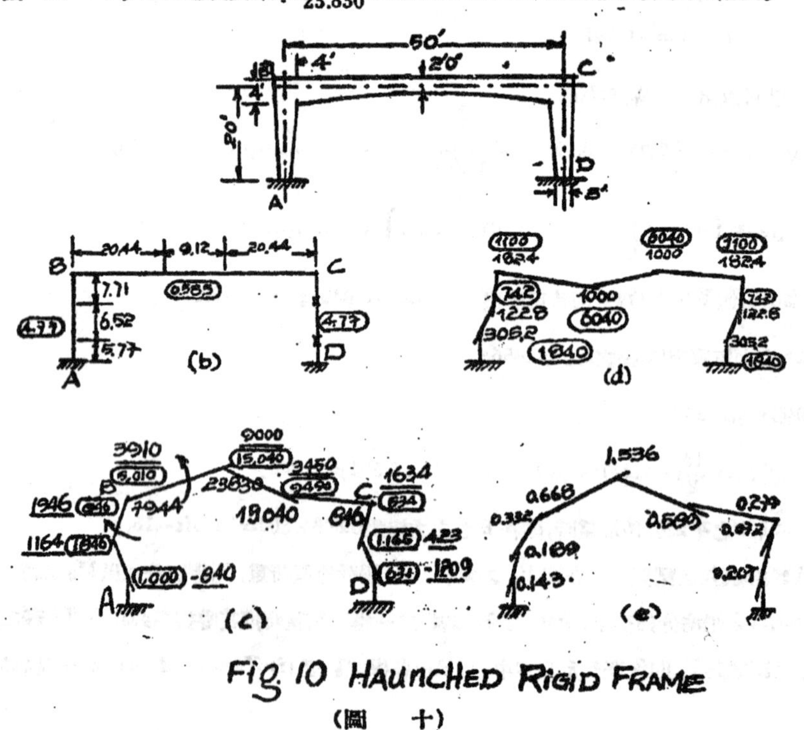

Fig 10 HAUNCHED RIGID FRAME

（圖　十）

（表五）

表五　BC梁性能之計算

Segment	x_n (ft)	dx	d^3	$\frac{M}{E/12}$	$\frac{M}{EI}\Delta x$	$\frac{M}{EI}\Delta x x_n$
1	1	3.84	56.7	980	34.56	34.6
2	3	3.55	44.2	940	42.44	127.4
3	5	3.28	44.1	900	51.24	256.2
4	7	3.03	28.2	860	61.02	433.6
5	9	2.82	22 3	820	73.50	663.0
6	11	2.63	18.1	780	86.16	949.0
7	13	2.46	14.8	740	99.90	1,297.0
8	15	2.42	12.4	700	112.90	1,692.0
9	17	2.20	10.6	660	124.50	2,116.0
10	19	2.12	9.39	620	132.10	2,508.0
11	21	2.05	8.59	580	135.10	2,838.0
12	23	2.01	8.07	540	133.90	3,079.0
13	25	2.00	8.00	500	125.00	3,125.0
14	27	2.01	8 07	460	114.10	8,082.0
15	29	2.05	8.59	420	96.80	2,808.0
16	21	2.12	9.39	380	80.90	2,504.0
17	33	2.20	10.6	340	64.10	2,016.0
18	35	2.32	12.4	300	48.34	1,691.0
19	37	2.46	14.8	260	35.12	1,301.0
20	39	2.63	18.1	220	34.30	948.0
21	41	2.82	22.3	180	16.14	662.0
22	43	3.03	28.2	140	9.94	427.8
23	45	3.28	35.1	100	5.70	254.6
24	47	3.55	44.2	60	2.72	127.8
25	49	3.84	56.7	20	0.70	34.4
					1,711.18	35,075.4

（表六）

表六　AB桿性能之計算

X_n ft	dx	d^3x	$\frac{M_B}{E/12}$	$\frac{M_B}{EI}\Delta x$	$\frac{M_B\Delta x}{EI}X_n$	$\frac{E_A}{M/12}$	$\frac{M_A}{EI}\Delta x$	$\frac{M_A\Delta X}{EI}(20-x_n)$
1	3.95	61.5	950	30.86	30.8	50	1.62	30.8
3	3.85	56.8	850	29.92	89.8	150	5.28	87.6
5	3.75	52.3	750	28.66	143.2	250	9.56	143.4
7	3.65	48.5	650	26.60	187.6	350	14.22	187.6
9	3.55	44.2	550	24.88	223.8	450	20.38	224 0
11	3.45	40.8	450	22.06	242.8	550	26.96	242.4
13	3.35	37.4	350	18.72	243.4	650	34.74	243.0
15	3.25	34 0	250	14.70	220.4	750	44.10	220.4
17	3.15	31.0	150	9.68	164.4	850	54.84	164.8
19	3.05	28.1	50	2.56	67.9	950	67.70	67.6
Σ	………	………	………	209.84	1,613.8	………	279.62	1,611.6

在沿垂桿中之未平衡剪力爲$1010\frac{E}{12}$，在圖10d中已知B節點上之水平分分解答。由對稱關係證明BC梁之導綫角必等。假如此等角爲1,000，則各導綫可繪出如圖所示，在圖10C中之水平力爲 $167.5\frac{E}{12}$，在圖10d中指出更正之角，此號角係爲圖(10C)之相等角之代數和，此乃旁傾側數之更正數．在圖(10C)中有一劃者卽爲其更正數也。在B節點之旋度爲1946+3910=5856。爲使各旋度爲整數，故各角均以$\frac{1}{5856}$乘之，此等角均在圖10C中示出。BC梁彈性曲綫(感應綫)可由表7中求出，在感應綫所包括之面積爲 $\Sigma y_n \triangle x$ 或 $122.58\times2=245.16$，而1.5kips/ft所生之力矩爲367.4 kip－ft。

（表 七）

表七　偏差曲線之計算

x_n (ft)	$\frac{M_B(1-\frac{x_n}{50})}{E/12}$	$\frac{M_C(1-\frac{x_n}{50})}{E/12}$	$\frac{M_n}{E/12}$	$\frac{M_n\triangle x}{EI}$	θ_n	$\frac{\triangle x_{n-1}+x_n}{2}$	$\triangle y$	y_n
0	－0.899	0	－0.899	………	0.668	……………	………	0.000
1	－0.881	0.007	－0.874	－0.031	0.637	1	0.668	0.668
3	－0.845	0.020	－0.325	－0.037	0.600	2	1.274	1.942
5	－0.809	0.304	－0.775	－0.044	0.556	2	1.200	3.142
7	－0.733	0.049	－0.724	－0.052	0.504	2	1.112	4.254
9	－0.737	0.062	－0.675	－0.060	0.444	2	1.008	5.262
11	－0.701	0.076	－0.625	－0.069	0.375	2	0.888	6.150
13	－0.665	0.089	－0.576	－0.078	0.297	2	0.750	6.900
15	－0.630	0.104	－0.526	－0.085	0.212	2	0.594	6.491
17	－0.594	0.117	－0.477	－0.090	0.122	2	0.424	7.918
19	－0.554	0.131	－0.426	－0.091	0.031	2	0.244	8.162
21	－0.521	0.145	－0.376	－0.089	－0.058	2	0.062	8.224
23	－0.486	0.159	－0.327	－0.081	－0.138	2	－0.116	8.108
25	－0.450	0.172	－0.278	－0.069	－0.208	2	－0.278	7.830
27	－0.414	0.186	－0.228	－0.057	－0.265	2	－0.416	7.414
29	－0.378	0.200	－0.178	－0.040	－0.305	2	－0.530	6.884
31	－0.342	0.213	－0.128	－0.027	－0.332	2	－0.610	6.274
33	－0.306	0.227	－0.078	－0.015	－0.347	2	－0.664	5.610
35	－0.270	0.341	－0.029	－0.006	－0.353	2	－0.694	4.916
37	－0.234	0.255	0.021	0.002	－0.351	2	－0.706	4.210
39	－0.198	0.268	0.070	0.008	－0.343	2	－0.702	2.508
41	－0.162	0.282	0.121	0.011	－0.332	2	－0.686	2.822
43	－0.126	0.296	0.170	0.012	－0.320	2	－0.664	2.158
45	－0.090	0.310	0.221	0.012	－0.308	2	－0.640	1.518
47	－0.054	0.324	0.270	0.013	－0.295	2	－0.616	0.902
49	－0.018	0.338	0.320	0.011	－0.284	2	－0.590	0.312
50	0	0.345	0.345	………	………	1	－0.284	0.000

例題九：簡單支持連續樑

在圖(11)中之梁可應用間接法及力矩分佈法繪其感應綫，在圖(11)中，同時分析在B及C節點中

25730

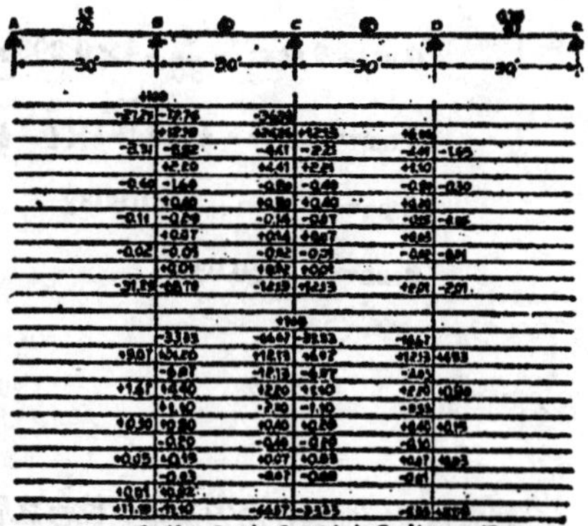

之（+100）未分配力矩之影响。（在此例題中，爲便利起見，以前所成之符號畧有更正，力矩依時鐘方向爲正號，至于其他計算與前所用符號相同。）

求在 B 支點上之力矩時，設一單位旋度在 B 節點下，而旋度只在 A B 跨度上有作用，則定端力矩M_{AB}爲 3EK，X 因其硬度 EK爲 2,則 $M_{BA}=3\times2=6$,

依圖（11）之分佈情形分配此定端力矩，則 $M_B=4.120$，$M_C=-0.729$；$M_D=0.121$，又末端之旋轉亦在下列算出：

$$\theta_{AB}=-\frac{4.12}{6}\cdot\frac{1}{2}\qquad=-0.34$$

$$\theta_{BA}=+\frac{4.12}{3}\cdot\frac{1}{2}\qquad=+0.69$$

$$\theta_{BC}=-\frac{4.12}{3}\cdot\frac{1}{4}+\frac{0.729}{6}\cdot\frac{1}{4}\qquad=-0.31$$

$$\theta_{CB}=\theta_{CD}=\frac{4.120}{6}\cdot\frac{1}{4}-\frac{0.729}{3}\cdot\frac{1}{4}=0.11$$

$$\theta_{DC}=\theta_{DE}=-\frac{0.121}{3}\cdot1\qquad=-0.04$$

$$\theta_{ED}=+\frac{0.121}{6}\cdot1\qquad=+0.02$$

Fig.11 Simply Supported Contiguous Beam.

（圖十一）

又在圖（11）中，在 B 之支點反抗力上有一單位偏差，故其定端力矩必需應用：

$$M_{BA}=\frac{3EK}{l}=+\frac{3\times2}{30}=+0.200$$

$$M_{BC}=\frac{6EK}{l}=+\frac{6\times4}{30}=+0.800$$

$$M_{CB}=\frac{6EK}{l}=-\frac{6\times4}{30}=-0.800$$

用圖(11)之分佈狀況，則最後結果之力矩爲：

$M_B=+0.298$，　$M_C=-0.194$，　$M_D=0.032$

及其相當之旋度爲：（$\delta_B=-1$）！

$$\theta_{AB}=-\frac{1}{30}-\frac{1}{6}\times0.298\times\frac{1}{2}\qquad=-0.0583$$

$$\theta_{BA}=\theta_{BC}=-\frac{1}{30}+\frac{1}{3}\times 0.298\times\frac{1}{2} \qquad =+70.0164$$

$$\theta_{CB}=\theta_{CD}=+\frac{1}{3}\times 0.194\times\frac{1}{2}-\frac{1}{6}\times 0.032\times\frac{1}{2}=+0.0296$$

$$\theta_{DC}=\theta_{DE}=-\frac{1}{3}\times 0.0032 \qquad \doteq 0.0107$$

$$\theta_{ED}=+\frac{1}{6}\times 0.032 \qquad \doteq 0.0053$$

結 論

(Conclusions)

感應線可依照其作用力及單位位移之彈性曲綫計算。此等計算法必需具有：

(1)結構物在單獨位移時之靜力分析。

(2)在彈性曲綫上計算所必需之各点。

有時此法會比簡單靜力分析工作爲多。不過靜力分析只係包括沒有荷重之結構物，使在靜力分析時用任何方法均可，或可間接用在每個節点上分佈其未平衡之力矩，然後依照比例，視其所需之偏差，取一端固定力矩與之相加。彈性曲線可應用幾何法或其他幾何原理求出，(分析法在此章暫不討論)。

有時在很多情形不需要完全彈性曲綫，如爲綫形桿時，極大效力之荷重，可用方程式(12)算出，如爲平均分佈荷重時其在感應綫所包括之面積可照方程式(14)算出可也。

重力壩之應力分配法

BYO.C.Zieukiewicz.PH.D., BSc.

——何思源譯——

引言

普通在重力壩設計之古典原理中，往往假設在平面部份上之垂直應力分佈爲縱長者。但據本世紀前一般人之觀察，則此假設與壩底基礎附近地域之彈性原理（Theary of Elasticily）互相抵觸而致矛盾，更且，由此假定，眞正產生應力之值常與計算之值不大相符。

在此文中，上述之假設是經過一種應力分析之新數學方法來審定，而此方法則概說於下列之(Appendix I.)

附錄一：

此研究工作之進行方在倫敦之帝國科學及技術專門學院中。

符號法

X, Y,　表示卡達兒（Cartesian）坐標

r, θ,　表示極向坐標

X_x, Y_y, X_y　表示在X,y系統內之應力平面應變分力

p_1, p_2　表示主要應力

S　表示在任何平方面上之剪力

X, Y　表示在 x 及 y 方向，邊界應力之分力

M　表示在邊界各力上之力矩

s　表示邊界之長度

p　表示壩與壩基材料之比重

w　表示水之密度

g　表示因地方吸力而發生之加速度

φ　表示 Poissons 之比率

α　表示 Airy's 應力之功能

R　表示每單位闊度（在壩底上）之總反應力

H　表示壩壩基之水高度面

L　表示壩基底之長度

F_o　　表示在O点之剩餘

a　　表示網眼之長度(Mesh Length)

x　　表示在主要應力平面上之傾向

隄墻問題之敘述及其假設

每一隄壩必有壩基，扶壁等以構成其體積，而此結構之眞正分析又爲一絕對之難題。但在，一長，直之重力壩上，此問題則可減爲兩面體積之平面應變問題。假設此壩係一延續及壩基齊次均勻之隄壩，其結構物質又均爲齊次均勻而絕對彈性者。但此假設非係絕對眞確，不適長之應力應變關係，在連續角重下之崩潰，及彈性係數因含水成份而有之不同值，此種種均爲混凝土之特有性質，更且建築壩基之材料往往與建築隄壩之材料不同.

上述因素應力之彈性分配所發生之影响究爲若干，現尙爲懸案，有待未來實驗紀錄始能加以解答及證明；但據甚多作者之見解，則此彈性原理已能給予與眞正應力之分配值互相接近結果，故迄至現在，此理論確爲設計隄壩之合理根據.

產生在壩之應力包括：—(a)由外界作用力成本身作用力所產生之應力，(b)與上述外界作用力不同之內壓應力，由收縮性及溫度變遷種種而產生者。

上述(a)，(b)所生之應力應個別加以應付而其影响所附加其上云.

於此文中，祇論及流體靜力學之壓力及重力壓力，而(b)所發生之應力則不加以討論或分析，因其每因建築方法及其他因素而畢其值者。總之殘餘應力(Residnal stresses)可因適當之建築方法而減至最少值，而在普通設計中常對此不加以任何之考慮云.

歷史性之調查

自從本世紀之始端，已有不少對於此應力彈性分配之研究，但其中頗能成功者則爲根據實驗所得之結果，於1908年Wilson與Gore曾用橡膠模型法從實驗得來美滿之成績.其後各方多用此法在試驗中，但其最近者則純粹應用"Photo-elastic (technigue).

在堅硬壩基上之三角形側面圖可用數學方法計算及分析而得一準確之值，但在普通一彈性壩及彈性基上，種種困難之問題常用之而生。最有効之試驗則爲1908年中由人F.Richarason所做者.

故此文以下所得之結果乃完全根據Richardson試驗中所得之可靠方法而得來.更且於其計算當中，亦常採用Sonth well's method of Relaxation.之方法云.

用以分析之側面圖

側面圖用以分析者有下列三種：—

第一側面圖：—此乃一206.'呎高隄壩之側面圖，根據Wegmaun Method之古典假設法而設計者，其設計之條件如下：

壩頂之寬度	20呎
堤頂高度距離 最高水位之距離	5呎
在上流面之最大壓力	18000 磅/平方呎
材料之比重	2.30

第二側面圖：——三角形側面圖，其壩底基線之寬度為其高度之 $\frac{2}{3}$。

隄壩材料之比重 ： 2.25

當儲水池滿水時其作用力之綫又恰在其橫面之 Middle third.

第三側圖面：—此乃一長方形之側面圖，其壩底寬度為壩高度之三分二.

隄壩材料之比重 ： 2.25

當儲水池滿水時其作用力之綫又恰在其底邊之第三點(Third Point)

所得結果

圖表一至圖表十五表明當儲水池滿水或空缺時所分析第一側面圖而得之結果，及當儲水池充滿時之第二側面圖。而曲綫乃表示應力之分力方向分佈法。第一側面圖之主要應力亦已計算清楚，而曲綫表示相等應力及極大剪力者亦已印出於下列圖表中。

第二側面圖之應力詳細分析已於圖表十六至圖表十八中表示清楚，其使壩尖圓滑所得之影响亦加已說明：

結果之討論

分析應力所得之結果能撮要於如下：

當儲水池滿載時之應力。

(1)垂直應力： Y_y

(a)在分析隄壩上部之側面圖中，由壩高度之三份一至四份一，應力之分佈往往是縱長的分佈在橫面上，此正足以符合古典之假設原理。

(b)在壩基底之橫面部份，此應力之分佈多非為縱長者。第一與第二側面圖之分佈法大致相同，在壩之中部應力通常比由縱長定律 (Linear Law) 所得來者為大，但又比在壩面所得者為少。每當下流面之坡度太大時所應注意之條件有二（1）在上游面各應力在縱長分佈 (Linear Distribution) 之下者之距離往往圖是較少(2)在下游面之應力往往斷趨增大，尤以在第三側面圖中其應力分佈在縱長分佈之上。

試將第二及第三側面圖之應力作一比較，後者之體積雖為前者之两倍，但其所能負荷之張力，根據古典原理亦祇能抵受同等之張力而已。然事實上說來，根據彈性應力分佈 (elastic Stress disfribution) 第三側面圖確爲比較安全穩重。

（2）剪力：X_y

（a）在隄壩之上部，剪力分佈在橫面部份之表現爲一拋物綫形之分佈法。在第三側面圖時，此剪力分佈之表現又爲一時稍之拋物綫，而此拋物綫在三角形之第二側面圖中又可能變爲一直綫。

（b）在壩底之橫截面上，此剪力之分佈可能脫離古典原理所規定之分佈法而成凸形之曲綫，其極大値在壩面中。

（3）水平應力：X_x——在壩上游之垂直壩基及在壩內之若干距離中，上述三種側面圖之水水平應力可能變爲張力。此皆因隄壩與堤基之牽电性作用有以致之。其他在壩之各部份內，其水平應力多係壓力性者，此壓力增加之方向乃純向壩趾而發，因此壓力之量常與垂直應力之量相等。

（4）主要應力：p_1，p_2 及極大剪力.

（a）在壩面下游發生之壓力性主要應力通常多向着壩趾而增加.

（b）在上游壩基之背面，張力性之主要應力將可能發生於任何件情形下，而與X_x力連貫.

（c）最大剪力放射體之軌道幾爲一直線，此正足證明普通以水平應力抗抗剪力之學說.

（5）在隄壩趾尖脬頂之應力亦足研究。因在壩脬，倘其轉角爲急切者，其兩者之主要應力均可銳趨向變爲一極限大之張力，甚至其剪力亦能趨向至極限大。此使壩尖側滑方法所得之影响并不可靠，或賴以改進應力之分佈云.

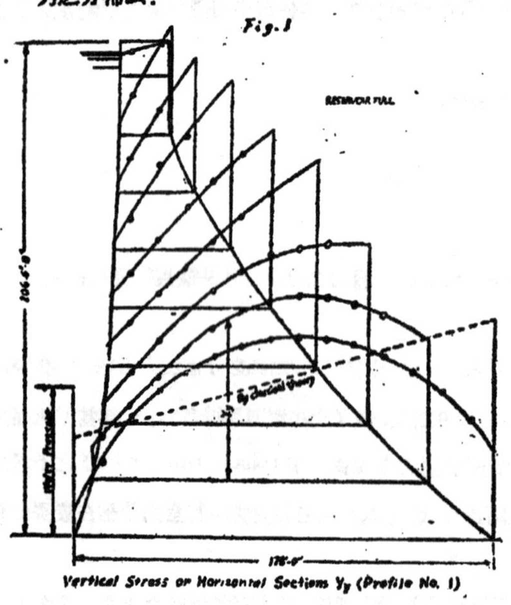

Vertical Stress on Horizontal Sections Y_y (Profile No. 1)

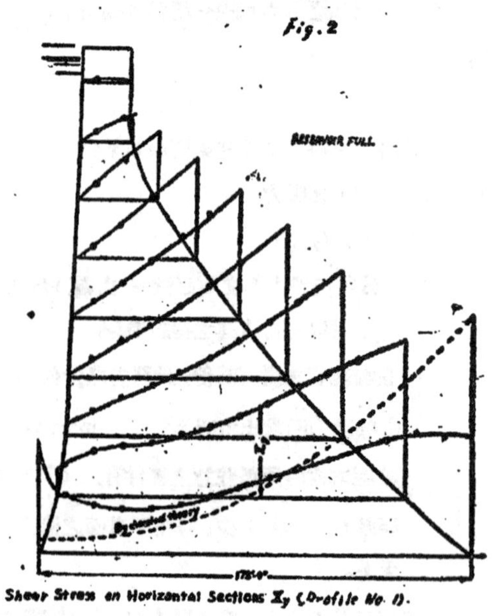

Shear Stress on Horizontal Sections X_y (Profile No. 1).

25736

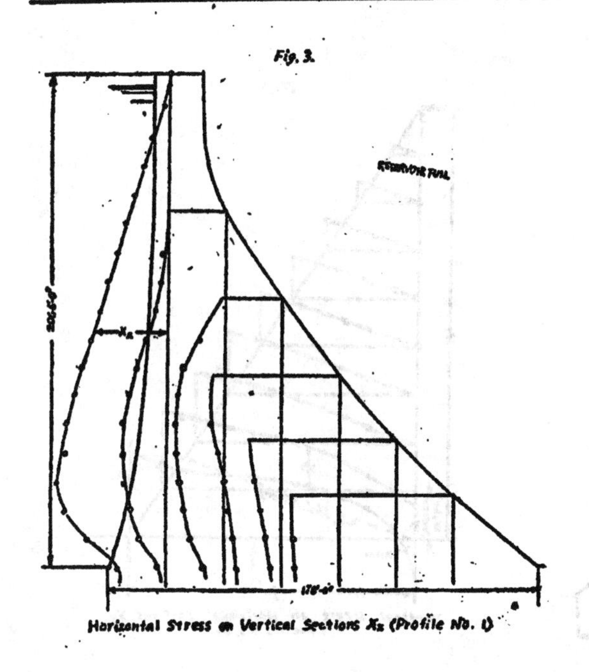

Horizontal Stress on Vertical Sections X_x (Profile No. 1)

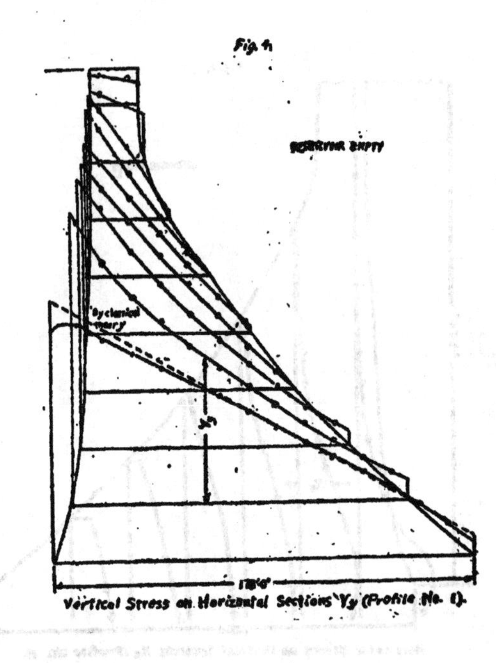

Vertical Stress on Horizontal Sections Y_y (Profile No. 1).

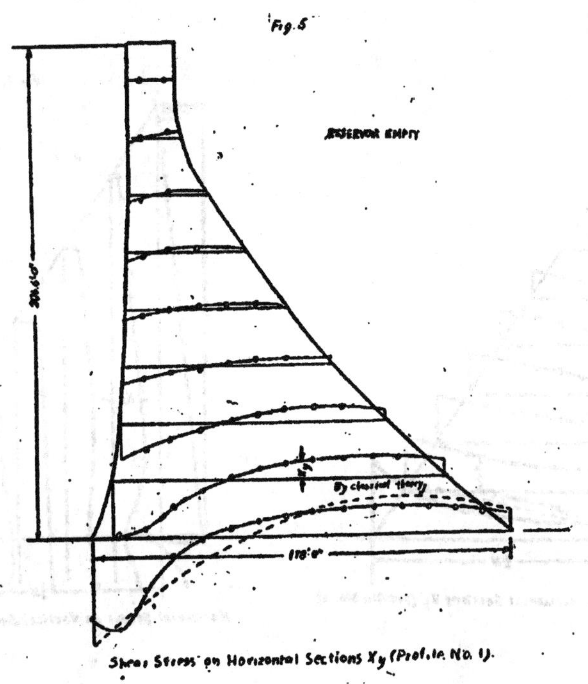

Shear Stress on Horizontal Sections X_y (Profile No. 1).

25737

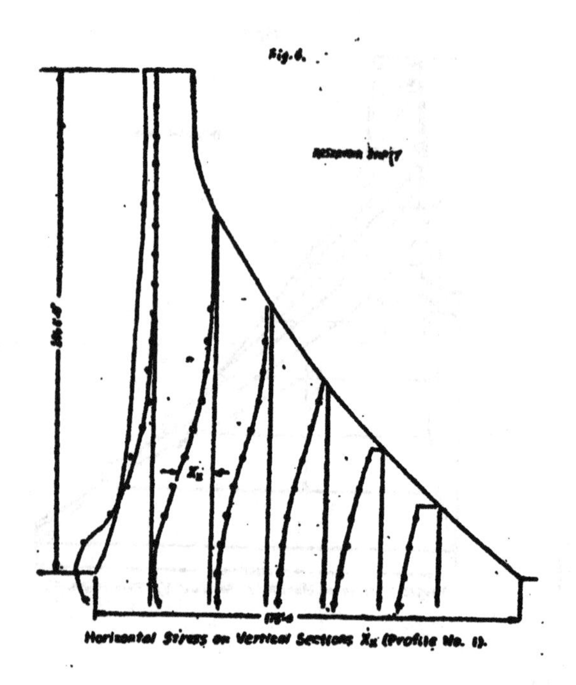
Fig. 6.

Horizontal Stress on Vertical Sections X_x (Profile No. 1).

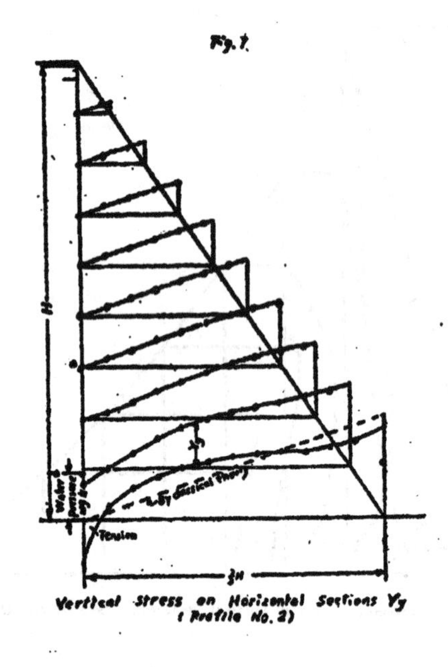
Fig. 7.

Vertical Stress on Horizontal Sections Y_y (Profile No. 2)

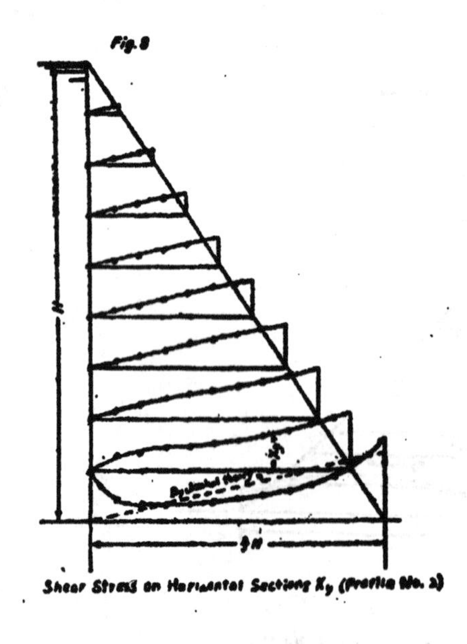
Fig. 8

Shear Stress on Horizontal Sections X_y (Profile No. 2)

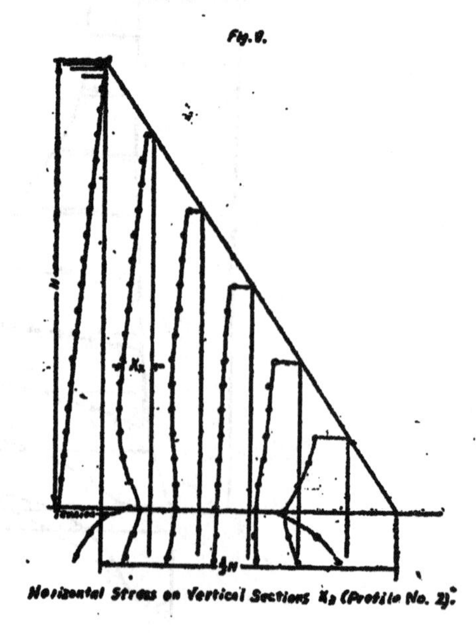
Fig. 9.

Horizontal Stress on Vertical Sections X_x (Profile No. 2).

Fig. 10

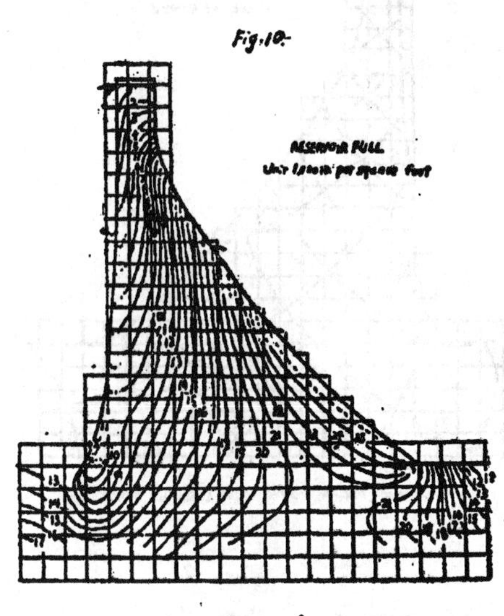

Lines of Equal Major Principal Stress (Profile No. 1)

Fig. 11

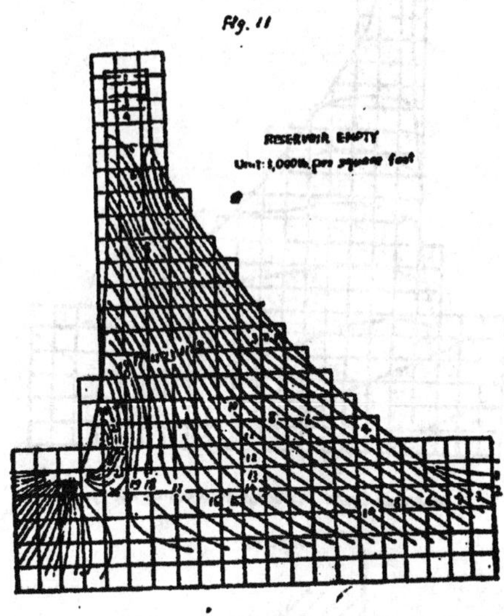

Lines of Equal Major Principal Stress (Profile No. 1).

Fig. 12

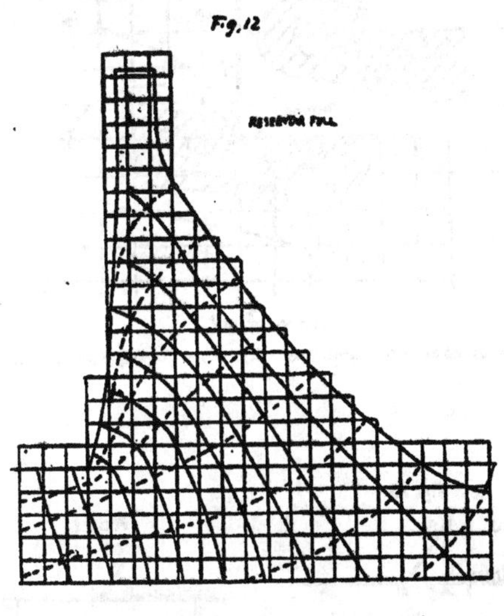

Principal Stress Trajectories (Profile No. 1)

Fig. 13

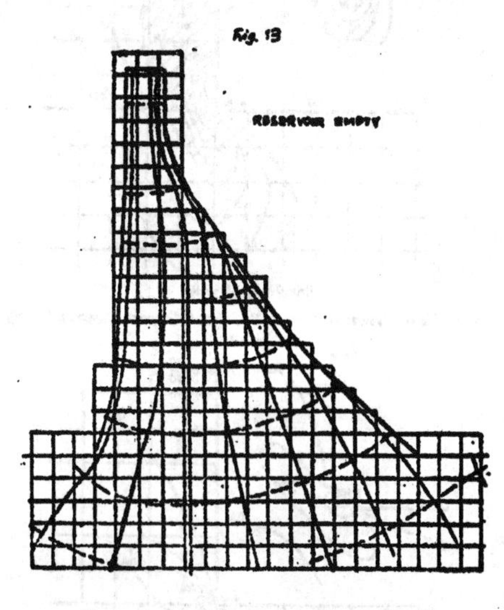

PRINCIPAL STRESS TRAJECTORIES (PROFILE NO. 1).

Fig. 14.

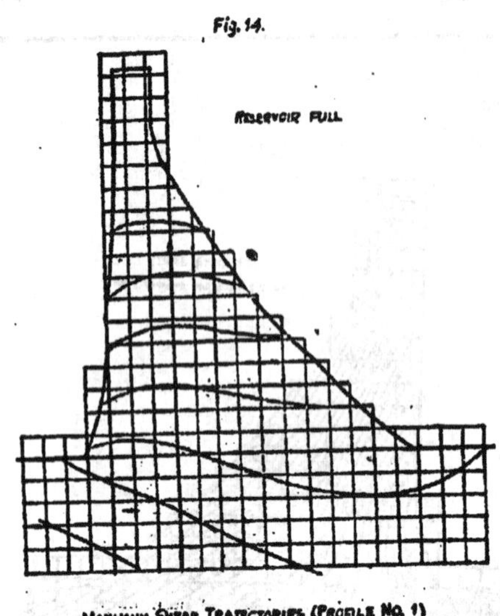

Maximum Shear Trajectories (Profile No. 1)

Fig. 15.

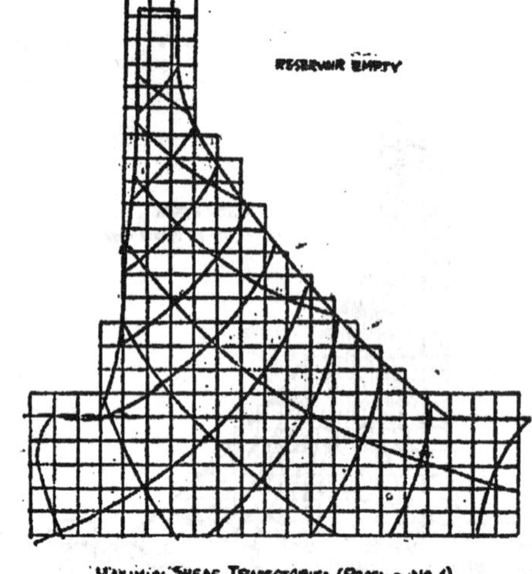

Maximum Shear Trajectories (Profile No. 1).

Figs. 16

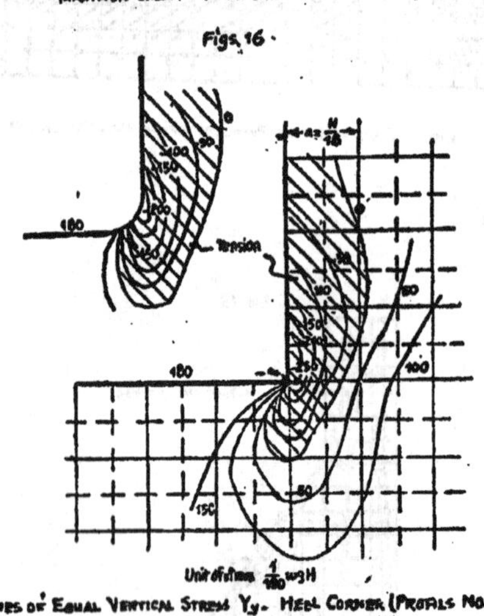

Lines of Equal Vertical Stress Y_y. Heel Corner (Profile No. 2)

Figs. 18.

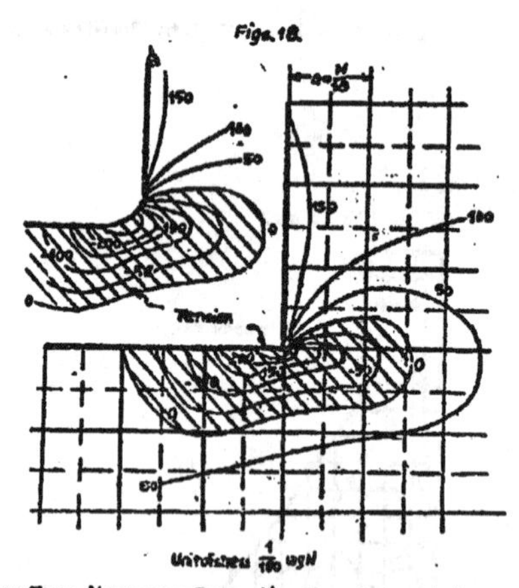

Lines of Equal Horizontal Stress X_x. Heel Corner (Profile No. 2)

Figs. 17.

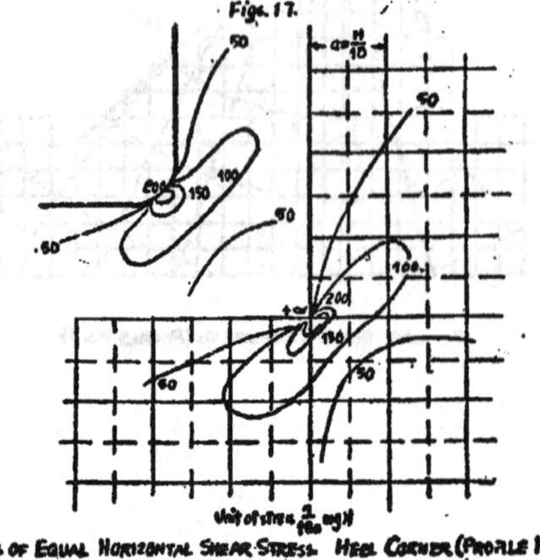

Lines of Equal Horizontal Shear Stress. Heel Corner (Profile No. 2)

Figs. 20.   Figs. 21

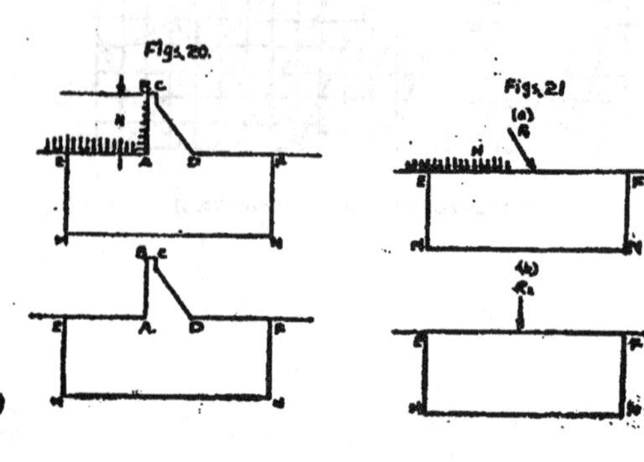

當儲水池空載時之應力.

(1)垂直應力：Yy——垂直應力在水平面上多爲縱長，雖近壩基底邊之處亦然。此可用如下之理由來解釋因此垂直應力常有使其應力相等於該點上所用材料重量之趨勢，而剪力則純爲一再分佈之媒質者(Re-distributing agency)．

(2)剪力：x_y——在水平面部份上所得應力之表示又爲一拋物綫形之分佈

(3)水平應力：x_x——所有在垂直部份之水平應力皆爲少量之壓力

(4)主要應力，p_1p_2及最大剪力.

(a)最大主要應力之產生常在上游之踭角.

(b)無張力性之主要應力產生

(c)最大剪力之值常超越水平剪力之值

(d)最大剪力之放射體軌道多爲斜線而與垂直成45°角。然此又爲最可能當儲水池空載時所能使壩崩潰主要因素之一.

下列 Table 1 所表示者乃根據古典及彈性原理而將各種應力加以比較。假設壩角多爲圓滑以避免無限大之應力，但仍可發生於壩角之應力極大值乃從普通應力分佈曲綫而得之云.

Table I

當儲水池滿載時

	古典原理	彈性原理
在壩基部份之最大垂直應力	18000磅/平方呎在壩趾尖	16700磅/平方呎在壩底邊之中部
在壩基部份之最少垂直應力	8000磅/平方呎在壩之踭背	2000磅/平方呎在壩之踭背
在壩基部份之最大主要應力	41200磅/平方呎在堤趾尖	26500/磅平方呎在壩之趾尖

當儲水池空載時

在壩基部份之最大垂直應力	22000磅/平方呎在壩之踭背	20000磅/平方呎在壩之踭背
在壩基部份之最少垂直應力	1800磅/平方呎在壩之趾尖	1800磅/平方呎在壩之尖趾

結　論

由上述所得之結果，關於設計採用古典方法之安全與否，其總論所得如下：

如上言，根據彈性原理，當儲水池滿載時，在壩中所產生最大之彈性應力往往不能超過由古典原理所得來之值。此壓力性與剪力性之應力常產生於壩之下游面。在此面中，主要應力中有一爲零，其材料恰爲應用於單軸壓力（Uni-Compression）之上，如普通之試驗焉。又因此限制應力常被採用於普通壓力試驗之情形下，故不須要考慮採用任何材料所能引起之失敗因素。此說早已在 1933 年前當 Kelen 氏根據古典原理，設計一三角形壩并應用 Mohr's 氏之失敗因素探討之理論所得結果中，加以如上之判斷及肯定。

普通在忽略總共剪力之阻力當中，最少剪力失敗之安全係數往往產生於受高度壓力之点上。更且，吾等在設計隄壩時，壩底綫之阻力及摩擦力亦應詳加討論及計算，因混凝土與石之聯結力甚不完滿云。

又其最大困難則係如何能使壩內絕無張力。

根據最近分析所得之結果則謂無論在任何情形下，當儲水池滿載時，張力將可能在壩之上游面基底下產生，此雖能引起壩基崩裂，但此亦不能影响壩之穩定性。其最危險之地帶乃在壩之脊背，因爲 Middle Third 之規定往往不能免却張力之計算，而此又爲特別適合於計算之角形之例面圖者。雖然將壩之轉角處加以圓滑之施之，但此祇能避免無限大之應力，倘欲全無張力時，則計算方面須由古典原理分析而加以一相當壓力。同理，當上浮力存在時，吾等亦得由古典原理方法分析而加上一相當壓力以平衡之。除此之外，別無他法。且除上述之地帶外，其他壩之任何部分均無張力產生。

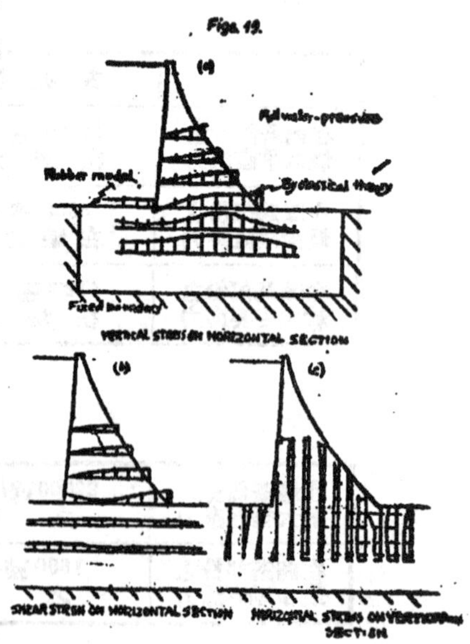

因古典及彈性原理互爲調洽，尤以能適用於設計壩之上部，故普通設計多如上法。

而古典原理已有適當之方法而解決壩內所生之各種應力，尤以當儲水池爲空虛時。又當向上力綫能在（Midde Third）之內時，則張力更難產生而普通採用壓力性應力多較計算出者爲大，但其量亦祇爲甚少甚少而已。

在第一側面圖應力分佈所得之結果正如 Wilson 及 Gore 以同等種類之隄做實驗所得之結果相同。但此非証明彈性原理可能適用於眞實之隄壩上，惟其祇足以爲例証說明此文所述之分析方法爲一可靠之計算重力壩應力方法。又此方法之根據純粹以材料之彈性效能爲依歸云。

附　錄　1

應用在解答問題方法之概說

問題之叙述：吾等須決定在邊界綫EABCDF (Fig 20) 下之半極限地域上每点之之種應力分力。圖表 20 (a)表示出當儲水池滿載時邊界荷重之種類圖表 20(b)表示出在儲水池空載時。

設將壩沿 AD 截開兩部，而壩基則採用圖表 21(a)及(b)所表示出之荷重系統。然後，由 Saint-Venant 之理論，在長距離之應力相等圖表 21(a)及(b)所指示者。倘吾等使沿 EMNF 綫上之應力相等於圖表 20 及 21者，則問題可被簡化。

假設 EA=DF=EM=AB，其在壩本身內應力之　誤少過百分之一．

應力功用：一設 x與y 爲某一點之 Orthogoral 坐標，Xx，Yy及 Xy 則爲該點之三應力分力，圖表 22 指示彼等皆爲正方向(Positive directions)

設A爲 x 與 y 之原點(圖表 20)，x量向右，而y向上應力可寫爲

$$X_x = \frac{\delta^2\phi}{\delta y^2} \quad \cdots\cdots\cdots\cdots\cdots\cdots\cdots\cdots\cdots (1)$$

$$Y_y = \frac{\delta^2\phi}{\delta^2 x} wpgy \quad \cdots\cdots\cdots\cdots\cdots\cdots\cdots\cdots\cdots (3)$$

$$X_y = -\frac{\partial^2\phi}{\delta x \delta y} \cdots\cdots\cdots\cdots\cdots\cdots\cdots\cdots\cdots (3)$$

w　　表示水之比重

g　　表示由地心吸力所產生之加速度

p　　表示材料之比重

ϕ　　Ariy 應力功用，此仍用於平面應變及平面應力上問題甚廣．

爲求適應應變方程式之合理，ϕ 必需適合於下列之方程式

$$\frac{\delta^4\phi}{\delta x^4} + 2\frac{\delta^4\phi}{\delta x^2 \delta y^2} + \frac{\delta^4\phi}{\delta y^4} \equiv \triangle^4\phi = 0$$

倘 ϕ 之功能能適應於 EABCDFMN 邊界綫上之應力情形，同時，(4)方程式能在該邊界綫之限制內，則每點之應力可能由(1)至(4)方程式中求出

邊界值：——倘欲求 ϕ 之值以便利上述之計算時，則需先行從應力學識方面求得邊界綫上每點之值此爲一計算數字之必要步驟．而使(4)方程式爲唯一解答之方程式

沿 AEMNFD，應力功能之值可由普通公式根據圖表 21 之荷重而求得之。此皆因係在圖表 23(a) (b) 及 (c)中所表示之三種荷重所致：

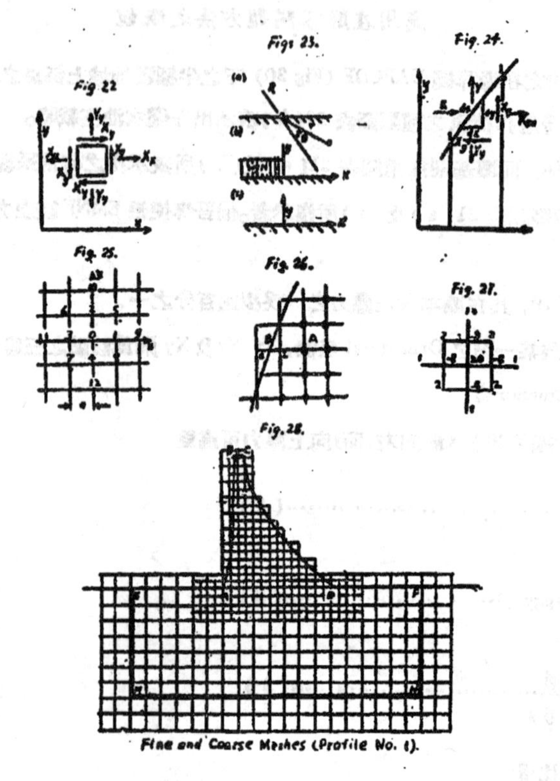

Fine and Coarse Meshes (Profile No. 1).

圖表 23(a)：——傾斜集中之荷重綫R

$$\phi' = -\frac{R}{\pi} r\theta \sin^2\theta \cdots\cdots\cdots\cdots (5)$$

r 及 a 均爲極限標座，由荷重之作用綫量出來者

圖表 23(6)：——均佈荷重之壓力 WgH，由A點至無限大者

$$\phi'' = -\frac{wgH}{2\pi}\left\{ (x^2+y^2) + \tan^{-1}\left(-\frac{y}{x}\right) + xy \right\} \cdots\cdots\cdots\cdots (6)$$

圖表 23(c)：一壩基材料之重量

$$\phi''' = -\frac{\sigma}{1-\sigma} pwg\frac{y^3}{6} \cdots\cdots\cdots\cdots (7)$$

σ係 Poisson's 比率

因方程式(1)至(4)均爲直綫縱長(Linean)者，故沿AEMNFD應力功能之總值取決於 ϕ'，ϕ''及ϕ'''之和.

當儲水池空蕪時，$\phi''=0$；在ϕ'中之爲垂直荷重R現倘有ϕ須待取決者

設X與Y代表邊界綫每單位長度之邊界分力，同時，又設一被dx，dy，及ds所環繞之表面之平衡狀態.

$$Xds+X_x .dy-X_y .dx=0 \quad (8)$$

$$Yds-Y_y .dx+X_y .dy=0 \quad (9)$$

代入(1)至(3)

$$X.ds+\frac{\partial^2\phi}{\partial y^2}dy+\frac{\partial^2\phi}{\partial x\partial y}dy=0$$

$$Y.ds-\left(\frac{\partial^2\phi}{\partial x^2}+wpgy\right)dx-\frac{\partial^2\phi}{\partial x\partial y}dy=0$$

因爲
$$\frac{\partial^2\phi}{\partial y^2}d_y+\frac{\partial^2\phi}{\partial x\partial y}dx\equiv d\left(\frac{\partial\phi}{\partial y}\right);$$

及
$$\frac{\partial^2\phi}{\partial x^2}d_x+\frac{\partial^2\phi}{\partial x\partial y}dy=d\left(\frac{\partial\phi}{\partial x}\right)$$

所以
$$Xds=-d\left(\frac{\partial\phi}{\partial y}\right) \quad (10)$$

$$Yds=d\left(\frac{\partial\phi}{\partial x}\right)+wpgydx \quad (11)$$

由I点至II点之積數
$$\left[\frac{\partial\phi}{\partial y}\right]_I^{II}-=\int_I^{II}Xds \quad (12)$$

$$\left[\frac{\partial\phi}{\partial x}\right]_I^{II}=\int_I^{II}Yds-\int_I^{II}wpgydx \quad (13)$$

由此，沿邊界上每点$\frac{\partial\phi}{\partial x}$及$\frac{\partial\phi}{\partial y}$之值皆可因其已知於A而決定。在(12)方程式中，右方純爲代表沿邊界上由I點至II點之總水平水，而(13)方程式中，右方則代表因上述在邊界點上所產之應力及在此邊界下部材料之重量而產生之向上力之總值.

倘欲決定ϕ由I點至II點之變化，首先應計算M一時鐘向之邊界力之總力距

由圖表24.

$$dM=X(y-y_{II})ds-Y.ds(x-x_{II})$$

或由(10)與(11)方程式

25745

$$dM = -d\left(\frac{\partial\phi}{\partial y}\right)(y - y_{II}) - d\left(\frac{\partial\phi}{\partial x}\right)\cdot(x - x_{II}) - wpgy(x - x_{II})\cdot dx \quad \ldots\ldots(14)$$

用 I 至 II 點之限度將(14)方程式之一部份積分所得

$$\Big[M\Big]_I^{II} = \Big[\frac{\partial\phi}{\partial y}(y_{II} - y)\Big]_I^{II} + \int_I^{II}\frac{\partial\phi}{\partial y}dy + \Big[\frac{\partial\phi}{\partial x}(x_{II} - x)\Big]_I^{II} + \int_I^{II}\frac{\partial\phi}{\partial x}dx$$

$$+ \int_I^{II} wpgy(x_I - x)dx$$

試因 $\int_I^{II}\frac{\partial\phi}{\partial x}dx + \int_I^{II}\frac{\partial\phi}{\partial y}dy = \int_I^{II}d\phi = \Big[\phi\Big]_I^{II}$

$$\Big[M\Big]_I^{II} = \left(\frac{\partial\phi}{\partial y}\right)(y_I - y_{II}) + \left(\frac{\partial\phi}{\partial x}\right)(x_I - x_{II})\left(+\Big[\phi\Big]_I^{II} + \int_I^{II} wpgy(x_{II} - x)dx\right.$$

可再調整爲

$$\Big[\phi\Big]_I^{II} = \left(\frac{\partial\phi}{\partial y_I}\right)(y_{II} - y_I) + \left(\frac{\partial\phi}{\partial x_I}\right)(x_{II} - x_I) + \Big[M\Big]_I^{II} + \int_I^{II} wpgy(x - x_{II})dx \quad \ldots\ldots(15)$$

因此，ϕ 之變化及沿邊界ABCD上之各值均可決定

因 $\partial\phi$ 及 $\frac{\partial\phi}{\partial y}$ 之值經已求得也.

同時，在(15)方程式之後項，乃代表因邊界各力作用於 I 點至 II 點及在邊界下部之材料重量所產生作用於 II 點之時鐘向力距

如邊界由 I 點至 II 點爲一直綫則 (12)(13)(15) 方程式便因之而簡單化，倘邊之曲綫橫部劃分爲一連串之短直綫，則 ϕ 在每點之值可用漸次逐步方法 (Step by Step method) 而求得之，而此等值已於圖表 20 在A點及 D 點表示出來，因此，如須作一覆核，則可由A點起，求沿邊界綫上之各值，覩其與否符合在D點之值.

最後差別計算法：—— 旣已知沿邊界綫上各點之值。則可着手求解 (4) 方程式，如圖表 25，先作一平方形之網，然後計算在該網上所註之各點，以O點爲原點，則 ϕ 可伸展爲一相重泰萊連串數(double Toylors Series)：

$$\phi = \phi_0 + A_{1,0}x + A_{0,1}y + A_{2,0}x^2 + A_{1,1}xy + A_{0,2}y^2 + A_{3,0}x^3 + A_{2,1}x^2y + A_{1,2}xy^2 + A_{0,3}y^3 + A_{4,0}x^4$$
$$+ A_{3,1}x^3y + A_{2,2}x^2y + A_{1,3}xy^3 + A_{0,4}y^4 + A_{5,0}x^5 + A_{4,5}x^4y + A_{3,2}x^3y^2 + A_{2,3}x^2y^3 + A_{1,4}xy^4$$
$$+ A_{0,5}y^5 \quad \ldots\ldots(16)$$

An'm 之繫數及代表 $\frac{1}{nm}\frac{\partial^{n+m}}{\partial x^n y^m}\phi$

25746

不理X與Y之第六項及其他高次項，而代入以適當之坐標則可得：—　　　　a 係網之長度

$$\phi_1+\phi_2+\phi_3+\phi_4-4\phi_0=2a^2(A_{2,0}+A_{0,2})+2a^4(A_{4,0}+A_{0,4}) \quad \text{(17)}$$

$$\phi_9+\phi_{10}+\phi_{11}+\phi_{12}-4\phi_0=8a^2(A_{2,0}+A_{0,2})+32a^4(A_{4,0}+A_{0,4}) \quad \text{(18)}$$

$$\phi_5+\phi_6+\phi_7-4\phi_0=4a^2(A_{3,0}+A_{0,2})+4a^4(A_{4,0}+A_{0,4})+4A_{2,2}a^4 \quad \text{(19)}$$

（4）方程式可化爲

$$\Delta^4\phi=\frac{\partial^4\phi}{\partial x^4}+2\frac{\partial^4\phi}{\partial x^2\partial y^2}+\frac{\partial^4\phi}{\partial y^4}=24(A_{4,0}+A_{0,4})+8A_{2,2} \quad \text{(20)}$$

採用方程式(17)(18)與(19)

$$\Delta^4\phi=a^4\left\{20\phi_0+2(\phi_5+\phi_6+\phi_7+\phi_8)+(\phi_9+\phi_{10}+\phi_{11}+\phi_{12})-8(\phi_1+\phi_2+\phi_3+\phi_4)\right\} \quad \text{(21)}$$

此乃對方程式(15)之最後差別計算。如欲得一滿意之結果，則可設a網長爲一較少之值；而錯誤祇在a^6之次項而已。同理，網上各點之方程式亦可同樣寫出及求出。

但由 Professor R. V. Southwell 所發明之 method of Relaxation 則計算此等方程式更形便捷。

Method of Relaxation：—

在聽度中選出ϕ之數值若干個，而方程式(21)可化爲

$$20\phi_0+2(\phi_5+\phi_6+\phi_7+\phi_8)+(\phi_9+\phi_{10}+\phi_{11}+\phi_{12})-8(\phi_1+\phi_2+\phi_3+\phi_4)=F \quad \text{(22)}$$

而F_0與0不同，F_0被稱爲在0點之餘數，此餘數對網上各點之值計值甚爲容易，倘ϕ更改而減一，則在0點之餘數更改而減20，則在圖表27所示環繞各點之餘數均更改而減一少數量，此數量普通被稱爲 Relaxaliton Pattern. 倘此更改餘辦法將繼續依次進行，則結果數值將漸趨準確，有時，數點得可同時更改，以增解答之效能進行方法：

（a）先劃一網形包含所求之地帶。(圖表28)

（b）沿ABCD邊界綫上與網相交各點之值可用積分方法而求得之．假設邊界爲一直綫與各點相連。則R之方向，數量，及用點皆可求得，同時，ϕ中A點至D點之值亦可計算。

（C）地網點沿AE及DF之ϕ' ϕ''及ϕ'''之值均可計算，同時，沿EM，MN，及NF之網點亦可計算以決定ϕ之值．

（d）花邊界點外之點，與其在內之點，有方程式(26)以連貫之．

（e）假設在邊界綫內ϕ之初值，而以 Relaxation Method 計算之，則可得一近似值．

（f）將（d）與（e）之方法繼續進行，直至各點之餘數均可忽畧爲止．

（g）在幼細網上各點之應力均可採用最後差別計算法以方程式(1)，(2)及(3)而計算之

$$X_x=\frac{\partial^2\phi}{\partial y^2}=\frac{(\phi_2+\phi_4-2\phi_0)}{a^2} \quad \text{(23)}$$

$$Y_y=\frac{\delta^2\phi}{\delta x^2}+wpgy=\frac{(\phi_1+\phi_3-2\phi_0)}{a}+wpgy \quad \cdots\cdots(24)$$

$$X_y=-\frac{\delta^2\phi}{\delta x\delta y}=-\frac{(\phi_5+\phi_7-\phi_6-\phi_8)}{4a^2} \quad \cdots\cdots(25)$$

（b）從此上求得之應力分力，主要應力及其方向可於如下普通之公式求出

$$P_{1,2}=\frac{X_x+Y_y}{2}+\sqrt{\frac{(X_x-Y_y)^2}{4}-X_y^2} \quad \cdots\cdots(26)$$

$$S_{max}=\frac{P_{1-2}}{2} \quad \cdots\cdots(27)$$

$$\tan 2a=\frac{2X_y}{X_x-Y_y} \quad \cdots\cdots(28)$$

α 爲一主要平面傾向軸所成之角.

在解答中之差誤——對於解答應有之近似計算爲如下：

(1)距離隄壩一穩定距離，及沿 EMNF 綫上之應力皆已肯定。

(2)最後差別之方程式非係完全解答，故常有小部餘數存在。

(3)最後差別不能準確代表微分方程式

此差誤之根源可因採用幼細之網形及距離隄壩極遠之 EMNF 而減少。

在此上所述之解答，除去失銳急切之隄壩彎角所引起之差誤不計外，其最大可能之差誤率爲滿載水壓力之百份六，其可能差誤（Probable Error）則又爲此數目之一半。

『交通工程 (Traffic Engineeing) 應用於鄉村公路之重要』

By Charles E. Conover, M. Am. Soc. C.E.

——何思源譯——

普通在公路工程上,如欲堆砌一路面及路基使其適合於接駁一曲線彎道,而其半徑又在五百或六百英尺以下者,根據此文作者,則謂此殊不可能。而在路面接近彎道之右方,紅色內耀之標誌往往對於小心之車輛駕駛者有絕大之幫助,反之,站在公路工程者立場上來說,在普通較爲急轉之彎道上,特別刺眼之標誌常常能使胆大之駕駛者觸目驚心而警惕危險,不致因疾駛而肇禍,凡此皆爲公路設計及建築之工程者應必加以考慮之條件也。

轉接曲線 (Transition Curves) 雖然在普通公路工程上仍未見大量採用,然此對於彎道曲線其半徑在千五英呎以下者,則甚有利。因據作者之見解,則謂普通欲得一工程上理想之彎道,甚爲困難,尤以在不甚繁榮之地域,山嶺地帶或鄉村公路等,則更爲難得。

在有彎道之公路路面,其堆砌寬度應比原來與彎道相切之切線爲大,但據筆者週遊『密西西比河』(Missisippi River) 東部經歷所見,則此『加寬』實爲不甚重要而可謂之多餘。

筆者又謂,普通車輛失事之地點,十份之九之機會可能發生於彎道上,因往往在坡度6%之雙行車道之公路上,笨重緩慢之貨車多小心駕駛,速度較低,故甚安全。反之,燥暴胆粗之駕駛者,常因好勝心理而欲使其輕快之車輛越頭,因之而妄顧安全在彎道上將行車速度加速,意外失事遂因此而生。據筆者之見解,則謂此種失事之補救方法,非如上述將路面加寬便能澈消此種危險或減低此失事機會之可能性,據其經驗所得,則謂如能將笨重緩慢貨車之速度加强,使之與普通輕快客車之行車速度相等,使車輛依次行駛,不能超越別車,此未嘗非一治本之補救方法。故換而言之,各種車輛構造之設計;其重要性實比工路設計爲大,此皆站在公路交通安全上之觀點來說,而以盡量減低車輛失事之機會爲原則,故於近年來此政策亦多爲各繁榮都市公路交通設計者所採用云。

筆者又謂,普通車輛失事危險之可能性,常多發生於公路相交處,尤以於車輛作左轉之時爲最多,故在市鎮或人口稠密之都市中,常規定禁示車輛作左轉,而應以右轉代之。但此雖使車輛交通多增累贅及麻煩,而倘爲安全計,則此實爲一合理之規定。然此禁止車輛作左轉之規定祇能適應用於市鎮及都會而已,因在鄉村公路或山嶺地帶之公路上,普通,車輛應有之左轉常因不能繞道故而不能以右轉代之也。

又如有二公路相交,其一爲車輛叢集,交通頻繁,反之,其一爲路靜人稀,交通鮮落者,則往往有標誌之設備,而此標誌之安置,應以平向後者公路爲佳,更且,特別電器之裝置,使每一車輛經過交通鮮落之公路而將近相交點時能將此標誌轉換,而停止交通繁忙之公路來往之車輛,則此又更爲科學化及合理之公路交通上所應有之必須裝置設備云。

報導

嶺南給水系統

劉益信 陸景文 鄺國樑 何逢康 潘世英

假如你是一個從未到過嶺南的客人而第一次和朋友進來遊覽，多數是喜歡從水路來的，因爲水路是比陸路清閒優雅得多，浩浩的珠江，四周的景色，對于一個在城市住得厭的人，簡直是一盆清水，什麼也會洗得一乾二凈的；船行不久，你的朋友便會遠遠地指着灰色的一點告訴你，這就是嶺南了，這灰色的一點，便會吸引着你的視線，使你不厭疲勞地望着，一直等到清楚地看見。這就是嶺南景物中的水塔了。它襯托于綠樹紅牆之上，直指藍天，顯出那未來底英雄；到過嶺南的人，沒有一個不喜歡它的，所以遊客和同學們，總喜歡替它照一個相留爲紀念的，或者和它合拍一個，多少顯出自已一點軒昂；它可算是嶺南景色之一，又是本文所述給水系統的中心了。嶺南的給水系統是頗現代化的，現在讓我在這裡佔些篇幅，作一個簡單的介紹罷！

在碼頭的旁邊，有一條小河，這裡有一個三合土門，這就是嶺南用水的進口了。河水經過短短的三合土管，便到沈澱池（Setting Basin），它是這樣的清澈無波，從珠江走進來的砂石，便慢慢地沈下了；這個池塘，長着嫩綠的水草，活潑底游魚；初到嶺南的人，也許會以爲是點綴風景的魚塘呢！其實，池中的一芥小草，都是這系統中重要的一環，缺一不可的，水草的工作，是供給氧氣，使食水較爲清甜，活潑的魚兒，是用來吸收過剩的氧氣，使不致不利于鐵水管的持久性。在池的末端，有一張鐵絲網，這也不是廢物啊！它是用來阻隔什物的，水由它的小孔穿過，流入一條十八吋直徑的水坭管，然後再流到機房的生水池來（New Water Basin），便由池邊兩架七匹半馬力的低壓離心力水機（Low Pressure Centrifugal Pumb）推到八角噴水池去了。這裡有廿五枝管子，把水噴射到空中，如巫山的烟霞，如黃梅的細雨，使水和空氣接觸的面積加大，那些不潔的有機氣體，便完全發散出空中去，同使使水接受太陽的紫外線而完成初步的消毒。水由空中落到八角池來，由于水位的關係，流到混合池（Mixing Basin），加入五磅的白礬（Aluminum Sulfate）（以每日用三十萬加侖水計算）和些小的石灰，使變成膠狀的氫氧化鋁（Aluminum Hydroxide）便流到凝結池去（Coagulating Basin），凝結池有兩個，每個又分四格，水到這池便慢慢地靜下來，于是氫氧化鋁就和坭污凝結成塊沈到池底，水經過這一步驟便有百分之八十潔凈了。水經沙濾池（Rapid Sand Filter）的進水管，便流到沙濾池來；經過過濾之後，就從出水管流入淸水池（Treated Water Basin）了，和沙濾池連接的水管有五條之多，除了進出兩管外，一條叫洗滌管，一條叫排水管，一條叫廢物管，都是用來清理池底泥污的。淸水池在機房的池底，水到這裡來，經過消毒後，就變成完美

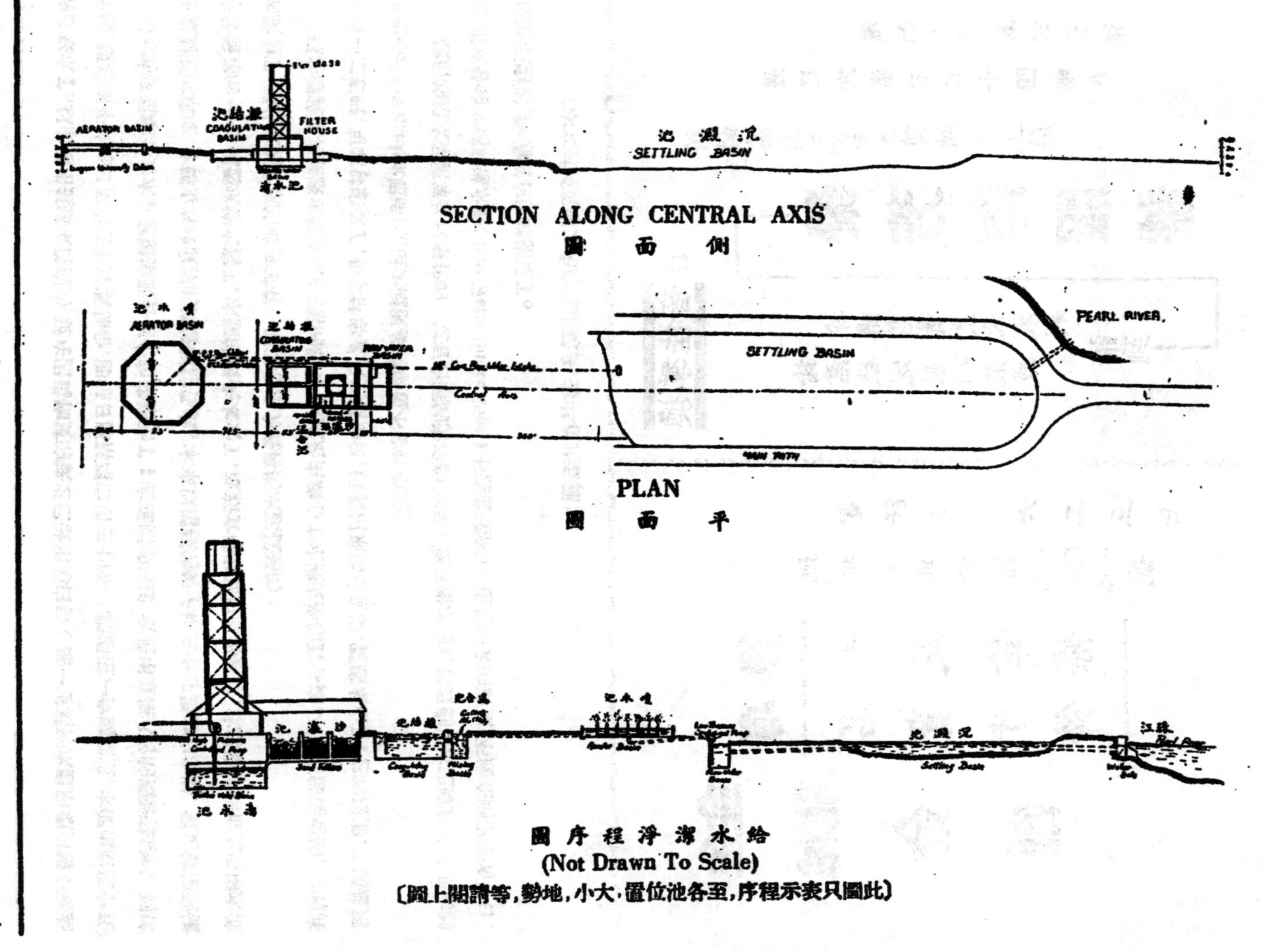

SECTION ALONG CENTRAL AXIS

側面圖

PLAN

平面圖

給水潔淨程序圖

(Not Drawn To Scale)

(此圖只表示程序,至各池位置,大小,地勢,等請閱上圖)

的自來水了。以前是用氯氣來消毒的，現在因爲物質的缺乏已改用漂白粉（每一百萬磅水應用漂白粉十磅至廿磅，現嶺南平均每日用水三十五萬加侖，即每日應落漂白粉三十磅。）經過這一步處理後，水便由兩架十五匹馬力的高壓離心力水機直接推到各宿舍和建築物了；在離開機房起，水是用直徑八吋的鐵管輸送的，到懷士堂對開的草地，才轉由六吋直徑的鐵管轉往各處。本來以這樣的管子和三十匹馬力的水機，實足够推到嶺南最遙遠的一角，但因水管太舊了，水頭的損失太大了，故新女學和新中學等區域，水已幾乎無推上頂屋水池的壓力了，因此在新女學的地窖下有一個離心水力機再把水頭提高的。

現在讓我來再談談水塔罷！在正常的狀況下，水是由離心力水機直接推到各處，不需再經過它的，它的高度爲一二五呎，可算够宏偉了，但它的容量只三萬加侖，但若用來供給全校，實感太小了，它的用處，不過是用來洗滌水池和加大水壓而已。這大概很多同學也不會知道罷！

嶺南的給水系統算是科學化的，它有合理的潔凈和消毒，更有着水壓計和自動用水記錄計等輔助的器械，但水管的太舊和進水管（Raw Water Intake Pipe）不能直達河心，這是較大的缺点；不過在剛復原的今日，而能有這樣的設備實在已經滿足了。

（本文材料黃郁文先生供給殊多，玆特在此致謝）

工程零訊

滾動！滾動！世界向前進步

兩個人的氣力能拖動一在水平軌道上重400噸在停止狀態中的火車龍頭，這是因為在鐵馬車軸上的鋼棍軸承。用了這種軸承，比用老式的平軸承，起動的阻力要減低百分之八十八之多。關于這種減低磨阻力的軸承，新近 Los Angeles 還有一個有趣的報告說有一所房子在地基上裝了這種軸承，地震時竟可向任一方向滑動6吋，藉以避免房子爲地震所撓屈；每一隻裝在這房子底下的軸承，裝配後共重600磅，而所支持之負載竟達250,000磅之巨。——工程界—4/36—

日光取熱爐的威力

在法國 Mendon 天文台中，特洛默博士（Dr Jrobme）等裝置了一具日光取熱爐（solar furnace）這只爐子的心臟是一個拋物鏡；直徑約六呎半，焦距約三十三呎半；這是裝在一個舊的軍用探照座子上的。據說所接收到的熱能有 3 瓩，强度可以達到絕對溫度 5,200 度。即使是幾種稀土金屬，也會被這只爐子熔融過。在絕對溫度 3,500 度時，石墨被它從固體變成了氣體。一片重約一盎司的鐵，在十秒鐘內就被它燒掉，其中一部分變成了鐵的氣體 —— Science Digest ——

舖砌磚牆機

芝加哥哈維公司（E. L. Harvey, Chicago）設計了一部機器，能够在八小時內舖砌 100,000 塊磚頭。這部機器重 200 磅，可以在鐵軌上前後移動。磚頭用皮帶式運送器取起，送經裝在一根向外伸出扛臂頂端的排整機（arranging device）後。于是砌磚器（layes head）便把磚頭撿起砌入膠泥（Mortar）之中。膠泥是經過一組水管用壓力輸送到牆上的。運用這部機器需要十個工人。

三小時45分鐘架成新橋

因爲每天有56列火車來往，Reading 公司的紐約小火車綫要拿一座150噸的新橋去代換那座不適當的 7 噸橋時，必須顧慮到時間的重要性 那座新的 65 呎雙軌下承鈑橋先架建在舊橋旁邊的鷹架上，舊橋的另一邊也搭了鷹架。然後把舊橋用絞盤機舉提起來，在座墊下置墊軌，并把路軌擡高到新橋的高度。

舊橋被滑車起重機舉高32呎，放到還就的鷹架上。起重引力是用二架鋼抓，以壓縮空氣來發動的。然後移轉起重機，把新橋舉到橋座上，再用絞盤把它提高，移去墊軌，置上座墊鈑及鉚釘，然後放下到舖露面的地位。新舊橋的更換只費掉了三小時又五十四分鐘矣。—— Mc Graw-Hill Digest ——

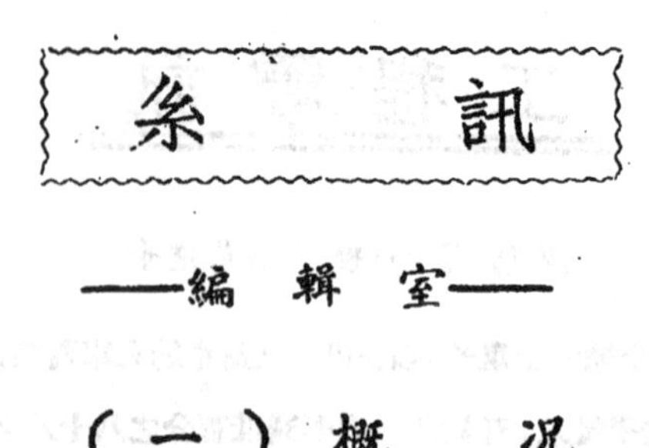

——編輯室——

（一）概況

本系之創立，遠在民國十八年的鐵路部部長孫哲生先生因鑒于華南一帶各大學尚未有工學院設立，工程人才，頗感缺乏，故委托本校開設工學院，以訓練土木工程人才，從事建設工作．是年哲生堂落成，繼之水力試驗室，金木工廠亦次第完成，儀器設備陸續由外國運來，日臻完善，學生人數與年俱增，研究工作亦多進展．第一任院長爲我國工程界先進胡棟朝先生，繼之者爲李權亨先生。至民廿七年與理學院合併而稱爲理工學院，自茲而後，教學與行政之效率益見增高，而各系間課程與研究工作及各方面之聯繫亦愈見密切，但不幸是年秋日軍侵入廣州，學校不得已遷移香港，借香港大學上課，另賃屋宇爲各系實驗室之用，所有設備除由廣州原校運到應用外，在港添置者亦復不小；在此期間內，設備雖不若往時之完善，但教學及研究成績仍斐然可觀。及至民卅年冬香港失陷，學校內遷粵北，以儀器設備一時無法內運，抱寧缺毋濫之旨，乃決將理工科停辦，迨至民卅四年秋，日敵投降，本校重返原址，系內圖書儀器幸未完全洗刼破壞，大部份尚得保存；三年以來，整理殘缺，補充遺失暨陸續添增設備，原有教授亦陸續歸來，而新聘者亦復不小，學生人數亦增至一百廿餘人，系主任一職原由桂銘敬先生担任，後桂先生因付湛江就來湛鐵路局副局長兼湛江港務局長職，暫時向學校請假，系內行政由馮秉銓博士兼理，師資充實，儀器添置不遺餘力，漸次恢復戰前水準矣．

（二）設備

(A) 材料試驗室概況

本系材料試驗室之設立，一方爲增進學生實習之經驗，一方擬利用所設備者爲大規模及精密之材料試驗，以期貢獻于科學界及建築界，自戰前購置各種材料試驗機後，規模略備，試驗成績，尚屬準確，經美國全國材料試驗會承認本會爲會員，凡該會試驗結果概寄本會以資比証，惜以經濟困難，未能極度發展，所有專門人材之聘請機械之添置，各種材料之搜羅及試驗。均待于經濟籌措，方克次第進行。

（一）木材之試驗，本省北江及西江杉木及山樺等木材用于工程者極多，獨惜對于此等木材，每種向無加以力量之研究，故工程界對此甚感困難，考外國出產之鋼鐵木材每種必加以精確之試驗；如每方寸可受力若干。是以本國或外國工程師或商人均以此等材料有精確可靠之受力標準樂于購用，反顧我國對于此等材料無人加以注意，誠爲可惜！本系有見及此，擬將西北兩江木材分別採集試驗，如將來結果良好則對于工程界固多一貢獻，而對于本國木材原料之出口當自激增也。

（二）士敏土之試驗：我國近日從事建設事業，士敏土的出產日增，對于土質雖經試驗，然關于士敏土之力量試驗尚鮮注意，致工程界之選擇竟毫無標準，以故使用時在經濟上及安全上均發生很大問題。因此多採用外貨士敏土而小敢採用本國出品。但倘若能將每種士敏土均加以精細之力量試驗，証明其受力之數目大小，則對于國產品自必增加信仰矣。

（三）土磚之試驗：粵省素產磚，年來從事建設磚廠更應運而生，其出品名目繁多，如青磚，明企紅磚，白沙磚，通心磚，舖路磚，火磚……之類，不勝枚舉；惟商人只知牟利，對于質地方面甚小研究。本系有見及此，以爲每種構造不同則性質亦因之而異，故每種施以受力耐熱吸水種種試驗，藉以決定某種構造適合某種用途，則工程各界當受益不淺也。

（四）築路材料試驗：築路材料以石沙爲最大宗而瀝青等亦爲重要材料。石之種類繁多，沙亦種類不小。普通上一般人對于這類物品性質均很小注意。本系現正設法搜羅各地名產，從事試驗，俾可決定某種材料適合某種路面，使材料經濟兩均用得其宜，庶有裨于建設。本系對此所以尤爲注意者也。

（五）竹之試驗：我國產竹不下千數百種，就本校農學院試驗中者亦有數百種之多，考竹之用途極鉅，舉凡傢俬用具棚廠等等莫不可以竹爲之。近者美國工程家且以竹代三合土內之鋼筋，其用途之廣，未可限量，若再加以研究，不難續有發明；此種實驗實不容漠視也。

（六）鋼鐵之試驗：在南中國土洋鋼鐵向無試驗處所，土產鋼鐵力之試驗固屬闕如，卽外來所謂精良出品，亦無由爲之實地試驗。本系有鑒于此，戰前特在美國運來 Universal Machine 一架，專供鋼鐵拉力壓力彎力之試驗，其他三合土木材、磚：竹……之拉力壓力彎力均可試驗使可澈底清楚材料之力量，而適合經濟條件，一破昔日盲目使用之糊塗矣！

茲將本試驗室備有儀器登記于后：

MATERIAL TESTING LABORATORY ROOM.

(1). **Instrument:**

(A). **Machines:**

(1) Universal Testing Machine.
(2) Vickers Hardness Testing Machine.
(3) Riehle Cement Tester.

(4) Hydraulic Cement Tester.
(5) Soundness.

(B). **Accessories to the Machines :**

(1) Universal Testing Machine.

(i) Steel Accessories.

a. Extensometer.
b. Shear Attachment.
c. Multiplying Extensometer.
d. Cold Bend Apparatus.
e. Toughness Apparatus.
f. Micrometer.

1″ – 2″ .... 2 set.
2″ – 3″ 1
Outside and inside diameter 1

(ii) Concrete.

a. Compressometer.
b. Moulds :

4″ diameter × 8″ cylinder 1
6″ diameter × 12″ cylinder 6
6″ diameter × 48″ cylinder 1
6″ × 6″ × 12″ cube 1
6″ × 6″ × 48″ cube 1

(2) Vickers Hardness Testing Machine.

a. Microscope.
b. Diameter Indenter.
c. Steel Ball.

1 mm. diameter 1
2 mm. diameter 1

d. Table of Hardness .... 1 Copy.

(3) Riehle Cement Tester.

a. Briquettes (1″ sq. cross-section) 8
b. Drop Shot ....12 lb.
c. Drop Shot Reciever 1
d. Steel Table 4′ × 6′ 1

(4) Hydraulic Cement Tester.

a. Moulds:

2″ diameter × 4″ cylinder 2

2″ × 2″ × 2″ cube 5

b. Steel Table 4′ × 8′ 1

(5) Soundness.

a. Steam Chamber 1

b. Vicat Needle 2

c. Gilmore Needle 1

d. Sieves with cover and pan No. 200 to 1.05″ series...20

e. Le Chatelier Specific Gravity Bottle 2

f. Thermometers

– 10°—105° C. 1

– 10°—100° C. 1

– 15°—150° C. 1

– 5°—200° C. 1

– 5°—360° C. 1

g. Cylinders

200 cc 2

500 cc 1

50 cc 1

h. Graduated Flask 500 cc 1

i. Funnels........................ 4

j. Glass Plates

1/4″ thickness × 2′ × 2′ 1

1/4″ thickness × 4″ × 8″ 10

1/4″ thickness × 8″ × 14′ 1

k. Glass Rod........................ 1

l. Distilling Flask 2

m. Balances with Weigths........................ 3

n. Trowels 2

o. Electrical Heating Boxes........................ 2

p. Electrical Heater........................ 2

q. Slump Tester 2

r. Pans 5

(C). The Uses of Machines:

(1) The uses of Universal Testing Machine.

A 50,000 lb. machine which with the appropriate accessories could perform the main tests on most of the engineering materials.

The main tests that could be performed are:

(i) Steel and Metal.

a. Compression.
b. Tension.
c. Shear.
d. Bending.
e. Cold Bend.
f. Toughness.

(ii) Concrete.

a. Compression.
b. Transverse Bending.

(iii) Timber.

a. Compression.
b. Tension.
c. Horizontal Shear.
d. Bearihg.
Transverse.
Direct.

(iv) Brick.

a. Compression.
b. Transverse Bending.

(2) The Uses of Vickers Hardness Testing Machine.

The machine is generally constructed for the hardness of various materials but can be made to test the hardness of gear teeth.

The hardness number of any material is measured by the size of the indentation made by a diamond or steel ball.

By use of a microscope, these indentation could be as small as possible eliminating the waste of test specimens.

(3) The Use of Riehle Cement Tester.

This machine is constructed to test the tensile strength of cement briquettes.

By use of small size briquette whose cross sectional area is only 1″ square, the tester is compact so that specimens could be easily made and tested.

The assumption is that the cohessive or adhesive strength of cement is a measure of the value of the cement.

(4) The Use of Hydraulic Cement Tester.

The machine is a hydraulic operated press whereby cement mortar cubes and cylinders could be tested by compression to faiure.

The machine, the cement tester (tension), is a compact machine for the testing of cement mortar usually being a better evaluation of the quality of the cement.

(5) Soundness.

This steam chamber is for the testing of cement under steam conditions to simulate action of cement that might occur in later life.

Un-soundness is indicated by the cracking and warping of a cement pat placed in to steam for five hours.

(D). The Uses of Accessories to the Universal Machine.

(1) The Universal Machine.

(i) Steel Accessories.

a. Extensometer.

For use in the measurement of determation of a steel bar in tension.

b. Shear Attachment.

For use in double shear of various metal rods.

c. Multiplying extensometer.

For measuring the deflection of a beam.

d. Cold Bend Apparatus.

For the test of bending of various steel bars to check on ductility.

e. Toughness Apparatus.

For the purpose of measuring toughness of metal plates.

(2) The Accessories for Testing of Cement.

(i) Vicat Needle.

For testing of the normal consistency of cement and to determine its time of set.

(ii) Gilmore Needle.

For testing the time of set of cement.

(iii) Sieves.

Used for the mechanical analysis of sand, cement and aggregates to determine their suitability for use as a concrete aggregate.

(2). **Rooms:**

(A). **Meterials Testing Laboratory.**

(B). **Equipment Room.**

(C). **Moist Room.**

(1) Curing Room.
(2) Trough.
(3) Specimens Racks.
(4) Thermometer.

(B) 測量試驗室概況

（一）測量爲工程計劃之基礎，用途繁要，在工程上所佔之重要地位，自無容多敍，測量之學習須以方法與理論爲基礎，而以技術爲依歸；就以工程建設而言，缺乏理論知識即難獲進步，若無技術，必失之空虛。是故本系測量一科，對其有關之知識與經驗均極注重，俾於縝密研究之中，收融會貫通之効。

（二）本科包括各種測量儀器之施用校正及原理；測量之方法，結果之計算及製圖原理；小三角網觀測，精密水準測量，地形測量之重要，選讀測量者每週須作兩下午之實習，二三年級並得利用暑期假日至工地實地觀摩，實地操作，務使在技術上更多體驗，更多增益。

（三）本系測量儀器設備戰前尚稱完備，中日戰事發生後，零星設備散失甚多，復員至今測量儀器力求恢復舊觀，並事擴充改進，惟因限於經濟，雖已擬就各節充實改進計劃，而全部之實施，則尚有待於愛護本系者之指示與協助焉，目前測量儀器粗告不缺，茲臚列如下：

Name：名稱	Quantity：數量	No. 編號	Manufacturer 製造廠
經緯儀及三脚架	1 副	(001)T—01	
經緯儀及三脚架	,, ,,	(002)T—02	Watts & Son-London
經緯儀及三脚架	,, ,,	(003)T—03	Tamaya-Japan
經緯儀及三脚架	,, ,,	(004)T—04	K. & E.
經緯儀及三脚架	,, ,,	(005)T—05	Engene Dietzgen Co.
經緯儀及三脚架	,, ,,	(006)T—06	Stanley, London
定鏡水準儀	1 副	(008)L—02	Breithaubt & Sohn
精密水準儀	,, ,,	(009)L—03	K. & E.
定鏡水準儀	,, ,,	(010)L—04	Engene Dietzgen Co.
定鏡水準儀	,, ,,	(011)L—05	
活鏡水準儀	,, ,,	(012)L—00	Germany
手提水準儀	5 副	(013)L—06至(017)L—10	
定鏡水準儀(新到 1947)	1 副	(152)L—11	K. & E.
六分儀	1 副	(018)S—01	
六分儀	,, ,,	(020)S—03	New Eltham, London
六分儀	,, ,,	(021)S—04	
平板儀連三脚架及繪圖板	2 副	(022)P—1至(023)P—2	Germany
平板儀連三脚架及繪圖板	,, ,,	(024)P—3至(025)P—4	Dietzgen Co.
導線平板儀連三脚架·照準儀·繪圖板·羅盤儀及垂錘	9 副	(026)P-t-1至(034)P-t-9	China
照準儀	5 副	(035)P-t-10至(039)P-t-14	
平板羅盤儀	1 副	(040)P-t-15	
Tubular 羅盤儀	1 副	(041)C—01	Cail Zeiss
測量羅盤儀及 iron rod Stand (A.B.)	2 副	(042)C-02至(043)C-03	Dietzgen Co.
手提羅盤儀及 iron rod Stand	1 副	(044)C—04	
流速儀及 Steel rod	2 副	(045)C.M.-01至(046)C.M.-02	A. Ott Kempten Babaria

Name; 名稱	Quantity; 數量	No. 編號	Manufacturer 製造廠
彈簧秤	2 副	(047)M-01 至 (048)M-02	
Pocket balance	,, ,,	(049)M-03 至 (050)M-04	Germany
Rod Level	4 副	(051)M-05 至 (054)M-08	K. & E.
放大鏡	5 副	(055)M—09	
Binoculus (8 × 30)	1 副	(056)M—10	
測量氣压計	2 ,,	(057)M-11 至 (058)M-12	K. & E.
Passometer	2 ,,	(059)M-13 至 (060)M-14	
面積儀	3 副	(062)M-16 至 (064)M-18	Dietzgen Co.
Higgin's 繪圖墨水	12 副	(065)M—19	
平行尺	1 ,,	(066)M—20	Ref. to Zeiss Level
Folding 平行尺	3 副	(067)M-21 至 (069)M-23	Dietzgen Co
T'parent Celluloid railroad Curve	1 ,,	(070)M—24	,, ,,
面積儀		(071) M—25—(072)M—26	,, ,,
工程計算尺	2 副	(073)M-27 至 (074)M-28	Engg. Slide Rule Co
Surveyor's duplex slide rule	1 ,,	(075)M—29	K. & E.
Phillip's 計算尺	3 副	(076)M-30 至 (078)M-32	Dietzgen Co.
Gern union Protractor	1 ,,	(079)M—33	,, ,,
Wricolettening pen	5 副	(080)M—34	,, ,,
Soft ink erasers	3 ,,	(081)M—35	,, ,,
繪圖儀器	1 ,,	(082)M—36	Reisszeng, Germany
Flat archi!ect's scale	15 副	(083)M 37 至 (084)M-38	Dietzgen Co.
2 Straight edge rule	1 ,,	(085)M—39	Universal drafting Machine Co. U.S.A.
Straight edge rule	2 ,,	(086)M-40 至 (087)M-41	,, ,, ,, ,, ,,
Lettering pen	1 ,,	(088)M—42	Dietzgen Co.
Tape Mending outfit	1 ,,	(089)M—43	K. & E.
三稜尺	11 副	(090)M—44	Dietzgen Co.
白邊三稜尺	3 副	(091)M—44	,, ,,

Name: 名稱		Quantity: 數量	No. 編號	Manufacturer 製造廠
Hook Gauge		3 „	(092)M—45	
Universal drafting Machine		1 „	(093)M—46	Universal drafting Machine Co U.S.A.
Suspended Pantograp		1 „	(094)M—47	Dietzgen Co.
丁字尺		17 副	(095)M-47 (T_1) 至 (111)M-63 (T_{17})	
三角膠板		9 副	(112)M—64	
Franch Curve		16 副	(113)M—64	
標準法碼		4 副	(114)M—65 至 (117)M—68	
大小木製模型	For Descriptive Geometry	17 副	(118)M—69	
模型	For Descriptive Geometry	2 „	(119)M—70	
Wooden crystal models	For Descriptive Geometry	60 副	(120)M—71	
Detached natural crystal	For Geology	50 副	(121)M—72	
Rock collection specimen	For Geology	65 „	(122)M—73	
Mineral specimen		150 ,	(123)M—74	
自讀水準尺(英尺制)		8 副	(124)R—01	British
自讀水準尺(咪尺制)		3 „	(125)R—02	
覘牌水準尺(咪尺制)		7 „	(126)R—03	
覘牌水準尺(英尺制)		4 „	(153)R—15	
三咪尺長測竿		4 „	(127)R—04	
2-meters long rule		1 „	(128)R—05	
Invar tape		5 副	(129)R—06	
鋼尺		4 „	(130)R—07	
皮尺		6 „	(131)R—08	
測鍊 (Surveying Chain)		3 „	(132)R—09	Chesterman Sheffield England.
Lead depth pumb with ropes		2 „	(133)R—10	
測深竿		1 „	(134)R—11	
測竿		44 „	(135)R—12	
測針		102 „	(136)R—13	
木竿		9 „	(137)R—14	

Name; 名稱	Quantity: 數量	No. 編號	Manufacturer 製造廠
繪圖板 (2′×2′8″)	31 „	(138)B—01	
繪圖板 (1′×1′5.5″)	8 „	(139)B—02	
繪圖板 (2′×2′6″)	4 „	(140)B—03	
繪圖板 (2½′×3½′)	19 „	(141)B—04	
Copper astronomical globe	1 副	(142)M—75	
Big Wooden Compass	1 „	(143)M—76	
(8½×11) drawing paper graph Sheets	2 „	(144)M—77	K. & E.
小型木製模型	1 副	(145)M—78	
Transit boxes	3 „	(146)M—79	
小型竹篋	3 „	(147)M—80	
Hatchet	3 „	(148)M—81	
Wooden frame for rods	2 „	(149)M—82	
„ „ in 8 squares	2 „	(150)M—83	
木箱	2 „	(151)M—84	

(C) 電機實驗室概況

本系現時所開電工學一科，爲本系三四年級必修課程之一，物理及化學系學生亦得選修之。每年之上學期爲直流電路及電機，下學期爲交流電路及電機。每學期爲四學分，每週課堂討論三小時，實驗室工作三小時，以期選讀同學獲得系統理論之訓練外，并能熟悉各式電機之性能與實際工作。

査本實驗室成立於戰前，全部設計係出於馮秉銓博士之手，內容甚爲充實，戰時本校遷移內港話，機件笨重，無法搬運，淪入敵手，幸該時敵僞未加破壞，惟多年未加使用，機件稍欠靈活，去年上學期得本系金工廠及物理系之協助，已先後修復，開始實驗工作，成績甚爲圓滿。

現時實驗室備有交直流電機十部，包括電動機，發電機，感應電動機，Repulsion Motor 諸類，并有大小變壓器，電動機，電阻器及自製電阻水箱等十數件，各機俱配有適宜之底架皮帶及耦合裝置，可以隨時安裝需要之組合，使全部電機使用具有最大可能之伸縮性。

至於測電儀器，現有大小交直流電流電壓計、瓦特計及時計（湯姆生整流式及感應式）總共約三十餘具，并有電力頻率計及精細電計等其中有電表十餘具係本年新自美國訂購寄到者，關於較精細之測量，本校物理系給予極密切之合作，尤使本室得到非常之便利。

全年之實驗約可排出二十餘個，本期交流實驗前部偏重交流儀器之使用及電路理論之証明，後部注重電氣機械在工業上之運用，近正擬安置一電動工率計 (Electrodynamometer) 一座，以代替通常所用之制動器測量工率。可望測之知較準確之電機工作多曲綫。

本課程之教科的 Dawes 之電工學第一二冊爲藍本，在適宜之情形下加添 Langsdorf 及 Lawrence 之補充或較便之方法。實驗則以 Karuppetoff 及 Tucku and Ricker 之實用電工學爲主要參攷書。

(D) 金木工廠概況

從哲生堂工學院的北門望去，還不到一箭之地，我們可以見到兩根淡紅色的柱子，和一條灰白色的橫額的牌坊，上面塗上了幾個灰黑色的字，可是被一部份的白粉蓋上了，而至濛糊看不清，當仔細看去還可以隱約辨得出"工學院實驗區"幾個大字。牠這座古舊的牌坊，含有歷史性底遺跡。恰好跟哲生堂遙遙相對。爲了被風雨所浸，就不多大給人們所留意和關心了。在這牌坊的西北角底一隅，；這是一塊較爲低陷而崎嶇不平的大地，在牠底面上蓋上了幾幢小平房，同是淡紅色的牆，上面蓋上了一大塊被雨露所氧化了而發出棕紅色。略帶銀灰色的鐵板。這是紅灰底徵象吧。

假如你是局外人，我是說假如你們並不是工學院的同學的話。你們就很少會留意到這座小平房，或許甚至不會發覺牠們，因爲牠們是那末渺小而且位于偏僻角度，當然就不會被人家重視和關心。不過世上有不少轟烈的好事也是多從無名中發掘出來。這就是由于我們的前土木系主任黃教授毓文沉默寡言。刻苦實幹而不好大肆宣傳的人，所以牠就沒人去欣賞了。但他却是土木系的同學們精神寄托和樂園，我們也會消磨不少時光在這樂園中。沒有嘗試過這滋味的人是沒法領畧其中的奧妙。兩座相連的就是木工室和金工室，是由一度小小的門便把牠們連接起來了。後面另一座長涌形的是水力實驗室，不過我們須要談的還是木工室和金工室這座。每天下午。假如你行道經過這區域。你可以聽到機器軋軋的聲音，發電機的呼號。大皮革緊接着大輪子。大小輪齒的互相緊貼中軸的旋轉如磨擦！金屬柱子受了鑽子如鋒利的鋼刀。所發出來的悲鳴，而成了調和而有節奏的美好旋律。你會感到這是單調嗎？不，一點兒也不感到，我會感到這是一條條美好而有曲綫美的肌肉，一點點一滴滴的汗珠，發電機的底音調。金屬柱受了車牀上發出比女高音還高的音韻。皮革的節拍。組成了一個美好的交响樂，一首廿世紀的『工場交响樂』就在這班刻苦的工佬們演出了，這是工佬們的樂園，可不是嗎，我們不少的時光都在這靜寂中充實起來了。

在木工室裡面有一副皮革鋸。(Bend or Jic Saw) 是用來鋸曲綫板的，一副木刨牀是用來把木板刨平的另有風車牀和四副木車牀，在金工室裡算是一個較爲充實的機房，有三副大車牀。一副機刨牀，一副鑽牀，一副 milling machine 和一副火石牀，另有無數的大小配件。每一副機器都由發電機拖動。牠們的用處很廣而且沒有一定方式，我們的產品也不少。例如大小各式各樣的螺絲釘最大的有四五寸長直徑將及一寸的大釘。和造了很多物理實驗室，尤其是化學實驗室裡所用的三脚鐵架支持燒瓶用的銅圈。一些 Vise ，(老虎鉗)，和一些零碎的機件都是我們這小小工場所出的，雖然這是一所小小的工場，可是，假如我們好好善用這些機器，我們可以製造機器和實用的工具，而在這工學院裡是甚够用的了，聞說現在還有一批機器由英美輸運途中，將來這工場的發展是更無可限量的。

(E) 水力實驗室之概况

從水塔那邊有一支水管直通至距離其二百呎左右的一間建築物內，那便是我們的水力實驗室了，往內一探，最初是看不出有些甚麼儀器，特別注目的，因為它不像金工實驗室，材料試驗室等機器林立的啊，但仔細察看，我們便可看出各實驗部門了。入室後最初使我們注意的便是水管環繞全室，這些水管都可連續貫通的，而有水掣可以定水的流向，各實驗部門所用之水便是由這些水管運輸去的，至於實驗後之用水，可出地下管運至一水窖，或再由動力機將水運至室頂旁的蓄水池，當需用時又可以再取用的，所以室中的水可循環使用而不至有所消耗。關於各水管的佈置，可由下圖看見。

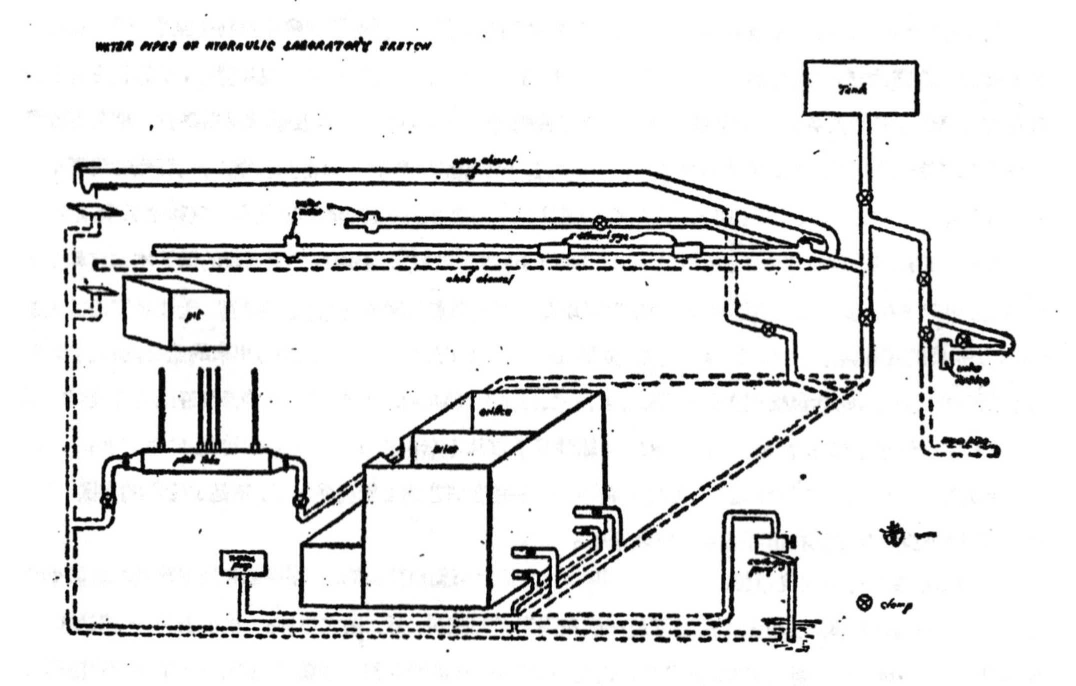

圖一

現在根據圖中所編的號數，我們將各實驗部份詳細觀察罷：（1）在號數（1）所指的地方，有一水平的『雲慈理氏計』（Venturi Meter Tube），這是由兩半徑不同的水管連接而成，在不同半徑的水管上連接有二水銀量壓管，是用來量度水管兩點間的壓力差的；整個儀器的構造大約如下圖所示：

究竟我們拿這儀器來實驗甚麼呢？最基本的是求這水管中的流速係數。我們知道，當我們拿着實驗得來的紀錄來求水流速度時所得的數值是比眞確的流速度為大的，這因為管內有磨擦阻力，以至水流被阻而速度變慢的原故。設理想流速為 V_i 眞確流速為 V_a 二者間的關係可用下列公式簡單表示之：

$$V_a = CV_i$$

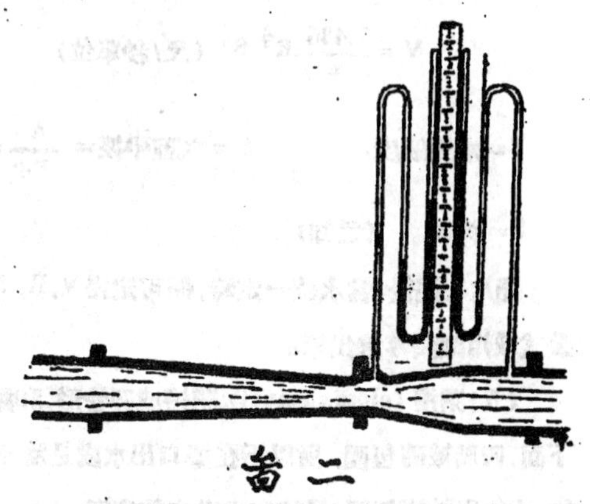

圖二

式中的 C 即流速係數，亦是我們需要知道的數值，當我們得了這數值後，那麼以後用這管做實驗時，流速便可一算而得了。至於 C 值的求法，我們先通水入管中，從水銀管中定出 A，B 兩點間的壓力差 h；又水從管流至管口，我們用盛水器量度單位流量 Q，所以在 B 點的眞確流速 $V_2 = \frac{Q}{\frac{\pi}{4}(D_2)^2}$ 應用下列公式：

$$V_2 = C\sqrt{\frac{1}{1-\left(\frac{D_B}{D_A}\right)^4}}\sqrt{2gh}$$

各值已知，所以 C 值便可算計而得。

（2）從同一管綫向左望去，我們也可以看見有二半徑不同的水管連接在一起，在管上有水銀量壓器，這裏是實驗當水由大管突然進入細管時，水頭突然降落的情形。

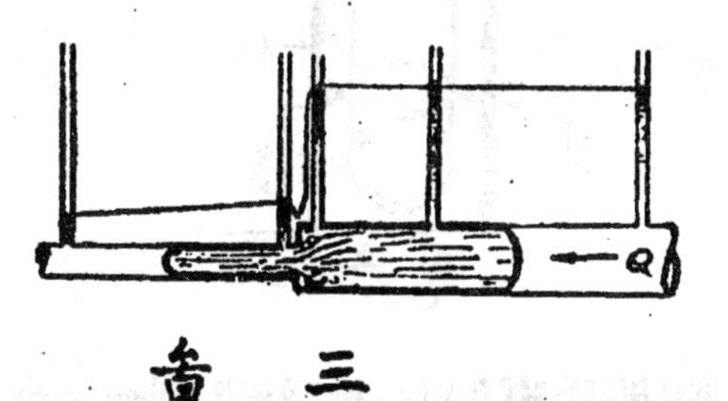

圖三

在這裏又可看見同一管內二點的水頭高低有別，拿了記錄我們便可求得因水管摩擦而耗失的水頭數值了。

（3）在同一管上再向左望去，我們常會看見有些量水計安置在管上，那裏便是檢驗量水計處，那些量水計多是從校中市塲的商店拿來檢驗的，檢驗的方法很簡單，先說下量水計中所指示的數值，然後通水入管，經一已知時間後，停止水流，再記下所指示的數值，前後數值值差額若與從水管口流下的水量相等，那麼那量水計是正確無誤，否則便要看差額多少而修改了。

（4）當看完這一連三個實驗部門後，我們抬頭向前一望，便可以看見一露天的用混凝土造的長槽，這便

是敞槽 (open channel) 了；它的横截面是一抛物綫形，全槽對水平綫畧爲傾斜。

關於敞槽的各種實驗，最基本所需要知道的條件是它的斜率和糙度 (roughness)；斜率可以用簡易的方法求得，但糙度便要用實驗求出來，所用方法可用下列公式而定：

$$V=\frac{1.486}{n}R^{\frac{2}{3}}S^{\frac{1}{2}}\text{（尺/秒單位）}$$

V＝水流速度.　　R＝水理半徑＝$\frac{A}{w.p}$＝$\frac{\text{水在渠的横截面面積}}{\text{横截面的周界}}$

S＝斜率　（已知）

通水入渠後，當水位一定時，便可定得 V, R, 值，所以糙度可用上列公式或用圖表等求出來。

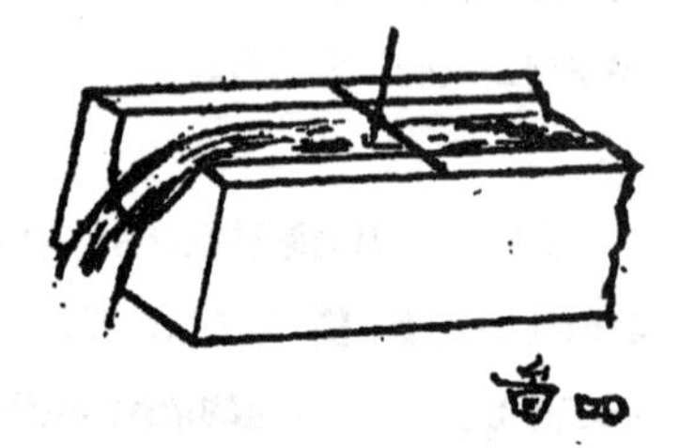

圖四

（5）閉槽 (close channel) 閉槽成圓管形，和敞槽平行，在上述水管下面，四周被磚包圍，所以不在渠口出水處是看不到的，至於實驗的目的，大約與敞槽相同，目的在求出它的糙度。

（6）回轉頭來，在入門處右邊，我們可看見一轉輪那便是水渦輪 (water turbine) 了，水渦輪的功用是將水位能轉變成動能，所以最重要的是測定它的功率。水渦輪的構造大約如下圖所示，

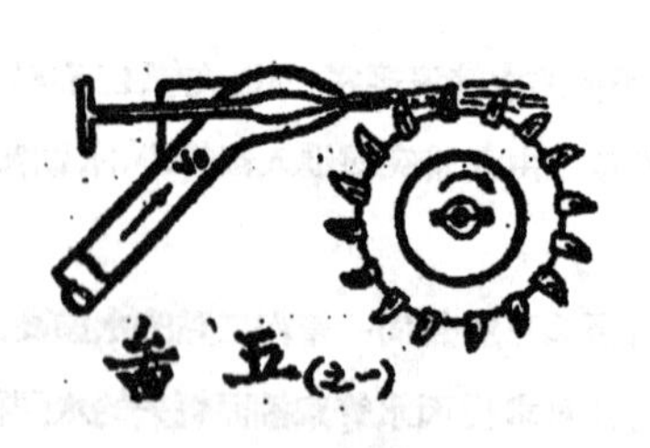

圖五(之一)

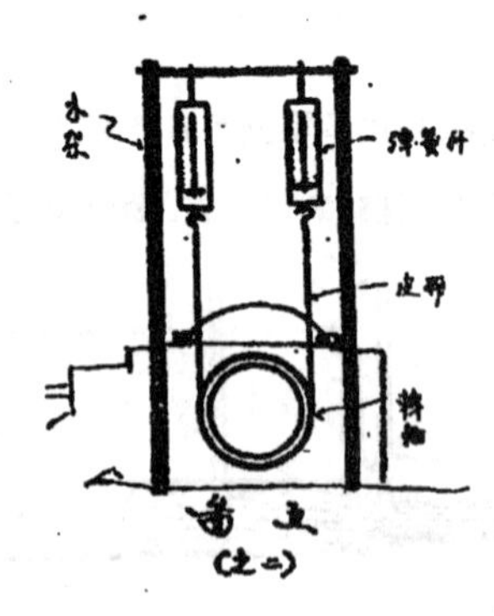

圖五
(之二)

（7）水渦輪前面有二大磚造的水箱互相接粘，右面那一個是用來有關『孔』(orifice) 及短管 (Short tube) 的實驗的。當水從箱旁一孔流出時，我們常以爲它的單位是 $Q_1=AV=A\sqrt{2gh}$

A＝孔的面積.　　h＝水面至孔心的高度

但實際上我們把流出來的水量度一下，便知道眞正的流數 (Q_A) 比理想的爲少，這因爲有流速係數和收束係數 (contraction coefficient) 所致，所以我們有下列公式表示

$$Q_a = CQ_i \,, \qquad C = \frac{Q_a}{Q_i}$$

式中的C是這孔的流量係數，這亦是我們對孔實驗最先要做的一個。

關於短管的實驗，和孔大致類似，不在這裏重言了.

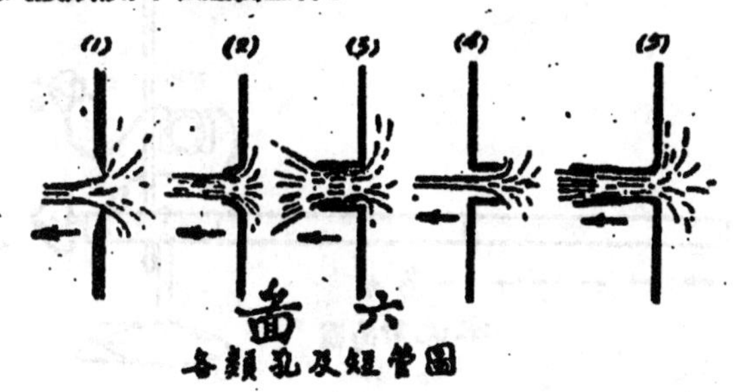

圖六
各類孔及短管圖

在左邊的一箱，是用來有關堰（weir）的實驗，堰有四方形，三角形，梯形幾種，如下圖所示：

圖七

這部份最主要的實驗和上述相同，在求堰的流量係數.

（8）在這二水箱前面，我們可看見有一大圓形管凸出地上，管上有細玻璃長管，在表面來看，各管的形狀是一樣的，但我們如透視內部，那麼便可知各管形狀不一的了，有些是直的，有些伸至管中心向左或向右成直角的彎曲，這是實驗壓力頭（Pressure head）和速度頭（velocity head）對於水管放置的關係的

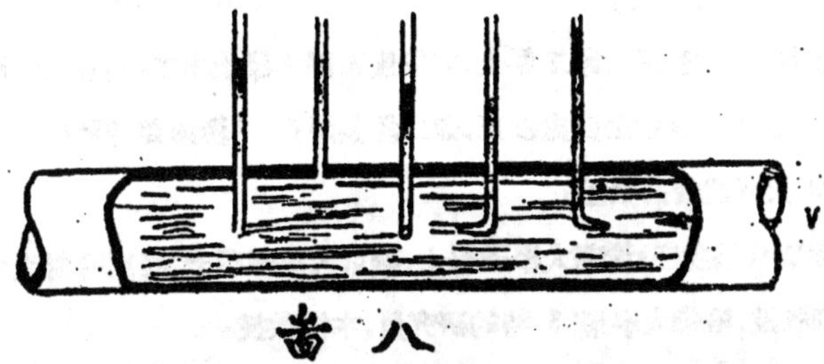

圖八

此外室中看完畢的儀器後，我們走出來，不要以為這就完了。還有一水壩模型，這是根據最近在開平建築的水壩造成的，這會另有文說明，這裏不再贅言了，我們還要實驗室後去看一看呢；在那裏我們看見一大水槽，濶約四呎左右，約比實驗室長一倍，兩旁有鐵軌，上有平板小車一輛，在槽的一端，安置有一電動機，用鐵纜管理小車的前進或退後，這是甚麼玩意兒？是水流計的實驗啊！當實驗時將水流計裝置車上，使其中的轉舵

放在水中，並且附加電綫乾電池和感應器放在小車上，於是利用電動機使車行動，轉舵便因小車移動而轉；每五轉感應器响一次，記錄所响次數，乘以五，便得轉動次數，並記錄車行時間與距離，於是用下列公式：

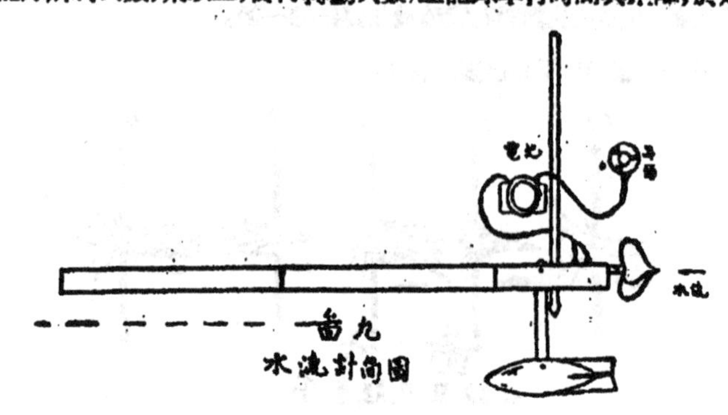

圖九
水流計簡圖

$$V = a + bn$$

V＝速度，　n＝每秒所轉次數　（a,b 爲常數，各轉舵不同，）

可計得車行速度　又速度＝$\frac{\text{車行距離}}{\text{車行時間}}$，

所以如二數值相等，則表示無錯誤．

這是在靜水中實驗，以車行代水流速；若在流水中，我們只將水流計放在水中，看轉次多少，運用以上公式便可計算水流速度．

最後，謹向各位致謝，費了這許多時間來觀看呢！

（三）本系教授資歷表

桂銘敬教授兼系主任在假：交通大學工學士，美國康奈爾大學土木工程科碩士，歷任廣東大學講師，中山大學教授，廣東省建設廳公路處技正，兼工務課課長，粵漢鐵路株韶段正工程師，湘桂路天成路副總工程師，本校教授兼系主任。

馮秉銓教授兼本年度代系主任：清華大學理學士，燕京大學理碩士，美國哈佛大學科學博士，歷任本校助教，講師，副教授，哈佛大學講師，特約研究員，本校教授。

黃郁文教授：美國亞麻工程專門學校工學士，歷任咪吔洋行，富新機器製造廠，謙信洋行等工程師，廣州工業專門學校教授，本校副教授，教授。

梁健卿副教授：嶺南大學工學士，美國麻省理工大學工程碩士，歷任美國公路總局實習工程師，交通部公路總管理處幫工程師，軍委會滇緬公路監理委員會專員，軍委會滇緬運輸局駐緬副工程師，軍委會運統局公路總處副工程師，美國軍部正工程師。

韓約瑟副教授：美國麻省理工大學工學士，曾任美國瑪麗諾大學教授。

(J.A.Hahn)

王銳鈞講師：嶺南大學工學士，歷任軍委會技師研究室研究員，技佐，技正，交通部西北公路局幇工程師，

廣州市工務局技士。

凌鉄錚講師：湖南大學電機工程學士，交通大學電信工程碩士，曾任上海廣播電台工程師。

林炳華助教：交通大學工學士。

馮啓德助教：交通大學工學士。

劉載和副教授：嶺南大學工學士，歷任交通部第三區公路局副工程師，正工程師，鐵路工程師，設計組組

長，本校副教授，廣州大學教授。

黃發瑤兼任副教授：復旦大學工學士，美國愛歐華大學碩士，歷任交通部公路總管理處幇工程師，西南公

路局副工程師，交通部公路標準委員會技正，廣東省建設廳技正。

譚沛雷講師：美國畢玆堡大學工學士。

江開鑛講師：美國加州大學工學士。

（四）本學期開設科目表

科目	學分	年級	每週實驗時數	教員
投影幾何	2	1	6 小時	馮啓德
工廠實習	1	1	3	黃郁文
材料力學	4	2		韓約瑟
機動學	2	2		黃郁文
平面測量	5	2	6	林炳華
材料試驗	1	2	3	江開鑛
初級結構學	3	3		王銳鈞
鋼筋混凝土設計	2	3	6	王銳鈞
結構計劃	2	3	6	韓約瑟
鐵路測量	3	3	3	譚沛雷
土石及基礎學	3	3		梁健卿

交流電機工程	4	3	3	馮秉銓
				凌鉄錚
水文學	2	3		黃發瑤
鐵路工程	3	4		劉戢和
合約與規範	1	4		黃發瑤
高級結構學	2	4		梁健卿
水力設計	2	4	6	梁健卿
道路工程材料試驗	1	4	3	劉戢和
道路設計	2	4	6	劉戢和

（五）研究工作

1. 數年前本系曾致力于研究用竹竿代替鋼鐵包含于三合土內，並研究以三合土凝結而成之磚柱作爲屋厦中之棟樑，其結果雖相當完滿，但牽涉之技術問題尚多，此當有俟將來之繼續研究。
2. 最近爲研究排洪道之流量，將開平縣梁金山水壩造成模型一座，備高年級生作研究之用，本系雖無研究所之設，但學生人數爲全院各系之冠；四年級生論文工作甚爲注重。

會聞

三年來工學會同學動態

——謝國光——

在這三年間，本系的活動，大致可以分爲兩類。第一類是關于學術方面的；例如參觀本市的鉅大工程，邀請工程界先進演講等。第二類是屬于同學間的友誼的聯絡。自嶺南復員以來，在本系肄業的同學，經常在一百二十多人以上，佔全大學人數的十分一强，是土木系有史以來未有過的好現象。可是，問題來了，因爲人數的衆多；同學們所選科目的參差，和遼遠的宿舍距離等等；要打破彼此間的隔膜，實在不是一件容易的事。惟一的方法，那就是增加我們課餘活動。

第一類，關于學術方面的。

我們常常舉行學術演講，講者都是工程界中知名之士。其中更爲值得注意的，有麥蘊瑜先生的「原子彈與城市設計」一文。對于城市內各區域的設計，例如工業區，商業區等等的分佈；及城市與城市間的建築，都有詳細的說明和解釋，怎樣才可以將敵人的原子彈的威脅，減少到最低程度。例如城市內的工業區不應該集中一點；城市的建築，以小型都市爲最佳，作串珠形的連貫狀爲最妙等等。這些都打破了工程上的習慣，是近代建築學上的新理論。(原文見同期本刊)其次是容祖誥先生演講中印公路踏勘經過，是把他個人所見的所聞的告訴我們，內容備極充實。

我們曾經先後參觀過西村士敏土廠兩次，分別由鄺正文及江開嬌兩位先生領導，士敏土在近代工程上所佔位置的重要，恕我不再饒舌了。在製造的過程中的每一個步驟，蒙該廠的負責人詳爲分析，使我們得益不少。此外，例如西村電廠，自來水廠，飲料廠等，我們都也曾到過。

在去年海珠橋大修的時候，在中段橋面的構成物都拆卸下來，祇剩下構成橋架的支撑架，桁架等。這是難得的機會，由王銳鈞先生帶我們前往觀光，本系的舊同學李文泰先生担任解釋，怎樣將書本上的理論應用，使我們對于橋樑的結構，有深一層的了解。

粵漢鐵路的黃沙站，我們也去過兩次。第一次是關于工程方面的，我們看見到它的修理廠，認識幾輛拆開的機車頭，最後我們看到了它的翻沙廠；這使我聯想起我們被敵人破壞了的翻沙廠，假若沒有這一場戰爭的話，我們的設備，至少比現在好十倍。第二次是選讀運輸學的同學，會同商學會一同參觀該路的管理法及未來發展計劃。隨後至沙河築路工程處，參觀機械築路大隊，我們看到許多聯總贈的築路新機械。

遠東第一大的挖泥船——新廣東號，在上月初，由黃郁文教授率領我們前往參觀，這是世界第四大的挖泥船，每小時的挖泥量可以達到六百立方公尺。對于珠江的航運，佔着一個舉足輕重的位置。其中的每一件機械的性能及運用，都經該船的負責人解釋清楚，而其中惟一遺憾，那便是因爲該船的消耗燃料量太大，他們沒有法子作一試驗給我們看看。

還有黃埔建港工程及海軍船塢的參觀，這是劉誠和先生的領導，很可惜的是我們祗是見到日本人築成的數百尺堤岸及新近完成的三合土道面。事實上，因爲種種關係，黃埔建港的工程，直到如今尚未有顯著地開工呢。在海軍船塢裡，我們祗看到一些機器，據該塢負責人說，一月後美國贈給我國的浮動船塢運到後，那時相信會有些對于我們有幫助的資料。

關于課餘的研究，在梁健卿教授指導下，高年級的同學，對于梁屋山水庫作一個實地的研究。他們在水力試驗室內。根據當日在該地所獲得的寶貴資料，採用相當的比例，造成了一個與當地一樣的模型，研究着排洪道渦流對于土壤的冲刷狀況及其流量及流水速等關係。目的是作一個試驗，看看此書本上所獲得的智識和理論，是否用與實地情形相吻合。(按關于該壩的報告見上期南大工程)

兩年一度的暑假野外測量，是選讀測量同學的必需實習。在土木系中，可以算得是一件大事。這類的野外實習，在許多學校是沒有開設的。在平時，我們已把校園測遍了。這次更進一步，將我們的智識，應用到廣大的原野去。一方面是爲地方上作一有意義的工作，而另一方面是可以增加自己的經驗。一舉兩得，又何樂而不爲呢。復員後的第一次暑期測量隊，是在去年七月初舉行。地點是新會六區潭岡，由梁健卿何育民教授率領，工作時間共三星期多，參加的同學有三十六位，該地的平面圖，地勢，土方等都有詳細的結果，成績相當滿意，現在的測量室內，還有當日的圖樣存在。

攝影術對于我們工科的同學有很大的幫助，同學們有見及此，遂有攝影社之組織。因爲這是共同的嗜好，參加者極爲踴躍。最可惜的是冲晒放大的地方不大，以致常有人滿之患，我們希望對于冲晒室的地方加以改善。在工學會日各部展覽的照片，那便是攝影社成績的一斑了。

第二類，關于學術方面

每學期之初，我們都有一個盛大的歡迎新教授及新同學的聯歡大會。雖然是一個清一色的團體，可是出席的人數，却有百分之九十五以上。因爲全體同學聚集在一起的機會實在是太少了，所以我們每一個都珍惜這一個集會，輕易不肯放過。雖然不是什麼晚會，但是情況的熱烈，都是遠非晚會所能比擬。桂銘敬主任在離別前，曾在他府上舉行一盛大話別會招待我們，會中充分地表現出師生間友誼的密切。

此外，舉辦分組的球類比賽，是聯絡同學間感情的妙法。這樣可使遠居新疆(魁業堂一帶)及中原(陸佑堂一帶)的同學玩成一片。因爲全系都是男孩子,故此極易進行。人數是太多了,然而沒有問題,分成八九組比賽更加熱鬧。普通是採用單淘汰制度,這樣一來,便有許多職業球員應時運而生了。例如甲隊被淘汰出局後,一部分的球員便加入了另一隊。結果各隊競招職業球員精彩百出,而冠軍隊的出場球員,已一半是屬于職業性的了。同學中架眼鏡的不少,有因躍高頂球而眼鏡跌下者,有因踢球而波鞋飛天者;而呂九之滾地葫蘆射門姿勢,大鼻之蜂式衝刺,及姚孟岐之琵琶絕技,觀衆莫不歎爲觀止。

去年中秋,工學會舉辦海上賞月會,是一個別開生面的玩意。

在昨年三月及本月,我們曾分別舉辦電影及話劇籌款。目的是爲着南大工程的印刷費,第一次是在廣州戲院,獻映名爲「鳳樓魂斷」。第二次是敦請留穗劇人公演曹禺名作「日出」。由于同學們的努力與協助,兩次的籌款,可以說達到預期的理想了。使南大工程能够問世,我們能够在這荒蕪的學術園地裡,發出一點兒聲響。

我們也曾和教育系的同學,聯合開一次野火會,這是去年冬天的事情了。我還記得,這晚上的天氣很好,並沒有刮大北風。每一個參加者都玩得很高興。我們團團圍着那光明的火燄邊,象徵着南大一家親的精神,野火會的最後一個節目是燒東西吃,然而教育系的同學却是太不濟事了,過不慣野人的生活只吃小小便喊飽,而我們這一羣却感覺到物資的可貴,結果,幾乎每個同來都嚷着肚子痛咧!

潭江三個星期內的測量生活,參加者都認爲饒有趣味。其中甘苦參半,至今猶有餘興,可見其印象之深刻。據云每晨天甫露曙色,大隊長以身作則,在廣場上吹動銀笛,催促同學起床。但是同學們每晚都要整理,較正和計算當日野外工作的所獲得的資料,以備繪圖之用,工作常至午夜。清晨時候,正是好夢方酣之際,不料被此尖銳笛聲喚醒,同學們莫不命此笛曰「無情笛」。

在每天的午間,例有夏季的驟雨。最初的數天不知道,在野外的同學,惟有除下身上的衣服,遮蓋着我們的經緯儀或水準儀;屹立山頭,憑着一股工作的熱誠,與大自然相抗。因爲在潭岡一帶,都是童山濯濯,除此以外,人有什麼法子可想呢。以後漸漸有些經驗,那在宿舍附近工作的,都寧願多走兩次,在下雨前將儀器搬返大本營,那距離較遠的,惟有繼續逗留在山頭,靜候風雨的來臨。

午餐是在野外吃的,有一件趣事,菜肴中的湯,往往是飲之不盡。午間的驟雨,在潭岡的廿一日內,祇有三天沒有光臨。晚餐在大本營內進行,這是鄉公所隔鄰的一所大屋,規定八人一桌,有時因欠人而同桌七人不能用膳者;有時兩桌各有七人而須呆候者,于是俱竭力施展招徠方法,務使本桌能够及早用膳,而趣事以此發生矣。派信的時候,必定是我們剛剛在用早餐。郵差一到,每一個人連吃飯也置諸腦後,一擁而上,看看自己有沒有信。收到信的同學便興高彩烈,得意忘形,躲在一旁慢慢欣賞,否則便吃多兩碗飯以補償損失。據某權威統計家云,收信最多者是巢公與何兜,是否真確,則非筆者所知,僅照錄而已。

飯後多數是在門口納凉,這是一天中最舒適的時候。斯時也,在一片花話之後,例由康道黃領導唱歌,一本嶺南歌譜,由校歌而運動會歌而戰歌都唱遍了,接着又是民謠。等到天色將黯,便在室內架起大光燈,整理一天工作的結果完畢後，已在十一時以後了。

為着要使各界人士，對于工程學上有較深切的認識。發生些興趣，和聯絡本系校內及已畢業的同學與外間的工程界先進及同學起見，我們作一個大胆與嘗試，暫定五月十六日為工學會日。是日將工學院的材料試驗室，水力試驗室，電機試驗室，測量儀器室，金工廠，木工廠等實驗室全部開放，加以詳細的說明和解釋；並將我們實驗的成績展覽出來。所得到的結果是怎麼樣，我們是虛心地祈望着和接納各界先進的批許與指教。我們祇是盡力地去幹，希望發生小小的作用，在嶺南及外界的靜寂學術園地裡，激起共鳴來，而另一方面則招待畢業多年的舊同學歸寧母校，希望先進者給予我們一些指導，和聯絡彼此間的友誼。

很久以前，校外及校內的同學，都感覺到大家應有更密切的聯絡，故本月初，遂有「私立嶺南大學工科同學總會」之組織，會章等經已草就，理事監事刻在選舉中，在五月十六日，乘着校外校內同學齊集一堂的時候，便宣告成立，相信以後同學與同學間，校內的同學與已畢業的同學間的聯絡，益形緊臻了。

以上所述，僅就筆者所知，聊盡報道之責，倉卒完成，錯漏難免；尚希讀者指導與原諒。

臨別贈言

——別　者——

「天下從無不散之筵席」，現在，我們是要別了——這就是說，不祇是離開曾經和我們憂患與共的母校——嶺南，更兼我們是要暫時的離開這一群同窗共硯友愛得像兄弟們的仝學。朋友！大概你們也會曉得人類是情感的動物，眞情的透露和友誼的可貴常常是特別的顯著在離別的當兒．也許你們過去也有着或咀嚼過「別」的經驗和滋味，在你們高初中的畢業期間裡，驪歌已經隱隱地奏起的時候，你們快要離開這可愛的母校校園底懷抱，離開這一羣多年敘首的可愛同學，和那些永遠是孜孜不倦像春風時雨地教誨我們的師長，在這刹那間心情的轉變和難過，確非這支禿頭的筆所能形容於萬一．可是，我們正是準備着來再度嘗嘗這難過的離愁別緒哩！

不過，雖則是我們現在是快要分離了，但這祇不過暫時性的，我們將來仍多着聚首的機會，而且我們的精神還永遠聯繫在一起。說不定我們將來聚首的機會比我們同窗共硯的時候還要多，因爲我們始終是「同學」和「同行」，但切不要記着「同行如敵國」這句話，這麼說來，假如我們每位之學院的同學都是學而致用的話，那將來見面和合作，互助，切磋的機緣委實是不少，但願我們每位同學都牢記着這一個念頭，盡量利用我們所學的來貢獻給國家社會，正如像我們德高望重的系主任桂銘敬先生所訓示過的「做一位頂天立地和立已立人的工程師，爲國家與建設，爲社會謀福利」，更且，爲着我們是要繼承過去先驅和先進們的豐功偉績，我們現在工學院的同學也許是將來在社會裡的仝事要團結的聯繫在一起，互相扶持，因爲團結就是力量，有了智識還須要有了力量才能使我們的工作穩定，發展和利用我們的材能，因此，體系的團結是我們永遠不能輕視或忽畧的。同時，我們還希望這些團結的觀念和習慣應該是不單祇是爲了將來的工作才產生或養成，這并不是說爲了將來的出路我們才肯去這樣做，我們都知道學校是一個社會的縮影，學校裡的社團及學會是構成「社會關係」的橋樑，是羣體生活訓練的組織，像我們青年時代的學生，更是應該積極參加這種社團及學會的生活，如使我們在德羣智體四育方面多所進益，而以學會集會爲其聯繫的工具，可是在現在嶺南的工學會本身來說，許多同學對學會的興趣是不大濃厚，學術研究的風氣也許可以說得是罕有得像鳳毛麟角，或許這是學會的負責人的一點忽畧，不過，同學間對於學會的活動漫不關心的現象是不容否認的，反之，很多的同學們特別是關心那些所謂交際集會。假如，我們能够分析一下這種偏岐的享樂心理，或許是研究用怎樣方法來轉移這個心理到研究學術方面去，這未始不爲工學院的福氣。同時，我們知道青年人的模仿性是極大的，近朱者赤，近墨者黑，如果我們不提高我們對生活的鬥爭性和警覺性．遏制我們對沉迷享樂的心理，來接受那些有益身心陶冶修養的活動，我想這效果正和我們臨別所期望的恰是相反，不過，我以爲工學院的同學們都是理智特別清醒和具有高度自覺性的，我所說的和替他們担心的說不定祇是博得同學們的嘲笑，或許是說我多餘而已．

說了這許多「多餘」的一大堆廢話，筆光又應轉回我們的身上來了。本來，在我們這一羣懷着快要畢業和走出校門跑進社會的心境裡，我們是有着「一則以喜，一則以悲」的感想，喜的固然是我們的學業，說是完成了一個階段，悲的是我們撫心自問，我們所學的是有若干，我們不敢說是已經把自已好好地充實，反之，我們祇徒覺得我們是依舊空虛，沒把握，尤其是當我們準備跑進社會裡去掙扎的時候，我們的境遇眞有點像迷途的羔羊，徬徨於十字街頭的叫化子，究竟我們爲什麼要這樣徬徨逡巡呢？這一點我想不用說人們也會清楚，歸根到底的答案也是，我們雖則是經過了好幾個年頭的埋頭苦幹，但是，我們始終沒有自信的感覺到已經把自已充實。固然，這些責任的攸歸是自已負上的，不過，我們却不能否認這未嘗不是學校在抗戰期間受着戰事的影响而顚沛流離三徙其居所給予我們的賜與。無疑的，這是抗戰期間整個戰局的變動問題，那一間在大後方的大學學府也是不能例外的避免這些偌大的波動和轉變，間接上來說，我們這一羣的大學初期的基礎也隨着它而起了動搖，固然，我們對於這些不幸的影响是不能埋怨任何一個，但是，到底這也是我們命運中的不幸，使到我們到了現在還是沒有很堅強的信心，暫時更談不着怎樣的來立已立人了。

朋友，你們現在有的是優越的環境，多材多藝和德高望重的師資，能够互相切磋砥礪的同學。你們是優秀者，因爲你們德羣智體四育的培養都有那些優良的背景來幫忙它的孕育，你們是應該盡情地去享受和吸收這些環境所能給你們的賜予，利用它把自已堅強起來，不要自暴自棄的放過這個良好的機會．而不盡量的充實自已，去秉承我們工學院已往曾經發出過的那亮到發紫的光芒，這樣你們所得的結果就不會有得着別人的輕視和冷語，熱諷和冷嘲，假如你們（也許是我們整個工學院的全學），到了現在還沒有急切的反省和熟視着環境的需求，那又怎能怪我們的工學院一間給別人的印象是那麼淺陋，低能，而致各方面對它也持着一種輕視和漫不關心態度，有如一個高度權貴的社會裡對待一個私生子一般。

友愛的同學們，你們得體諒我們這一羣別者的撩亂情緒。語無倫次的牢騷，說不定我們說了這一大堆的廢話而引起了的反响是「多餘」，但是，到底我們是要胡亂地吐出來，正如「骨鯁在喉，不吐不快」一般。我們不敢奢望這幾句話所能意外的引起一種特別良好的反應，不過，我們始終是懇切的期望着，你們這一羣優秀者是有把握地把工學院一向所給別人的印象洗脫過來，同時，你們還得有體系的團結在一起，不論在學校裡在社會裡也會苦幹下去，眞眞正正的做一位頂天立地和立已立人的工程師，正如　蔣主席在中國之命運裡所訓示的一樣。

本系歷屆教授畢業肄業同學姓名及通訊處

姓名	服務機關	電話	住址	電話
(1934)民廿三年級				
崔兆鼎	本市叢桂路九號三業	14771	東山啓明大馬路	
	粵漢鐵路廣州鋼樑廠		漢園	
劉俶和	本市河南嶺南大學		本市河南嶺南大學	
李文泰	珠江水利局	16697	本市多寶路多寶街51號二樓	
吳慶鳴	本市官祿路36號裕泰建築行	14474	第六甫木排頭二號	14576
潘祖芳	東莞糖廠			
彭震東			東山保安南路二十號地下	
劉　登	香港德輔道中交易行四樓			
許寶漢	黎樹仁			
(1935)民廿四年級				
陳景洪	太平南二一號僑聯行	10094	廣德北路沖園一號四樓	
陳國柱			東山岡嶺北一號地下	
陳銘珊	官祿路三六號裕泰建築行	14414		
高永譽	官祿路三六號裕泰建築行	14414	西關昌華大街廿三號	14651
林榮棟	市政府工務局第二科	14504		
李廷汾	粵漢鐵路十一總段	10229	長壽東路高基大街高興里二號之一	
梁卓芹	滇江湘桂黔鐵路工務局來滇段粵境工程處			
梁寶琨	達成營造廠	12551	迴龍路龍橋新街五號	
韋金信	本市白雲路珠江水利工程總局	16697	海珠中路杏花巷四十八號二樓	
容永樂	達成營造廠	12551	迴龍路龍橋新街五號	
鄒漢新　黎均縣　李文遠　文鑑堯				
(1936)民廿五年級				
李卓傑	珠江水利工程總局	16697	白雲路一一六號	

姓名	服務機關	電話	住址	電話
龍寶鎏			本市德政路設飛坊四號	
莫佐基	資源委員會水力發電工程總處滃江勘測隊現暫調廣州電廠	10721	法政路五七號之一三樓	
黃樹邦	惠愛東路301號嘉頓公司	11244	惠愛東三〇一號四樓	11210
黃惠光			長堤二八六號華盛頓餐廳	12962
趙宗武	星加坡大坡馬路二一二號 馮强膠廠星州分行			
趙長有	Cihu Chsng. You c/o Chiu Mei Lai 924 Kusaug Loop Manali. P.I.			

陳秉淮　陳書聲　陳元力　夏進興　王宏猷　陳守勤　崔世泰

(1937)民廿六年級

陳士較			本市增正路三號地下	
沈錫琨	湖南零陵冷水灘工務總處轉			
曾德甫	廣西柳州柳江橋工所			
鄒煥新	彭奐原			

(1938)民廿七年級

梁健卿	本市嶺南大學			
吳潤楷			香港山光道十號地下	
黃文海	湛江市政府			
王銳鈞	本市嶺南大學			

陳振泰　陳棣廉　陳自仁　林守就　溫兆明　余廣才

(1939)民廿八年級

夏傑榮			龍津東路二二五號	
林文賢	市府工務局第三科	17789 轉32號	東皋大道二橫路13號三樓	

陳樹賓　馮葆鎏　馮佑曾　黎辛才　李卓平　李梁材　伍瑞明　黃寬昌

(1940)民廿九年級

陳德泰	湛江湘桂黔路來湛段號境工程處			

25782

姓名	服務機關	電話	住址	電話
區錫齡　黎廣杰　楊啓疇　余槐欽				

(1941)民三十年級

姓名	服務機關	電話	住址	電話
陳尚武			19 Buckingham st. Penang	
朱錦池　李世能　盧炳煌				

(1942)民卅一年級

姓名	服務機關	電話	住址	電話
陳乃鼎	市府工務局第三科	14519	維新路一五六號二樓	
何耀波	珠江水利工程總局	16697	河南仁喜橫街十二號	
何育民	湛江湘桂黔路來湛段粵境工程處			
賀喜定	粵漢鐵路衡陽段			
林漢中	珠江水利工程總局	16697		
馮漢邦	王啓祥			

(1943)民卅二年級

姓名	服務機關	電話	住址	電話
程聯興	市府工務局第四科	14441	一德路四三號	12360
許銓錦			本市大新西路四五八號永泰行	
劉柏江	粵漢鐵路工務十二總段三十分段	12597 轉三十分段	大塘鄉四七號	
梁鼎燊			2E Lorong Slamat Penang	
張炯翰	珠江水利工程總局	16697	十六甫東四號一八號二樓	
甘光儀	新志利洋行		香港大道中法國銀行四樓	
阮樹鈞　陳世柏　符和丰　黃炳禮				

(1944)民卅三年級

姓名	服務機關	電話	住址	電話
梁百衡	廣西柳州柳江橋工所			
甄賓光	市府工務局			
余進長	粵漢鐵路工務十一總段二十八分段	10229	黃沙本處內	
關肇明　黎恩炘　林永泉　馬錦源　徐承熾				
余伯長			中華南路小市東街十四號	

姓名	服務機關	電話	住址	電話
潘祖芬			多寶街五八號之一	
黃炳禮			大南路一三四號三樓	10441
張錦華			龍津東路洞神坊觀蔭坊四〇號	
符和豐			維新北路南朝新街九號	
關蓋明			新荳欄上街二十一號轉	
趙礪艮	粵漢鐵路十一總段	10229		
朱士賓	同上	同上		
李偉廉	同上	同上		
楊克剛	珠江水利工程總局	16697		
梁光能	同上	仝上		
冼楠勳	同上	仝上		
林穎夫	同上	仝上		
容永文	上海京滬鐵路工務處			
陳光明	市府工務局			
李滋深	仝上			
黃華照	粵漢鐵路工務十三總段	11710		
黃昆昌	農林部水利工程隊		南朝新街九號四樓	
陳榮士	大原營建行		惠福東路惠新東街一二號	
劉文漢	太平南二一號僑聯行	10093		
伍慶常	光復南路八一號			
蕭漢輝	沙面肇和路101號三樓	11390		
卓華耀	太平南路一一號四樓	11725		
楊啓讓	一德中路374號二樓瑞昌行	15207		
陳賡暢			南關麥欄街一號	
文耀芳			長壽東路咸嘉巷14號	
黃国原	市府工務局第一科	14504		

姓名	服務機關	電話	住址	電話
何耀波	市府工務局第三科	14319	西湖路小馬站二二號	
嚴寶琨	電力管理處工程課	11710	西關華貴路厚福大街六號	
黃寬昌	農林部第五工務隊		維新北路南朝新街九號三樓	
王啓祥	西村士敏土廠	10270		
張炯湖	珠江水利工程總局	14763	第十甫七十一號	
黃蘇熙	廣九鐵路工務課	11710	長堤201號四樓	
梁廷俊，阮其江，歐允文 李立賢，王槑威，朱懋澄 陳景洪，譚　殼，黃文海			湛江 湘桂黔鐵路工程局來湛段粵境工程處	

教　授

胡棟朝：本市東山合衆路九號棟園

馮兆端：本市東山新河浦四横路十號二樓

桂銘敬：湛江湘桂黔鐵路來湛段粵境工程處

黃湘湖：廣西柳州湘桂鐵路工務處

黃錫九：本市維新南路朝新街七號三樓

梁綽餘：貴縣，湘桂黔鐵路貴縣橋工所

劉耀鋼：本市啓明四馬路三號二樓

林鴻恩：本市連新路二十九號四樓

羅石麟：Lo Shih Lain

Fung Keong Rubber Mfg. Co.

1 st. Mile Kapar Road, Klang Selengor F.M.S

王叔海：唐山工學院礦冶系

黃郁文：嶺南大學

林逸民　葉穉安　李文邦　區東

因爲時間的匆迫各師長各同學的地址未能詳細調査清楚，如有錯漏或遺漏的地方，請卽函本市河南嶺南大學南大工程編輯室更正爲盼。

——編輯室——

25785

在校同學通訊處

第一年級

陳興華	番禺縣塘步鄉陳明德堂
陳熾琮	台山潢村三合宮步站
陳紹強	廣州高第路大榮街弍號
鍾汝楷	香港般含道十九號
詹忠達	c/o Ban Cheang Co. 6, Mac Arthur St. Kuala Gumper
方奇芬	香港軒尼詩道三百號
符致橋	104, Pekan China Alor Star. Malaya
許 魁	新會，古井，網山鄉
黎錦洲	香港興漢道十三號
黎順康	廣州西關十六甫翁源巷四號
黎懷德	香港軒尼詩道六十七號四樓
林漢寧	廣大路弍巷三號三樓
林汝恒	沙面復興路二號三樓
劉良駒	廣州德政北路雅荷塘六十九號
劉貽光	廣州龍津西路逢源中約六十七號
李建勳	香港加咸街二十一號三樓
李宗浩	汕頭信箱一百號
李偉文	廣州百子路東平路六號「新壘」
梁學誠	澳門沙梨頭五十二號E寶泰公司轉
凌耀光	86 King St. Kingston Ja. B. W. I.
羅寶勳	香港利東街四十二號三樓
麥家平	香港機利文新街順華祥
伍卓漢	香港荷李活道四十六號三樓
倪汝霖	台山冲蔞益元堂

吳志道	廣州芳村明心里
潘成南	香港西摩台四號
岑譜芳	越南堤岸古都大道四十四號
譚建輝	台山白水市郵局交月山村
徐良佐	香港禮頓山道七十二號四樓
王紹高	廣州維新路醬巷街五十二號
黃作新	Sun W. Wong 524 N.W.6th-Ave. Portland 7 Oregan U.S.A.
黃　英	星加坡大坡大馬路一七二號
楊維正	揭陽棉湖南山墟
胡炯群	台山縣斗山墟光豐寶號(轉交西山柵堡安村)
余炳沛	香港永樂街五十二號
戴祖真	12 Logan Rd. Penang
陳　湘	廣東台山新昌區石龍頭郵局

第二年級

歐陽剛	香港中環區多利街十一號
區元侃	廣州沙面肇和路卅一號三樓
陳星明	33 Paradise, Rae Town, Rings Ton, Jamaica
陳世華	廣州太平南路四十六號二樓
張啟良	西貢打冷街八十八號
周毓民	開平沙洲墟瑞華壹寶號
顏文學	上海愛多路一四六二弄一〇七號
鄺國榮	廣州高第路十五號
郭日維	香港永安公司
林卓斌	廣州文德路文德坊二號四樓
林崇義	廣州東山梅花村十七號
林春寧	揭陽蛟龍宮後
林敏初	香港文咸西街四十二號二樓

劉君厚　澳門新馬路永隆號
劉烱坦　安南堤岸嶺街二八三號
劉宏閑　中山石岐大平路四四七號
梁家本　西關逢源路寶盛沙地二號之一
梁權鉊　香港上環永樂街一四九號
梁　榮　香港摩利臣山道五十二號四樓
盧民福　上海王原路華村二衖五號
老洪基　香港皇后大道中十號宏興行
龍建昌　下九路文瀾巷十號
鮑維綱　東山龜岡三馬路二十號二樓
岑悦芳　廣州逢源路寶源中約五十八號
沈潤棠　廣州六二三路新興街二十號
鄧翰儀　西關逢源大街十四號
鄧國樑　澳門草堆街三號
謝國光　澳門亞素平街三十二號
蔡耘耕
尹行賢　廣東東莞萬江鄉永安坊卅三號
温衍智　順德大良八閘一巷七號
王　詗　汕頭永泰街二十號信興行
黄朝澤　汕頭永安街卅一號大利行
黄兆歡　廣州十六甫東街一號
黄耀華　廣東台山縣潮鏡墟和合號
余文傑　香港跑馬地觀馬台十號
余炳沐　廣東台山荻海中和路廣福堂

第三年級

陳子浩　香永樂街七十六號
陳錫元　本市文明路定安里二十三號

周公海	本市寶源路八十四號三樓
鍾福華	廣東新會城南里六號
趙光鴻	台山海晏街廣綸
何達康	廣州十七甫懷遠驛二十七號
姚寶照	香港般含道光景台二號
關學海	香港跑馬地黃泥涌一五三號二樓
劉　鉛	澳門白鴿巢前地三號B
鄭國良	九龍旺角新填地街三〇三號三樓
林思進	香港南北行街四十二號
劉維寶	廣州抗日西路一壹二號
劉益信	廣州寶華路十五甫
李蔭康	香港般含道育賢坊七號
梁耀顯	本市東山新河埔三橫路四十一號
陸景文	廣州東山梅花村十九號
龍頌漢	順德大良新路福聯社龍寶善堂
馬信輝	廣州海珠南路三府新街十四號
吳秉俠	文昌縣羅豆市后山村
保端納	(Bowman Donald) Linn, Kansas U.S.A.
潘世英	南海西樵沙瀛
潘演強	香港跑馬地成和道六十九號二樓
潘應標	香港西摩台十四號二樓
譚條靈	廣州市海珠北路一七五號
鄧錦榮	香港大道中六十二號同興公司
曾文達	澳門荷蘭園二馬路二十七號F
徐保羅	澳門柯高馬路卅四號
辛基球	廣州市一德西路四八三號
黃康道	香港灣仔中華循道會
黃國強	廣東台山縣潮境墟廣信號
黃文添	香港銅鑼灣金龍台二號
鄔境厚	香港渣菲道一三五號三樓
鄔堅柏	番禺南村南街
袁天煦	廣州將軍前瑞南路二十四號號二樓

第四年級

歐陽謙	順德江尾倉華鄉
陳潤初	香港般含道五十九號三樓

梁永崇	嶺南大學
張悅楷	廣州寶源路六十號
趙浩然	高要二區上蓮塘鄉勝益號
何思源	香港彌敦道四九一號三樓
何　慧	香港跑馬地觀馬台十四號
林壽庚	湛江市赤坎
李克勤	香港乍畏街廿九號美德洋行
李小覺	嶺南大學
羅明享	河南寧興中北約四號
呂惠炎	香港干諾道中六十號
黃漢基	本校
黃煥林	廣東新會江門水南鄉德馨里十五號
楊民安	順德陳村廣教鄉

編後話

——浩　然——

在還未有決定出版這本特刊以前，我們曾經考慮到經費和内容這兩大問題：經費方面：因爲目前幣值的低跌，紙張人工…的飛漲，都在我們預算外跳躍着，同時因爲時間的匆廹，只不過一個月的籌備，稿件和編排當然會有很大的辣手，尤其是在嶺南這校園裡，考試和習作是這樣的多，同學們多半時間消磨在應付考試裡，那裡能夠找得出充分的時間來搜集材料研究課外的問題呢！可幸經過了大家同學的努力，本着知其不可爲而爲之之精神向前邁進，同時更得到各舊校友教授及各界人士的熱烈指導和幫忙；在徵求廣告上得到林文贊先生，陳潤初君，呂惠炎君，黄文添君與各舊同學的幫忙與愛護更且得到留穗劇人在青年會演出曹禺先生名作：社會諷刺白話劇——日出——爲我們出版基金籌款，替我們解決經費上的疑難，這是萬二分感激的。在内容上得到廣州港工程局陶局長述曾，三區公路局王局長節堯，湘桂黔鐵路局副局長來港工程處處長——我們的系主任——桂銘敬先生在百忙中爲本刊撰文，使本刊增光不小。同時更得到本校老教授劉雅錮先生，舊同學甘光儀先生和劉義和先生林炳華先生的指導，使本刊能完滿出版，我們是感激萬分的。

這本小刊物因爲籌備的時間太廹促，中間又因稿件和經費的阻折，未能依時在五月九日出版，加以編者的才力能力薄弱，很多地方未能遵循和達到各位師長和同學的意志和理想的，更且會有遺漏錯誤之處，這是要請大家原諒的。

五月六日晚于工學院

介紹

本刊此次出版時間忽迫，交由廣州市教育路十六號蔚興印刷場承印，蒙予協助，依期出版，且工作優良，特此介紹。

——南大工學日紀念會——

（港）

軒鯉詩道一百九十七號　電話：弍伍叁玖弍號

承辦海陸工程

德聯建築公司

（粵）

清平路一十三號　電話壹弍陸捌肆號

Jarman
SHOES FOR MEN

美國馳名「嗜文」鞋

•大帮新到•

款新•料美•價廉

永安公司

Farnsworth

豐富牌收音機

GT—669—WAZ

BANDSPREAD

案頭型

五波帶

收音機

ET—651

小型玻璃膠質收音機

此兩種收音機俱具
有通用變壓器可適
用國內各地電流由
一〇五至二二〇電
壓且能接收全世界
播音聲音清晰柔和
式樣大方美觀

FARNSWORTH-KNOWN FOR TONE

先施公司

總代理

太原營建行

惠愛東路惠東新街十二號

電話：一一七六二

穗港滬梧邕柳

建基行

經營業務

代理亞細亞火油公司出品

出入口貿易

信託業・各種企業

廣州市六二三路一〇八號

電　話：一〇三五七

電報掛號：〇五九四

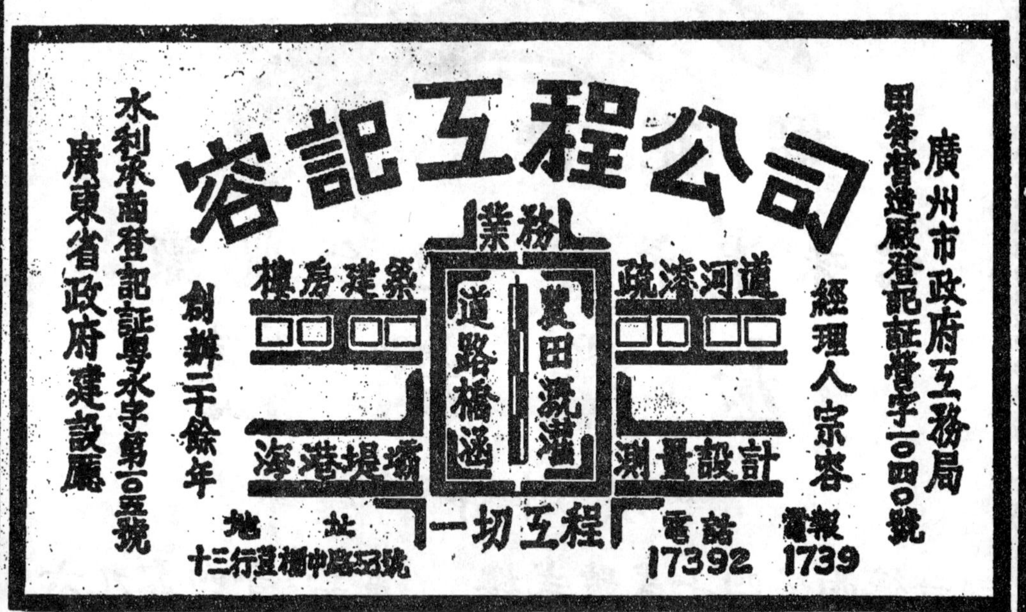

宏業東莞製糖廠出品

味濃氣芳·色白質潔

宏業糖廠

廣州辦事處

一德西路四七弍號弍樓　電話：壹四零九六

南大工程

桂銘敬

康樂再版 第三期

月刊通訊

主編：潘世英

編輯：吳秉俠 何逢康

目錄

嶺南大學工學會刊印

民國三十七年十一月一日出版

寫在編前

潘世英

在多難的世紀，在紛亂底中國，這樣安靜的學習機會，是值得珍惜的。若我們不好好地利用它，簡直是太對不起國人，太對不起自已了。嶺南的同學，能够屹屹窮年，是很值得誇耀的。但因教育方針的未盡完善，與乎功課的繁重，以致各同學終日在圖書館裏苦鑽，也鑽不出課本的範圍，這可算是美中不足了。是課本的知識，巳够滿足我們底求知慾嗎？答案當然是否定的。大學的教育，與中學的教育不同。它是一種專才的訓練，重在自由的發展和興趣底培植。課本所給我們的，不過是些基本的知識，建成了堅固的屋基，而不築上大厦，是很可惜的。而且，一個人能終生力學不倦，能在百忙中不忘學習，每由于在大學時，曾經養成了堅强底讀書興趣之故。若在求學的時候，天天只是在課本裏邊鑽，鑽到精力疲乏，感到木然無味，則他日出到社會，還能刻刻手不離卷，力求上進，恐怕是百中無一了。南大工程的出版，主要的作用，就是希望它能給我們一個翻書的目標，引起我們閱讀的興趣。我相信在功課的重壓下，去探求課本以外的知識，一個現成的目標是須要的。其次，一個人的精力有限，圖書館裏的書籍，汗牛充棟；有些是應該放入文獻館，供考古家的研究，有些是「超乎時代」的創作，在中國還用不着。我們欲利用那僅有的瞬息課餘，在琳瑯滿目底書堆中，找尋我們的「寶島」，眞是像在荒漠裏尋找甘泉一般的不易了。在學術環境普遍染上了貧血病的今日，想在四年的大學生活中，得到很多的心得，簡直是一種奢望，由於上面的兩個原因，我們是要分工合作，集腋成裘了。我們很希望這少少的刊物，能盡我們知識交換和積聚的一部份責任。我們更希望教授們的心得，能在這裏賜給同學，獻予社會。

最後，我應該說說本刊籌備的經過了。

本刊是學術性的刊物，一向是期刊或年刊的，何以現在改爲月刊呢？這有下面幾個原因。

（一）爲要配合各人寫作的速率——嶺南各刊物出版的時候，編輯是很少不爲找稿而彷徨的，若因稿件不足，勉强塞責，致淪於爲出版而出版，（校長語），則未免太浪費物力了。我們自度能力，改爲月刊，增加期數，減少篇幅。就是爲要與同學們的學習和寫作速率配合之故。

（二）爲要配合同學們閱讀的興趣——當我們接到一本厚厚的刊物，往往是不知要從何處讀起的。當閱得未及一半時，這本刊物又會被遺失了。雖然，知識像佳肴一般，不會被厭多的。但多桌佳肴，要一次吃完，那總不及分多次吃的好！分期刊出，也是爲了這同一的理由。

（三）經費較易籌措——復員以來，本會曾多次向外籌款。下學期的工學會日，又必須籌一大筆的數目。故本學期「南大工程」的經費，我們不想再煩擾同學太多了。唯一可走的路，就是多招廣告。但廣告是不易找的。多出幾期，改爲月刊，相信對這一問題的解決，亦較容易。

由于以上的三個原因，我們就決定改爲月刊了。但何以月刊下面又有「通訊」兩字呢？理由很簡，單因爲工程是決不能離開實地底經驗的，這裏不用多講了。我們都是學習，不完善之處，務希指正

工學會學術講座

第 一 講

「大學生就業應有之認識」

主講：桂銘敬系主任（在假）

十月二十日
吳秉俠筆錄

兄弟今日得機會回校，謹以至愉快之心情與各位相見。本人離校多年，然無日不緬懷於嶺大。蓋以此地環境幽靜，際此遍地烽烟之日，可謂絕優之境，諦望各同學能充分利用，而致力於學問工作也。

今日所欲向諸君研討之問題，乃一極普通而爲衆所欲知之問題。此即「大學生就業應有之認識」。

兄弟於各工程機關中服務多年，對各青年工程人員接觸特多，彼輩初爲社會服務，對於此問題之了解尤感迫切，每以此而見詢，故今特擇此題目與諸位共商，蓋諸位來日亦不免爲此問題所困也。

大學畢業同學初入社會，其工作階層必從低而達於高，事業發展乃自小而趨於大。其中必經艱辛之奮鬥，尤達成功之境。尤以吾等工科同學，在校雖學習各高級之設計課程，然一旦置身社會，初必無發揮此等學問之機會。非自低級工作起始而達於高位後，難於應用也。是以切勿目視背負經緯儀者爲可恥，而手持軟皮尺者爲不屑。蓋事業之成功，乃基於根基之鞏固也。

猶憶抗戰期間，對於從事工程建設應有之知識，諸如物價起跌，金融動盪，包工等各問題，皆非書本中可能獲得者，惟有不諱卑職，窮力研究，方能獲得經驗，乃可爲來日擔當重任之基本。此即「以犧牲自我，爲他人服務」之信條爲旨，而拋棄職位高低之成見，以服務人羣爲前題也。以上所述，乃諸君應有之第一點認識。

第二點欲告諸君者，即本國乃一工業落後之國家，雖迎頭苦幹，尤恐不及歐美，然回顧目前國勢之混沌，不無悲觀之意。加以現日手捧之書本，多皆過時之學識，因而灰心意冷，然此皆極端錯誤之舉，吾人務須凡事樂觀，以書本之知識爲基本，而輔以目前新異之科學知識。以爲國家效勞。

其次，吾人所習之土木工程，乃始自帝國羅馬時代。其時只有軍事工程（Military Enqq）。及後國勢平定，乃分有土木及軍事兩種工程。再後隨時而異，而有今日之機工，化工，建工等門系。是以乃各形成一狹窄之學習範圍。以吾輩土木系而言，每以爲熟習各土木知識後，即可應付各問題。然此乃錯誤之見。若以築路爲例，如不稍具機械之知識原理，則遇有開山鋪路之機車時，即無從駛用，而陷工程於停頓。因而吾人必需以土木爲主，而兼習他種工程。尤以本國工程分門之未見明晰，更須擴濶學習範圍。

末點所欲告及者，即工程人員必須具有如下之志趣：「以工程爲終身事業」。普通一工程畢業生，每爲待遇薪餉問題而感徬徨。或就高薪而從之，如是則對工程志趣極不符合。吾人必須以工程之知識奉獻國家，以求科學之進步。而切勿以環境之舒適爲前題，以生活之享受爲目的也。

廻視今日艱難之國家，交通最感困難。以美國二十萬里之鐵路較之吾國只萬多里鐵路，實太稚弱。目前只京滬路爲完整耳，如浙贛鐵路，粵漢鐵路，皆破碎未全，而湘桂黔一線，亦未齊劃。故吾人必需立心，爲國完成此等工程。

括而言之，即各青年工程人員，必須以勞苦自任，小心學習，擷取經驗，以便來日身肩重任，不至遇事煩惱。尤憶抗戰期間，政府曾選派一稍有工程服務經驗之大學工程生赴美實習，彼輩於美最初之工作，即爲敲擊碎石。由此可知，工程之工作實無階層之分。吾人必須視任何之服役爲學習機會，切勿妄自菲薄，輕視低級工作。而應認之爲工作起點，而極力獲取經驗。

更有甚者，即諸君於校內所處理之工作，多無需肩負責任。然一旦置身社會，則舉凡一切工作，皆有責任存在。責任之輕重隨工作之大小而定。故屆時務須悉力爲之，方不負上級所托也。

今特敬告諸位，工程人員之工作乃最爲勞苦者，諸君若眷戀於舒適之享受，則請早日回頭，另尋興趣，免至畢業之日，即輟志之時，則徒虛耗時光乃矣。

至本人目下之工作，常爲各新舊同學所關懷，今謹藉此時機，略爲一告。本人担負之鐵路工程，爲自廣西之來賓而迄於廣州灣之湛江。此段路線若一旦通車，則可達來賓，柳州，再而桂林而接衡陽。如湘贛路完整，更可直達南京。西行方面，可通昆明。現目所缺之材料，全爲外洋之鋼鐵橋樑，炸藥等。若國家局面安定，經濟平穩，即可解決。

全段路線之初測與實測，皆告完竣。目前只候材料及款項到達，即可施工，若施工，於一年之內，即可通車。

至廣州灣之海港建設，其較重要之問題爲碼頭及倉庫之設置，河道之改良等。目前因經濟問題，只從事測量設計工作乃矣。

建設實非輕易之問題，吾國非一重工業國家，然舉凡建設，皆不離重工業，故多仰自外洋。然吾等切莫灰心，應堅志趣而力學習，以爲來日事業之根基。

末後，望諸君必須立定「以工程爲終身事業」之志趣，以「犧牲自我，爲他人服務」之信條，方可担負大任。尤憶抗戰期間西北之工程人員，爲鐵路綫之測量而直達新疆及蒙依靠，奔馳於沙漠地帶，其艱苦情況，非吾輩所可推想者。彼輩所以身受艱辛而不怨，風餐露宿而無苦，皆因有不泯之志趣而犧牲之信條也。

本人對各同學至感關懷，謹以過往之經驗，與諸位共同切商，希相互勉之。

氣體分子對於電磁微波之吸收

凌鐵錚

溯自二次世界戰爭以來，物理學家對於極高頻之電磁波研究不遺餘力，其中最著成效者，厥爲雷達之成功，不但有助戰事早日結束；且戰後民間採用，保障海上及空中交通安全，頗著成績。此種高頻波又稱微波（Micro—Wave），波長在100公分以下者屬之，因欲有別於通常所指之短波也。

微波最近另一應用，爲研究氣體分子之結構及狀態，一如以光波研究原子或分子內部之結構然。觀察原子之放射或吸收光譜可以印證數學之推理是否正確。此種光譜研究即光譜學是也。光譜學今日業已衍成一帙卷浩繁之科學，其所用波長，自X一射線至紅外線，約在數Å至數千Å之間。（Å爲Angstrom之縮寫，爲一公分之億分之一）今以微波研究分子吸收，方法與前者類似，不過頻率降低而已。

故微波分子研究可視爲光譜學範圍之擴張。但微與熱波之間，迄未見有研究報告，蓋因電子管之應用，尙不足產生如許高之頻率也。

約三十年前，吾人業已知道任何氣體，若具有電偶子者（Electric Dipole）均可以吸收電磁波之能量。

設有某氣體分子，具有兩個相等之慣量矩，則其本身之頻率與外來之電磁波諧振時，產生吸收。此種吸收可稱爲吸收頻率，可以下式表之。

$$f = \frac{n\,h}{I\,4\pi^{2}}$$

式中h爲浦郎克常數，n爲正整數，I爲分子之慣量矩。（圍繞以重心與對稱軸垂直之軸而旋）。故f可爲多個且互爲倍數。不對稱之分子，吸收波長之決定較爲複雜。通常大小之分子吸收均在微波範圍以內，其波長自一英尺至•02公分不等。

以往研究分子旋動之譜帶大多利用白熾之固體作爲光源，通過擬研究之氣體，然後以紅外線光譜儀分析之。一九三三年密歇根大學始有以電子管振盪器代替光源之嘗試。目前1.25公分之電磁波產生及檢波已臻完善，故微波研究分子工作日益衆多。

應用電子振盪之波測驗分子之吸收譜帶係自 C.E. Cluton 與 N.H. Williqms 開始。兩氏證明銨氣在1.25公分顯示甚寬之吸收譜帶。

1. 檢驗吸收譜帶之儀器

一九四六年牛津大學學者數人將銨氣灌充於空穴諧振器中。電磁波在此器中往返振盪，因銨氣之吸收而漸次失其能量。若以晶體檢視能量遞減之程度即可測出吸收之强弱。

另一種測驗方法爲 Cluton 與 williams 兩氏及其他美國物理學者所用的方法。其儀器包括一磁性電管之振盪器與導波管，大概佈置情形略如第一圖。

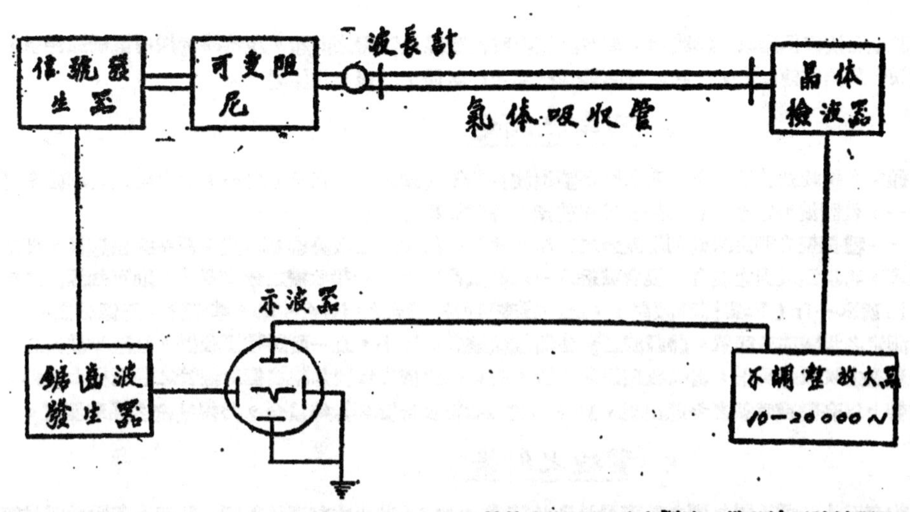

圖中所示吸收式之波長計，係一空穴諧振器，與導波管之間有極疏之耦連。當此波長計被調整至諧振時，自導管中吸取少量之能量，吾人可自檢波電計之指針向下突降得知。空穴之大小有綫旋之螺絲調節，并刻以業已校準之頻率度數，俾一視便知其頻率多寡。在導波管中之氣體被電母所製之窗門封閉於內，若波長在十餘公厘左右，則通常用1/4×1/2英寸之導管，管壁爲0.04英寸之銅或銀片製成。

導波管之選擇與頻率有直接關係，若管口過小，衰減太大，若管口過大，雖衰減較低，但可能發生多種波型傳送，尤有進者，設在晶體檢波之一端如未有適宜之終端調節在管中又足以產生反射及駐波現象，均足以增加觀察之困難與錯誤。乃免去多種波型及駐波之困難，吾人必須選用較小之管及調頻之電波，現時最令人滿意之方法當推 R.H. Hughes 與 E.B. Wilson 兩氏之方法，茲將各種不同佈置所得之吸收譜之形狀示如第二圖。

第二圖　如何削減吸收譜帶中之駐波

A.—譜帶中駐波甚大

B.—施用慢調頻方法以後

C.—施用差變放大路線以後

D.—當振盪頻率較波段爲小時之情形

E.—施用二次差變路線以後

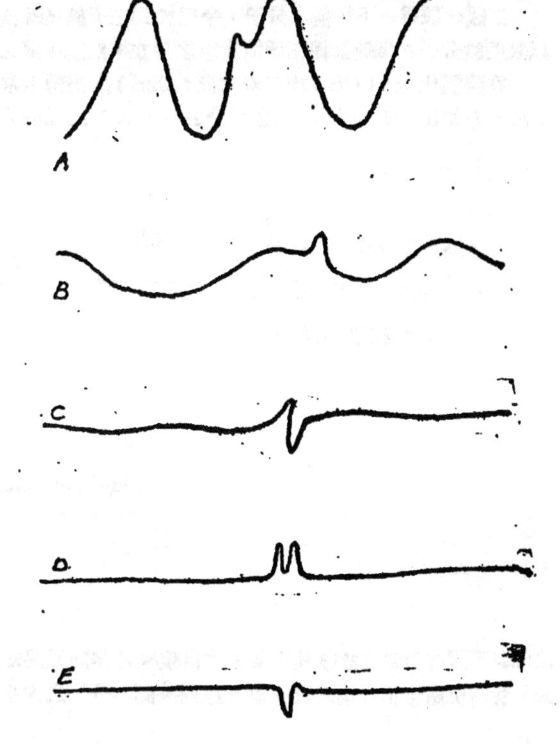

測驗之技術困難除上述諸項，尚有晶體噪雜信號混入譜帶之問題，現時學者採用電橋路線以減低之，即電波分爲兩路傳輸之，一路經過氣體，一路不經氣體，然後合併檢驗之。

2. 頻率之測定

通常之吸收式波長計自刻度上直接讀出波長若干，精確可至五千分之一，空穴較大者準確度可至五萬分之一，惟因波型較多之故必須在使用前預知各種波型之頻率。

另一種捷便之測定法爲利用波差之方法，係以一種穩定之振盪器（石英）經多次倍週後，產生一標準之頻率，再以源波與之混合，混合波送入一逕訊式接收機，若接收機之頻率爲f，則可知源波之頻率必爲Fo十f或Fo—fp（Fo爲標準頻率），此種頻率可與標準局之信號合校準，準確度可至億分之一。

測定之步驟亦殊簡單，將源波之發生器緩緩變動其頻率，且一面調整接收機，直至示波器之屏幕上顯示一頭峯之吸收線爲止，當吸收線出現之後，再以接收機之信號加諸陰極射線管之强度柵上，俾使拍頻在示波器上以亮點或暗點之姿態出現，於是調整接收機使此點與顛峰重合，乃得精密之頻率測定。

3. 實驗之結果

約有五十餘種不同氣體之電磁微波吸收綫業經測定。其中大部屬於對稱一類者，其吸收譜較簡單。如第三圖A所示者爲碳氧化硫（COS）之吸收譜，四條吸收線距離均勻，第五綫爲虛綫，係測定者適缺少10000至20000倍週間之振盪器，未能測定。

對稱分子之另一例爲氯甲烷，其分子爲三個氫原子排成直綫垂直於對稱軸，一氯一炭排於對稱軸上。此分子之吸收譜每綫相距23000兆周。

事實上分子多數構造不爲對稱。如第三圖B所示吸收線之距離極不規律，此圖爲二氧化氯在低溫時之譜帶，若在高溫或較重之同素體組成之二氧化氯，其譜帶更爲複雜，且各綫之間不甚明顯。

二氧化硫爲一不對稱之分子，今已得到三十餘條吸收綫，吾人雖尚不能一一加以滿意之解說，然亦足以表明該分子在低速旋轉三個慣量矩之存在，解說困難之點在於該分子高速旋轉時可能之變形不易估算。

在碳氧化硫（COS）或二炭化硫（C2S）之譜帶中常發現不少微弱之線，此等弱綫係由於硫之同位素之故，（如16S34. 16S33 約含百分之一，二）此羣弱線恆移數數百兆週，蓋慣量矩改變之故也。因原子

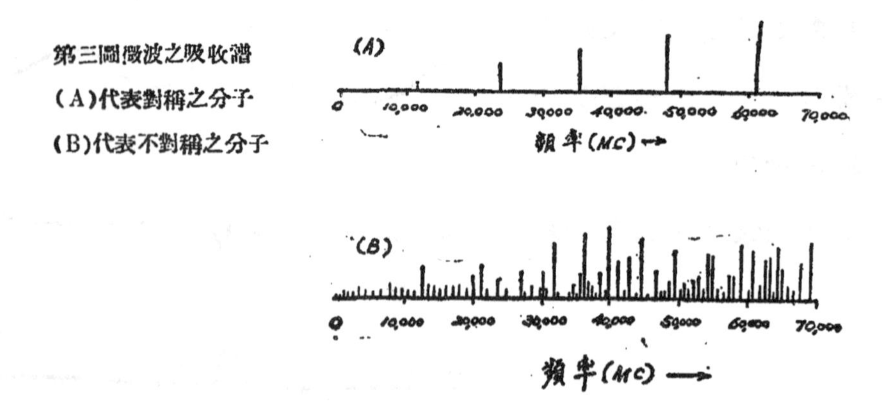

第三圖微波之吸收譜
（A）代表對稱之分子
（B）代表不對稱之分子

之間幷不因同位素之故改其距離，故自頻率之移動可獲知原子間之遠近。Gordy Simmons與 Smith諸氏，曾研究氫素化合物之吸收線。茲列其測出分子之大小如下：

鹵素甲基（Methye Chloride）化合物之分子大小

分子	C—X之距離（10—8cm）	C—H之距離（10—8cm）	HCH之角度
CH_3F	1•38$_4$	1•11$_2$	110°0'
CH_3Cr	1•779	1•109	110°0'
CH_3Br	1•936	1•104	110°05'
CH_3I	2•139	1•100	110°08'

4. 原子之自旋與細微結構

以上所討論僅及於分子旋動之能量換轉，若原子在分子中具有自旋，情勢自然有別。在多數情形之下，原子核恆有自旋，此自旋之角矩量與分子之軌行角矩量（Nuclear spin momentum and Orbital Angular momentum）合成一新向量，此新向量將圍繞一軸心作進動。於是整個動量準位遂分裂爲複帶，此種分裂在譜帶中將稱爲細微結構（Hyperfine Structure）若在氣體外圍施以電場或磁場研究其能量分裂情形，其產生之效應分別稱爲 Stark 與 Zeeman 效應。

茲以篇幅之故無法詳述。要之，由 Stark 效應中可推知分子之電偶之强度與其方位。自 Zeeman 效應中可推知核之磁偶也。

綜之，物理學家研究物質之構造已自分子，原子以至核心，所用工具自最短之X—射線以迄電磁微波，惟引爲遺憾者。迄至目前微波與熱波之間倘有甚大之空隙，若吾人能於不久將來，築成此一橋樑，使吾人之工具臻於完善，則對人類理解自然或求知活動，無疑爲一大貢獻也。

本篇材料多取自 D.K. Coles 微波分子譜帶一文，特此申明，以免掠美。

參攷資料：

1. Electromagnetic waves of 1.1 cm Wave-length and the Absorptim Spectrum of Ammonia, C.E. Cleēton 4 N,H, williams, phy.Rev. Vol.45 1943 P 234.
2. Ammonia Spectrum of 71Cm wave-luyth region, B. Bleaney, R.P. Penrose, Nature (London) Vql.157, 1946 P,339
3. Micro-Wave Spectra - The Hyper-fine Structure of Ammonia, W. Gordy, M. Kessler Phy. Rev., Vql. 71 1947P 640
4. A New Elutronic System for Detecting Micro-Wave Spectra, W, Gordy, M. Kessler, phy.Rev. Vol 72 1947 P. 6.44
5. A MicroWave Spectrograph, R.H. Hughes, E,B, Wilson, phy.Rw. Vol.71, 1947 P562
6. StarkqndZeeman Effctes in the Inversim of Spectrum of Ammonia, D,K, Coles, W,E. Good, phy.Rev. Vol 70, 1946 P. 979
7. Micro-Wave determination of Molecular Structure and Nuclear Coupling of the Methyl Halides, W,Gordy, J,w,Simimons A,G.Smith, Phy. Rev., Vol 74 1948 P. 243

台灣公營鐵路概況

劉載和

暑假間劉教授曾作台灣考察行，搜獲各方材料至多，奈限于篇幅關係，迫得分期刊出，殊引為憾。又劉教授於歸途間，得與化學系同學梁元溥，陳瑞勳，周寶芬，李啟濱等相遇，雖素昧生平而蒙承招待，足見南大尊師重道之精神，吾等殊表謝意。

編者

(一) 沿革

凡到過台灣的人，必往台北市，凡到過台北市的人，必參觀博物館。在博物館外園邊近馬路旁之處，放着兩個舊火車頭。這兩個火車頭，除了附近露宿街邊的乞丐，初由鄉間進城的行人，和少數自遠方到來旅行參觀的之外，普通市民，對牠們多不加以注意。露宿街邊的乞丐對牠們注意的原故，因為在大熱的天氣裡，牠們是最舒適的涼床。初由鄉間進城的行人對牠們注意的原故，因為在小便急而找不到廁所的時候，可以做他們最好的掩蔽。牠們雖然被人們遺忘，但牠們對於台灣的功績，則不能被抹煞。因為在這兩部火車頭當中，有一部叫做「騰雲」，當台灣最初修築鐵路的時候，曾向德國購買火車頭八部，在鐵路上行走，而這八部火車頭的第一號，就是這個「騰雲」。所以牠曾帶給台灣以文明，牠曾帶給台灣以繁榮，在台灣交通史上，填上第一頁的空白，所以牠對於台灣的功勞，實在不小。想起牠以前在鐵路奔馳的時候，叱咤風雲，萬民驚畏，每至一處，行人辟易，那種威勢，誠是不可一世。現在則矮小的身材，因為年事已高而有點偏僂了，堅實的鐵膚，亦因受雨打風吹而有點剝落了。被棄置在公園的一邊和乞丐為伍，度其悲慘冷靜的餘年，撫今追昔，益增遊客憑弔之情！

論鐵路建築，國內比台灣為早。國內遠在同治五年(1866)已開始修築淞滬鐵路，至光緒二年（1876）年完成，但因為受當時官民兩方的熱烈反對，卒至把經已築好的鐵路拆去，所以築等於沒有築。台灣計劃築路雖然是光緒十二年（1886）的事，但經巡撫劉銘傳氏於光緒十三年（1887）四月奏准在台灣敷設鐵路之後，即大舉開工，至光緒十七年十月，已把基隆至台北間一段完成，並向南展築。但不幸得很劉氏因事去職，由邵友濂代任巡撫。此公見解，和前任完全不同，認為新築以南的工程困難，奏請停築，所以到光緒十九年（1893）十一月台北新竹間鐵路竣工之後，台灣在未割讓給日本之前的鐵路，亦僅有這麼多而已。台灣鐵路雖然因為邵友濂之奏請停築而不能繼續進展，但已後來居上，比之國內勝一籌。所以現時到過台灣的人，每每說台灣的建設比之國內為勝。其實這並非現在如是，以往亦如是，好像命裡注定是台灣勝過國內的。因為交通為各種建設之母，交通有辦法，其他各項建設亦必有辦法。台灣在交通建設方面比國內跑快了一步，所以台灣各項建設亦比國內為優。

台灣割讓給日本之初，因為台灣民衆反日思潮甚盛，所以當時抗日的戰事，遍及全島，鐵路運輸，亦因之停頓。後來日本為運兵作戰計，首先恢復台北基隆間的鐵路運輸，並以軍費十萬元，充作縱貫鐵路之調查及建築出口港澳調查費用。調查大致已完，設計及編製預算正在進行的時候，適有人提議鐵路私營，在東京並有人設立「台灣鐵路」向總督請求准許自行修建。至1896年十月二十七日，總督果然准他們的要求，並由政府贊助他們的事業，給他們種種便宜。但因為當時各地抗日運動仍然很激烈，加以經濟又遇不景氣，所以卒不能實現。適逢當時基隆與台北間的運輸，非常不便利、運費又特別高昂、日本乃計劃從速敷設各地鐵路，實行官營。並依據調查所得的結果，草擬十年計劃，分從南北着手。北部以改善原有鐵路為主，南部則以高雄為基點，漸次向北敷設，至1908年基隆高雄間四百餘公里之南北縱貫鐵路，始全部通車。

除了縱貫鐵路之外，尚有支綫多條，或修築於縱貫鐵路完成之前，或修築於縱貫鐵路完成之後，玆將台灣各公營鐵路修築沿革列表如下：——

線段	里程(公里)	建築者	建築年份	附註
基隆—新竹	102.70	中國政府(巡撫劉銘傳)	1886—1893	
新竹—高雄港	299.16	日本政府	1899—1908	
台北—淡水	22.40	,,	1900—1901	
高雄港—九曲堂	17.42	,,	1907	
花蓮港—玉里	87.27	,,	1910—1917	
九曲堂—屏東	7.22	,,	1911—1914	
基隆—台北	28.62	,,	1912—1919	添築雙軌
屏東—潮州	22.34	,,	1917—1920	
八堵—蘇澳	94.96	,,	1917—1924	
竹南—王田	85.99	,,	1919—1922	
玉里—關山	41.45	,,	1921—1926	
台東—關山	42.01	台東拓殖株式會社	1922	由政府收購
二水—外車堤	29.70	台灣電力株式會社	1927	,, ,,
台北—竹南	97.10	日本政府	1927—1935	添築雙軌
台南—高雄港	46.70	,,	1927—1935	,, ,,
三貂嶺—菁桐坑	12.92	台陽礦業	1927	由政府收購
潮州—坊寮	18.00	日本政府	1937—1941	
社邊—東港	6.08	,,	1937—1940	
田町—高雄	6.87	,,	1937—1941	添築雙軌
高雄—九曲堂	13.74	,,	1937—1942	,,
民雄—嘉義	9.24	,,	1938—1942	,,
台南—新市	11.43	,,	1939—1943	,,
花蓮港—東花蓮港	4.04	,,	1938—1939	
新竹—竹東	16.60	中國政府	1946—1947	

(二) 現狀

(1)組織——

台灣公營鐵路，全由台灣省政府交通處之鐵路局所管，該局下設車務，機務，工務，材料，會計，總務，等六處，及秘書室，員工訓練所，高雄辦事處，花蓮港辦事處等十個單位。在此十單位之下，復分爲若干小單位，如工務處下設有工務段，車務處下設有車務段車站等，機務處下設有機務段修機廠等，材料處下設有材料庫等，全局員工共有15,731人。其組織系統如下：——

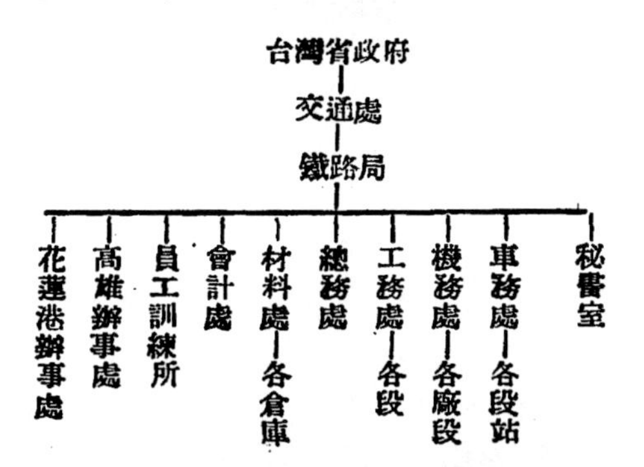

(2)路線——

台灣鐵路，分爲東西兩部，凡在台灣的東部的路線，稱爲東部路線，凡在台灣的西部的路線，稱爲西部路線。宜蘭綫雖然繞至本省的東北角，但因和西部路線相連，所以亦歸入西部路線的範圍，台灣西部地面開濶，人口衆多，城市村落，都比東部爲多，所以在西部的鐵路路綫，亦比東部爲多，鐵路里程，亦較東部的爲長。現把西部路線和東部路線的名稱，起迄站和里程列表如下；——

路線名稱		起點	迄點	里程（公里）	雙軌部份 起點	迄點	里程（公里）
西部路綫	縱貫綫	基隆	高雄	408.5	基隆 民雄 新市 田町	竹南 嘉義 高雄 高雄港	125.7 9.3 58.1 2.4
	台中線	竹南	彰化	91.4			
	屏中線	高雄港	林邊	60.0	高雄	鳳山	5.7
	宜蘭綫	基隆	蘇澳	98.7	基隆	八堵	3.7
	淡水綫	台北	淡水	23.6			
	平溪線	三貂嶺	菁桐坑	12.9			
	集集綫	二水	外車埕	29.7			
	竹東支綫	新竹	竹東	16.6			
東部路綫	台東綫	東花蓮港	台東	175.9			

(3)軌距坡度灣道

台灣鐵路建築的標準，各條路綫都不一致。這和建築的機構，建築的時期，建築的經費，和路線所經的地勢狀況有關。就大概而言，西部路線建築標準，比東部路綫爲高。西綫軌距爲 1.067公尺(3'-6")，而東部路綫的軌距則爲 0.762公尺(2'-6")。西部路線的坡度，在縱貫綫上最大爲1%，在台中線上最大爲 2.5%，而在東部路線最大坡度爲 2.5%。西部路線的灣道，亦較東部路線爲緩和，在縱貫綫上的灣道的最小半徑爲 300公尺，在台中線上灣道最小半徑爲 260公尺，而在東部路線的灣道最小半徑爲100公尺。由此可以見到西部路線和東部路線建築標準的大概。

(4)鋼軌與枕木

台灣公營鐵路所用的鋼軌，輕重不一，種類繁雜。現時西部路綫所用的，包括每公尺三十公斤，每公尺三十七公斤，每公尺四十五公斤，每公尺五十公斤四種。四十五公斤至五十公斤之鋼軌，備用在隧道之內，隧道以外的路綫，則多用三十公斤至三十七公斤的。東部路綫所用的鋼軌，更爲複雜，有輕至每公尺十二公斤的，有重至每公尺三十公斤的，而十五公斤，十八公斤，二十公斤，和二十二公斤的鋼軌，亦很普遍採用。

枕木的尺寸，亦長短厚薄不同，普通枕木，在西部路綫，其長度爲 2.15 公尺，厚度爲 0.14公尺；在東部路綫則長 1.53 公尺厚度由 0.105 至 0.20 公尺。玆將東西路綫各種枕木尺寸列表如下：——

西部路綫		
路枕	橋枕	轍枕
m m m 2.15×0.20×0.14	m m m 2.20×0.20×0.15 2.20×0.20×0.16 2.20×0.20×0.17 2.20×0.20×0.18 2.20×0.20×0,19 2.20×0,20×0.20 2.40×0.20×0.14 2.40×0.20×0.15 2.40×0,20×0.16 2,40×0.20×0.17 2,40×0.20×0.18 2.40×0.20×0.19 2.40×0.20×0.20 2.40×0.20×0.22 2.40×0.20×0.23 2.40×0.20×0.24 2.40×0.20×0.25 2.60×0.20×0.20 2.60×0.20×0.23 2.60×0.20×0.24 2.60×0.20×0.25	m m m 2.20×0.23×0.14 2.35×0.23×0.14 2.50×0.23×0.14 2.65×0.23×0.14 2,80×0.23×0.14 2.95×0.23×0.14 3.10×0,23×0,14 3.25×0,23×0,14 3,40×0,23×0,14 3,55×0,23×0.14 3.70×0.23×0,14

東部路綫		
路枕	橋枕	轍枕
m m m 1.53×0.155×0.105 1.53×0.20 × 0.14	m m m 1.83×0.20×0.15 1.90×0.20×0.15 2.44×0.20×0.15	m m m 1.83×0.23×0.14 1.98×0.23×0.14 2.05×0.23×0.14 2.13×0.23×0.14 2.29×0.23×0.14 2.44×0.23×0.14 1.83×0.155×0.105 1.98×0.155×0.105 2.05×0.155×0.105 2.13×0.155×0.105 2,29×0,155×0,105 2.44×0.155×0.105

枕木舖設時的距離，東西路線，固然不同，單以西部路線而言，亦每每因其所在位置而不同。西部路線枕木的舖設方法如下：——

位置	軌長	每軌用枕木根數			
		一等路綫	二等路綫	重要側道	其他
在直道及緩和灣道上	9.144m	14	13	13	12
	10m及10.058m	15	14	14	13
在半徑小過400公尺之灣道上	9.144m	15	14		
	10m及10.058m	16	15		

東部路綫枕木舖設方法如下：——

位置	軌長	每軌用枕木根數			
			幹綫	側道	
在直道及緩和之灣道上	9.144m		13	12	
	10.m及10.058m		14	13	
在半徑小過400公尺之灣道上	9.144m		14	13	
	10m及10.058m		15	14	

（5）橋樑

台灣公營鐵路的橋樑，載重量約等於古拍氏E—33級（Cooper'sE—33）。跨度大過5公尺的，在西部路綫共有1003座，在東部路綫共有244座，東西路綫合計1247座，總長達33700 公尺。跨度小過5公尺的小橋和涵洞，在西部路綫共有4838座，在東部路綫共有521座，東西路綫合計5359座。東西路綫跨度在200 公尺以上的重要橋樑，共有三十一座，如下表所列：——

路綫		橋名	長度	位置	
西部路線	縱貫綫	1 新店溪橋	868 公尺	距基隆	32.8 公里
		2 第一大嵙崁溪橋	316 公尺	同上	37.4 公里
		3 第二大嵙崁溪橋	416 公尺	同上	38.8 公里
		4 鳳山溪橋	282 公尺	同上	99.4 公里
		5 紅毛田溪橋	345 公尺	同上	102.9 公里
		6 下頭份溪橋	317 公尺	同上	127.9 公里
		7 下後龍溪橋	317 公尺	同上	141.8 公里
		8 三叉溪橋	237 公尺	同上	146.4 公里
		9 下大安溪橋	914 公尺	同上	176.9 公里
		10 下大甲溪橋	1213 公尺	同上	184.1 公里
		11 大肚溪溪	585 公尺	同上	211.7 公里
		12 濁水橋橋	953 公尺	同上	253.1 公里
		13 林之頭溪橋	236 公尺	同上	263.2 公里

		14 牛稠溪橋	209 公尺	同上 293.6 公里
		15 八掌溪橋	236 公尺	同上 308,0 公里
		16 曾文溪橋	710 公尺	同上 336.4 公里
		17 二層行溪橋	237 公尺	同上 372.0 公里
	屏東綫	18 下淡水溪橋	1526 公尺	距高雄港 19.2 公里
		19 東港溪橋	236 公尺	同上 38.1 公里
	台中綫	20 頭份溪橋	315 公尺	距基隆 127.9 公里
		21 後龍溪橋	294 公尺	同上 138,0 公里
		22 上大安溪橋	634 公尺	同上 172,1 公里
		23 上大甲溪橋	379 公尺	同上 179,4 公里
	宜蘭綫	24 宜蘭濁水溪橋	783 公尺	距八堵 77,2 公里
東部路綫	台東綫	25 知阿米溪橋	395 公尺	距花蓮港 26,7 公里
		26 萬里溪橋	296 公尺	同上 38,3 公里
		27 馬大鞍溪橋	246 公尺	同上 42,7 公里
		28 清水溪橋	427 公尺	同上 89'8 公里
		29 新武昌溪橋	467 公尺	同上 120,3 公里
		30 鹿寮溪橋	278 公尺	同上 142,5 公里
		31 初鹿尾溪橋	332 公尺	同上 148,5 公里

（6）隧道

台灣公營鐵路的隧道，共有56座，總長18,145公尺。其長度在500公尺以上的重要隧道，有如下表所列：——

路綫名稱	隧道名稱	長度	單軌或雙軌	位置
縱貫綫	竹子嶺隧道	542 公尺	雙軌	距基隆 2.400至 2.948公里
	第二隧道	726 公尺	單軌	同上164.750至165.476公里
	第三隧道	511 公尺	同上	同上168.926至169.437公里
台中綫	第七隧道	1361 公尺	同上	同上170.529至171.791公里
	第八隧道	519 公尺	同上	同上174.052至174.571公里
	第九隧道	1369 公尺	同上	同上177.839至179.038公里
宜蘭綫	竹子嶺隧道	554 公尺	雙軌	同上 2.500至 3.054公里
	三貂嶺隧道	1849 公尺	單軌	距八堵 17.189至 19.038公里
	草嶺隧道	2166 公尺	同上	同上 34.816至 36.982公里
集台綫	第二隧道	805 公尺	同上	距二水 22.938至 23.742公里
	第三隧道	507 公尺	同上	同上 23.747至 24.254公里
台東綫	搭叭隧道	1116 公尺	同上	距花蓮港 70.953至 72.059公里

（7）地磅及轉車盤

台灣公營鐵路，共有地磅三座，一在羅東，載重量爲30噸；一在基隆，載重量爲30噸，其餘一座在樺山，載重量爲30噸。

轉車盤則共有十五座。宜蘭蘇澳，大里各有一座，皆爲上承式，各長15.41公尺。頂雙溪有一座，爲上承式，長18.29公尺。基隆有一座爲下承式，長18.29公尺。台北的轉車盤亦爲下承式，不過長度爲20公尺。新竹的轉車盤爲上承式，長18.29公尺。苗栗，台中，彰化，嘉義各有轉車盤一座，都是下承式，但長度則不同，苗栗彰化嘉義三座皆爲18.29公尺，而台中則獨長20公尺。台南的轉車盤爲上承式長18.29公尺高雄和高雄港的轉車盤均爲下承式，長20公尺，屏東的轉車盤則爲下承式，長15,24公　，和宜蘭的轉車盤相同。

（8）水站

台灣公營鐵路，對於路綫給水設備，非常注重。各路沿綫每隔若干公里，就設有水櫃和水鶴，以應機車入水之需。玆將各綫給水設備列表如下：——

站	名	水櫃容量	水鶴數目	水源
西部路綫	蘇澳	8,2立方公尺	1	自來水廠
	宜蘭	24,8立方公尺	1	自來水廠
	大里	26.0立方公尺	1	河水
	頂雙溪	24,1立方公尺	1	河水
	三貂嶺	19,2立方公尺	1	河水
	菁桐坑	12.3立方公尺	1	河水
	基隆		4	自來水
	台北	120.2立方公尺	4	自來水
	淡水		1	自來水
	山	46.5立方公尺	2	河水
	新竹	75.0立方公尺	4	自來水
	苗栗	84.0立方公尺	4	河水
	三叉	26.7立方公尺	1	河水
	大安	36.5立方公尺	1	河水
	台中	40.0立方公尺	2	自來水
	大甲	38.0立方公尺	4	河水
	彰化	65.4 〃	2	自來水
	二水	78.5 〃	3	河水
	集集		1	河水
	外車埕	10.5 〃	1	地下水
	嘉義	45.0 〃	4	自來水
	台南	41.5 〃	3	自來水
	高雄	91.6 〃	2	自來水
	屏東	25.2 〃	3	自來水
	車港	79.5 〃	1	自來水
	中洲	8.0 〃	2	河水
	東花蓮港		1	自來水
	花蓮港	21.0 〃	1	地下水
	豐年	4.8 〃	1	河水
	壽豐	9.4	2	河水
	鳳林	6.6,9 4 〃	2	〃
	台安	20.0 〃	2	〃
	百庄	9.7.8.7 〃	2	〃
	三民	9.4 〃	1	〃
	玉里	14.6 〃	2	〃
	竹東	2.6 〃	1	〃
	海端	3.9 〃	1	〃
	關山	3.9.5.0 〃	2	〃
	瑞源	3.1 〃	1	〃
	鹿野	5.0 〃	2	〃
	稻葉	5.8 〃	1	〃
	初鹿	3.1 〃	1	〃
	檳榔	3.1 〃	1	〃
	台東	21.2 〃	1	自來水

（9）煤站

台灣各鐵路，沿綫多設有煤站，而煤站的容量，則由煤台的面積而定。東西路綫各煤站站台的面積，如下表：——

西部路綫		東部路綫	
站名	煤台面積	站名	煤台面積
蘇澳	14.44 平方公尺	花蓮港	24.80 平方公尺
宜蘭	54.61 ,,	鳳林	13.50 ,,
大里	18.45 ,,	台安	18.50 ,,
頂雙溪	14.00 ,,	百庄	14.09 ,,
三貂嶺	18.49 ,,	玉里	39.25 ,,
基隆	162.00 ,,	關山	14.40 ,,
台北	222.50 ,,	鹿野	14.40 ,,
台北（淡水）	14.08 ,,	台東	11.30,,
新竹	97.00 ,,		
苗栗	114.00 ,,		
台中	111.34 ,,		
彰化	184.60 ,,		
二水	27.93 ,,		
外車埕	11.25 ,,		
嘉義	121.83 ,,		
新營	31.50 ,,		
台南	31.68 ,,		
高雄港	92.56 ,,		
高雄	95.70 ,,		
屏東	21.50 ,,		
東港	10.30 ,,		

（10）號誌

台灣公營鉄路的號誌設備，甚爲完善，雖然經過這次大戰之後，有些已經被破壞了，然亦較國內各路爲勝。茲將各路綫號誌設備情形列表如下：——

路綫名稱	號誌						聯鎖機			
	機械			電氣			第一種			第二種
	臂式	色燈式	燈列式	臂式	色燈式	燈列式	機械式	電氣機式	機械式	電氣機式
縱貫線	553	5	31	50	39	—	12	6	69	1
台中綫	112	—	—	8	1	—	1	—	13	—
屏東綫	84	—	—	—	7	—	—	—	13	—
宜蘭線	84	—	—	11	1	—	2	—	18	—
淡水綫	26	6	—	—	—	—	2	—	7	—
平溪線	3	—	—	1	—	—	—	—	2	—
竹車綫	8	—	—	—	—	—	—	—	2	—
集集綫	10	—	—	1	—	—	—	—	4	—
台車綫	45	—	—	—	—	—	—	—	23	—

（11）機車與車輛

在台灣公營鉄路上所行駛的車機和客貨車輛，現共有機車 253 輛，客車 464 輛，貨車 5792 輛。乃因使用年代日久，時有損壞，亦只有隨壞隨修，隨修隨用而已，所以在一年之中，每月可以行駛的車輛數目，並不一定相同。茲將民國 36 年內各月份機車客貨車的數目列如下表：——

月份		1	2	3	4	5	6	7	8	9	10	11	12
機車	合計	246	246	246	246	246	248	253	253	253	253	253	253
	完好者	140	144	144	144	147	149	148	139	140	137	134	148
	損壞待修	102	98	98	98	96	99	105	114	113	116	119	105
	損壞不堪修理	4	4	4	4	3	—	—	—	—	—	—	—
客車	合計	497	497	497	497	497	465	465	465	465	465	464	464
	完好者	356	353	358	353	365	356	361	364	375	376	376	377
	損壞待修	109	112	107	112	100	109	104	101	90	89	88	87
	損壞不堪修理	32	32	32	32	32	—	—	—	—	—	—	—
貨車	合計	5942	5942	5942	5942	5941	5793	5793	5793	5794	5793	5792	5792
	完好者	4979	4975	4979	5062	5041	5080	5108	5094	5098	5123	5145	5201
	損壞待修	791	193	789	706	726	713	685	699	696	670	647	591
	損壞不堪修理	172	174	174	174	174	—	—	—	—	—	—	—

（12）運輸概況

台灣鐵路局所轄下的公營路綫，每日平均行駛列車凡 351 次，計 19,485.2車公里。每日客運約 121,498人，貨運則爲 9,425噸。列車行駛速率，若以卅六年十二月份爲標準，客車平均速度爲每小時 42公里，貨車平均速度爲每小時 32 公里。依照當時基本運率，客運第三等爲每公里 1.79元，貨運第五等爲每噸公里 4.95 元，而全年客運收入佔全數百分之六十，貨運則佔百份之四十，所以客運比貨運爲多。

（三）　結語

當我開始執筆寫這篇文章的時候，我原定的題目，爲「台灣公營鐵路概況，及以後改善計劃」。因爲當我在台灣旅行的時候，我感覺到台灣的鐵路，有幾點應當改善的，我見鐵路局工務處樊處長的時候，我也曾將我的意見提出過的。後來因爲南大工程的編者催稿甚急，時間太促，而本人又忽爲私事所阻，沒有辦法抽出較多的時間，從事寫作，因爲談到改善問題就牽涉到水力發電，台灣煤，台北市的擴充，環島鐵路的修築等問題，範圍既廣，篇幅必多，實非目前我的時間所能許可，所以後來就把題目削去一半，只留上截，而變成現在這題。

美國鐵路橋樑最近之一個研究問題 陸能源

在最近這幾年裡，美國許多大鐵路的單跨度架橋，發生了一種很有趣味的損壞(Failure)這事發生在差不多七十個架橋上的同一地方，引起了不少工程師的研究興趣，美國鐵路工程協會(American Railway Engineering Association)特地爲了這事組織了一個委員會來研究這損壞的原因和補救的方法，筆者最近曾在這委員會工作過，感到確是個很有趣味的問題，故特地來報導一下，雖然目前尚未有確切的研究結果。

事實：這種損壞發生在單跨度架橋的横樑(Floon Beam Hanger)上端與夾鐵版(Gusset Plate)連接處，損壞的地方多半在夾鐵板上最低卯釘線上(Lowest Rivet Iine)一切損壞都在横樑吊的本身(Main Material of the hanger)。而非在夾鐵板上，有這種損壞的架橋的年齡都在二十年與四十年之間，故包含新型的橋和古舊的設計，這種架橋包括單軌，雙軌及三軌，Riveted 及Pin Connected，有些是斜角跨度(Skew Bridge)的每個横樑吊的損壞，有發生在向河處(Side away from track)有發生在向軌處(Side near track)有發生在向河及向軌兩處，但都是一種微少的裂縫(Crack)損壞發生時，多半未經察覺，而是由工程師後來檢驗橋樑時發覺，但亦有於損壞發生時車輛受其影响而突然搖擺，立刻檢驗而察覺。

研究對象：這種損壞原因可能複雜，可能簡單，可疑之原因大概是Impactand Vibration，Stress Concentration and futigue, secondary stress。

研究方法：這種損壞的研究方法包括三點(一)理論上力學的計算，包含一切可能之力學影响(二)實際上力的量度選擇兩三個能代表的橋樑，用儀器度量整個横樑吊的力的分佈，(Stress Distribution)包括動力(Moving load)及靜力(Static Ioaa)的影响，(三)試驗室的工作，包括各種模型研究及力的量度。

25817

最新型之中央支撑橋樑

何達康

(譯自 Engineering Newms:Reord Sept 1948)

由於戰爭使然，德國之橋樑工程師已將其智能運用于戰後重建之事業上，其最先值得注意者乃為在Cologne之新萊茵橋(Rhine bridge)，在其中一長650呎之板距支持其兩端相連之二樑距，並負一特重之道路石灰板。最近更有一個偉大之提議，則為運用一鉄架，一拱形骨幹或一鋼纜建于橋之中線，將道路從其二旁伸懸出來，此類新型橋樑之優點基于下列二點，一為節省鋼鉄及基座之材料，二為對於高速度之運輸，此類橋樑實較舊型者為佳，蓋其將車道自然分開，並消減足以阻礙行車者視綫之左右兩旁之鉄架也。

(圖一)

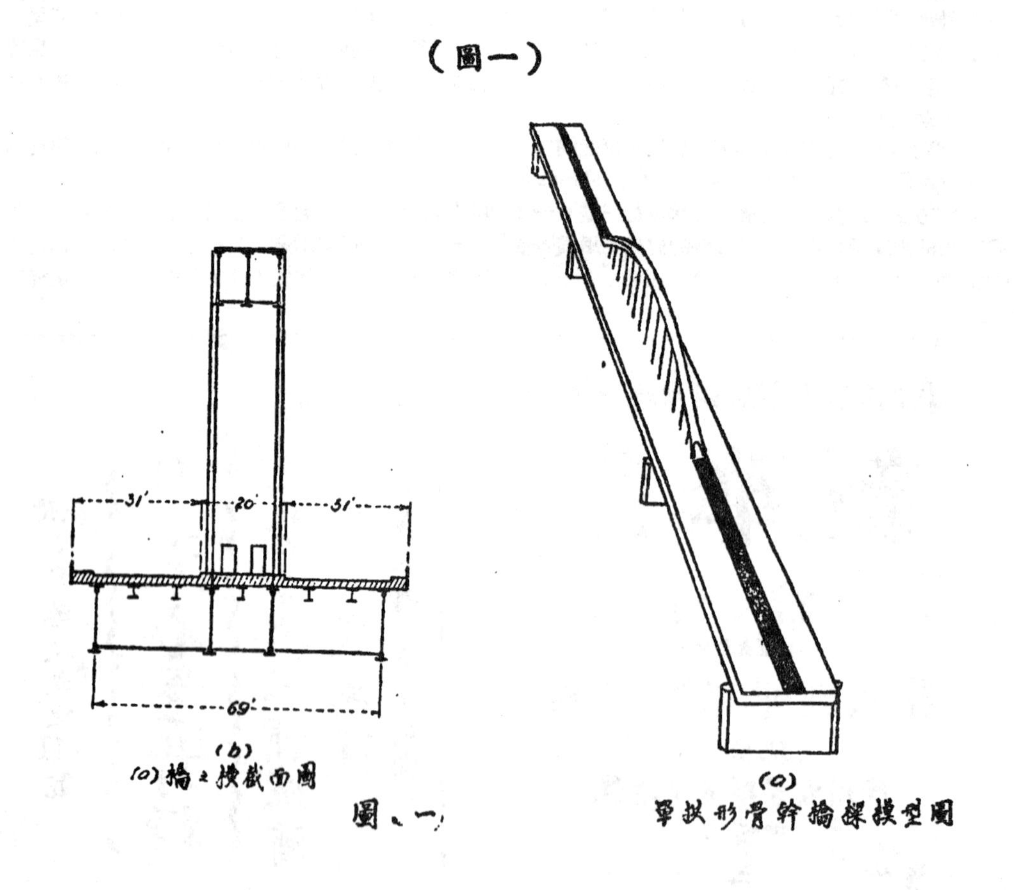

(b) (a) 橋之橫截面圖

(a) 單拱形骨幹橋樑模型圖

圖、一

第一座「中央支撑」橋樑已在德國之西部Ostertal 河上建築起來其 距長330呎。橫截面如圖二所示。

設計中央支撐橋樑所須特別注意者，則為須計及強大之扭轉阻力，且對風及運輸重荷偏于一方時之安全率須充份擴大，關於此二點之應付，用三弦之三角形式架為之，可得良好之効果。

圖二 中央支撐橋樑之橫截面圖

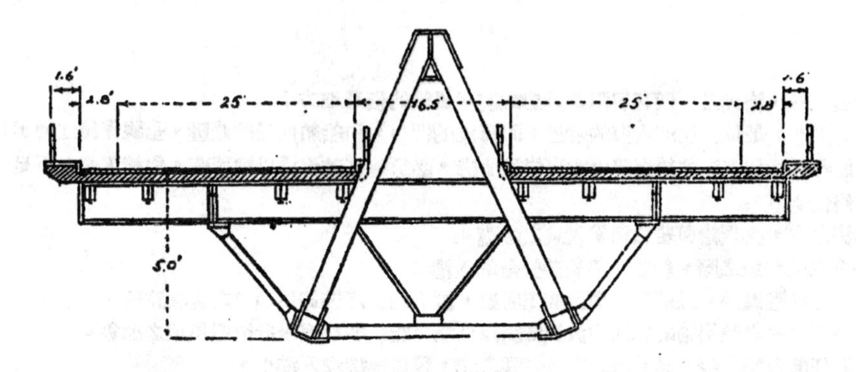

從旁看來，中央支撐橋樑與舊型橋樑之外表差別甚微，惟工作方面則輕便得多，蓋在車道之上只有一主要支撐處，此若從斜方觀察之，則為一末條無異此類新型橋樑最初是以鋼鐵建造，後研究結果，若用鋼筋混凝土為之，在實用上亦相等，且在鐵路方面，此橋樑之功用較與公路方面所用者亦一般優越。

築海港之大要

黄康道

海港之定義：——海港是一個有屏障蔽護船舶避免風浪的優良地方：

優良的海港，是位于江河入海的附近，而海港的門戶，和船舶所受的蔽護，全視保持江河出口的工程而各異。如海港位於海灣中或海岸稍曲處，必須建築破浪堤以增蔽護，而補天然之不足，建築海港之重要條件：

1. 港海之中，必須給與船舶相當優良之蔽護。
2. 即有巨風和浪之際，船舶，仍能甚安全的入港口。
3. 不只是躲避風浪，且須爲通商及商業而設，應有適宜之裝卸出進口之貨物設備。
4. 建築波浪堤必須適適的把海中波浪截斷以便港中能收容所有來此港避風浪之船隻。
5. 在避狂風來襲之際，船舶須能在風來到之先，得從容疾駛入港中，
6. 港海在可能時必有避風港，與停船處，而停船處不可能離避風港太遠。
   例如過香港的航洋巨輪船，全停在對岸尖沙咀及九龍對岸，但是香港的避風港，是在銅鑼灣及旺角避風塘。
7. 港口或海港的門戶，必須位於正與最大風浪之方向相對。
8. 因令船隻能易於進入港口者，海港門戶須濶但波浪亦因之而隨入，因所以由港口至停泊之處，須有一相當距離，以便波浪得而展開，故可以漸減其高度，
9. 兩個破浪堤端所對峙成港口處，須有相當深的堤基及厚度，以免最大的波浪經過時，不至破碎。
10. 在可能之際，破浪堤須與最險惡之波浪之方向成正角，即波浪平行，欲以最短的破浪堤以得最大的蔽護。
11. 如海岸原來是甚直，則破浪堤須漸曲向海岸，不能與波浪成平行者，但兩破堤間所夾之角，不可過小，以便波浪進入港中後，得到相當的展大，而波高漸減
12. 爲避免衝撞堤端的危險，所以兩堤端峙處之港口，須有相當寬度，即在最險惡的風波中，稍受漂驅，亦能安全的駛入港口，而不至衝撞堤端。
13. 如過受潮港面甚小之時，港口亦緊窄，以維持相當深度。
14. 在河口甚大的地方，應宜先注意能引入大量的潮流以便改善上游河道
15. 其次須顧及來往此港船隻之大小，及其流入波浪所激起之紛擾，須減至最低限度
16. 如受潮港而甚少者，而港口沙洲由平行束堤以一致劃深，當風向岸之際，該港即難於安全進入。如由兩堤環轉而成之港，港面較濶，內港即無需要也
17. 在大湖中之商港，多由平行束堤以濬深河湖出口而成，爲避免大風浪時具有破壞性之波流入內束堤以前，概築以破浪堤壩，如支加哥（Chicaqo）可雷夫蘭（Cleveland），泊發盧（Buffae）蓋束堤外破浪堤內之區域，旋闢爲外港，
18. 故此在繁盛之商港，初先建築破浪堤以阻波流進港，而漸闢破浪堤以爲外港
19. 河湖口之商港，在風平浪靜之時，即有相當航行深度，往往在大風浪時，波槽之驅進，大減航行深度，甚至過進口淺灘，波浪至於破裂，而露海底，在此次要港，往往亦須築設破浪堤，
20. 爲航行安全起見，港口之寬濶度，以船隻吃水深度而定，在吃水十九英尺至卅英尺的船隻，其最小寬度爲四百英尺，至六百英尺港口之最小寬度約在來往此港船隻吃水深度之廿倍以上。

21港口過狹，固失厥航行之安全，而港太濶，又招致波浪之進港，如港面較大時，港口亦可較濶，故海港設計雖適當，則以三百五十英畝之港面，可使波浪于甚短時間內，減殺其勢。卽如在港口有十英尺高之波浪，亦不至於影响於對面碼頭前所停泊之船隻。倘若港面較大，港口濶度可由六百英尺至一千英尺，可酌量增加，

22在大的海港，有時爲增加駛進便利及安全起見，而增置兩個港口，但其異方向。有時說得犧牲進港之便利，而保持港中之寧靜如錫蘭島上（Ceylon）的哥崙布（Colombo），港和法國的卜龍（Boulogne）港。甚至港口外再築多一層破浪堤以蔽護之物希替港（Citte）

23有時水流及風浪之關係而帶着沙泥而來故築兩堤，非皆爲截浪之用，若築一堤爲拒絕泥沙流入港內而築建的此種情形之下，通港水道往往沿其一堤，而波浪亦斜衝進，浪力之減殺，遂極有限。故必築二堤而一缺口并在其後方置一有斜披之小小水塘，則波浪可在此塘展開，而波高得以大減，此爲之蛸塘如第裂（Dieppe）和哈弗（Havre）兩皆具有蛸塘。

前屆畢業同學近況

編輯室

前屆畢業同學只有五人，這就是本系復員後第一屆的畢業生由於在康樂園裏成長的友誼，他們出社會後情形如何，我相信各同學一定很想知道的，現在把他們的近況列後

陳潤初——畢業後東渡美國深造，在紐約洲立大學攻讀，對 Soilmechanic 一科甚爲努力。

趙浩然——他畢業之後，在社會的活動甚大，除在水利局任技佐外，幷與舊同學金澤光等，經營新湖建築行，進行十分順利。現在已生意興隆，承接的工程經有四五處開工了。

呂冠炎——他是嶺南的「大天二」。富服務性，頗得學校當局器重。畢業之後，被留在學校，負責監建幾件工程。新近屹立路邊的小學禮堂，和計劃中的小學膳堂等，都是他的得意之作。

歐陽讓——是本系的高生才，金鑰匙的獲得者。因身體關係，暫時不想出當艱辛的工程工作，而希望休息些時。故畢業之後，在本校附中任教，頗得同學的愛戴。

黃漢基——本是往廣州灣工務局工作的，但因該處沒有工程開工，閒坐沒有興趣。乃轉往香港國際救濟會任副工程師。他的畢業論文，是寫梁金山水壩的，近來曾往開平監理該件工程的開工。但因經費不足，該件工程未能如意展開云。

會聞

編輯室

九月廿五日，本會舉辦歡迎新教授暨新舊同學聯歡會，秩序甚爲豐富。日間除聚餐，雞尾酒會，及遊藝等外，夜間更於綿園舉行音樂欣賞會。紅燈之下，平台之上，花香鳥語，倩影雙雙不知周結了多少純潔底靈魂更不知激下了幾許真切的友誼。

九月卅日，本會與政治學會作足球友誼賽於小學球場。是戰精彩百出，緊張之處，令人叫絕。結果，本會以四對零獲全勝。是役我軍以梁副主席親自掛印中帥，「傻梁」居右，楷仔主左，前衛則爲體育部大臣，「靚 Mad 」及佐兄担當，後衛一職，交由「鍍記」供任。而大本營則由羅莘君座鎮，全軍脚法爛熟，奔走迅速，加以球路亨通，因是攻則銳不可當，守則無隙可擊，其勝實理之必然也。

十月十五日，學生會舉辦游泳比賽，本會參加各項，均獲錦標，內中尤以二百咪四人接力，獲得亞軍，其他如百五咪三式接力及四百咪自由式等項，皆有極優異之成績表現。據路邊社批評，工學會之得勝，啦啦隊之功尉偉云。

十月廿一日，桂銘敬系主任（在假）回校主持本會第一次學術講座講，題爲「大學生就業應有之認識」，席間除本題外，對同學之學習生活，更感關懷，各同學聆聽之下，至爲感動。

會際排球比賽已開始多日，本會人心激昂，三軍用命，連戰皆捷，每當比賽，助威者衆，據觀察家言，大有緊握霸座之概。

南大工程

桂銘敬

康樂再版　第四期

月刊通訊

主編：潘世英

編輯：吳秉俠　何達康

目　錄

嶺南大學工學會刊印

民國三十八年元月一日出版

廣州市惠愛西印刷工業合作社承印

如何學習工程

劉載和

修讀工科之同學欲知如何學習工程，必須先明白工程之意義。工程爲何？工程爲一種利用物質之機械性能，在各種建築及機器上，造福人類之科學與技藝。由此可知工程包含兩點，一爲科學，一爲技藝。前者注重智識，後者注重經驗。智識之尋求，須靠學，經驗之獲得，則靠習。故讀工程，須學而時習之，不可有所偏廢。

學生學習工程，必須有師長教導。無教則所學進步較緩，無導則所習易流於誤。故學習與教導，需相伴而行。且須互相配合，苟一方面學習甚勤而另一方面則教導無方，學生固不能得益，苟一方面諄諄善教而另一方面毫無留心，教授亦徒費氣力！

工程之主要原素，既爲科學之智識與實際之經驗，則不論學習或教導方面對此二者，皆應並重。若專注重科學之智識而缺乏實際經驗，則每每成爲閉門造車，不切實際，並不符合工程學之原理。如專注重實際之經驗而忽視科學之智識，則每每墨守繩法而日趨落後，亦與工程學之原意相違。故學習工程者必須以科學之頭腦，尋求經驗；教導工程者，必須以實際之經驗學理分析；如是則授者固不至於流爲空談，受者亦不至於流爲幻想。否則言者口沫橫飛，聽者靜坐遊埠，言雖無害，而聽者則毫不獲益矣。

學習工程時，探討學理，須靠書本，尋求經驗，則靠試驗。惟工程書本之講授，與其他學科不同。因工程學理，離不開事物。故教授講解學理之時，除使學生明白一事一物之學理外，仍須使學生腦中，構成該事該物之形象。在一毫無印象有如白紙之腦中，專憑口講，而能繪上某事某物之完整形象，實非易事。普通講解之時，乃在黑板上繪畫草圖，以補言語之不足。但遇結構精復之物體，內容繁雜之事件，則無所施其技。且黑板之面積不大上課之時間有限，繪畫草圖，亦有一定之限度。其最良方法，乃爲詳細之標本及掛圖，使學生能一目了然。工程試驗與其他學科之試驗不同，因工程試驗離不開現實。故從事試驗之時，除注重其結果外，對於試驗之實用，亦應留意。欲求試驗之結果優良，則須有精確之儀器；若欲試驗能合乎實用，則試驗須與將來實際工作相切合。故學生學習工程困難，教授講授工程更難，而學校設立工程學科亦不易也。

學生學習工程，其所學之學理，雖處處不離事與物，然仍爲書本上之工程智識。其所習之試驗，不論與現實如何接近，仍爲學校內之工程經驗。若欲增加學生智識使不限於書本上，擴大學生經驗使不限於學校內，則必須到各大工程機關參觀其工程與工作，則必須聽各工程前輩講述其經驗與見識。如是則可採取他人之經驗，增加自己之智識，觀察別人之工作，增加自己之經驗，既可補充在校內在書本上之學習，且可增進學習之興趣。由本期所舉行之學術講座與參觀粵漢路黃埔支線兩點觀之，則見各同學對於參觀演講之注重及獲益之多。本人因種種關係，對於各同學之學習工作，未能盡最大之努力，深表歉意。希望各同學下學期對於演講出版參觀等學術工作，能繼續努力，使能配合課程，增加自己學習範圍，則所厚望焉！

第十五屆工程師年會在台北

離校多年，對於母校狀況，常在懷念中，前接閱各期「南大工程」，藉悉母校之近況與學術成就，至感愉快，迨月前赴台參加第十五屆工程師年會後，慶承老同學黃煥林君來函，囑爲文將年會經過刊登，爰不揣鄙陋，將見聞所得，記敘如下，幸各位師長同學，有以指正。

譚君乃本系舊同學，畢業之後，在工程機關工作多年，經驗豐富。是次出席十五屆工程師年會，在百忙中亦撰文寄回，足見對本會之關懷，誠值得新舊會友之學効也。——編者

（一）年會紀要

中國工程師學會

中國工程師學會的前身，原爲我國工程界先進詹天佑氏所創辦的中華工程師會，及凌鴻勛，陳體誠二氏所組織的中國工程師學會，因爲兩者宗旨相同，俱在於研究工程學術，協力發展中國工程建設，乃於民國二十年在南京舉行聯合年會，議決合併組成今日的中國工程師學會，以後曾在許多不同的地方，舉行年會，抗戰及復員期間，曾先後停辦三次，直至去年，才於南京繼續舉行。

第十五屆年會

本屆年會，是在台北市的中山堂舉行，會程是自十月二十六日開始，一連四天，舉行各種集會，然後自十月三十日至十一月三日分組參觀台省各地工廠，秩序十分豐富。大抵國人對光復後的台灣，及其建設事業，多感覺興趣，是以本屆年會，參加的會員及來賓，濟濟不少，計共有一千五百餘人，打破以往年會紀錄，其中一部分會員，還是從廸化，瀋陽等地，遠道而來的。開會期間，由年會籌備會辦事處，負責招待，大部分會員，是招待於台北附近的草山及北投溫泉區居住，那裡離開市區有十多公里，但是有汽車及火車聯接，班次很多，而且早晚還有工程師專車來往，是以交通尚覺方便，各會員於開會勞頓之餘，遨遊於青山綠水間拍拍照，洗洗溫泉浴，倒也頗有意思。

年會第一日

十月二十六日上午，中山堂便擠滿了人，台省多數機關，都派員在分發小冊子，其中有價值之報導頗多。上午九時，會員及來賓，陸續蒞臨會場，濟濟一堂，頗極一時之盛，在樂隊齊奏聲中，年會宣告揭幕，主席茅以昇會長，首先致詞，代表全體會員，對台省各界人士的招待與幫忙，表示感謝，并對台省官民在建設方面的復員成績，表示敬佩。最後申述其對台灣建設的意見，希望：（一）台灣建設，應與內地建設配合。（二）台灣建設，要使台灣得到眞正的繁榮。（三）台灣建設，要以台灣人爲推動幹部。（四）台灣建設，要成爲中國工業的重點云云。繼着國防部何應欽部長、台省魏道明主席、工商部陳啓天部長等，相繼致詞，祝年會成功，下午繼續舉行大會，先由顧毓琇總幹事報告會務，略謂中國工程師學會本年會員人數，已達15.817人團體會員爲197單位，分會達49處，比之于民二十六年之2.994人17單位，已大有增加，足見學會在日漸發展中。其後各分會代表，相繼報告及發言，都說在這個年頭，要舉辦一些建設事業，是如何的困難，而各地工程師的境遇，尤其是北方各地，又是如何的艱苦云云。本來工程人士，素來是比較上吃得苦的，現在也異口同聲地訴起苦來，則其苦況可知。最後茅以昇會長報告三十七年度會長及董監事選舉結果：當選者，會長沈怡、副會長趙祖康錢昌祚、董事茅以昇、朱其清、周鳳九、譚炳訓、張延祥、吳必治、楊簡初、薛傳儒、汪胡楨等。會畢，是晚台省魏道明主席，假座中山堂光復廳，舉行雞尾酒會，公讌全體會員，賓主盡歡而散。

年會第二日

二十七日的叙會，主要的是上午的專題討論及演講，下午則爲各專門學會會務討論。是日的專題討論是，「中國建設投資問題」，由前交通大學校長程孝剛先生、交通部次長凌鴻勛、徐恩曾二先生等相繼發言討論，都認爲建設事業，首重資金之籌措，而資金之籌措又必需以利潤爲依歸，政府如能保證投資可得利潤，則投資者不致因經濟環境影响而有所卻步，比方在可能範圍內實行保護關稅與酌量減低原料之進口稅及有計劃地發展關鍵工業如交通與電力，不直接干涉民營

企業之經營等等，又如果某一建設事業爲社會國家所需要，政府應該實行保護政策，則建設事業，即有開展之希望云云。

年會第三日

二十八日上午繼續專題討論，題目爲：「台灣建設與大陸之配合，」及「台灣建設發展之可能性。」討論要點，大致是：（一）台灣之建設大陸之配合，絕對需要，因爲台灣天然資源，並不豐厚，工業製造過程中的自給不可能，非此不足以使工業建設順利與圓滿地展開，譬如以台灣之煤及水泥之類的物資，換取海南島的鐵砂等等。（二）台灣建設，應確立政策，而政策之確立，應與日人在台時之採取政策有別。（三）台灣之工業建設，决不是將過去日人所經營的事業，予以復員，而是從全國的立場上，建設爲中國的工廠，但是發展台灣工業，應當着重輕工業抑重工業或是兩者兼顧呢？這問題發生了不同的意見：（一）主張着重重工業的，所提出的理由是台灣動力發達，工資低廉，環境安全，宜於發展重工業。（二）主張介於輕重工業之間的，理由是台灣既無鐵礦，又乏煤藏，着重重工業比較吃力，還是二者兼顧的好。（三）主張發展農產加工與輕工業的理由是中國沒有海軍，將重工業建立在海島或港口，一旦發生戰事，很易爲人襲擊，而台灣農產品豐富，倒不如發展農產加工與輕工業爲佳。

工業建設千頭萬緒，地域上的輕重本末，程序上的先後緩急，即有不同的觀點與主張，但向前推進的目標則一，惟是我國正在多事之秋，對於各項建設措施，常覺有心無力，良堪歎惜！

下午會務討論，通過議案三十條，其中一條是下屆年會在廣州舉行。晚上土木、市政、衞生、土壤工程學會舉行聯合叙餐，餐後放映工程電影，有錢塘江大橋工程，美國橫貫東西部石油管之裝設，林肯隧道工程大要，及基隆港沉箱特輯等四卷，并由茅以昇會長對錢塘江大橋施工經過，作一詳細之解釋。

年會第四日

十月二十九日，是大會最後的一天，全日宣讀論文，本屆收到論文一百餘篇，特編就「論文摘要」一冊，分發各會員，其中以有關台灣之論文爲多。

晚上舉行年會宴，席間并由孫立人將軍等致詞，强調今日之工程師，應特別注重國防工業。宴會後在中山堂舉行遊藝大會，由台省交響樂團等演奏名曲，并由台灣高山族人表演歌舞，節目相當精采，大會由是告一段落，自十月三十日起，便開始分組參觀台省各地工廠。

（二） 工廠參觀雜記

十一月一日清晨，筆者所加入的小組，由台北乘火車出發，參觀工廠，行程是第一日往日月潭，第二日往高雄，第三日返台北。每組雖然只有45人，可是我們這一組，倒也人才濟濟，論系別，是土木、機械、建築、電機、礦冶、化工等都全，論學校，則國內南，北方大學及留美留英的都有。四面八方，合攏而來，正是有緣千里能相會，一路大家談談笑笑，玩玩橋牌，空氣十分融洽輕鬆這一天的午飯，是在火車上吃，每人「便當」——飯盒一份，吃完便將盒子連筷子扔掉，眞是「便當」而衞生。車行九小時，沿途所見，都是稻田蔗田，麥浪隨風，一望無際，台灣眞不愧爲一寶島，下午四時，抵達台中縣屬的集集鎮，再換乘十分鐘汽車，便到達日月潭下的鉅工發電廠。

台灣的發電廠很多，數個爲一組，名曰發電廠羣，以發電廠而分別名之曰羣，則其電廠之多可知，我們要參觀的鉅工發電廠，便是屬於日月潭水力系發電廠羣，此外尚有火力系發電廠羣等等。

日月潭的發電系統，大致是先在日月潭的水社及頭社兩個主要出水口，建築水壩，將日月潭的水位提高70公尺，然後以鋼筋混凝土的隧道，明渠及暗渠等，將水引送至18.5公里外之大觀發電廠附近山上的調節池，導供該電廠發電之用，用畢之水，再以隧道，水路橋等引送至其下五公里外之鉅工發電廠，再供發電之用，然後才將水導往別處，灌漑農田，流入河川，此系統中，尚有萬大發電廠一所，亦是以同樣方法送水發電，聽說整個日月潭水力發電系統，是於十八年前由著名的Dr. Salvage設計的。

鉅工發電廠，建立於十年前，需時兩年半，方才築成，所用有効水頭高（Effective Head）爲123公尺，計有發電機二座，以前最大發電量爲43,500 KW.，但因戰時曾遭破壞，目前只可發電35,000 KW.

在鉅工發電廠參觀完畢，便繼續乘汽車上山，晚上抵達日月潭邊的涵碧樓，該樓建築頗爲清雅別緻，憑樓遠眺，日月潭景色，一覽無遺，地點甚佳，各人於整日奔走勞頓之餘，得此幽靜清潔之住所歇宿，倍增舒適之感。

十一月二日，在涵碧樓吃罷早飯，便乘汽艇遊湖，是日也，雖非惠風和暢，但亦天朗氣清，乍寒還暖，仍是遊山玩水的好時光，於明媚溫暖之太陽光下，斜倚汽艇欄杆，流覽日月潭之山光水色，頗有心曠神怡之感，日月潭四面皆山，中間是一座名叫光華島的小島，青山綠水，景色甚佳。船行約半小時，便抵達有名的高山族人聚居的魚池鄉，捨舟登岸，土人乃聯羣表演歌舞，以示歡迎，當下各人并與該族兩位公主合拍照片，以留紀念，然後返回涵碧樓，換乘汽車下山，直奔大觀發電廠。

大觀發電廠建於十多年前，費時四年，方才築成，所用有效水頭高爲320公尺，較鉅工廠大一倍多，有20,000 KW. Y型發電機五座，最大發電量爲100,000. KW.，但亦因曾遭破壞，目前只能發電 80,000 KW.云云。

十一月三日的程序，是參觀高雄港、鋁廠、機器廠、及孫立人將軍的練兵營等。

早上乘坐高雄港務局的汽艇遊港，高雄港之一邊，是一長達十二公里的防波堤，另一邊是市區及工廠區，航道水深十公尺，可泊三千噸至一萬噸級船舶共百餘萬噸，每年之吞吐量，最高時達三百餘萬公噸，各項海港設備如乾船塢、挖泥船、給水船等等，都相當齊備，實不愧爲台南第一大港，而港內各種輕重工業建設，都相當發達，倘能善加利用與改進，則對國家資源之交流與充實，當有莫大的裨益。

遊港歸來，便換乘汽車，前往參觀鋁廠。是日天氣頗熱，但是在鋁廠喝着冰凍的檸檬茶，細細研究一下該廠的圖表及工作程序，然後逐步看看機器，倒是很有意思的。該廠目前年產鋁錠四千噸，氧化鋁一萬六千噸，全部運銷上海，但原料則來自福建及國外。

高雄的機器廠，規模頗大，產品頗多，如各種製糖機械、工作母機、柴油機、窄軌機車貨車等，都有出產，最近爲了擴充業務，還向賠償委員會配領到不少日本賠償物資，但據該廠職員云：日人狡猾異常，該項賠償物資中，每種機械，一定缺少一些重要零件，或是各種配件零亂裝箱，以致裝配時極感頭痛，此種現象，非設法防止不可云云。

孫立人將軍的練兵營，是在高雄附近的鳳山鎮，佔地頗廣，營地之一角，裝設了一些障礙物，用以訓練學兵障礙跑及越野跑，該營訓練程序，係採用美國式，術學科并重，頗爲有系統而科學化。

是日下午，乘火車北返台北，年會會程，由是圓滿結束。

本屆年會，不在別的地方，而在台灣舉行，委實有着其深長之意義，正如台省魏道明主席於年會開幕時所說：「台灣雖然可以說是一個寶島，但是這寶島的天然條件，并不十分優厚，要靠人力奮鬥的成份還多，……」在這烽火遍全國的今天，就只有台灣這一隅，比較安定，而台灣各項工業建設，雖然缺乏足以自立的基礎，但是還算粗具規模，今後應如何刷新整頓與改造，以配合內地各項建設，使成爲我國重要工業之一環，尚有待於全國人士之認識，研究與努力，本屆年會之在台灣舉行，原因即在於此，會期中專題討論，特別着重於台灣建設，也是爲此。倘能因本屆年會在台灣舉行，而能普遍引起國人對台灣之認識與注意，進而共謀配合台灣工業，與大陸建設，向前發展，以爲復興建國之本，則其對國家人民之貢獻，誠未可量也。

25828

竹在航空上的用途

余仲奎

竹除採用于各種手藝工業外，其可取性能，本可擴及更大規模之應用。是以數十年前，歐美及日本即有竹筋混凝土之試驗，結果皆有紀錄，本系現亦進行是項工作。近年中國航空研究所，更研究用層竹代替鋁片，製造飛機及各種航空器材，成績美滿，本文作者，乃中國航空研究所所長，為是項試驗之主持人，現任本系教授。今撰是文，以饗讀者。 ——編者——

一、起頭語

南大工程的編輯向作者索稿多次，因時間關係，無法寫出。最近又來催稿，不得已將本年九月間在廣州扶輪社的講辭，加以整理繳卷。

這篇文字是介紹性質，講述抗戰期間竹在航空上的幾種重要用途，所討論的問題均是輕描淡寫的。

本來竹與中國文化有着密切的關係，古人削竹記事。紙發明後，竹又用為原料，現今西南各省仍有保持土法用竹製造紙漿。我們所見的日常用品，有不少是用竹作成的。

竹在工程上的利用，極難找到正確的紀錄。一九一二年周厚坤在美國麻省理工大學用竹替代鋼根作加強混凝土的試驗，算最早的科學試驗，此後有關竹之強度試驗，亦有不少文字發表，但甚少詳盡的研究。

抗戰期間航空研究院在成都因缺乏鋼根，曾利用竹代替鋼根作一個受一千匹馬力的地基及風洞試驗平台。美軍亦曾于上次大戰中，在菲律賓用竹代替鋼根作防禦工事。上述用途，均無數字發表，以表示其優劣點。

二、抗戰時期竹材的研究工作

抗戰期間，交通被敵人封鎖，所有由外輸入之器材，極難進入內地。是時作者在成都航空研究院負責航空器材研究及尋求國產替代品。竹材試驗是其中之一重要部份。

此項研究所包括的問題：（一）何種竹最好？（二）如何將之利用？（三）如何防腐？在此期間試驗的竹種有十餘種，其有詳盡之試驗的有數種，（見後文參攷資料）。楠竹與慈竹的試驗合共有二萬餘次。

由這種試驗結果證明竹青部份的強度可以超過低炭鋼或鋁合金的強度，而牠的比重僅為鋁的三分之一，鋼鐵的九分之一。

舉例來說，慈竹的部位抗張強度（見參攷一）在10%的壁厚位置其無節部份之平均值有六萬五千磅每一平方吋；有節的部份亦有三萬三千磅每一方吋。

選擇航空材料的主要條件為「強度——比重」比率，即每一比重的強度，鋼鐵的強度比鋁合金強，但鋁的比重僅為鋼鐵的三分之一，如果兩種材料的強度是相同，用鋁質的材料來負担荷重，其重量自然比用鋼少小三分之二。

航空木材的強度比鋁更低，但牠的比重約為鋁的六分之一，從這個觀點說，用木材作成的構架，來負担鋁質構架所負担的荷重，可以不須增加很大的重量。

木材既然可以作飛機的主要材料，竹的強度比鋁還好，自然可以利用來作飛機。不過問題不是竹的能否利用而在如何利用。竹和木是有機的物體，有機物第一缺點是組織不均勻，這樣會影响到牠本身的強度的變異，結果這一部位的材料與另一部位的材料不同；這一種強度與另一強度不同，上文所舉的抗張強度，有節部份僅為無節部份的一半。

抗張強度是竹的最優的強度，其他如抗壓，抗剪，抗彎等強度等不甚高，所以利用竹作工程材料，應利用牠的優點。但是竹的利用不是一件容易的事，牠的桿壁是很小，中間空洞，又加上節疤，不比木材容易施工，許多人看見竹的缺點，就感覺到它不能做工程上的材料。事實上如此，如果竹不經過加工處理，這是不能直接利用的。

所謂加工處理是如何將低強度的部份配合高強度的部份；如何將小徑的竹變為整個的材料。下文就簡單的叙述一些加工的方法，和用途。

三、層竹

「層竹」是一個新創的名詞（見參攷二），是代表一種經過加工後變成如層板一樣的竹材。簡單的說，牠是用多層的「單竹」用膠黏合復經大壓力機壓製出來的薄板。

（圖一）　竹材製成的教練機

「單竹」是層竹中的一層，牠的構造頗似市面上所售的竹簾。所不同的地方，編織所用的竹條，是經過選擇。取材有一定的部位，長、寬、厚等度數均有一定的規定。此外在編織時，有節部份的竹與無節的部份配合使每整張成品的部位強度趨于均勻。

製造的步驟可分為：（一）選料，（二）製竹條，（三）編織的方法，（四）層竹的壓合和其他有關問題。選料包括竹種和膠合的材料。竹條是由鋸成長度適合的竹筒劃出的。牠的位置是由表皮外面量入百分之三十壁厚的部位。牠的長度約為二公尺，寬度四公厘，厚度半公厘。編織是用手工的，耗費時間甚多。一度擬改用半機械化方法如(織布)並且製成一小型的編織機，試驗頗為滿意。後以製造大機較困乃中止進行，仍用手工。普通製造一張廿五方呎面積的單竹，熟練技工需要十八工時。這是大量生產的障礙。

（圖二）　竹材製成的滑翔運輸機能載十四個全副武裝的降落傘部隊

壓合的膠料是用酪膠，此種膠是從牛奶中提煉出來的。酪膠的濕性強度僅及乾性的一半。牠的防水和防腐能力不及酚醛膠，可是前者可以冷壓，不用許多設備。

膠壓的壓力，每一平方吋要用一〇〇至一五〇磅。有時亦可以用較小的壓力。

木材與竹材的直紋和橫紋方向的強度，是相差很遠。單竹是編織而成的竹「布」，可以不像木材須受紋向的影响，在壓製為層竹的時候，又可以使每層的「紋」（竹條）與另一層的竹條成一指定角度，結果可使竹材每一方向的強度，趨于均勻。

層竹的試驗前後共作了一萬七千次，其中包括抗張，抗剪，抗壓，抗彎，彈性係數及曝晒等試驗。由此驗試的結果決定此種材料適合於製造飛機。

（圖三）竹材製成的V字尾飛機尾部之結構頗為奇特

四、三種竹製的飛機

層竹的主要用途是用來替代鋁合金的薄片，和層板。抗戰期間用此種竹材所製成的飛機有三種。第一架是基本教練機，如第一圖。這架飛機的內部構架，是用木做的，外面的是用層竹蒙上作成一受應力蒙皮的機體。

第二架是滑翔運輸機能裝十四個全副武裝的降落傘部隊。（如第二圖）第三架是一V字尾的飛機（第三圖）。

第一架作成之後經過科學試飛，獲得滿意的性能。這飛機在飛機場上露天放置了五年，仍可以飛行的。第二架在勝利那一年完成，亦經過科學試飛，獲得滿意的結果。第三架因尾部爲V形與一般飛機的構造不同，製成之後，因人事的變遷，未得上級機構的同意，至今仍未有試飛，誠屬可惜。

上述三架飛機設計的主持人是王助先生。

五、竹質外掛副汽油箱

飛機上之汽油箱分「固定」與「外掛」兩種。外掛的油箱皆造吊在翼或機身的底下，用來增加飛機在空中活動的時間。作戰的飛機出戰時必須攜帶，如遇過敵機，將之卸擲應戰。

竹材在航空上大量的應用是製造此項外掛油箱。抗戰期間總共製成了九千五百枚。容量最大的每一油箱可以裝一百〇五加崙汽油，最小者廿五加崙，製造最多者百七十五加崙，約佔全數的百份之五十。

此項油箱不特供給中國空軍，且供給當時陳納德所統率之美國志願隊（飛虎隊）。

油箱之製造（見參考三）分爲「膠壓式」及「編織式」兩種。前者的製造方法是用流綫型的陰陽模加壓力機壓而成，後者用陽模用手工依模編織而成。第四圖表示此類汽油箱。

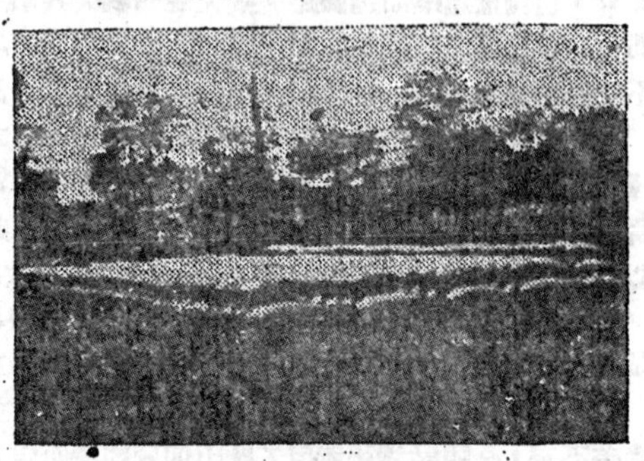

（圖四）　竹材製成之油箱抗戰期間曾供中美空軍之應用

防止滲漏汽油，並不是一件很容易的事，油箱內部除裝了數百磅的汽油所產生的壓力外，有的飛機還須將油箱內加以每方吋五磅的壓力，使汽油能由機底下升入發動機內。膠壓式的防漏比較易辦。

防漏方法是將箱的內部塗上數層的塗料，這種塗料不特要有防漏的作用，且要有防止汽油溶解的作用。中國的生漆，經過長期的試驗，是國產塗料中最好的一個，可是她的薄膜甚是脆弱，及乾燥很慢。減除脆弱可以用漆將沙紙或綢綢裱上，油箱內部是裱上約四層的漆紙，爲要使箱外光滑減少空氣阻力，外面又用生漆混合石膏填塞不平的面。然後再打磨使之光滑。

生漆乾燥與空氣的水份及溫度，有很大的關係，雖然可以加催乾劑。不過成效不甚顯著，加高溫能使漆在很短時間乾燥，但大量生產，須有大型的烘爐及操縱的設備。

五、其他竹的製品

除上述的用途外，尚有飛機上的無綫電定向儀的天綫環外壳是用竹做的。（如第五圖）有一批美國製的飛機的天線環外壳，是用一種塑膠壓成的，因爲過於靠近發動機，到了中國使用不久就變了形，結果將天線環阻住不能轉動，後來換上竹製的外壳，故障乃得解除。

另一種用途是用層竹作成各種「層竹管」。在此種研究進行時適值中英文化委員會的李約瑟博士（Dn .J Neeaham）來參觀，並要了些大小不同口徑的層竹管運回英國供應部（Ministry J Sufrlg）作火箭炮發射筒的試驗。

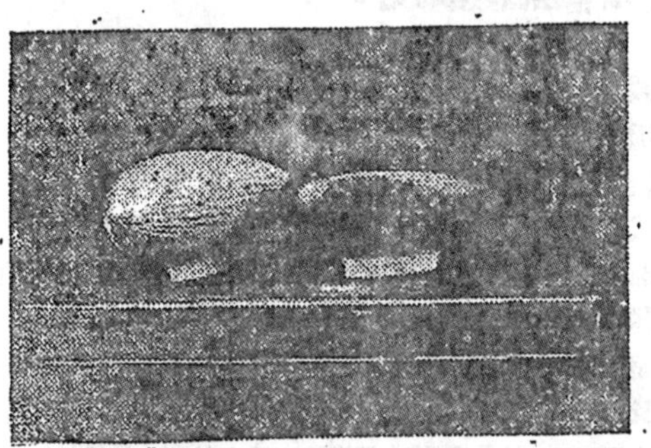

六、結語

竹在中國有極豐富的產量，且生長極速，在供給方面是不成問題的。根據以往的經驗，可以推斷此種材料除航空之外，亦可適應其他工程部門之用。

近代塑膠工業（PIastics ludvstry）進步，更可以開闢竹材爲加强料的新園地。如果我們能不斷的繼續研究，利用竹材的優點以補償塑膠的缺點，相信在不久的將來，總可以見到很多其他新的應用。

參攷資料——

1. 余仲奎等：「川產楠竹性質之研究」，航空研究院報告第十號
2. 余仲奎等：「川產慈竹性質之研究」，同　上　第廿八號
3. 余仲奎等：「層竹之創造」，同　上　第廿六號
4. 余仲奎等：「竹質外掛汽油箱」，同　上　第卅號

工學會學術講座 第三講

主題 鐵 路

主講 梁永槐先生

講題：從事工程數十年來之失敗經驗

日期：十一月二十四日 吳秉俠筆錄

梁先生現爲粵漢路十二總段長，曾任廣梅鐵路局局長，對鐵路擁有豐滿之實際與基本經驗。以廣九路而言，自測量建築而迄養路，梁先生皆親與其事。故實際之經驗至富，梁先生不特具有豐富之工程經驗，而更有良好之工程道德。對一般藉工程處以升官之工程人員，多鄙視之。梁先生爲人忠厚慈善對於下屬，不論員司或工人皆一視同仁，本身亦毫無架子，誠爲一不可多得之工程老前輩也。 ——編者——

凡從事工程之人員切勿奢望過高，務需實事求是。職位之高低，不足以移其志，薪津之厚薄，不足以降其格。對於社會，工程人員乃眞正之創造者，對人類生活，影响至鉅。其爲眞正之創造者，故應以勤勞爲自勵，以寬大自慰。而放棄物質之享受，苟如此，則凡事認眞，不辭求捷徑，工作實際，不好高騖遠，以服務社會爲樂，時刻滿足本身之崗位，則偉大之工程人格成矣。

從事於工程之人員，首重經驗，經驗之獲得多由於失敗，失敗爲成功之母。失敗愈多，成功愈大。因是本人謹以自身從事工程數十年來所經歷之失敗披歷於諸君之前而勉之。

（一）中國鐵路之概況：一中國之鐵路以前並無一定之標準，每隨地而異，建築方法與所用材料，亦參差不齊，全視該鐵路所用工欵由何國商借而定。茲依照其建築方法分述如下：

（甲）英國式 我國採用英國式之鐵路主要者有下列各綫：

A.京奉綫 爲中國最先完成之鐵路，建築動機乃在採運開灤及撫順之煤礦。

B.津浦綫 借英國欵項築成者。

C.京綏線 完成京綏綫之工程爲詹天佑先生，彼乃參與京滬線之工作者，是以平津綫亦採用同一格式而爲英國式。

D.滬寧線 雖同爲英國式，然比之京奉綫，較爲新穎。

E.湘鄂綫北段

（乙）法國式：

A.平漢綫 是綫乃借法國欵建築而用法式。

B.隴海綫 此綫與平漢綫互交於鄭州，作十字形，亦爲借法國欵築成者。

（丙）德國式

A.膠濟線 乃德人佔領青島後由德人建成者，故全爲德國式建築。

由上述各點觀之，則全國路綫共有三種格式，而各種格式皆具有獨特之優劣點，概言之爲德式輕巧，美觀，惟强度，則較弱。法國式堅固而穩實。英國式則健全而耐久也。

在英國式之鐵路線中，因其建築時間之遲早，故格式亦有差異。滬寧建築最遲，故爲最新格式。廣九綫爲英國工程師所建，而依照滬寧線之格式建築。然大部份工程師曾在京奉鐵路工作有年，故廣九路可謂包含滬寧與京奉路兩種格式。廣九線於一九〇七年開始測量，一九〇八年施工，而完成於一九一〇年。

（二）廣九鐵路之失敗教訓 廣九鐵路經數年之室內野外工作，順利完成，當時至感歡欣，然一經正式通車即發覺弊病百出，一切皆非學理所能預測者，比如林村至平湖一段之九十五號橋之橋墩，竟爲大水所冲毀。又一百號橋，每遇江水上漲，即遭掩沒而無從通車。考其失敗之原因，實由於以下三點：——

A.忽畧小節——事前未能顧及各足以妨礙通車之小節 。例如各鄉人常於小橋之上游築壩阻水，用以灌溉，水滿之後，乃越壩而過，其勢甚洶，冲刷橋基，至使小橋傾斜。本人經手修理此等橋樑甚多。

B.橋基太弱——河流至橋樑位置，河面因受橋墩影响，而至變狹，流速驟增，每每將橋基冲刷成空洞。初期補救方法乃以石塊填阻。及後發覺此非善法，蓋在深水下投以石塊，並不能將橋基空洞填實，橋基仍甚弱。故改於橋墩上游阻絕河水，然後以唧筒將橋基之河水抽乾，再行將橋墩澈底修理，此法雖善，惟用費甚大，非經費拮据者可採用，當阻絕河水之時，上游之水，仍繼續流至，水位增高，勢將水壩增高，水越流越多。壩愈築愈高，將無止境，為避免時將壩增高之費用，乃用大木槽引河水越橋而過，由此得到處置河水之經驗不少。

C.孔寬不足——因孔寬過少，每每招致橋樑之破壞，九十五號橋即其一例。本人於抗戰期間，曾負責天水蘭州一段鐵路之踏勘工作，對於各橋樑之孔寬，甚為注意。當決定其孔寬時，每於實測之紀錄中加大一倍。此於學理方面似覺輕鬆，惟經驗所示，必須如此方可避免後患。故橋樑孔寬，寧失之多，莫失之少也。

廣九鐵路創始之初，虧本纍纍，後經不斷之努力與掙扎，方達於戰前之黃金時代，其時廣九路行車速率，全程平均速僅次於京滬線之下，若以每段之平均速而言，則佔全國鐵路之第一位。當時之全程速率為二小時五十五分。惟本人猶未滿足，而欲縮短為二小時三十分。當時之計劃乃將全線七十餘道行徹尖釘牢，使列車通過時仍可保持高速度，而將全程縮短七十分鐘。對於各站傳遞路牌之方法，亦擬改以自動機傳遞。使列車經過車站無須降低車速以便傳遞也。由於上述之改善，則二小時三十分之時間，即可自廣州直達九龍。後因局內各人意見參差不一，故此種改善計劃，未見實行，殊為可惜。

目前之廣九鐵路，雖極力改良，然實難恢復戰前狀況。蓋人力不濟，工人技術標準降低；物力不足，使與器材缺乏；地方環境惡劣，員司毫無朝氣有以致之。故欲在十年內恢復戰前之盛況，實屬疑問，實則廣九一綫，以正常而言，二小時即可完畢其旅程。蓋目前之路軌，多為八十磅者，若全部換為九十五磅軌，則二小時之速率當可到達，本人對此段車速之改進，所獲得之經驗甚多，成功與失敗俱在，上述種種即為一例，相信諸君當樂於一聞也。

現日之工程人員，可分為穩健與急進兩派：

穩健派——凡事以慎重態度處置，方針既定，絕不為流言所動，全部計劃以安全為重。

急進派——只求時間上之縮短，金錢上之節省而忽視日後之禍患，對於任何之工程，只求其能於最短時間內用最少金錢完成，而不計較其耐久性幾何，安全性大小，及日後損壞時所耗之修補費用。此派對於整個工程計劃，根本缺少堅定信仰，易為別人所動搖。

本人從事工程，向持慎重態度，設計路線之時，絕少貪圖捷徑而採用一大於1%之坡度，惟天水至蘭州一段內，因山坡過多，而有採用 1.5% 之坡度者。然亦實出於不得已而為之。

此外則有本人計劃廣梅綫之時，初以為已意甚合，及今思之，亦覺錯誤。蓋廣梅線之起點問題，廣九路英籍工程師認為應在廣九路石龍站以下開始，惟本人極力反對，理由為廣九之最初三十英里乃廣潮之平原，次三十英哩為江河汎濫之區，此段橋樑至多，最後之三十英哩，則在寶安境內，山嶺起伏。當時已入戰事時期，橋樑易被炸燬，苟廣梅線之起點在廣九路石龍站以下，則一旦石龍大橋被炸，則廣梅線雖完成，亦無可置用矣。因而力主自石灘為起點。結果卒照本人意見開工，然時至今日，始知此種建議，乃屬不當。蓋其起點應在廣州而不應設於廣九路之半途，因所省有限而對於運輸管理，皆有甚大之影响也，然此皆非當時所能覺察者。故對於一計劃必須預料及微小之可能錯誤。再舉例言之，月台有高低之分，普通貨台約為三尺六寸，客台則為二尺三寸。廣九沿綫車站，初多設二尺三寸之站台。後有人主張改高，本人亦以為然。乃一律將之改高。及至完工之後，始發覺此等站台，不特對於檢查油箱輪軸之工作發生阻礙，而旅客上落，亦不安全，故有再行改低之必要。諸如此類之錯誤，非經實地觀察，每每未能發覺也。

（三）工程人員之道德觀念——凡從事工程之人員，切勿眷望過高，務需實事求是，職位之高低不足以移其志，薪津之厚薄不足以降其格。對於社會，工程人員為真正之創造者，對人類生活影响至鉅，其為真正之創造者，故應以勤勞為自勵，以寬大自慰。而放棄物質之享受。苟如是則凡事認真，不尋求捷徑，工作實際，不好高騖遠。以服務社會為樂，時刻滿足本身之崗位，則偉大之工程人格成矣

末後欲提及者，則工程師應立堅定之志趣，具遠大之目光，否則隨波逐流必一事無成也。

（本篇錄未經梁先生過閱如有錯誤筆記者自負）

主題：公　路

第四次學術講座（十二月二日）主講黎傑才先生　　何進康筆錄

講題：築路機械與機械築路及其在我國之關係

講者黎傑才先生，現任交通部機械築路第二總隊總隊長。抗戰期間，曾任湘桂越鐵路橋樑隊隊長，丹竹呈貢等機場工程處長，與保寗公路工程處長等職，勞苦功高，學術講座第四講，得先生主持，誠各同學之幸也。　——編者——

說到築路，從前覺得它的範圍是很小的，政府也不很重視，但由于時代的進展，運輸日加繁重，而且因爲要工作效率迅速之故，所以政府現在不單注重築路，且亦注重機械築了。

先說說築路機械，普通來說，如起重機，挖泥機，及凡能用來築路的機械均可稱爲築路機械，若將其分類，可有下列三法：（一）築路機械以其產生之工作各有不同而分成各種類。（二）因其裝備不同而分類。（三）因其發生原動力所需要的原料不同而分類。

現在我們來談談爲什麼我們要用機械來築路呢？這問題的解答可有下列數點：（一）由於人類生活逐日進化之使然；從前人類是用自己的力量來產生工作的，其後便用牛、馬等畜類替代了人力，現在更進化至用機械來代替畜力了。（二）因個人的力量是有限的，遠不及機械的力量。（三）由於工程需限期完成，在要迅速完成的條件下，於是便要運用機械的力量了，由上述數點，我們可知道工程必須要機械化才可追上人類生活進化之途徑。

機械築路成功的要件要工作的數量相當大，因機械隊一流動的機械廠，所以必須要給予它一龐大的生產工作之機會，它才可表出它偉大的効能，如工作太小，則效果無從示出，至於我國用機械築路，則困難的問題特多，因（一）我國無重工業，各物及機械所用的零件也須由外國運來。（二）機械所用的液體原動力原料我國產量不敷，（三）由於我國現況不定，且爲入機械化之初期，所以人才缺乏，欲維持一批工作人員及提高水準不易。（四）工作發展受人力的阻礙。（五）工程界方面有反對者，因爲上述種種的困難，所以我國欲成功運用機械築路障礙很大，但我們要知道，我們不能因困難而退縮，須知工程必須機械化才可跟上進化的途徑，所以我們只有努力的幹；經過最近數年來的努力，可証明用機械是可成功的，並且在工程方面佔有很大的價值並取得地位，換句話來說，人力有人力的地位，機械有機械的地位，兩者相並齊行，是一樣重要的。

在這裏說一說，國家經濟這樣的動盪，預算無從根據，那麼機械築路怎會成功呢？以過往經驗來說，一機械所需要原料的價值約等於總工作價值十份之三至二份之一，故用機械築路，對於預算可有三份一至二份一之準確，所以對工程進行較易，若用人力而爲，則無預算了。還有一點，就是管理機械比管理工人爲易，機械吸收原料後所發生之能力，比人力所生之能力爲大，且保養及修理機械皆較易，費用方面亦較人力爲宜。

以上所說的都是欲表明工程須機械化，把人類的生活水準提高，使能跟上時代的進展，使國家踏入一新的偉大的途徑。

舊事重提

魁北大橋的倒塌 陸能源

當時驚動了全世界工程界的魁北大橋的倒塌（Failure of Que bec Bridge）已是四十多年的事了。筆者在此重提舊事是因爲這重大的損失曾給予工程師們一個寶貴的教訓，欲予未曾注意這個故事的人們一個參考。

魁北大橋的倒塌發生於一九零七年八月二十九日下午。這橋是加拿大政府建築於離魁北（Que bec）城約九英里橫跨約半英里闊二百呎深聖羅蘭斯河（St. Lawrence Ruer）的一座當時世界在建築中最大的懸臂橋（Cantilever Bridge）。倒塌時該橋全部工程已完成一半。倒塌那天，在橋上工作人員，八十五人中死亡者達七十四人，在僅僅的十五秒鐘的時間內，一座完成過半的大鐵橋立刻變成一大堆曲斷了的廢鐵。該橋的跨度等可由下面簡圖明之。

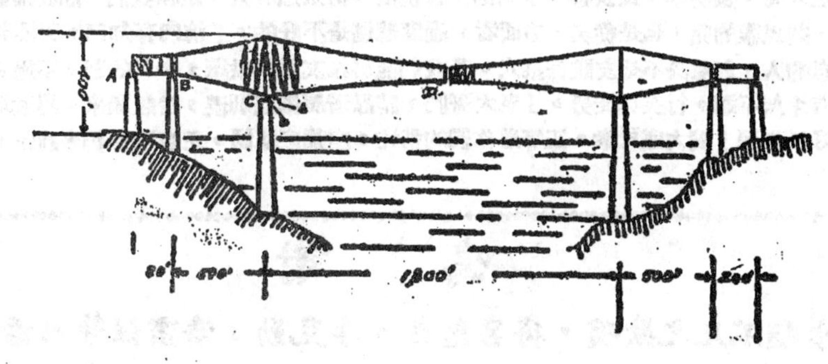

該橋之大小可說於下列諸事項：

1. 該橋係雙火車軌道，雙電車軌道，雙汽車道及雙行人道的大鐵橋。
2. 該橋鋼料共重38.500噸。
3. 該橋最重之 member達100噸。
4. 該橋最長之 member達105英呎。
5. 該橋所用鋼栓（Pin）之直徑爲24英吋。
6. 該橋鋼栓所繫最多之鋼條（eye bar）爲56條。

這座大橋於橋墩建成後由A處（上圖）起建，經B.C.到達D處時即突然倒塌。

關於這橋突然倒塌的原因，根據各方研究結果，可簡述如下：

1. 該橋倒塌起於近橋墩C處（上圖）之 Compression Member a b 之突然損壞。
2. 檢驗這條支柱的損壞，可知係由於 lattice bars太弱未能支持所發生之剪力（Shear），根據後來試驗大小等於原柱三分之一的模型的結果可確知之，檢驗這已試驗後的模型，可証實 lattice bars及其卯釘（Rivet）因剪力過大而損壞，故此模型支柱之損壞非因 Buckling 或 yielding所致。此點與原橋支柱之損壞情形相似。
3. 雖設計時對Dead load之估計稍低及所用 Working stress 稍高可能爲其損壞之原因，但主因仍要歸究於次要設計（Design of secondary Part of Structural member）之未臻完善，以致latticing 太弱。當時工程界對於鋼柱（Steel Column —— Compression member 即一條

Column)之知識及其受力質性未徹底明瞭亦有以致之。

自從這座大鐵橋倒塌之後，就引起了許多人對鋼柱之研究和試驗，這種研究一直到現在仍在進行中，設計鋼柱之規範書(Specifications)亦日在改進。

根據上列事實，我們可以有下列的感想：

1. 工程知識，有時要用極大的代價去獲取。
2. 如果設計一件空前偉大的工程，其成功失敗，可能爲一極小未及注意之事項之是否完善而影响。
3. 凡一種工程理論，其中必有多少假設(Assumptions) 其理論之是否正確可靠，應該根據實驗室中對此理論之試驗的結果。每設計一空前少有的工程，其所根據的理論，應該加上試驗結果作爲一種可靠的復核(Check)

別——給畢業的老大哥們　　一凡

現在，我們又看着一班老大哥，離開了我們，踏進茫茫底人海去了。說起別，誰人不感到傷感，何况是同畢業的老大哥，羅明享、黃煥林、李克勤、張悅階、雷永鼎五人，是和我們一起在搖籃裏長大，曾經一起同甘苦，共患難的呢！但是從另一方面看，過度悲傷是不好的。「海內存知己，天涯若比隣。」有共同目標和意志的人，別離决不是友誼底墓穴。若我們能緊拉友誼的鐵纜，今日的別，不過是日後合作的先聲而已。「有才無不適，行矣莫徒勞。」老大哥們，請忍着感傷底別淚，抬起頭來，勇敢地踏出校門，做我們日後底好榜樣罷！廣大底民衆，正等着你們的帮忙，荒蕪底祖國，正等着你們來建設。

鳴　謝

本期不足之款項，得呂惠炎、李克勤，潘濱强等四君捐助謹特此致謝

土製士敏土之應力

江[illegible]作
鄺曉厚譯

是文本用英文寫成因印刷與其他關系特譯成中文　　——編者——

1947年春季，筆者得指導材料試驗一課。在工程上的一種重要材料，是混凝土，這種材料，能够在實驗室中製造試驗者。

這科所佔的時間很短，每一組學生，能搜集關于混凝土性能的知識很少。甚至七組集合，試驗之所得，對本地產品的强度和其他特性，所知亦不多。欲在其性能方面得一清晰之認識及結論，就已非這科目的範圍了。因爲製造混凝土有多種因素與混合之可能，筆者對于其他從事這種試驗者所用之方法與結果，甚感興趣。筆者希望此文能引起討論之興味，故所有批評，均表歡迎。

這科的範圍：混凝土成分的試驗，是根據ASTM 標準的。士敏土沙和石的試驗，主要目的是欲製成一種足以代表商業上應用的混凝土。此種混凝土之骨材（Aggregate），多由未經處理之沙石合成。

士敏土試驗所用之士敏土，是西村士敏土廠五羊牌。試驗凝固時間（time of set）用，Vicat needle和Gilmore needle。初期凝固（Initial set）小于三小時。Soundness test 用灰漿餅（Pat）置於100°C 蒸氣內，經過五小時而無徵細之裂縫，細度試驗（ Fineness Test）用200號篩來篩，少過20%留於篩內。Normal Consistence 用Vicat needle 測定爲27%至30%。標準灰漿之張力與壓力試驗，用7日至 28日之標本，1：3之灰漿之張力，經過7日爲194 PSi。28日爲211Psi。壓力之結果，7日爲2292 Psi，28日爲3232P.s.i。骨材（Aggregate）：試驗所用的骨材爲本地產物，沙由本地商店購取，而商店由河取得，試驗結果，知道此種沙不含有機物，但含有小量淤泥（Silt）。用篩分析之結果，知道細度係數（Fineness Modulus）多爲 2.43 及 4.19。沙與石之等級，亦如想象中之惡劣，因爲牠們沒有分類，沙包括太多低過48號篩之徵細沙，而碎石包括太多平石塊。

混凝土（Concrete）：影响混凝土强度之主要因素，爲水分士敏土之比率（Water cement ratio），和單位體積混凝土所含之士敏土之多寡，至於稠度（Workability）亦要顧及。稠度能因變更沙與石之量與乎變更骨材的粗細而變更。1：3：6：之混凝土，其水與士敏土之比率爲 8 加侖與低落爲 2 吋者，經18個6 ×12吋之圓柱模形，試驗其壓力强度如下：

	7 日	28日	yield
平均	787 psi	1190 psi	7.8 cuft
範圍	650 至 880 psi	900 至 1420 psi	

至於通常所用以建築房屋之 1：2：4 混凝土，其水與士敏土之比爲 7 加侖者，爲

	7 日	28日	yield
平均	1238 psi	1570 psi	4.54 cuft
範圍	1140 至 1375 psi	143 至 1650 psi	

故根據整個試驗結果，產品之强度，是以根據 ASTM的規範。但只是少數試驗模形的結果，不是整個的結果。

普通螢光管之光量設計表

潘應椽

螢光管之用途日益見廣故其光量設計爲吾工程界所應注意者，茲以其普通光量設計表及設計程序附錄于後

此表以白色螢光管爲標準，如用日光管，則將瓦特數量增加1770；如用柔白色光管，則將瓦特數量增加40%									
最大燈距	1.5 ×地面至天花板高度(呎)			13. ×地面至裝置高度(呎)			1.1×由地面裝置高度(呎)		
瓦特(附屬品在內)	瓦特每平方呎面積			瓦特每平方呎面積			瓦特每平方呎面積		
房間之比例	W=H	W=2H	W=4H	W=H	W=2H	W=4H	W=H	W=2H	W=4H
學校及辦公室 5-15 10Ft-c. 議會室 食物店 應接室 走廊 講堂 洗手室	1.0	0.8	0.6	1.0	0.7	0.5	0.8	0.5	0.4
15-30 20Ft-c. 課室 圖書館 實驗室 臨時寫字室 普通審記室 文件室	2.0	1.6	1.3						
30-75 50Ft-c. 精細工作室 繪圖室 打字速記室 簿記室 會計室 閱晉室	5.0	3.9	3.2						
75-150 100Ft-c. 繪圖室 商業機器室	10.0	7.8	6.4						
5-15 10Ft-c. 儲物室 洗手室				1.0	0.7	0.6	0.8	0.5	0.4
15-30 20Ft-c. 間乎儲物與普通資物室	2.0	1.6	1.3	2.0	1.4	1.1			
30-75 50Ft-c. 普通貨物店	5.0	8.9	3.2	5.0	8.5	2.8			

5-15 10Ft-c.	器具室 倉 固定工作區 起貨區 船務區 汽鍋室 車房							0.8	0.5	0.4
15-30 20Ft-c.	粗糙室 製造廠 木工室 紙機室 剪機室							1.6	1.0	0.9
30-75 50Ft-c.	普通會議室 顏料室 打寫室 縫紉室 臨時試驗室							4.0	2.5	2.1
75-150 100Ft-c.	機械室 精細試驗室 顏色配合室 精細工作室							8.1	5.1	4.2

W＝室闊　H室高　　　　Ft-c.每燭光（Foot. candles）

以上設計表爲每平方呎面積所需之光量，例如有某工廠，其面積爲50呎×50呎由地面至天花板高度爲12呎而此廠用爲做粗糙工作者，在設計表上可知所需每呎燭光（foot-candles）爲20室闊約爲高度四倍，故在表上查得瓦特平方呎面積爲0.9由地面至天花板高爲12次故裝置高度約離地10呎而每管相隔亦爲10呎，10呎數量小過1.1×由地面至裝置高度（即10呎）故適合，所以每管所佔之面積爲 10×10＝100 平方呎而每平方呎有0.9 瓦特，故該面積共有90瓦特，用雙管每管40瓦特之光量已够，如此其總瓦特連損失在內爲90瓦特，故在此工廠內需用5個40瓦光管在100平方呎面積內用一對光管其餘一管單獨裝置。

啓發天然力中的中國

余文傑

(譯自：Engineering News-Record Sep. 16 1948)

當第二次世界大戰結束後，中國資源委員會，(NationalResources Comission of China)曾經考慮過復原後工業化中國的計劃。這會在1935年四月由政府組成。他的主要目的如下：

(1) 推進基本工業

(2) 發展主要鑛業

(3) 增加天然力的利用

(4) 推進其他政府有關業務

現在中國最需要的莫如動力。因爲它是工業的始源。他預算計劃中，先發展蒸汽發電840,000KW，水利發電280,000 KW.，而且還希望在五年中能够把這個數目增加兩倍。

因爲這個原故，中國資源委員會於是成立了一個水利發電工程會(National Hydro-electric Engineering Bureau)，這個會曾經在近三年來很勞碌地從事於設計，測量，及推進不少急需的工程事業。但，可惜得很由於經濟及政治上危機的阻阻，許多計劃都遲緩起來。雖然如此，這個水利發電工程會還安然努力工作。在中國各區考查和測量，希望光明日子的來臨。

最堪注意的就是上清源洞(Upper Tsing Yuan Tung)水壩。位於楊子江中的一支流名曰龍子河(Lung Chi River)。

這幾個動力區域都集中於四川省的東部，約距重慶六十英里。龍子河長約八十英里，這裏地勢，瀑布和急流都很適宜於水利發電事宜，成爲重慶主要動力的中心。這四個區域一共能够發動64,000 KM，替代現在的16,000 KM。

第一個區域就是下清源洞(Lower Tsing Yuan Tung)，始建於1937年，落成於1942年。這裏有兩大渦輪，每渦輪工率爲1,000馬力，一個是在英國製的，其他一個是在中國製的，能推動一1,800 Kva發電機。

第二個區域就是在許龍濟(Hui Lung Chai)。現在希望能够在最短期內建設2,000 KN。這裏的渦輪和發電機都計劃完成，裝設於不久的將來。

在籌建中的15,000KM第四個區域，就是獅子流(Lion Rapids)水塘及發電廠。但可惜由於經濟缺乏，工作還未開始。

龍子河的第三個區域就是上清源洞。現在要利用它一方面來冶金，另一方面來建築獅子流水壩。這個壩是建於河的上流，發電廠位於下。這樣便可以把壩的高度減低了許多。此處河床都是由沙岩石做成，此大部份石都被河水冲去，所以對於掘挖工作只不過是一件易事而已。

這裡工程的展開是由包工做的，由包工者僱工人，工人的薪水多數是用米來做標準。每件工程假如包工承接後，不論贏虧都要做完，這樣便可以減去通貨膨漲的顧慮。

有些小工程，例如鋪石，它的價錢以每立方米特而定。又有些包工喜歡以「日薪」計算的，但因爲通貨膨漲的原故，「日薪」要常常更改。在工程中所有　料、器具，例如洋灰機鑽等都是由水利發電工程會供給的。

水壩完全是用人力做成的，　料是用砂岩石和洋灰漿。壩高19米特(62英呎)，長120米特(394英呎)。連支台長23?米特(76?英呎)。這壩的預算在最大水時流量可達130.000立方呎/秒。

剛才曾經說過整個壩都是用沙岩石砌成的，可是關於成本多少問題，還要着重於洋灰的價錢來決定。至於技術方面，兩來數世紀，中國石工的巧妙，還是很著名的。

下面就是1947年九月所調查上清源洞的最高預算价格表。說明用石灰漿和洋灰漿砌石的比較值。這些比值只限於本地，因爲，例爲在南京及上海，勞工的價錢或會比這裏高五倍，而洋灰的价錢又或會高兩倍之多。

上清源洞壩所需材料及其每立方米特的比較值

	每立方米特計						
	洋灰（桶 bbl.）	石灰（磅 lb.）	砂（厘米 c.m.）	碎石（厘米 c.m.）	塊石（厘米 c.m.）	勞工（人日 man-days）	比值（%）
全用三合土	1.30		0.5	1		5.0★	100
洋灰漿砌石	0 33		0.2		1	3.6★	49★★
石灰漿砌石		55	0.2		1	3.6★	17★★

★包括養病津貼，惡劣天氣，和盈利。

★★包括所有砂岩石價錢，例爲採石，修碌，運輸，安置等。

石之大小，約0·4×04×1米特。這個尺寸剛適宜於搬運。石的採取完全都在河的兩邊，所用的器具不過是斧，鑿，劈而已。

工作的展開完全在建築商的手中。他們在合同中規定了每立方米特的價錢，然後每隔兩月依米價更改一次，岩石打好後便用電車或船搬到壩邊。

在石未安置之前，地脚首先要修平，然後用1：3洋灰漿砌上。但是因爲在中國洋灰漿價貴，而勞力價平，故此岩石的接口可用1/4"的濶度。每層石的安放須隔36小時。這樣便可以來得比較穩固。

普通水高約28米特（92英呎），最高爲110英呎，最低爲87英尺。在發電廠裹平均流量約1,080立方尺/秒。這裏發電廠包括兩個單位：9,000和4,000Kua。

發電廠通常在地底，在設計時已經預算了地下最大的水壓，同時亦裝置了抽水設備，至於空氣流通方面，亦有很完善的布置。

會　聞　　編輯室

十月卅日，本會曾舉辦參觀黃埔港旅行。是日天氣清和，一行三十餘人，乘專車直達，得該港工程處人員之講解，對黃埔港建築計劃及水利情形，得一眞實之認識。隨即赴建築地點參觀。時一鋼筋混凝土之大倉庫在建築中，故對結構學又得一見習機會。各同學對于各問題均興趣甚濃，發問頗多，直至下午二時，始乘興而回。

十一月十八日，本會學術講座，第二講，在工學院102課室舉行。講者爲劉載和教授，講題爲「工程政治學」。劉教授乃本校舊同學，畢業之後，服務工程界十多年。抗戰期間，足跡遍于中國，人生經驗更富。「工程政治學」乃劉教授經多年經驗著成，內容充實，給吾等對工程學另一新觀念。

十一月廿四日，學術講座第三講，由梁永槐先生主持。梁先生年近古稀，但精神飽滿，直講點多鐘，面無倦色。詞句亦諧亦莊，由始至終，坐上百多人，無不興緻十足者，足見梁公口才之佳，與內容之充實。

十一月廿五日，本會又舉辦一廣九綫旅行，以便利選修鐵路工程各同學，得一實習機會。該日得廣九路總段長，先備工程專車與卡車一架，自廣九路出發，沿途過有橋樑及與鐵路工程有關之處，均下車參察，使從書本中得來之理論與實際，作一比較。其後又轉入黃埔支線，重赴黃埔，先參觀黃埔鐵路上路軌之扣件及轉轍器等，然後再入建築區參觀，至三時才乘原車返廣九站，總括此二次旅行，對於土木工程所包括之水利、結構、鐵路、橋樑等，均有所觀及，是誠不可多得之機會也。

十二月二日，學術講座第四講，請得黎傑才先生主持，黎先生亦爲本校之舊同學，對母校愛護備至。爲人和藹可親，無架子，坦白眞誠，誠紅灰精神之表徵。抗戰期間，歷任工程要職，曡建功勳，所得勳章至多，誠工程界之功臣也。

十二月十七日，本會於西大球場舉行冬季野火大會，會中秩序繁多，興味極濃，卽年過半百之教授，亦參與同學之踢火球，捉迷藏，歌詠及舞蹈等玩意。直至十一時方宣佈散會，同學之戀，其趣者，大不乏人。

十二月廿三日，學術講座第五講，講者國民大學工程學院院長張建勳先生，講題爲美國加省水利建設，張先生旅美十二年，曾在美政府各級工程機關服務，對加省工程之機構，更爲熟悉，其講詞不獨使吾人如身歷該地，耳聞目見，且提起吾人對水利之興趣，使吾人想起中國未來之遠景，誠堪珍惜也。

一月六日，在泰山歡送畢業同學，到會之人甚多，足見南大親切之精神。

編　後　　編　者

在學會中負責學術，乃最艱難者，吾等資質愚魯，更感工作進行之不易矣。幸各教授與同學不棄，時予指導，工作始不致停頓。此點吾等應向各位致謝者也。

吾等爲學生，一切首重學習。本刊之出版，主要之目的，亦不外增加學習之興趣。故學術股之工作，除出版外，知識之進修，尤關重要。故本學期之工作重心，由出版而移至學術講座與參觀，卽是故也。

學術性之刊物，非重在篇幅或期數。稿件之獲得，亦非易易。若稿件不够，勉强充數，則不出版尤妙。故本學期南大工程本欲出版三期，今只出兩期，卽是理也。

至各寶號惠登廣告一節，本訂明之期連登惟鑒于縮短期數之故一律于本期內將各廣告照原訂面積增大，以補損失。加之初本訂定每期出紙一張，今則每期出版一本，此點亦可稍補少登一期廣告之損失，望惠登廣告者予以愿諒。

至學術講座，劉載和教授之講詞，因他暫不願在此刊登，故留于下期。張建勳院長之講詞，因時間關係，整理不及，未克在此期出版。今特在此致歉。

南大工程

桂銘敬

THE JOURNAL OF THE LINGNAN ENGINEERING ASSOCIATION

廉樂再刊 第五期

VOL. 5

MAY 8, 1949

本期要目

嶺南大學工程學會刊印

中華民國三十八年五月八日出版

Spurning the spell of distance, cleaving across the land,
Binding mankind's far outposts to all the human scheme,
Cutting the shifting landscape, the highway's level band
Winds through the mountains' passes and over the sluggish stream
That curls beneath the arches of a bridge's rhythmic span
Down to the mighty gorges in the valley far below
Where the river's surging torrent is tamed by the will of man,
Balked by the Titan masses of the dam that stems its flow.
The dam that leashes power to turn the distant wheel,
That makes lights glow in cities far—the road that links all near;
The bridge that spans the chasm—all are monuments that seal
The triumph over nature by the Civil Engineer!

by Eric Fleming, M. ASCE

南 大 工 程

康樂再刊 第五期

目 錄

工學總會第一屆理監事職員表

監　　事

桂銘敬　黄郁文

王叔海　馮秉銓　鄺正文

理　　事

主　　席：劉載和

副主席：高永譽

文　　書：吳乘俠　鄺國良

財　　務：黄銳鈞

福　　利：崔兆鼎　韋金信

交　　際：關學海

研　　究：梁健卿

庶　　務：曹文達　黄惠光

特輯

總會成立週年對同學說的話

——劉　戴　和——

嶺南工科同學總會，自成立迄今，已足足一年了。回憶去年成立的一天，曾開會慶祝，並出版特刊，以留紀念。是日到會參加的，除本校工科畢業和在校的同學外，還有工程機關的首長，校外來賓友校代表，以及校內的師長等，盛況空前，爲本校自設立工學院以來所僅見。今年爲總會成立一週年紀念，本應比以前更爲熱鬧。惟際此國家多難的時候，念及生活的艱苦，物質的缺乏，所以極力撙節，一切鋪張，固然完全避免，而所出特刊，亦改用較差的紙張。但特刊的內容，仍力求充實，開會的秩序，亦力求豐富，以紀念這一個富有意義的日子，相信到時情況的熱烈，必不減當年成立之時！

本校設立工科，遠在二十一年前。但在抗戰期間，因受戰事影響，曾經一度停頓，故工科畢業同學，迄去年止，僅得十二屆。且嶺南的傳統作風，爲貴精不貴多，故每屆畢業同學，少者四五人，多者十餘人，因此畢業離校的同學，前後共不過百餘人，與其他大學較，實相差甚遠。嶺南工科畢業的同學，人數雖然不多，但活動力則甚强，現除在本國者外，在歐美以及南洋各地，皆有嶺南工科同學，而以在美國及南洋爲較多。在美國的同學，有繼續深造者，有在美國工程機關任職者。在南洋的工科同學，則多在大商業機關和工廠工作。至留居本國的同學，在抗戰期間，活動最廣，在自由中國的每一省份，都有嶺南工科同學的足跡。他們以種種姿態，協助政府，從事抗戰，如造機場修鐵路築公路與水利等，皆能克盡厥職，効忠國家。抗戰之後，內戰繼起，嶺南工科同學爲避免無謂的犧牲，皆紛紛南返。現在除數位仍不避艱難，不辭勞苦，留居廣西，在湘桂黔鐵路繼續其未竟之工外，其餘多返回粵穗，此乃嶺南工科畢業同學移動之大概情形。

在嶺南畢業的學生，都抱有一種崇高的理想，但這種理想，須在一個良好的社會一個安定的國家始能實現。但環繞着他們四週的社會，是一個黑暗的社會；擺在他們面前的國家，是一個紛亂的國家，所以當他們一踏出校門，就覺到情形不對而失望。在抗戰期間，因在國家至上的號召之下，他們不得不捐棄其個人的理想，混在污暗的圈子裡工作。抗戰之後，國內局勢並不好轉，反而變本加厲，他們覺得這種犧牲是太不値，寧可轉行，亦不肯再留在原有的崗位，因是紛紛辭去政府的職務，或教書，或經商，或從事建業。其仍留在政府機關工作的，亦表現出一種超然的立場，廉潔的操守，優異的成績。這是嶺南工科畢業同學從業的大概情形。

嶺南工科同學的學問水準並不低，工作能力亦不弱，以往在各機關服務，都曾表現出優異的成績，這並不是妄自誇大，而都有事實爲証。目前因不滿現實，紛紛離開原有的崗位，轉就別業，並不是嶺南工科同學放棄他原來爲國家爲人民服務的高超理想，他們日夕盼望能在一個開明的政府領導之下，再演身手。所以環境一轉佳，社會一安定，他們必挺起胸膛，整隊向着建設國家，服務民衆的大道邁進！

嶺南已畢業的工科同學的處境既然如此。嶺南在校的工科同學又怎樣呢？嶺南有良好的讀書環境，嶺南有豐富的圖書儀器，所以在學問的尋求智識的獲得的一方面，目前嶺南學生的確比別校爲勝。但畢業後，找尋職業，則每每比別校畢業生爲困難。因此在畢業典禮舉行之後，畢業的歡樂，瞬即變成爲失業的恐怖。造成這種現象的主因，固然是國家政治不上軌道，此外尙有兩種原因，一爲我們沒有在政府裡任要職的政客做我們的教師，所以我們缺乏提拔的人；一爲我們沒有和社會發生密切的聯繫，所以我們陷於孤獨。我們想以後不再有這種痛苦，我們就要在這兩點下工夫。但我們並不歡迎政客跑入我們的圈子裡，所以只有從深入社會一方面着手。然而從那一方面深入社會，由那一角度接觸社會呢？最安全最妥善的方法，就是透過畢業的舊同學而深入社會，來接觸社會！

嶺南工科同學總會成立的目的，就是想聯繫着這班有能力有意志現在正等候着機會來大顯身手的畢業同學，來爲國家工作，來爲人民服務。就是想引導這班具有純潔的心靈，具有聰明的頭腦的在校同學接觸社會深入社會！

嶺南工科同學總會經已成立一週年了！在這整整的一年中，我們曾做過什麼事呢？我們的答案是一無所成。但我們並不以此爲恥，亦不因此而悲傷，因爲這個總會並不是一個政治黨團，亦不是一個工商會社，而是一種精神上的維繫，一種友誼的交流，所以在這種危機滿佈的社會裡，在這種物價飛漲的時候，我們實在沒有從事任何重大工作的必要！

同學總會雖然沒有做出什麼工作，但因這一個總會的成立，我們已意識到一種很大的收獲，就是這一百數十位工科新舊同學，一方面透過我們舊的友誼，維繫着我們新的意志；一方面樹立一種新思想，革除原有舊作風，前一點可由我們一班新舊同學的言語行動見之，後一點可由我們一班新舊同學的生活方式見之。以往社會給我們的惡名，說我們全是美國作風，洋氣十足，說我們是資本家的產物，養尊處優，現在工科同學已將它完全抹去。如不信，請聽聽我們的談話，請讀讀我們的文章，請瞧瞧我們的皮膚，請看看我們的工作！

工科同學們，時代巨輪現在急劇的轉進，我們應握實拳頭，咬緊牙根，急起直追。我們不特要追上它，還要越過它，在它的前頭跑。因爲在它的後面是汚暗崎嶇的深淵，在它的下面是悲慘痛苦的死亡，只有在它的前面才是光明康莊的大道！

工學會日感言

關學海

（一）我感到非常欣慰因爲我是工學會一員，——我雖配不上說老嶺南，但總算在曲江入大學，也曾跟轉隨着學校在顛沛流離中奔跑，在數年光陰中對嶺南總算有多小認識，和很多老嶺南談話間，尤其在停學走入社會的兩年當中，在各地碰到畢業同學，彼此都有相同的感覺嶺南人多數着重個人的享受和發展，有了成就的沾沾自喜，炫耀人前，環境困頓的，消聲匿跡，顧慮顏面，不願求助于人，畢業同學在社會很少能保持聯絮，互相扶持的。

在復學當初，我眞恐懼，在數年大學中，除了取点書本智識外，恐怕將來在社會上很難獲得幾個朋友。

這兩年來的觀察，嶺南人的社會觀念似乎已改變了，尤其是工學院學的「工佬作風」一片，坦白自然，毫無矯揉做作，敢罵敢笑，一言一行都是眞情摯誠的流露，在工作上無分彼此，很少推諉責任，這種團結一致的精神，無論在工作上或未來的事業上，都是不可或缺的。數年來憂慮，現在總算將到点安慰，那就是因爲我是工學會的一員。

（二）老敎授們辛苦了，同學們也忙够了，——工學院自太平洋戰爭爆發以來，一直在艱難困頓中，黃議文先生和一老敎授，胼手胝足，艱苦支撑到今日，可說已具規模，但因人數激增，每個敎授所担任的功課，實在是相當煩重，不論什麼時候，不論在辦公室或實驗室中總可常看到他們忙碌地工作。

另一方面同學們，經過校方一次兩次的加緊，由六十五分的標準而提高至七十分的標準，已一天比一天忙起來，平日故然都在工作，卽星期日也常是埋首伏案拿着計算尺對着一堆一團的數學或筆畫細致的圖，起勁地幹。

這種現實的表現是值得額手稱慶，正如敎務長在大學週說：「現在怕的不是同學不念書而是怕他念得太苦。」

爲了社會未來的建設，爲了充實我們學問，爲了保持嶺南光榮，我們的工作是應該的，是有價值的。但另一方面過去和各位先進，敎授的談話中，他們都有一個共通的結論「現行的學制似乎太繁重了，且與實際有點入脫節。」試細㮣四年辛苦所得的學分，幾許是有實際用途，似乎旁枝別節的多，本科的科目少，不錯！大學不是職業專門學校應該專而且博，但「博」應該是隨興趣，應該是自由選修自由發展，似乎並非目下的劃一化的注入方式方可稱爲之博。不錯，這是整個中國敎育制度問題，並非一個學系可以改進，但各科目風輕風重，則應該是我們學校或工院所可規劃，旣感覺制度不良，而改革又有心無力，那麼可能做到規劃輕重，何不盡力而爲,以圖補救。

當然校方要求愈高,學生自然工作愈勤,教授嚴格的要求卽同學未來的幸福。但倘要求能愈切合現實,則同學收益將愈大,反之,則難免分散精力。流于泛而不專,同學對功課之注意力,將不繫于個性之興趣,或本科實際之要求而繫于分數取得之難易,或教授要求之緊弛。倘如是則恐有乖敎育之原衷,而教授與同學四年光陰,勞評亦恐未能收其應有之効果。

想先進諸公當更比稚子明察,事之然否。

(三)工會日我們除了慶祝還該幹些什麼!——記得一年前的今日劉戢和先生在南大工程裡的一段話:「嶺南人多養尊處優,衣食固與人不同,對社會事業都抱一種超然態度和社會格格不相入……惟我獨尊……」這是很嚴格尖銳的批評,現實有一種很顯明的現象,那就是「個人尊嚴」的偏重,一個嶺南人眞不容易向別人先點頭招呼,多數感覺自己家裏有地位有財產,自已有辦法,用不着和別人交接。這種幾乎成爲嶺南人的潛伏意識,有意無意之間,日常言行自然表露。

這種思想在瞬息萬變的時代中已是達到被完全淘汰的程度,而對于一個工程界的人更是危險萬分,我們現在的「工佬作風」已沒有這種現象,但在今年「工院日」再來一次自我檢討,是否這種思想正根除淨盡,似乎也是相當有價值。

希望在這一年一度的工會日,除了應有的檢討學問成績外,我們能眞眞實地發揚「工佬作風」照坦白摯誠的態度來接待我們的前輩,認識各友校的先進和緊握我們的手。

我們工科同學日增,日下人數已超過以前歷年畢業同學的總額,將來發展,方興未艾,來日在社會上互相扶持合作的機會正多,望能一貫初衷,保持去年今日的好的開始,發揮「工佬」眞摯精誠硬幹的作風。

工學會日節目表

上午	9:30	報 到 地點：工學院
	11:30—1:00	慶祝大會 地點：工學院
下午	1:00	拍 照 地點：工學院
	1:30	叙 餐 地點：大學膳堂
	3:30—5:30	球 賽 地點：足球場
	3:30—5:00	茶 會 地點：鐘 亭
	7:30—11:30	遊藝會及晚會 地點：工學院

◀學 術 展 覽▶

上午 9:30—11:30

下午 2:00— 5:00

地 點：工 學 院

工學會日籌委會職員

一九四八——一九四九年度

主　　席：關學海
副 主 席：余文傑，姚保照
特刊總編輯：吳秉俠
總　　務：梁耀顯
學術展覽：林崇義，鄺國樑
生　　活：鄧翰儀
體　　育：鄧國樑
文　　書：尹行賢
財　　政：王　嗣
招　　待：黃兆歡
宣　　傳：曾大民
佈　　置：陳春雨
交　　通：李小覺
攝　　影：巢永棠，韋基球，梁謹紹

論著

紹蘭測距法

陳永齡

紹蘭英名 Shoran，乃 Short Range 二字所合成，為雷達（Radar）之一種。其最初設計係用超短波無線電方法自飛機向地面上兩已知點施測，以決定飛機之空中位置。至二次世界大戰結束後（1944年）美軍又繼續研究應用紹蘭測定地面控制點之方法。經多次試驗與改善，紹蘭測距之精度現已可與大三角測量之精度相埒。但三角測量中每邊最長不過五六十英哩，通常則在二三十英哩左右，今用紹蘭法測距，其長可達五六百英哩，且測量程序至為簡單迅速。故此法發展之後，不但便於迅速測圖之控制，更可使各大洲三角網間之聯繫變為可能，意義至為重大。本文擬將此新穎儀器與其方法作簡單介紹，最後并將此法對於大地測量之影响，加以闡述。

紹蘭儀器之構造

紹蘭儀器可分為空中站（Airborne Station）與地面站（Ground Station）兩部。空中站設於飛機之內，其主要設備有播送機（Transmitter），收報機（Receiver）及指示器（Indicator）等。地面站之設備有收報機，播送機及校核器（Monitor）等。工作時空中站由脈動器（Pulse Generator）發出固定週率之極短脈動（Pulses）約為每秒930個，其速度係由石英振動器（Quartz Oscillator）所控制，極為穩定。此脈動一方面通於陰極光管（Cathode Ray Tube）之指示器（Indicator）上，現出標記脈動（Marker Pulse），作為傳送時距之標記。另一方面再用兩種不同週率（Frequency）經由真空繼電器（Vacuum Relay）交替播出，交替之速度約為每秒十次，播出之脈動分別由左右兩地面站之收報機交替收取，然後再以相同之第三週率轉播。空中站之收報機於收取此轉播之脈動後，經過適當擴大（Amplifier）再傳於陰極光管之指示器上。於是指示器幕上共現出三種脈動：一為標記脈動，一為經左站轉播之脈動，一為經右站轉播之脈動。標記脈動與左右兩站播回脈動在指示器上所顯示之相角差，即分別表示電波往來於空中站與兩地面站間之時距。為測量此時距，空中站於收得兩地面站之脈動後各經過相當之變相器（Phase Shifter）或名量角計（Goniometer）。旋轉此變相器即可使兩站轉播之脈動在指示器上與固定脈動相脗合，由此量得兩脈動之相角差。相角差乘以波長即為時距，時距再乘以電波之速度即爲長距。因波長爲固定值，電波之速度亦爲已知之常數，故變相器之旋轉盤上，可不刻角度或時距，而逕刻距離，是爲里程計（Mileage Indicator）。此里程計係以英哩爲單位，應用遊標可讀至千分之一英哩（約爲五英呎或二公尺弱）。左站（或右站）播回脈動所經之里程計讀數，即爲空中站與左站（或右站）之雙程距。此即利用電波相位差測定距離之原理。惟須注意者，即電波經由輸送線及收報機播送機等線路，已有一相位差。此相位差須另先測定，而加相當改正於變相器旋轉盤（即里程計）之零點。空中觀測者須隨時旋轉左右兩變相器，使指示器上之左右轉播脈動與標記脈動常相脗合，然後兩里程計上之讀數即爲空中站至左右兩地面站之雙程距矣。

爲控制電波週率穩定，地面站之校核器內亦設有石英振動器，並經常保持石英溫度於華氏70°，故地面播回之脈動，其週率可視爲標準。反之空中站之石英振動器，因限於地位及重量，並無保溫之設備，其週率常受溫度之影响而稍不規則，必須與地面轉播脈動之週率比較，時時以容電器（Condenser）調節之。蓋當兩脈動週率完全相等時，指示器幕上即現均勻之圓圈，反之則呈波動狀。故空中觀測者必須時時調整容電器，使幕上之圓圈均勻，則空中站播出之週率即與地面站週率同樣穩定。

此外地面站之校核器內尚設有固定遲滯儀（Fixed Delay Net Work），用以測定電波經過線路及收發報機所發出之相位差。此相位差在正常時乃爲一常數，爲避免其對測距之影响，前已述及係將空中站之里程計零位加以適當移動。但此種相位差仍可能有小量變化，故觀測進行時地面站人員仍須隨時注意固定遲滯儀所示之相位差有無變化。如有變化，須在里程讀數上再加以改正。凡此均爲控制測距精度之設備也。

紹蘭測距之方法

紹蘭測距係用飛機橫航越線法，或簡稱爲越線法（Line Crossing method）。施測時地面上欲測距之兩點上須各置一地面站之設備，另以飛機載空中站設備，以穩定之速度及不變之航高，飛行於兩地面站之間。飛機之位置應時常保持與兩地面站約爲等距，其飛行方向則與兩地面站之聯接線垂直，約如圖（一）所示。圖中 B，C 爲兩地面站，AE 爲飛機飛行之方向，約與 BC 線垂直，H 爲飛機之航高。欲期飛行結果圓滿，須先由可供使用之地圖中求出 BC 兩點之概距及其方位角，並將飛機須飛行之方向約略繪於圖上，以作參攷。當飛機適在橫越聯接線時（即在圖中之 E 點），EB，EC 兩距離之和應爲最小。此兩距離即爲吾人所欲由觀測結果求得之值。

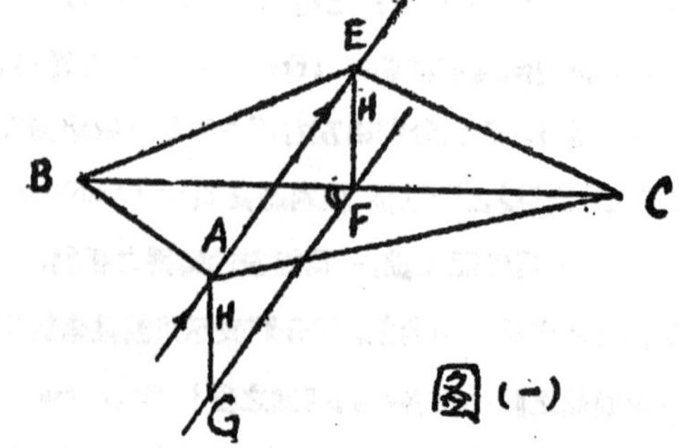

圖（一）

當飛機接近地面兩點聯接線之前十數分鐘，機上人員即開始播送短波脈動，并收取由地面站轉播之脈動。俟飛機接近路線四五英哩時，即正式讀數，直至飛過路線四五英哩後爲止。實際上里程之讀記全以攝影法行之。在飛機記錄板 (Panel Board) 之對面置一 35 公厘寬底片之電影機,每隔約 3 秒鐘自動攝取兩里程計之讀數一次,同時並將記錄錄版上之羅盤儀 (Fluxgate Compass) 高度表 (Altimeter) 氣溫表 (Air-temperature Gauge), 時鐘 (Clock) 及攝影號碼 (Counter) 一併攝入,以備參攷。此底片即爲將來計算之根據。每測一線距離,須如此横越數次至十數次,然後取其平均值,以求精度之增高。

觀測完畢後,計算人員將所攝得之底片洗出,按其攝影號碼每隔十號計其左右兩里程讀數之和,取其最低者。然後再將此最低者前後各 20 號之左右讀數和全部計出。其中最低之值即相當於飛機位置最近於聯接線頂上之時。爲求飛機位置適在横越聯接線時左右兩里程之應有讀數,即圖(一)中 EB, EC 兩距離,吾人自最低和數之號碼起,取其前後各十五號,一併用最小二乘法求之。其計算方法本文從畧。

速度改正及長度歸算

由前兩節所論紹蘭儀器及測距之方法,吾人可知其結果所得乃爲飛機於横越地面線時由飛機至地面站之距離。此距離之精度,一方繫於電波往返時距測定之精度,另一方繫於電波速度值之精度。前者可完全由儀器之設計控制之,後者則隨大地氣壓,溫度,溼度等情形而畧有微小之變化。根據光波在眞空中之速度歸化至海平面上乾燥空氣,標準氣壓時之速度爲每秒 186218 英哩。此值之精度約爲 1/50,000。但飛機之高度達於 30000 英呎以上,其空氣之密度,溫度,溼度等情形自與地面情形迥異,故其介質常數(Dielectric Constant)亦必不同。電波在此變化之介質中播行,其方向及速度亦必受相當影响。對於方向之影响,與距離關係甚小,將於歸算大地長度時顧及之。對於速度之影响,則必須用一改正式求之。目前應用之速度改正式係美國海岸大地測量局(U. S. Coast & Geodetic Survey) Donald Rice 氏所導出者,名爲 Rice Formula, 係以飛機航高,地面站高度及紹蘭量約之距離爲引數。式中之常數則根據標準氣層所求得,對於一般情形,已可敷用。若欲求更精密之改正數,則必須取用該地施測時氣層之已有資料另定式中之常數。

紹蘭觀測之距離乃爲由飛機至地面站之距離,實際應用時尙須將其化爲大地距離 (Geodetic Distance) 或名爲製圖距離 (Map Distance)。此化算公式係根據飛機與地面站之高度,化爲海平面距離,並顧及前段所述紹蘭測距線之折光改正。其式如下:

$$M = S - (2.3920\times10^{-8})\left[S(H+K)\right] - (1.7935\times10^{-8})\left[(H-K)^2/S\right] + (0.24848\times10^{-8})S^3 - (1.6083\times10^{-16})\left[(H-K)^4/S^3\right]$$

式中　M=製圖距離 (Map Distance) 以英哩爲單位

S=紹蘭距離 (Shoran Distance) 以英哩爲單位

H=飛機航高 (Plane Altitude) 以英呎爲單位

K=地面站高度 (Height of ground station) 以英呎爲單位.

上述化算公式與基線測量之改正式相類似。第一項改正爲化至海平面之長度,第二,四兩項爲坡度改正,第三項爲折光改正。此改正式導出時曾將高次項捨去，對於目前紹蘭所能達到之精度固已足用，若紹蘭測距之精度再加改進，則必須應用更爲精詳之計算式。

紹蘭測距法對於大地測量之影响

紹蘭測距法目前仍在發展之中。就其精度而言，已得之精度業能與普通大地測量所能獲致之結果相埒，而將來發展定能超過其精度殆無疑問。吾人目前所用之大三角測量方法，在測定地面上相距甚遠，兩點之長度時（如弧度測量），必須經過基線測量（長度最多不過十餘哩）水平角觀測等，然後以冗長之計算，始能求得。且今日大三角測量所用之儀器與方法，無論在基線測量或水平角觀測方面，因限於無可改善之觀測環境，其精度幾已達於極限，故紹蘭測距法之意義，不但在能增長直接觀測之距離，避免冗繁之計算，抑可更進而使更精密之測距變爲可能。

最初試驗紹蘭測距法之目的，原欲以最簡單方法求得較大面積之控制，俾可用航空測量而速完成大面積之地形圖。對於戰時施測戰區及平時施測未開發區域均有極大之功用。倘今後紹蘭之精度再能增高，則其功用尚不僅如此，蓋以其較高之精度將可作大三角測量之控制。吾人設想於每條大三角鎖之兩端均施以紹蘭測距，則此紹蘭距離即可控制三角鎖誤差之擴展。換而言之，吾人可以紹蘭測紹網爲大三角測量之骨幹。然此處必須注意者，即紹蘭網尚不能全部代替大三角測量。蓋在實用上，紹蘭點之距離過遠，對一般地面測量之控制尚不足用，必須在紹蘭點之間以大三角測量補充之。最多僅能免去大三角測量中之基線測量而已，至於在研究地球形狀方面，吾人如欲求各紹蘭點之垂線偏差 (Deflection of the Vertical)，除紹蘭線長度及紹蘭點天文經緯度之外，必須更知紹蘭線之大地及天文方位角，而此兩者因紹蘭線之距離過長，絕無法作直接觀測，勢非借重於普通大三角測量，然後推算得之不可。故在進行天文大地網之平差以求垂線偏差及地球形狀時，仍必須有大三角測量爲之輔助也。

紹蘭測距雖不能全部代替大三角測量，但可補助大三角測量之不足，蓋大三角測量因受邊長限制，無法越過較寬之水面，因此各大洲間之三角網始終無法聯繫，對於地球形狀之研究，障礙殊多，今後紹蘭測距法如能儘量發展，測越過海峽固屬不成問題，即越過較寬之海面亦非不可能者，蓋如在海面上每隔數百哩設一浮站，裝以紹蘭地面之設備站，則應用紹蘭測距法即可跨過矣。

滃江水力發電工程

莫佐基

莫佐基先生爲本校土木工程系廿五年級畢業生，肄業時從事工程工作，經驗至富，抗勝後一度任職於資源委員會水力發電工程總處滃江勘測隊，現則服務於廣州電廠。文中對滃江水力發電工程，作最詳盡之報導，想各讀者必嚮往之。 ——編者誌——

一、緒　　言

水力之利用歷史悠遠，而大規模之開發，則至十九世紀在歐美各國始行發達，蓋工業發展，動力之需求劇增，而成本之減輕則唯廉價動力是賴。水力發電之事業，遂應運而生。惜我國工業落後，對動力之需求無多，故水力發電之建設，亦無從談起而遠落人後。

廣東省內可以利用發電之水力地點甚多，在民國弍拾年時，省政府大事建設。然以煤產之質與量皆供應不足，故水力發電之建設亟覺需要，乃於廿二年由建設廳托請德國西門子公司到滃江踏勘，并搜集資料，盡量供給，擬成一初步開發計劃。建議在滃江口之上約 15 公里之黃崗附近築欄河壩，壩頂高度以不淹沒獅子口上游廣大農田爲原則，電廠緊靠欄河壩，水頭約 34 公尺，發電容量約 40,000 瓩。嗣以經費及政局之關係，未能實施，至 29 年又舊事重提，由資源委員會之水力發電勘測總隊派員再行作詳細之勘測，七七戰起又遭擱置，抗戰結束後，資委會水力發電工程總處于三十五年四月派遣滃江勘測隊由重慶來粵，進行水庫及壩址之測量，與繼續水文之測驗。更于三十六年十一月間派鑽探隊率同工作人員携同鑽探機械到所擇壩址實施鑽探，現在勘測鑽探均告一段落，而全部工程之設計及施工計劃亦已訂定。三十七年十月初資委會全國水力發電工程總處處長黃育賢，奉資委會孫委長命來穗晉謁宋主席，作確定性的商洽，結果完滿，將來由資源委員會與廣東省政府合辦。股本各佔若干，由資委會全國水力發電工程總處負施工責任，完成後另組機構管理，惜以時局變遷，籌備工作乃暫緩進行。

二、滃江形勢

滃江是粵北東部，北江之一大支流，上游分東北二支，於滃源縣屬之龍江合流，合流處成一大平原，西南下而會小北江于獅子口，至英德縣城滙入北江南去廣州 141 公里北去曲江 88 公里自獅子江以下两岸高山俯臨河邊，此段河流形成一長 24 公里之狹谷，平均比降爲 0.0015 直至距英德縣約 5 公里處始行開豁，其受水面積，有 5,200 平方公里。以前西門子計劃擬在狹谷下段距英德縣城 15 公里之黃崗築壩，其後由水力管電工程總處總工程師 Cotten 擇定黃岡下游 4 公里有樹坑地方爲壩址，以有樹坑爲中心，在 50 公里半徑圈內包括英德佛岡等縣，100 公里圈內包括翁源，曲江、從化、花縣、清遠等縣，150 公里圈內包括廣州，東莞、博羅、樂昌、順德、三水、四會等縣市，200 公里，圈內包括鶴山、寶安、高明、惠陽等縣，250 公里圈內則將香港、中山、台山、開平等縣市包羅在內，此等縣市有豐富而寶貴之資源如森林礦產等可資開發者，有工業源料充足可發展工業者，有交通利便人口稠密之商業區者，將來發電成功，用電市場固不成問題，而本省工商業之繁盛亦可拭目而待也。

三、工程概要

A. 雨量及流量

滃江流域之氣候水文測錄有二十年之時間，據測驗所得：每年平均雨量約爲1,600 公厘，其分配情形如下：

冬季(12月—— 2月)　12%

春季(3月—— 5月)　41%

夏季(6月—— 8月)　31%

秋季(9月——11月)　10%

最小流量——20秒立方公尺

最大流量——實測記錄爲 3,000 立方公尺/秒

設計用推算萬年週期洪水量爲 7,000 立方公尺/秒

平均流量——全年平均流量爲 160 立方公尺/秒

B. 工程佈置

全部工程計劃分兩期實施，第一期作 40,000 瓩配置，第二期再添 40,000 瓩容量。攔河壩分兩期建築，發電廠一次建成而分兩期安裝機器，輸電鐵塔則一次建築；現定計劃發電廠離開攔河壩，并由廠另築 5 公里長之鐵道與粵漢路接軌以利運輸。

(1)攔河壩——建壩地點原有 5 處，後經實地勘測及多方面比較，結果選定離粵漢鐵路 5 公里之有樹坑，經利用機械鑽探，每洞曾試以每平方英寸 40 磅之水壓，漏水尚不利害，施工時以高壓灌漿法，足以補救，惟兩岸地質尚待加續研究。

	第一期	第二期
壩高：	高出堅便基礎上 60 公尺	加高 20 公尺

落　差：約40公尺　　　　　　約60公尺

壩　頂：弧形洩洪門14座，可以宣洩滃江最大洪水，壩下設水墊以防冲刷，經南京水工試驗室模型試驗結果良好。

壩　型：混凝土拱形壩。

進水口：在壩之右岸以7公尺直徑隧道引入電廠。

引水管：鋼管4條引水入水輪分兩期建造。

蓄水庫：壩後蓄水容量有 610,000,000 立方公尺。

尾水道：水輪流出水量，在廠房處直接回到滃江。

（2）廠房——採半露天式（Simi-outdoor type），以省房頂造價，廠址在壩之下游，河之右岸，其下層爲水輪之出水道直接流入滃江，中層裝水輪及發電機，其引水管接于直徑7公尺之隧洞，此隧洞乃利用建築時之洩水隧洞改建者，上層爲起重機採露天式，每部機器之上均開井口以便吊放機器，平時用蓋罩密，運輸鐵道直達廠房，廠房大小以最後容量爲準，可容水輪發電機4套，每套容量爲 20,000 瓩，第一期先裝兩套，并預留地位爲將來裝置兩套之需。

（3）水輪發電機

水輪型式——法蘭西斯（Francis）直軸式與發電機直接連結。

淨水頭——第一期計劃39公尺，第二期計劃53公尺。

馬　力——第一期計 2 @ 27,000 馬力，第二期 4 @ 27,000 馬力。

轉　速——每分鐘 166⅔ 轉

發電機——交流之相同如式

容　量——第一期計劃 2@220,000 瓩，第二期計劃 4@440,000 瓩

電　壓——13.8 千伏（KV.）

週　波——每秒50週波

勵磁機——主副勵磁機各一具与發電機直接連結。

（4）輸配電系統——發電所設在有樹坑，再在英德及廣州各設一配電所。

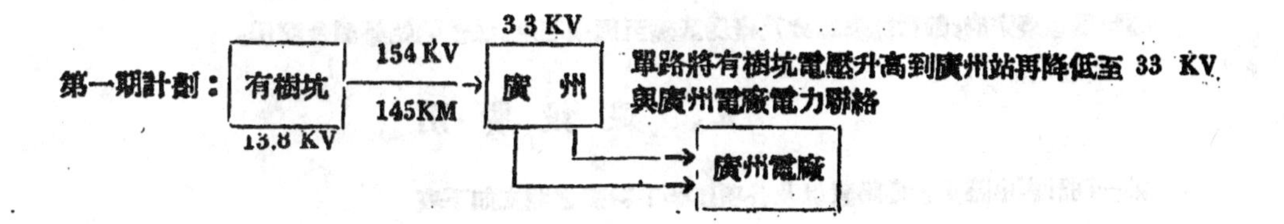

25859

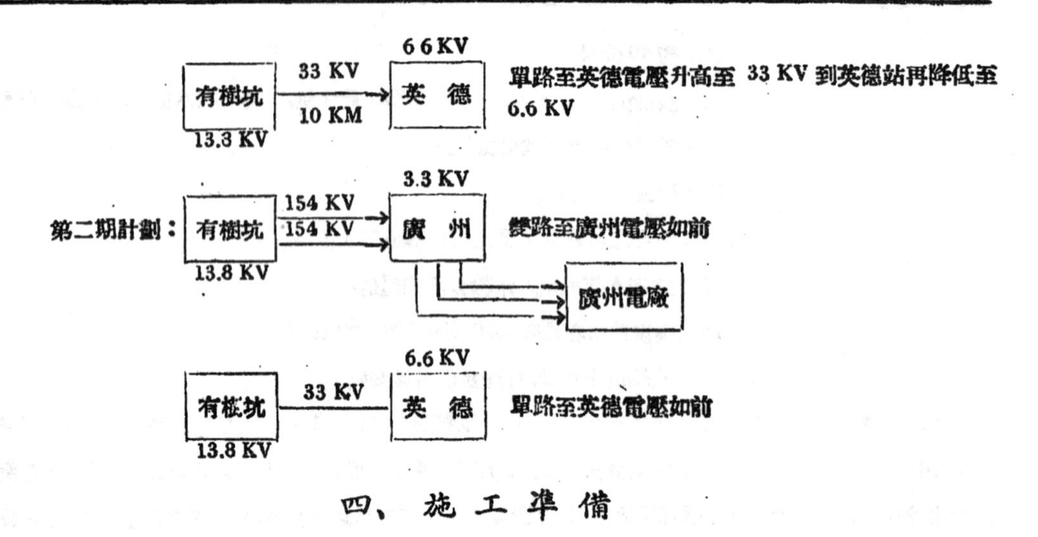

四、施工準備

本工程之規模在我國固史無前例，（日人建造者除外）在外國亦算中型，籌備工作相當繁重，其所需材料機械數量畧述如下：

(1)洋灰—— 250,000 桶，由鐵路運來後用風送法吹到貯藏塔。

(2)沙石—— 300,000 噸，採集於河身及山上各處再用傾斜鐵路拉上沙石場。

(3)鋼鐵—— 5,000 噸。

(4)各種營造機器—— 4,000—5,000噸。

(5)建築廠房材料共 10,000——20,000噸。

(6)磚瓦，木，石，灰——建築臨時斷流攔河壩及房屋所需材料其量甚鉅，類多就地採材，滃江雖位於粵北，但粵北一隅之產物，還不足以供給必須遠至各地，預早訂購，始不致中途不繼。

籌備工作開始，先行建造辦公房屋及數千員工之住所，與銜接粵漢鐵路之運輸綫路，同時派員四出洽訂及搜集各種材料，洋灰一項，將需協助西村士敏土廠加緊增產，外國器材亦同時訂購，分別緩急而分始交貨，務期配合工程之進度。

第二步：進行臨時攔河壩之修建，及洩水隧洞之穿鑿，同時修建混凝土拌合場，及鋼纜運輸設備，以為將來把混凝土運到壩身各部分然後傾例。

臨時壩建築完成，即行開鑿壩身及廠房基礎至預期石層，然後開始築壩及廠房。

五、建設費用

第一期計劃中關于設備購置以及各項建築工料費之開支如下表

項　目	美　元
水輪發電機兩套及各項水力機械電氣等附件	2,411,000
電氣開關及保護設護	130,000
輸電綫路及配電所	1,580,000
什項(包括水閘及引水鋼管等)	550,900
直接費用合計	4,671,900
*雜費	666,100
總計	5,338.000
營造器材購置(運費在內)	980.000
各項建築及設備(照37年3月份估計國幣值單位百萬元)	1,967,300

*此項什費包括間接費用5%,意外費7%,什支4%,建築期間所付利息3%。

本計劃全部工程用人最多時可有四千人,如各方面配合妥當,三年內可以完成,現如適應實情,暫擬定5年計劃,第一年造鐵路,倉庫,宿舍,洩水山洞,第二年起趁在枯水期建壩,所需機器交貨最快二年半,先交水輪,其他機器陸續交貨隨到隨裝。

六、輸出電能

假定廣州火力廠容量有30,000瓩全系統每年負荷因數達60%分豐水及枯水年估計輸出電能可如下表:

豐水年　　電能　　(百萬電度)

	水力	火力	合計	穩定	次等
第一期計劃 (水力40,000+火力30,000)	279.6	90.0	369.6	337.2	31,8
第二期計劃 (水力80,000+火力30,000)	508.3	72.5	580.8	522.0	58.8

枯水年

第一期計劃	186.3	174.6	360.9	337.8	23.1
第二期計劃	353.2	188.4	451.6	522.0	19.6

七、工程完成後之利益

廣州爲南中國政治經濟及工業之中心,電力之需求甚殷,而本省煤藏質皆不足以供應,且全賴火力電,實不合經濟原則，勢非建設廉價之水力電不可，故滃江計劃實現後，其電力足供本市之初步需求，而滃江本身亦可獲得其確定市場也，綜合此計劃之利益可如下列：

(1)水火相濟：第一期計劃完成 40.000 瓩，加上火力 30.000瓩，全系統共有70.000瓩，枯水期間，火力廠担任基本負荷，水力廠担任高峰負荷，豐水期間，火力廠担任高峰負荷，水力廠担任基本負荷，全年可供穩定電力 38.000.000 萬度，以每 5 年有一較枯年計，水電成本每年需金圓 1.5分，根據 20 年還本付息計算，每年利息以 6 厘計，輸電至廣州每度度金圓 2～3分，全系統（出水併計）每度需 7 分，各項損失估計 3%，每年純益可有金圓 13,000,000 元以 6,000,000元還本付息，可淨餘金圓券 7,000,000元。（金圓券未低折時價值）

(2)開發錫、錫、硫磺等礦產，粵北礦藏豐富，可藉廉價電力開發，還有造紙業亦可則用開發。

(3)製造肥料，有廉價之電力，到第二期計劃成功後，足以製造肥田料解決糧食。

(4)滃江高地可以利用電力戽水灌溉農田，增加糧產。

(5)因工程規模頗大，需要材料甚多，單以洋灰一項而論，其需要量即可促使現有士敏士廠之擴充增產，其餘機械修理廠亦可望增加。

(6)本工程需要人力最高達四千人無形中亦可解決四千人之工作。

結　論

全國可能利用水力發電之地點甚多，大者如黃河，長江三峽，四川大渡河等均以經濟人才各方面條件所限不易實現，中型者多屬資料不足無從根據，而滃江計劃資料既足需費不多，而交通便利接近城市，尤爲其他各地所無，故其開發價值極大，況其參與工作之人員所得經驗更可爲他日擴展水電事業之基幹，其利益實不可限量也。

此工程適位于本省境內，將來對吾人之關係甚大，凡我本省工程界人士皆當注意，惜以時間近促，搜習資料極少，又不暇整理，祇能概括寫來，拉雜無章，聊以引起各界注意而已，不敢謂對研究上有所裨益也。

重建柳江大橋墩座工程紀畧

曾德甫　梁百衡

本文作者曾德甫與梁百衡兩先生，爲本校土木系廿六年畢業生，現兩先生同時於柳州負責柳江橋之工程工作，來稿本附有圖片七幅，惟因經費所限，未能登錄，編者謹致歉意——編者誌

1. 柳江橋在湘桂黔鐵路之地位

湘桂黔鐵路爲西南大動脈，於三十四年冬合併湘桂黔桂兩路而成，後更成立黔境，粤境，桂境三工程處主持都(匀)筑(貴陽)及來(賓)湛(江)段新工事宜。試閱湘桂黔鐵路路線圖(見南大工程康樂再版第一期桂系主任之來湛鐵路)，以柳州爲中心，東北接粤漢鐵路之衡陽，西北達貴州省之貴陽，東南通粤南之湛江，西南聯桂越邊境之鎮南關，形如×，柳州適在交叉點中。柳江橋位於柳州西二公里之雅儒村，聯接南北兩岸，貫通全線交通，其地位之重要，至爲顯著。

2. 柳江橋初建及重建

柳江橋建於抗戰初期，完成於三十年春。當時以物資缺乏，且急於通車，故上下部結構均屬半永久性質，上部結構爲18孔30公尺皇后式鋼梁 (Queen Post D. P. G) 下部結構爲鋼軌塔5座，鋼軌架12座，活載重爲C—16級。迨三十三年秋，日敵入侵桂柳，始自行炸燬，勝利以還，本路當局以柳江橋關係全路交通，雖工程艱鉅，需欵孔多，仍應早日修復，爲維持永久計，决改建爲12孔49.25公尺華倫穿式鋼桁梁 (Through Warren Truss)，鋼筋混凝土橋墩座，橋墩座活載重爲C-24級，鋼桁梁活載重爲C-20級，全橋總長爲591.10公尺，(參閱示意圖)工程計劃，橋墩中之單號基礎(1.3.5.7.9.11)爲新建，雙號基礎(2.4.6.8,10)則利用原橋塔基兩端加長，墩高約23公尺，爲減輕靜重及節省洋灰，橋墩內部係空心設計。筆者於復工之初即參與其事，謹將施工情形畧述梗概於后：

3. 柳江河床水位及橋址測量

柳江爲融江及龍江之滙流，兩江皆發源於黔境，在柳州西北20公里之柳城滙合。沿江上游，山嶺叢叠，雨季時節，水位動輒日漲七八公尺，根據三年來紀錄，高低水位約差15.50公尺，水漲期間，流速爲18每秒公尺，流量爲 8.500每秒立公尺，河床爲沙礫夾卵石，厚度由二三公寸至四五尺。岩層爲石灰岩，但於短距離內變化頗大。

本橋復工之前卽施行測量，舊橋鋼梁鋼塔架雖被炸燬，惟基礎尙全部完整，且兩墩相隔 30 公尺以內，故用直接量法（Direct Measurement），再輔以三角測量，以備橋墩築至相當高度時測距中心距離之用。測量水平，則利用架基塔基作爲轉点，來回複測，尙屬便利，可省去兩岸對測。

本路在破壞後，材料工具損失綦重，尤以各種機具一時不易修理或添置，河床鑽探工作，以上述原因，未克進行，至屬憾事，所幸橋址爲利用舊址，從前施工，定經詳細勘驗及鑽探，江底岩層料不至有鉅大變化，故僅將較困難之第 7. 8. 9 橋墩用鋼軌打下河床試探，所得結果，證明岩層變化尙不算大。

4. 柳江橋工程之發包

本橋長度爲全路橋梁之冠，工程大而數量多，（見附表）工費鉅而歷時久，近年幣值日貶，料價工資，朝夕數變，爲管理工程及承包工程最感困難之問題。本路當局有見及此，乃將全部工程，委託中國橋梁公司上海分公司代辦。各項重要料具如洋灰鋼筋抽水機具油類等由路方供給，代辦制度（Cost Plus System），爲工程費外，另加代辦者管理費 15 %。抗戰後龐大工程，多採此項制度，誠爲物價波動中，辦理工程者辦法之一。

重建柳江大橋主要材料數量表

墩座號數	混凝土立公方	洋烊桶	竹節鋼筋 公噸						蔴袋個	防水堤片石立公方	備攷
			1"φ	7/8"φ	3/4"φ	1/2"φ	3/8"φ	總計			
北橋座	1.100	1.538		10.9	0.8	1.1	2.6	15.4			
第1號墩	880	1.228		6.8		1.5	0.4	8.7	360		
第2號墩	789	1.104		6.3		1.5	0.3	8.1	1.984		
第3號墩	978	1.361	4.1	0.3	1.9	1.6	0.3	8.2	5.729	59	
第4號墩	857	1.196	2.0	4.6	0.3	1.5	0.3	8.7	8.776	51	
第5號墩	1.073	1.493	0.8	6.5	0.5	2.0	0.1	9.9	1.992	40	
第6號墩	921	1.283	4.7	0.3	2.0	2.0		9.0	8.231	63	
第7號墩	1.056	1.468	4.7		2.0	2.0		8.7	7.794	179	
第8號墩	976	1.859	4.7		2.0	2.0		8.7	13.431	365	
第9號墩	1.027	1.428	4.7		2.0	2.0		8.7	12.676	618	
第10號墩	907	1.223	4.7		2.0	2.0		8.7	8.614	239	
第11號墩	1.020	1.419	2.3	3.0	1.4	1.9	0.2	8.8	165		
南橋座	1.162	1.650		6.5	3.6	1.3	2.7	14.1			
總計	12.746	17.750	32.7	45.2	18.5	22.4	6.9	125.7	69.802	1.614	

5. 橋墩座施工概況

(甲)第一期工程——第 1, 2, 4, 5, 10 號墩基礎 (36 年 1 月至 36 年 4 月淺水期間)

三十五年杪以訂約伊始，且工具材料，均付缺如，僅能作施工前籌備，乃由橋樑公司先行訂購河沙石子木料蔴袋船隻沉箱鐵脚等主要材料。

三十六年一月間，水位最低(標高 81.60)，便於工作，惟以炸燬之鋼軌枕木軌塔軌架及鋼樑，滿佈江中，殊多窒碍，爲配合橋墩基礎工程進行起見，須先將鋼件打撈或移開，方能工作，故各墩基礎施工之程序，往往未能計劃如意。本期工程祗能將容易打撈之第 1, 2, 4, 5, 10, 11 各號墩基先做，進行情形，大致皆用蔴袋裝土，在墩基地點圍以圍水堰用抽水機將堰內存水抽乾，挖清江底沙石後，卽灌注混凝土，俟高出低水位後，方暫停止。第 1, 2, 5, 11 各墩進行，尙相當順利。

第 4 號墩因利用舊塔基加長兩端，基邊片石纍纍，鋼梁碎件堆積甚多，且有深埋在沙礫下者清理不淨，影响防水工作甚大，緣以水沿鋼軌及壞梁部份侵入堰內，軌邊沙礫冲鬆，漏洞漸大，堵塞不易，雖集中四部抽水機之力 (當時祇有四部 5" 6" 8" 10" 各一部)，尙不能將水抽乾 (普通 5" 6" 各一部卽可濟事)。後用木板樁按墩基所需尺寸，預寬少許，打入河底，再用實土蔴袋及泥土將圍水堰至板樁之空位塡塞，并分格灌注混凝土，多費時半月，方克告竣。

第 10 號墩基礎之防水方法，與其他墩畧有差異。因利用塔基加長，且水深達三四公尺，故先在河灘參照河床高低，建築匸形木沉箱 (Wooden Cofferdam) 兩個，上下游各一，然後移正位置，徐徐沉入水內。箱壁內塡實泥土(墻厚1.5～1.8公尺)，塔基兩側，加圍實土蔴袋。當抽水時，發現箱底碎件頗多，底部滲水，無法抽乾，幾經設法補救方始見效，惟在大功將成之際，洪水突發，沉箱沒頂，致功虧一簣!

(乙)第二期工程——南北橋座基礎 (36 年 5 月至 36 月 10 月漲水期間)

新橋較舊橋增長 9.6 公尺，(見示意圖)，南北兩橋座位置，須往後推移，兩橋座於撤退時自動炸燬，施工之前，務須將殘餘混凝土淸除，方能開挖土方，是時水漲，利用墩基不能施工期間，進行橋座基礎工作。每橋座除能利用一部份原有基礎外，需另加方形鋼筋混凝土沉箱三座，形成品字，箱脚加鑲鐵跡，以防崩損。箱身分層澆築，以免過高易於傾斜。在挖土期內，雨水頻仍，土方常有坍塌，而橋頭附近，卽爲鄉村房屋，受地形限制，邊坡不能放大故工進頗緩，北座三沉箱於六月開始下沉，南座沉箱於八月開始下沉，箱內土方，全用人力分班輪流挖掘。起吊廢土，則人力絞盤，柴油裝吊機并用。每日約沉落五公寸，間或江水上漲，水從箱底滲入，則須用 3" 手搖抽水機或 5" 機動抽水機抽乾。此座約沉八公尺，南座約沉九公尺，始達岩層。將軟石鑿去後，卽行封底塡實空心，橋座基礎至是遂告段落。

(丙)第三期工程——第 3, 6, 7, 8, 9, 10 墩基及此橋座第 1, 2, 3, 4 墩身工程

(36 年 11 月至 37 年 3 月淺水期間)

鑒於上期鋼件碎片影响防水工作，此次水退，卽積極打撈，并派工潛入水內檢清碎片，故各墩基進行頗爲順利。第7,8,9墩爲主流所在，江水湍急，水位較深，圍築蔴袋防水堰，易爲水毁且蔴袋沉放，易爲水流冲歪，故在橋墩上游50公尺處，先築弧形片石防水堤一道，長約180公尺，以定水流。至防水堤之修築，端賴用竹籠滿裝片石，四五個紮成一排，疊堆爲堤基，方免爲急流冲去。第8號墩舊塔基原爲四圓形沉箱，上加口形鋼筋混凝土板，空隙滿塡片石，此次防水工作完成後，卽將片石沙礫挖淸，仍照各墩基尺寸灌注混凝土，將四沉箱及口形土板注爲一體。

第3.57.9各墩基礎無大困難。第10號墩基在水退後，卽整理防水木沉箱，加土塡築旁實，外復塡土圍繞，以免江水滲入，卒告成功。

在進行橋墩基礎期間，北座及第1,2,3,4號橋墩墩身施灌混凝土，均在36年12月底完成。最後一墩基礎第8號亦於37年1月中旬告竣。倘工欵及洋灰供應不乏，預計在大水期前（4月）可全部蕆事，奈以時局影响，欵料困難，工程時作時輟，37年8月間始繼續澆築第5,6兩號墩身，在加緊搶築之際，不料洪水早發，江水突漲五公尺餘，上游冲散木排，順流而下，兩墩脚手架及板摸，被冲毁殆盡，工作廹於停頓。

（丁）第四期工程——南橋座座身及第11號墩，墩身工程（37年4月至10月漲水期間）

洪水暑退，以所存洋灰仍足灌注南座及第11號墩座身墩身混凝土，第11號墩因靠近岸邊，水流較緩，縱水位稍漲，亦不易冲毁脚手架及板模。四月一月內卽將墩座同時灌注完畢，南北橋座俱爲T式（T-Abutment）。南座爲預留沿江碼頭鐵路綫穿過，在座身留有隧道拱形孔位，以備車輛通過。

（戊）第五期工程——第5,6號墩墩身工程（37年11月至38年4月淺水期間）

第5,6,7,8,9,10各墩墩身未完混凝土工程僅餘2,830立公方，佔全橋22%，預計50日內卽可完成，惟欵料紬缺萬分，無法接濟，本期祇將第5.6兩墩未了混凝土繼續澆築，祗三十天卽完，其餘四墩又不得不俟諸異日。綜計開工迄今共計完成82.5%

6. 結 語

本橋防水機具，除路局在港購有5"8"徑管帶動離心式抽水機各一部，6"10"自動離心式抽水機各一部外，在舊料中檢出五部修理使用，尙勉足支配。以能把握時間各墩基礎工程，尙無貽誤，惟橋址利用原有致打撈與新工變管幷下，尤須配合得宜方克爭取時間。利用舊塔基加長雖節省一部份材料，但防水工作不因是而簡便，反之每每增加困難，誠非意料所及。

墩座工程開工迄今，已歷兩年又四月，因欵料時斷時續，未能達到預定進展計劃，此乃抗戰後各項建設事業所常遭遇之困難。因時間之延長，致所備材料，來回搬運，腐爛散失，耗工費欵，爲數不貲，至堪惋惜！

本橋鋼桁梁12孔在加拿大訂製，大部份業已運抵工地，以工欵無着，未克動工安架。所有橋墩之完成，及鋼梁之安架，總在洪水過去，秋後方可繼續，倘有機會參加是項工作，再爲諸師長暨各同學奉告。

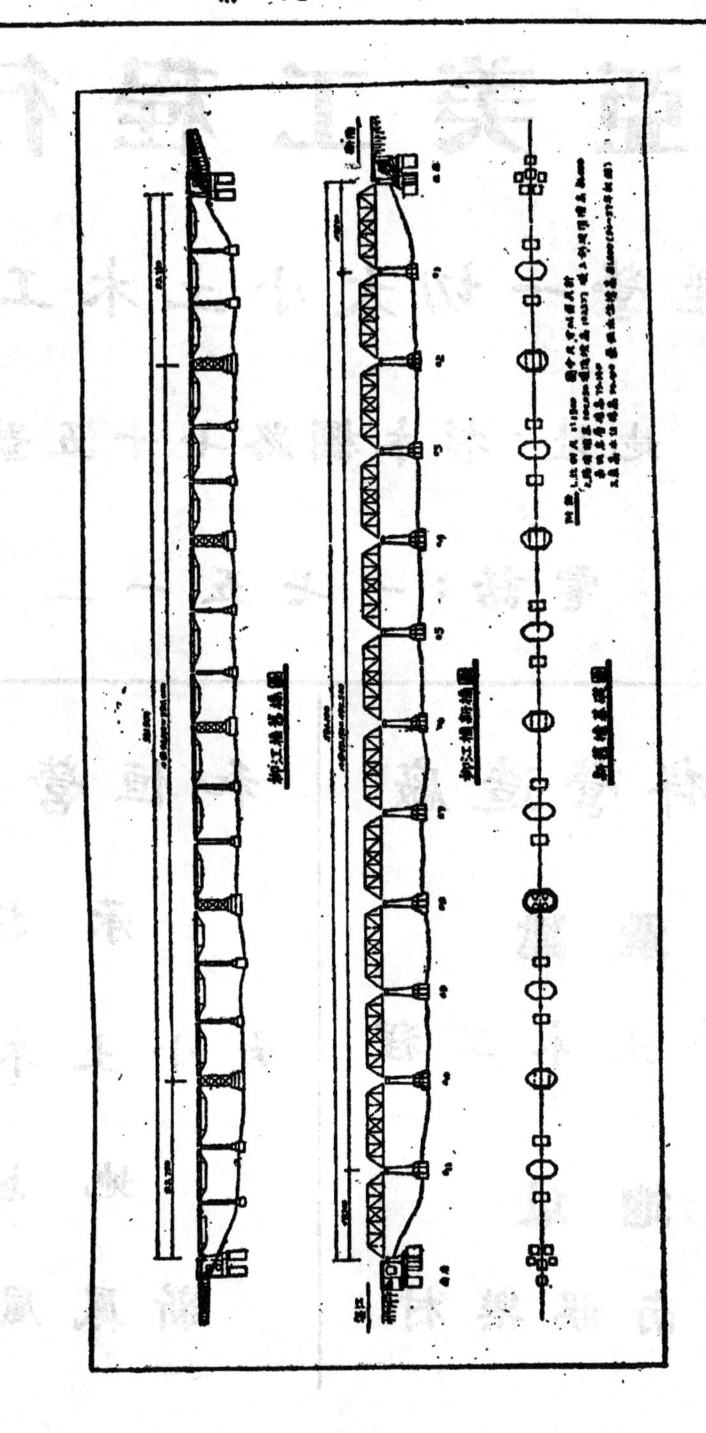

THE CIVIL ENGINEER AND SOCIETY

By

Rev. Joseph A. Hahn, M.M., I.A.S.

An engineer makes it possible for people to do the same things they would do anyway, but for more people to do them, to do them better, and to do them more quickly.

Without a bridge, people can still cross a river, but how relatively few would care to cross frequently and how much more effort and time would be spent. Without highways and railroads, people would still travel to distant inland points, but again this would be for only a chosen few. The great majority of people could not engage in any large scale trade or transportation unless the civil engineer makes it possible. Tall buildings concentrate a part of the population in large cities where their energies can be employed more efficiently than if they were scattered.

So the civil engineer is a catalyst which stimulates the various forms of enterprise that make a people prosperous.

Does he ever think of his responsibility in this regard? He gives the means to bring about the acquisition of great material wealth. He gives generously of his ingenuity and time, of his ability and wholehearted interest. But many times, though he himself succeeds in presenting to his fellow human beings the means, there is still something lacking.

Examine almost any large city in the world. There is an abundance of transportation, commerce and industry. Yet there always seems to be a fairly large portion of the population living in poverty. Why should this be? The civil engineer has done his share. His bridges are works of beauty. Highways and railroads are skillfully and plentifully intertwined among the tall buildings which show his handmark far into the distant countryside. You can hardly blame the civil engineer for this poverty in the midst of plenty.

Yet the civil engineer can do a lot. He is a real social worker. His works benefit countless numbers of people. While individuals help individuals, his work helps thousands, yes, tens and even hundreds of thousands. But if he makes a mistake, the detrimental effect is also very influential.

The civil engineer should insist that the fruit of his skill be properly used. He should attempt to help the greatest number of people, the poorest. Does it make him hadpy to build a road so that a few rich people can get richer, or a dam to supply light and power for those who can afford it? How much better to build highways so that all can use them, and supply power and light in such a way that the poorest can benefit from them. In erecting buildings and designing housing projects, he should insist on the greatest good for the greatest number.

The engineer should have the welfare of all the people before him. He ought to study the needs of the people and advise his employer in such a way as to influence him to help everyone.

The very nature of his profession makes the civil engineer a social worker on a mass production scale. Let him realize his tremendous influence and have the courage to sacrifice his personal interests for the sake of the greater number and the less fortunate of his fellow citizens.

ACOUSTICS OF BUILDINGS.

By Poon Ying Biu

If the intensity of heat and light can be controlled, why not sound? This was exactly what the engineers of the past had been trying to solve. Fortunately, by their incessant investigations and improvements, we are now able to control sound to whatever level we like.

Sound may be very annoying had it be not considered when designing a structure such as in a broadcasting studio. In order to separate the sound or noise from the street in a residential building, or to suppress the echo and reverberation in an auditorium, or to minimize the conversation or noise in a large office or workshop, a knowledge of acoustic control, acoustic material and their application in constructions are necessary.

Sound is a form of energy which behaves in many respects similar to heat and light. Sound waves are sent into the tangible media: air plaster, brick, etc by vibrating bodies, and consist of alternations of aerial condensation and rarefraction. Owing to the spherical spreading of the wave-front, the intensity of sound at any point—measured in energy units—is inversely proportional to the square of the distance of the point from the source. The intensity or loudness of sound depends upon the amplitude add frequency of sound waves. The decible unit is about the smallest change in the loudness of a sound which can be detected by the ordinary human ear. The decible scale bears a logarithmic relation to the amount of sound energy involved: tne number of decibles measuring the difference in level between two sounds is ten times the common logarithm of the intensity (energy) ratio.

At the bottom of the decible scale is the threshold of audibility, that is, the degree of loudness at which sounds become barely audible; at the top (126 db.) is the threshold of feeling, or the point at which sound vibrations begin to be felt as well as heard as shown in Fig. 1 below.

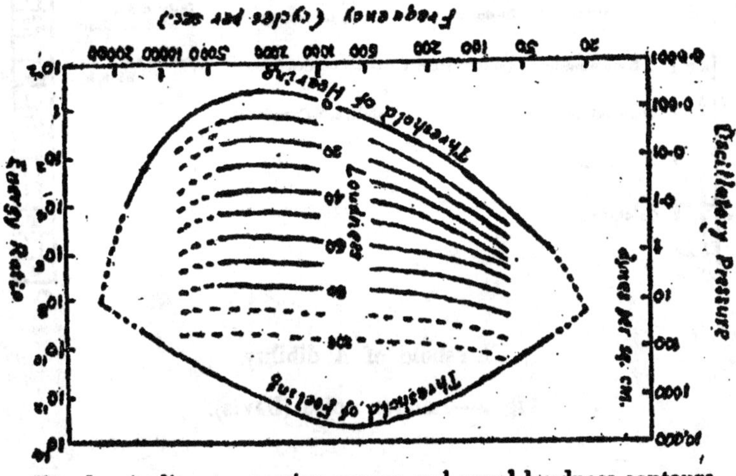

Fig. 1—Auditory sensation curves and equal loudness contours.

Before we step further into the acoustic control in a building let us study the course of sound or noise and their energy level. The measurements of acoustic in buildings have been made by many experimenters and the following diagrams and tables reveal some of the results or data of research and curves as references. Fig 2 represents a series of noise measurements made in London A.H.Davis, either by means of tuning-fork of frequency 640 cycles per second or by means of a Barkhausen audiometer of the same pitch.

Loudness Levels of Various Noises.

(Expressed in terms of the intensity of a standard note of equal loudness)

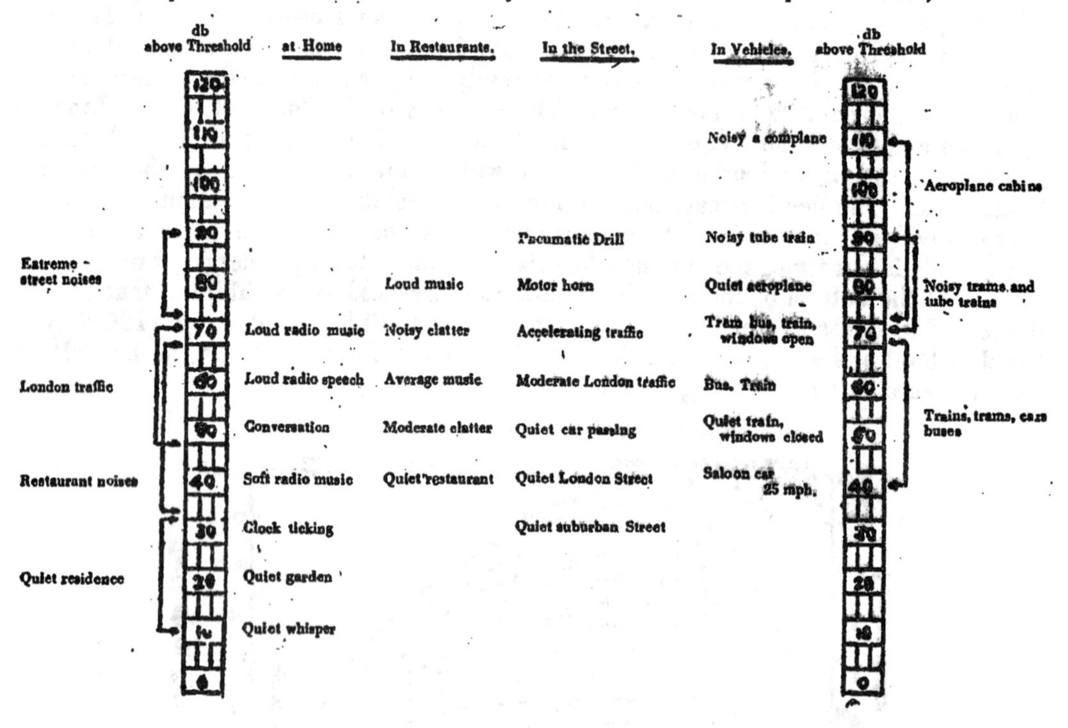

Threshold of Audibility.

Fig. 2—London noise (Davis).

Fig. 3 represents another measurement of noise in buildings from Joint D. & R. Subcommittee Survey—New York Data,

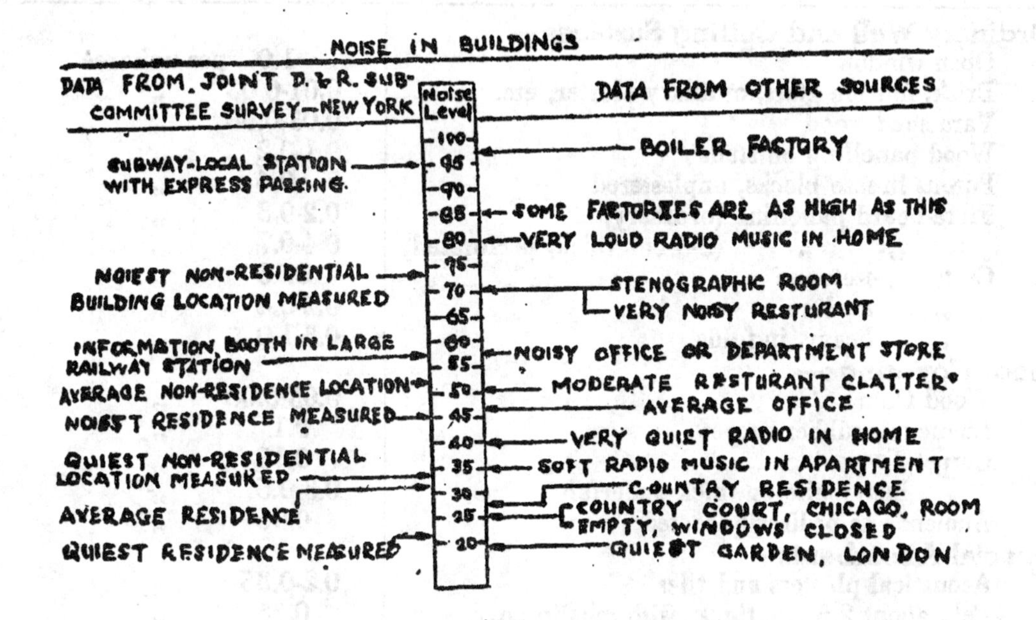

Fig 3

The waves of sound from a noise made in the room is partly reflected from the surface on which they impinge, partly transmitted througe the material of the structure, and partly absorbed. Any absorption at the point of reflection will tend to decrease the different components and minimize the characteristics of direct sound. Therefore by breaking the continuity of the material and the provision of layers of absorbent materials that the intensity of the transmitted sound waves is reduced. Acoustical treatment ane the shape of the room must be such that the reflected sound is thoroughly diffused. The coefficients of absorption of some of the common materials are shown in Table 1

TABLE 1— Table of Approximate Abrorption Coefficients for Frequency 500 Cycles per Second

Material	Absorption Coefficient	
Ordinary Wall and Ceiling Surfaces—		
Open window	1.0	per unitaree
Brick, marble glars, ordinary plarter, etc.	0.01-0.03	,,
Varuished wood	0.03-0.08	,,
Wood panelling on studs	0.1-0.2	,,
Porous breeze blocks, unplastered	0.4	,,
Fibre-board panelling (ordinary)	0.2-0.3	,,
,, ,, ,, (2-3 cm. thick, perforated)	0.4-0.7	,,
Curtains, cretonne	0.15	,,
,, medium	0.2-0.4	,,
,, heavy, in folds	0.5-1.0	,,
Floor Coverings—		
Wood floor	0.03-0.08	,,
Linoleum, rubber carpet	0.1	,,
Carpet	0.15	,,
,, heavy pile on thick underfelt	0.3-0.5	,,
Audience as ordirarily seated	0.96	,,
Special Absorbents—		
Acoustical plasters and tiles	0.2-0.35	,,
Felt, about 2.5 cm. thick, with muslin cover, distempered and perforated	0.75	,,
Fibre-board tiles, ⅝ in-1¼-in thick, perforated or slotted.	0.5-0.67	,,
Slag wool, wood wool, loose felts, ete., 1 in. thick	0.55-0.8	,,

(Continued)

Individual Objects	Absorbing Power per Object		
Wood seats for anditoriums, per seat	0.1-02	sg ft.	of complete absorption
Upholstered seats, per scat	1-2	,,	,,
Upholstered chairs, per chair	3	,,	,,
Audience, per person	4.7	,,	,,

Absorption coefficients depend upon the method of fixing—for instance, upon the distance of separation between the wall and the material. For this reason, materials are normally mounted for test in the manner which will be employed in practice.

Because of the tendency of sound to diffract, or spread, after passing through an opening, cracks and holes must be avoided in every type of construction if sound insulation is to be achieved. As a result of this tendency every crack or opening must be regarded as a source of sound, of importance almost equal to that of the original sound-producing agency.

In practice, the control of such sounds is best accomplished by

(1) reducing the initial intensity of the sound and damping the force of the impact by carpeting or providing a soft surface such as cork,

(2) isolatian of the floor or floor and ceiling from the supporting construction, and

(3) employing a construction with a high degree of resistance to airborne sounds.

The sound insulating properties, or resistance to sound transmission of a partition or floor construction are expressed in terms of decibles reduction. By this is meant the extent to which the loudness of a given sound, measured in decibles, will be reduced on passing through the partition. The transmission loss of and partition or wall is a physical property of that wall just as is its weight or rigidity, and depends only on the materials and method of construction used in erecting the wall and not on the loudness of the sound striking it nor on the size or acoustical properties of the room on either side ef it.

Below is the classification of various tested partitions employed by the U.S.Bureau of Standards providing valuable index of sound insulation value:

"Panels whose reduction factors are less than 40 sensation units (decibles)...... Conversation in ordinary tones heard through the panel is distinctly audible and intelligible.

"Panels whose reduction factors lie between 40 aud 50 sensation units...... Conversation in ordinary tones heard through the panel is quite audible but difficult to understand. If the voice is raised, it becomes intelligible.

"Panels whose reduction factors lie between 50 and 60 sensation units...... Conversation carried on in an ordinary tone of voice is reduced to inaudibility. If there is external noise in the listening room, a shout on the other side of the panel would be practically undeticable."

In practice, there are always present in a room certain sounds wnich tend to mask noises from adjoining spaces. It is therefore necessary to reduce the level of the transmitted sound to the persistent.

As a conclusion let us study the acoustic considerations in a radio studio room and an anditorium.

The Acoustics of Radio Studio.

To investigate the noise in a radio studio we need:

(a) To suppress at the source the course of noise.

(b) To confine the noise by enclosing the source.

(c) To exclude the noise from regions where quietness is desired.

Both from an a coustical and economical standpoint the cubical content of a playback or demonstration room should be consistent with the required end for which it is to be used. In this respect a suitable size would be approximately 8 feet wide, 8 feet long and 8 feet high, inside measurements. Observation windows should be constructed of double panes of glass set in felt or soft rubber. Weight, thickness, density and rigidity determine the "sound proofness" of any wall. Where weight and extreme thickness is impractical, special techniques have been worked out to make use of various combinations of free air spaces between independent walls. A few partitions and their average transmission losses are shown in Fig. 4. All cracks or openings which may exist around doors should be sealed by providing felt or rubber gaskets on the door steps at the head and jambs, and installing a special metal-bound, felt strip threshold-closer, as in Fig. 5. Efficient sound insulation in windows is required when windows are used for observation. These should be constructed with double panes of heavy glass with each pane isolated completely from the frame by gaskets of felt or rubber around all four edges. It is advisable to have the panes of slightly different thickness, to avoid resonance effects. Details of this construction are clearly illustrated in Fig. 5. In case of machine vibration the

Fig. 5

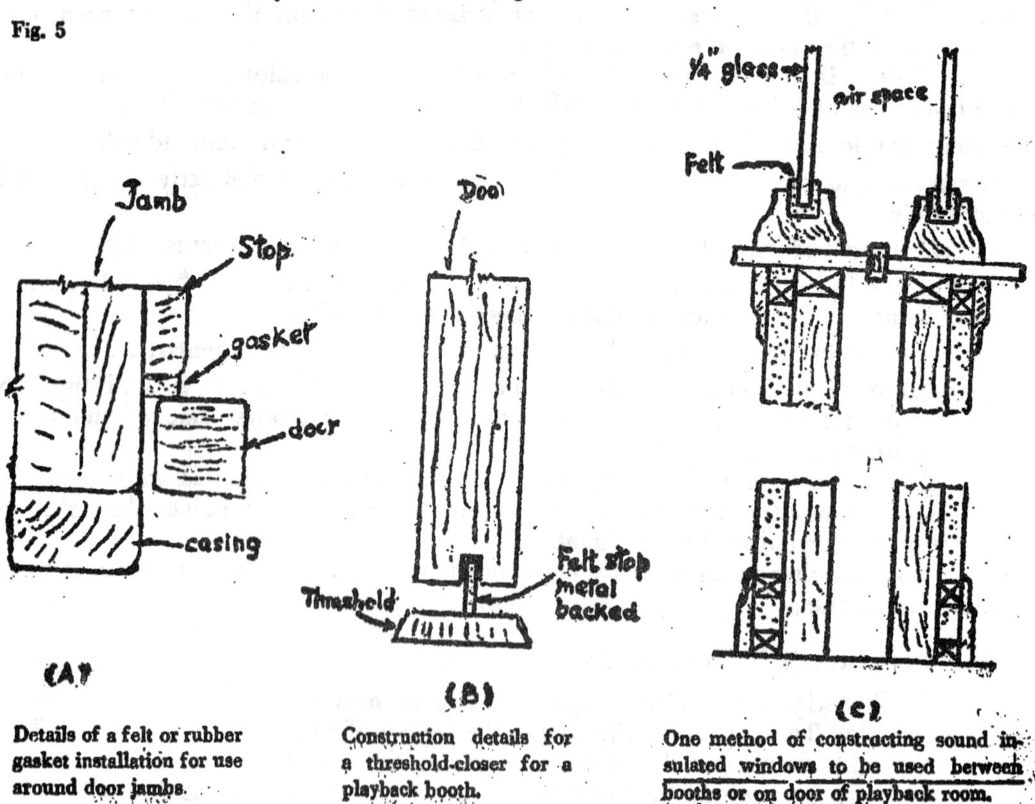

(A) Details of a felt or rubber gasket installation for use around door jambs.

(B) Construction details for a threshold-closer for a playback booth.

(C) One method of constructing sound insulated windows to be used between booths or on door of playback room.

simplest method of reducing it is to provide a resilient mounting which tends to cushion and absorb the vibration. Rubber, cork, or spring are used. Not to be overlooked in the process of construction is the ventilating system. The air ducts should be about 6″ × 12″ in size and should be about 12″ long. In case of fan or air noises conducted by ducts it will be necessary to line the duct with an absorbent material. This is also true of ducts leading to adjacent rooms.

Sound Insulating Details	Wt. in #/□	Transmission Loss in db	Rating
½" Plaster; ½" Insulating Lath; 2"×4" Studs Staggered 16" o.c. on 2"×6" Plate 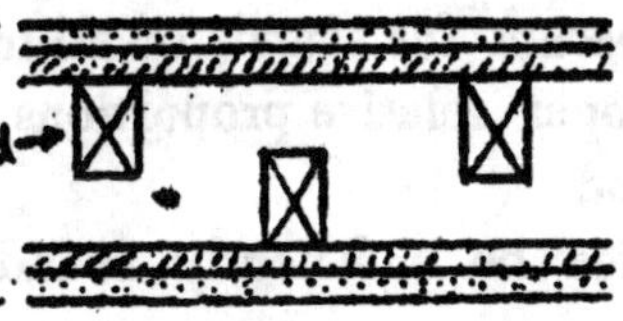	13.0	53.7*	Excellent
½" Insulating Board; ½" Insulating Board Stood loose, 6" lap; 2"×2" Studs, 16" o.c. on 2"×6" Plate 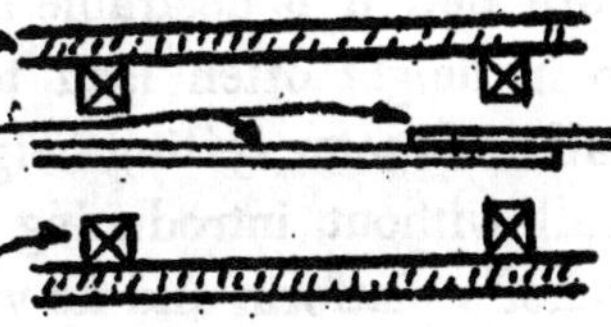	6.8	42.8**	Very good
½" Plaster; ½" Insulating Board; ½" Insulating Board Stood loose, 6" lap. 2"×2" Studs, 16" o.c. on 2"×6" Plate 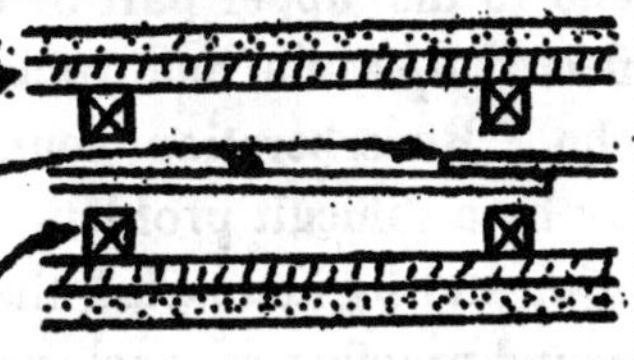	12.2	53.2**	Excellent
½" Plaster; Double 3" hollow gypsum tile 2" air space; ½" Insulating Board Stood loose, 6" lap. 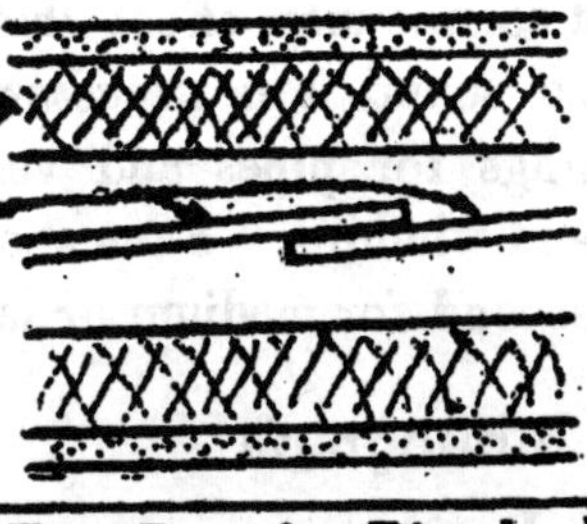	32.0	47.8*	(Fireproof) Excellent

* Test Data by Riverbank Lab.

** Test Data by Insulating Board Institute.

Fig. 4 Average transmission losses for different types of partitions.

The Acoustics of Auditoriums.

The chief conditions for good hearing in an auditorium are:

(a) The loudness should be adequate;

(b) That there should be no perceptible echoes or focusing;

(c) That there should be no undue reverberation, i.e., each speech sound should die away quickly enough to be inappreciable by the time the next is uttered;

(d) That where best music is concerned the hall should be non-resonant and suitably reverberant for sounds of all musical pitches, in order to preserve the proper relative proportions of the components of a complex sound; and

(e) That the boundaries be sufficiently sound-proof to exclude extraneous noise.

In general, where ceilings are flat, it is desirable to avoid a greater height than 40 feet. Ceilings 60 feet or so in height often lead to echo which are perceptible only in the middle of the hall. Greater ceiling height than 40 feet may be employed in the main body of the hall without introducing echo effects, provided that the ceiling is lower over the speaker's platform and its vicinity and is splayed to direct the reflected sound away from the front of the hall. Frequently it is useful to apply absorbents to back walls, and to the upper part of walls, and so to suppress sound which otherwise would be reflected upwards and returned to the floor of the auditorium later as an indirected echo. Reverberation, sounds being reflected to and fro without sufficient weakening, may be a difficult problem. It may be reduced by keeping the volume of the hall low, and by increasing the absorbing power of the surfaces of the auditorium. The sound-proofing of enclosures with a view to the isolation is of importance but of some complexity. The complexity arises from the variety of paths by which unwanted sounds may effect their entrance. To prevent the entry (or escape) of air-borne sound it is necessary to have walls sufficiently massive and rigid, to avoid openings for pipes and ventilators, and to exercise discretion in the location of doors and windows. Fans should be of slow speed, the tip speed not exceeding 55 ft. per second for medium or large sizes.

Bibliograph

Modern Acoustics: by A. H. David.

Sound Conditions: Celob Corporation. Nov. 1946.

Sound Insulation: Achitectural Forum. June 1937.

TESTING OF BOND STRESS OF BAMBOO

By

Lee Yum Hong *Chui Po Law*

What is bond stress? Bond stress is a kind of adhesion of two different materials, and is measured, say, by pulling a certain material such as steel out of the material such as concrete. Bond stress has not yet been investigated deliberately and what we should like to know, such as the history of bond stress, has also not been mentioned. So far we even can't get much results about bond stress in our references. In recent years, we think that this problem has been investigated, both in the laboratory and in practice, but we are very sorry that we can't get those results. Without much references, we have to start our test on this topic, and the majority of our work depends on our own thinking. We do not hope that we can give good results in our testing, but at least, we hope that we can find out some results that is of some use.

As there has much detail data in the field of the bond stress of steel imbedded in concrete, so what we emphasis here is the bond stress of non-metallic materials, such as bamboo and bamboo wire. The material that we emphasis most is bamboo.

Bamboo is a meterial of high tensil strength and it may be in certain cases used as a substitute of steel as reinforcement. So what, we are interesting in is the bamboo, such as, an original bamboo without knots, and those with knots, bamboo strips without knots and those with knots. At the same time, some of our specimens were made of bamboo strips with knots and those without knots, painted with asphalt. For steel and concrete, we found that the bond stress increases with time, that is as time increases, the concrete hardens and contracts, making a more effective clamping of the steel. But for bamboo, it is different, as bamboos are fibrous materials, which would decay with too much water content. From our result we know that as the moulds have been laid for two weeks, and four weeks, the bond stresses iucrease directly with time. But for the following weeks, the bond stress decreases as time increases. The painting of a layer of asphalt on the bamboo is to decrease the effect of the water to the bamboos and hoping that it can increase the bond stress. We have also noticed that, the bond stress of the bamboo strips are higher than that of rounded original bamboo, and the test were performed without any anchorage.

In this testing, the moulds were all made of bamboo sections, of 3 inches in diameter and 6 inches in length. It is very essential to know that the mould must be large enough that the volume of the concrete block can resist the cracking forces. The rims of the upper and lower side must be level so that the water content of the concrete must not leak out and affects the setting and strength of concrete.

The materials we used in this testing were bamboos of varies forms and some have treatments and stated as follows:

1. Original rounded bamboo with knots.
2. Original rounded bamboo without knots.
3. Bamboo strips without knots.
4. Bamboo strips with knots.
5. Bamboo strips with knots, with a layer of asphalt paint on it.
6. Bamboo strips without knots with a layer of asphalt paint on it.
7. Bamboo wire.

The original rounded bamboo is Asundinasia amobilis McClure (茶桿竹) and the Bamboo strips is Phyllostachys edulis A. et. C. Riv. (茅竹)

All the bamboos used in our testing must be washed and cleaned and be kep. in good condition. The surface of the bamboo were often rubbed with sand-paper. The tensile stress of the original rounded bamboo is around 28500 pounds per square inch. and those of bamboo strip is 37000 pounds per square inch.

We used portland cement, whose trade mark is Gladiator and made in Belgium This portland cement is just up to standard to be used. The water cement ratio, was 1. The ratio of nominal mix was 1:2:4. The time for setting of this cement seemed to be a little quicker. The compressive stress of this cement was found to be. 1000 lbs. per square inch. for 2 weeks.

The procedure of making specimens were as follows:

1. We clean all the bamboos and rub their surface with sand-paper.
2. Then we mix concrete with the ratio as mentioned above.
3. We use a right angle steel plate to adjust if the bamboo is vertial or not.
4. Then pour concrete in the mould. The imbedded part of the bamboo is about 4.5 inches.
5. We put the specimen into the humid room for 24 hours, then placed all the specimens into water with all the moulds removed, for four days. Then we take them out and let them set in the humid room, until we test them.

The machine we used for our testing is a Rieblé universal testing machine of 25 Tons.

We test the specemens every two weeks, four weeks, six weeks and eight weights, We haven't get all the results yet, and what we have now is just a minority. For original rounded bamboo, we bound out that its bond stress varies from 53 to 102 pounds per square inch for an age of two week, and for four weeks, the stress varies from 130 to 200 pounds per square inch of contact area. Then the bond stress for six weeks, was found to be from a value of 110 to 170 pounds per square inch of contact area, and we have noted that the bond stress was dropping. With those of original rounded bamboo, with knots, the value doesn't vary much, since the effect of the knot is not pronounced.

For bamboo strips, the value of bond stress has greatly increased, and the difference of the stresses between those with knots and those without knots is very pronounced. The bamboo strips without knots have a bond stress of, varying from 55 to 140 pounds per square inch for an age of three weeks, and for an age of six week, the bond stress have increased to a value of 110 to 120 pounds per square inch. With those with knots, for the age of four weeks, the bond stress varies from 120 to 230 pounds per square inch, and six week for a value of 115 to 220 pounds per square inch.

For bamboo strips with a layer of asphalt painted on them, we found a better result for the bond stress, for the asphalt tends to give a protection to the bamboo against water. The bond stress for those without knots varies from 150 – 170 pounds per square inch for an age of two weeks. And those with knots have a bond stress of 200 pounds per square inch.

For bamboo wire (竹䌫) we are very sorry to find that it is not a high tension material, and it often fails by tension around a load of 780 lbs.

Our unit bond stress is calculated by the following formula :

$$S_u = \frac{u}{P^1}$$

Where :

S_u = unit bond stress.
u = total bonds stress from testing.
l = imbedded length.
p = perimeter of the imbedded material.

From our testing so far, we obtain that the bond stress varies directly with time until about eight weeks, then the bond stress began to drop. Upon this fact, we think that this was due to the internal charge of the bamboo, such as the contraction of bamboo due to lose of water, and some other factors which we could not find out yet. So in taking this testing, we must notice that, bamboo is a fibrous material that would decay by making contact with water. The bond stress of bamboo also have a close relation with the surface of the bamboo. When the surface is rough, the bond stress would be increased, and perhaps with an treatment on the surface also would increase the bond stress, such as painting a layer of asphalt on bamboo. It is also essential to notice that the bamboo must be normal to the top plane of the concrete cylinder thus helping us to give a result of pure bond stress. In doing this testing work, we meet a lot of troubles when our testing work is proceding. But we should not afraid of those troubles, for what these troubles give us are not really troubles, but new discoveries. Thus, we find out that we would be taught to learn more in case we meet more trouble.

不定結構物之感應線

鄔境厚　　馬信輝

感應線之定義　感應線(Influence Iine)爲表示單位荷重(Unit Load)在結構物(Structure)上之連續位置所產生函數(Functions)之圖綫，其中包括反作用力(Reations)，力距(Moments)，剪力(Shears)，及應力(Stresses)等。

在簡單結構物(Simple Structure)上，感應線之求法甚易，且皆爲直線或連續之直線所構成，故毋需多述。然不定結構物(Indeterminate Structure)之感應線，較爲繁複。倘吾人能求得此等結構物之感應線，則對於分析及設計不定結構物，當有莫大之輔助焉。故此篇幅，全用於討論不定結構物之感應線。

一致扭歪之原理　(Principle of Consistent Distortion)

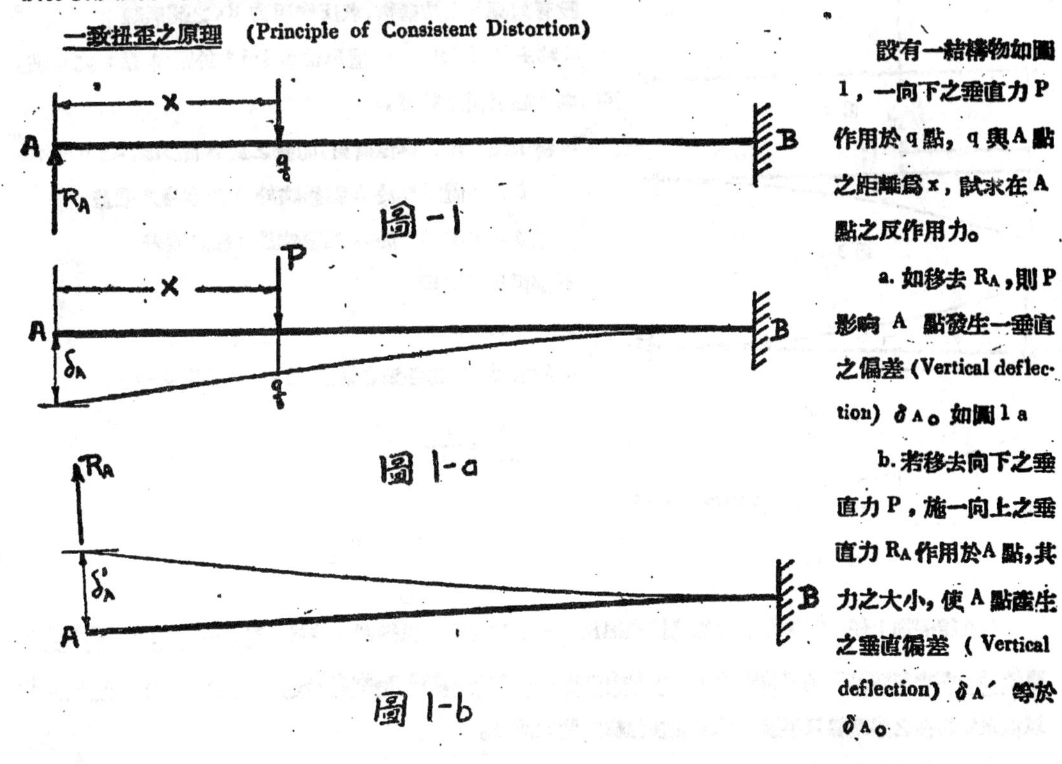

設有一結構物如圖1，一向下之垂直力P作用於q點，q與A點之距離爲x，試求在A點之反作用力。

a. 如移去 R_A，則P影响A點發生一垂直之偏差(Vertical deflection) δ_A。如圖1 a

b. 若移去向下之垂直力P，施一向上之垂直力 R_A 作用於A點，其力之大小，使A點產生之垂直偏差(Vertical deflection) δ_A' 等於 δ_A。(圖1 b)

c. 圖1 a與b之組合必與圖1一致，故此向上之垂直力 R_A 即爲所求之反作用力 R_A。

麥氏倒數定理　(Maxwell's Reciprocal Theorem)

1. 在一結構物上，一荷重 w 作用於 P 點影响於 q 點之徧差$\triangle_q$ 等於此荷重 w 作用於 q 點所影响於 P點之徧差 $\triangle_P$ (見圖 2－a)

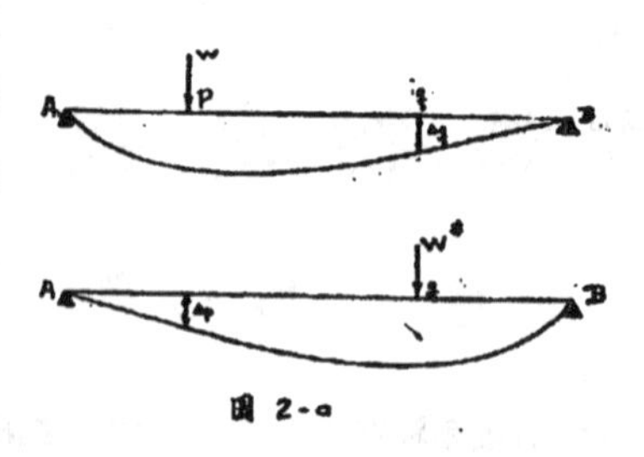

圖 2－a

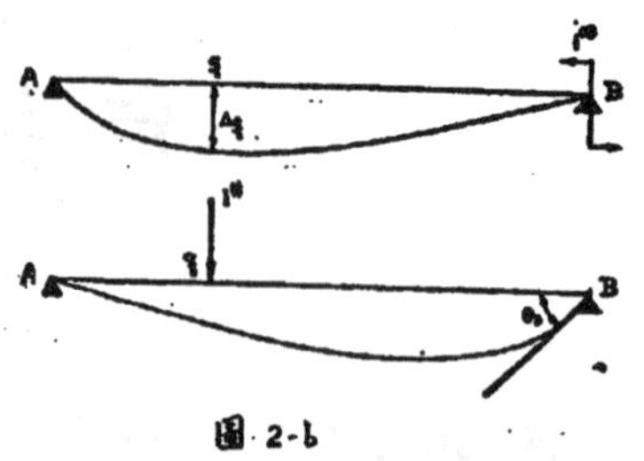

圖 2－b

將此定理推廣之得

2. 在結構物上單位力偶作用於鉸鏈 (hinge) B 點影响於 q 點之直線徧差 (linear deflection) $\triangle_q$ 在數目上等於單位垂直力作用於 q 點產生於鉸鏈 B 之角度徧差 θ_B (angular deflection) (圖 2－a)

感應線基本原理

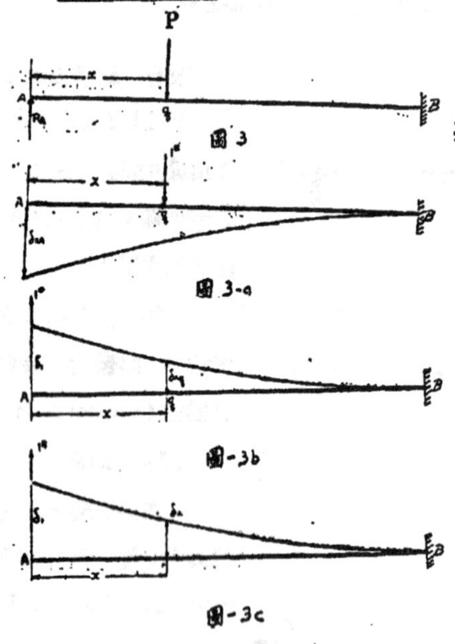

設有如圖 3 之結構物，求反作用力 R_A 之感應線

a. 移去 R_a 如圖 3－a，置單位重量 1# 於距 A 點 x 之 q 點。其影响 A 點之徧差爲 δ_{xA}

b. 移去在 q 點之單位荷重，而置之於 A 點如圖 3－b 所示。

δ_1 ＝置 1# 於 A 點影响於 A 點本身之徧差

δ_{xq}＝置 1# 於 A 點影响於 q 點之徧差

根據麥氏之定理

$$\delta_{xA}=\delta_{xq}$$

c. 如求得 AB 間各點之徧差 δ_x (圖 3－c)

$$\frac{1^{\#}}{R_A}=\frac{\delta_x}{\delta_1}$$

$$\therefore\quad R_A=\frac{\delta_x}{\delta_1}\times 1^{\#}=\frac{\delta_x}{\delta_1}$$

故在結構物上任一點反作用力感應線座標爲 $\frac{\delta x}{\delta_1}$ 由此我們可以得到一結論：結構物上任一點函數之感應線乃結構物在該點受到單位應變 (Unit Deformation) 所引起該結構物之屈撓曲線 (Distorted curve)。是以欲求結構物之感應線只不過變爲求屈撓曲線之問題而已。

求感應線之方法:

由於以上之原理，我們可以歸納一求感應線之方法，其步驟如下：

1. 移去吾人欲求之 Redundant。

2. 施荷重於 Redundant 之原有位置，使結構物發生扭曲。

3. 求得此結構物上各點之偏差及該荷重所在點之偏差。

4. 以荷重所在點之偏差除結構物上各點之偏差。

5. 以 4. 所得各點之商爲縱座標，連結之，所得之線即該 Redurdant 之感應線。

例：試求結構物 AB 之反作用力 R_A 感應線（圖 4）

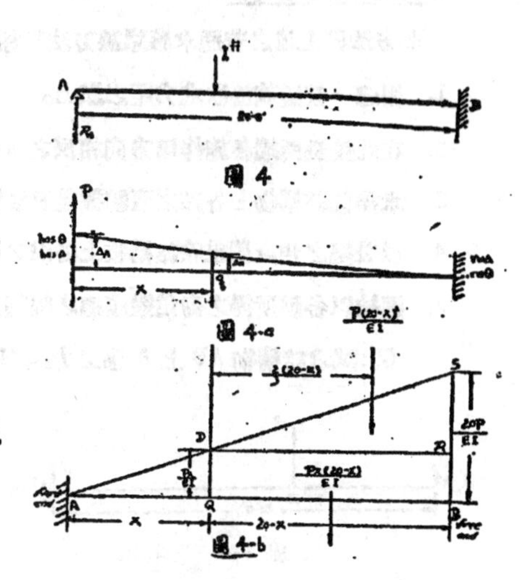

吾人利用共軛樑（Conjugate beam）以解決此問題，支撐 A 移去後，施以任何力 P 於 A 點。如圖 4—a

設 $\triangle_A$ 爲 A 點之偏差。

$\triangle_x$ 爲距 A 點 x 之 q 點偏差，根據共軛樑特性：

任何結構物上一點之直線偏差等於其共軛樑相當點之力距，由圖 4—b 之共軛樑上

$$\triangle_x = \triangle DSR \times \frac{2}{3} \times \overline{DR} + \square DQBR \times \frac{1}{2} \times \overline{DR}$$

$$= \frac{P(20-x)^2}{2EI} \times \frac{2}{3}(20-x) + \frac{Px(20-x)}{EI} \times \frac{20-x}{2}$$

$$= \frac{P(20-x)^3}{3EI} + \frac{Px}{2EI}(20-x)^2$$

當 $x = 0$，$\triangle_x = \triangle_A$

$$\triangle_A = \frac{\overline{20}^3 P}{3EI}$$

設 R_A 之感應線縱座標爲 Y_{RA}

$$Y_{RA} = \frac{\triangle_x}{\triangle_A} = \left(\frac{20-x}{20}\right)^3 + \frac{3x}{40}\left(\frac{20-x}{20}\right)^2$$

$$Y_{RA} = \frac{(x+40)(20-x)^2}{16.000}$$

繪 Y_{RA} 之方程式得圖 5.

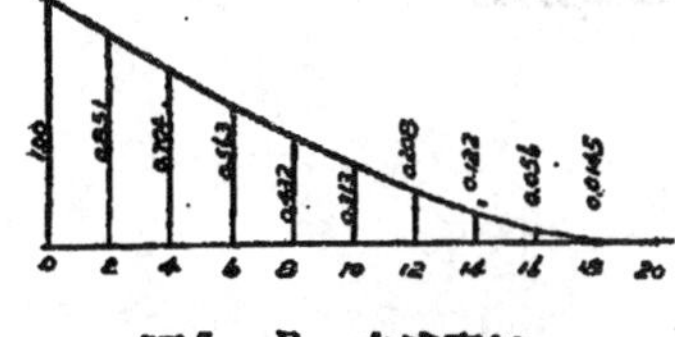

圖 5 R_A 之感應線

力距感應線之原理 欲求結構物上任一點 A 之力偶感應線，可假想該點有一鉸鏈，因在麥氏倒數定理內，偏差可爲直線偏差或角度偏差而荷重可爲力或力偶，再根據上述之感應線基本原理，

$$M_A = \frac{\theta x}{\theta_1}$$

M_A ——結構物上 A 點之力距。

25885

θ_x——單位力作用於結構物上任一點 p 所產生於 A 點之角度偏差。

θ_1——單位力偶作用於 A 點所產生於 A 點本身之角度偏差。

Δ_x——單位力偶作用於 A 點影响於 p 點之直線偏差由於倒數定理之 2. 我們知道 $\theta_x = \Delta_x$（在數目上）換言之 $M_A = \frac{\Delta_x}{\theta_1}$

求力距感應線之方法

此方法與上述之普通求感應線方法甚相似。

1. 假想一鉸鏈放於欲求力距之點上。
2. 在此鉸鏈兩端各施作用方向相反之單位力偶，使結構物發生扭曲。
3. 求得此結構物上各點之直線偏差及鉸鏈上本身之角度偏差。
4. 以鉸鏈之角度偏差除結構物上各點之直線偏差。
5. 連結以各點所得之商爲縱座標之線卽爲該點之力距感應線。

例：試求結構物 AB 上 A 點之力距 M_A 感應線：（圖 6—a）

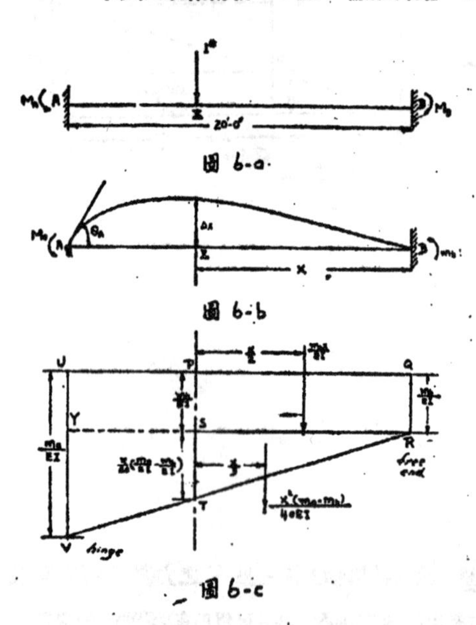

利用上節之方法，假設 A 點變爲鉸鏈 A，然後施以任意力矩 m_a 於 A 點使發生扭歪如圖 6－b。設 Δx 爲 B 在 X 點之直線偏差

θ_A 爲 A 點之角度偏差

m_b 爲固定邊 B (Fixed end) 之反抗力矩

$$\Delta x = \square PQRS \frac{x}{2} + \triangle TRS \cdot \frac{x}{3}$$

$$= \frac{m_b\ x}{EI} \cdot \frac{x}{2} + \frac{x^2}{40\ EI}(m_a - m_b) \cdot \frac{x}{3}$$

$$= \frac{m_b\ x^2}{2EI} + \frac{x^3}{120EI}(m_a - m_b)$$

又根據共軛樑之另一特性：任何結構物上一點之角度偏差等於共軛樑上相當點之剪力。

故 θ_A = Area QRUU

$$= \frac{1}{2} \times (\overline{QR} + \overline{UV}) \times \overline{UQ}$$

$$= \frac{1}{2} \times \left(\frac{m_b}{EI} + \frac{m_a}{EI}\right) \times 20$$

25886

$$= \frac{10}{EI}(m_a + m_b)$$

因鉸鏈A之力矩爲零,故$\triangle_A = 0$

$$\therefore \frac{1}{2} \times \overline{UQ} \times \square UYRQ + \frac{1}{3} \times \overline{UQ} \times \triangle YVR = 0$$

$$\frac{1}{2} \times 20 \times \frac{20m_b}{EI} + \frac{20}{3} \times \frac{10}{EI}(m_a - m_b) = 0$$

$$\therefore m_a = -2m_b$$

以 $m_a = -2m_b$ 代入$\triangle_x$與θ_A之方程式

$$\triangle_x = \frac{m_a x^2}{40EI}(20 - x)$$

$$\theta_A = \frac{10m_b}{EI}$$

$$\therefore y_{MA} = \frac{\triangle_x}{\theta_A} = \frac{x^2(20-x)}{400}$$ 此爲A點之力距感應線方程式

圖7　M_A感應線

機械方法(Mechanical Method)求感應線　用理論之方法以計算高次不定結構物之感應線,非常麻煩。倘吾人製成結構物之模形,應用前述求感應線之步驟,而用儀器屈撓之,以尺度直接量出偏差之值,以求感應線,此法謂之機械方法。應用此方法,非常迅速而且簡便,故甚適用於研究及分析複雜而新穎之結構物。

剪力感應線　剪力感應線可利用反作用力之感應線以求得之。

設有圖8之結構物ABC.今欲求其上一點q之剪力感應線。

荷重之位置若在q點之左,則q點之剪力$V_q = R_A - 1$

荷重之位置若在q點之右$V_q = R_A$

吾人先用前述之方法求得R_A之感應線(圖8-a)

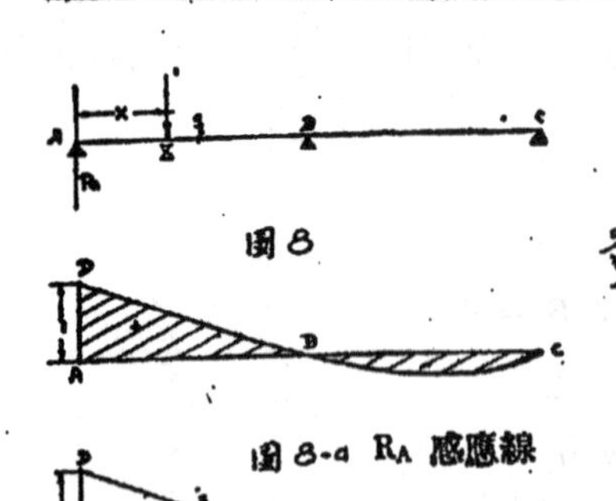

圖8

圖8-a R_A 感應線

圖8-b V_q 感應線

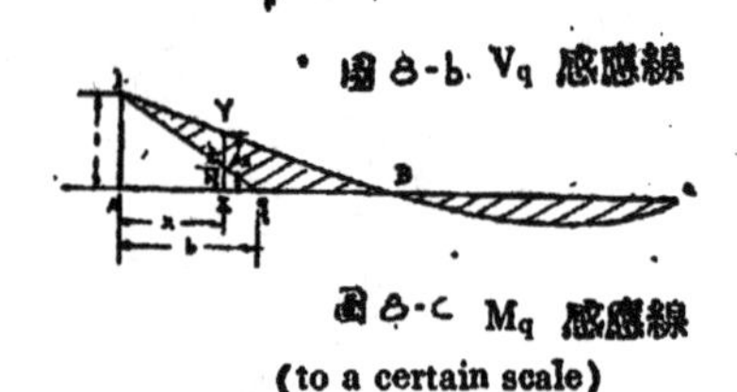

圖8-c M_q 感應線
(to a certain scale)

然後在 q 點作一垂直線，並作 AF//DE，則 AFEBC 爲所求之剪力 V_q 感應線。　(圖 8-b)

間接方法求力距感應線　力距感應線亦可根據 R_A 感應線以求得之。今求圖 8 結構物 q 點之力距感應線以明其法。

聯 Dq 直線，與 Aq 上任一點 x 之鉛垂線 XY 相交於 Z (圖 8-c)

設 $Aq=b$,　$XY=\triangle_x$,　$XZ=\delta$

$$R_A = \triangle_x \text{（感應線原理）} \qquad (1)$$

當荷重之位置在 q 點以左

$$\text{q 點之力距 } M_q = R_A \cdot b - 1\# \cdot (b-x) \qquad (2)$$

又由圖 8-c　$\triangle ADq \sim \triangle XZq$

$$\therefore \frac{b}{1} = \frac{b-x}{\delta}$$

$$\text{即 } b-x = \frac{b}{1}\cdot\delta \qquad (3)$$

以（1）及（3）代入（2）式得　$M_q = \triangle_x\, b - b\delta$

$$= b(\triangle_x - \delta) \qquad (4)$$

當荷重之位置在 q 點之右

$$M_q = R_A \cdot b = \triangle_x \cdot b \qquad (5)$$

由方程式 (4) 及 (5) 可知，在 Aq 段內，Mq 之感應線以 Dq 至 R_A 感應線之鉛垂距離及支撐 A 至 q 點距離之積爲縱座標。其餘之 qc 段內則以 R_A 感應線之縱座標及支撐 A 至 q 點距離之積爲 Mq 感應線之縱座標。

感應線之應用　玆舉數例以畧明感應線之應用。

例 1. 設結構物 AB 之長度爲 60 呎，試求 H-20 之車在 AB 上發生最大力距之位置及其大小。

吾人可利用圖 5 之 R_A 感應線，而以圖 5 之橫座標每單位代表 3 呎。

最大力距必發生於較重之車輪，故以車之後輪爲 q 點移動 H-20 於 AB 上以求 Mq 之值，經數次不同之位置後，即

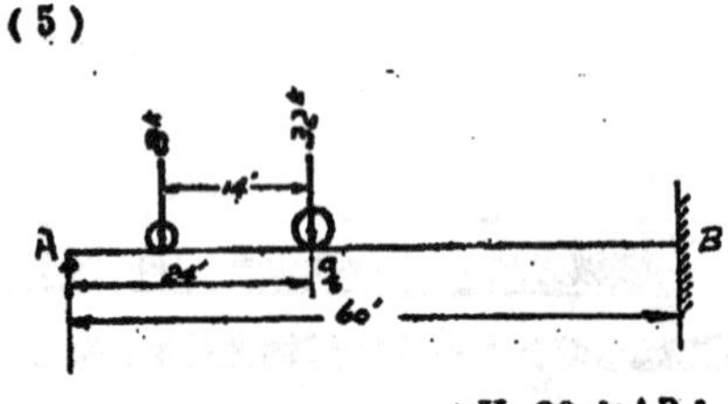

圖 9-a　H-20 在 AB 上發生最大力距之位置

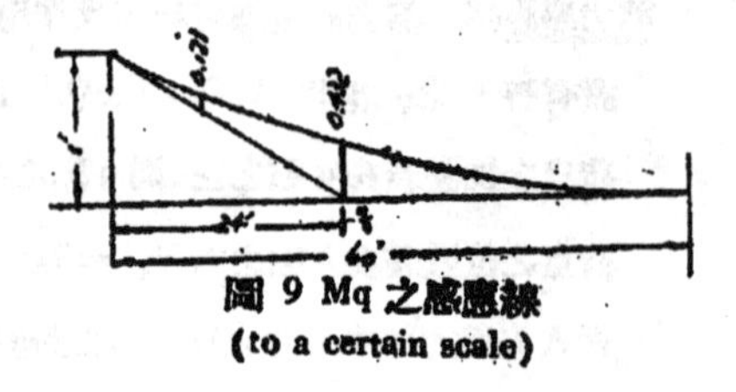

圖 9 Mq 之感應線
(to a certain scale)

25888

可找出其大約之位置。

a. 置後輪於距 A 點 30 ft 之處

$M_{30} = (32\times0.313+0.135\times8)\times30=332.4$ ft－k's

b. 置後輪於距 A 點 24 ft 之處

$M_{24} = (32\times0.432+8\times0.171)\times24=364.0$ ft－k's

c. $M_{23} = (32\times0.451+8\times0.165)\times23=262$ ft－k's

d. $M_{25} = (32\times0.411+8\times0.167)\times25=362.2$ ft－k's

觀乎以上各點之力距，M_{24} 之值最大，故所求之位置如圖 9-a 所示，而其值爲 364 ft－k's

例 2:－ 感應線在聯樑方面之應用：－

已知：－－一三跨度之聯樑其載重及跨度如下圖。

求　：－反作用力 R_B 及在支點 B 上力距 M_B 之數值。

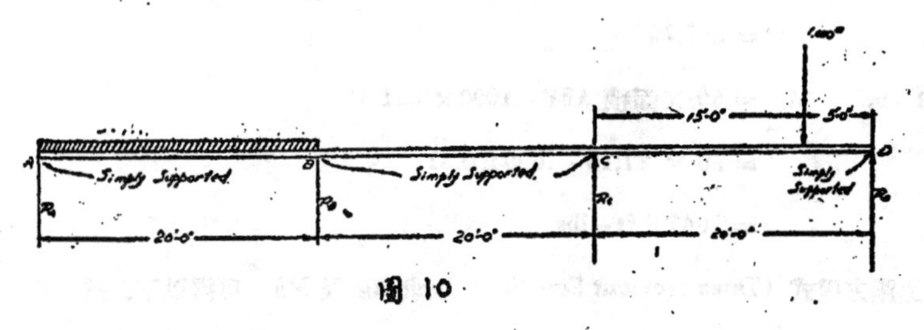

圖 10

由以前所敍之方法可求得支點 B 之反作用力及支點 B 上力距之感應線如下圖

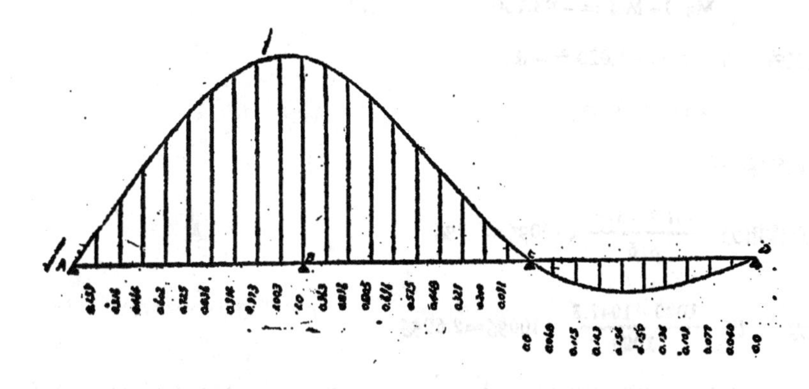

圖 10-a　　反作用力 R_B 之感應線

25889

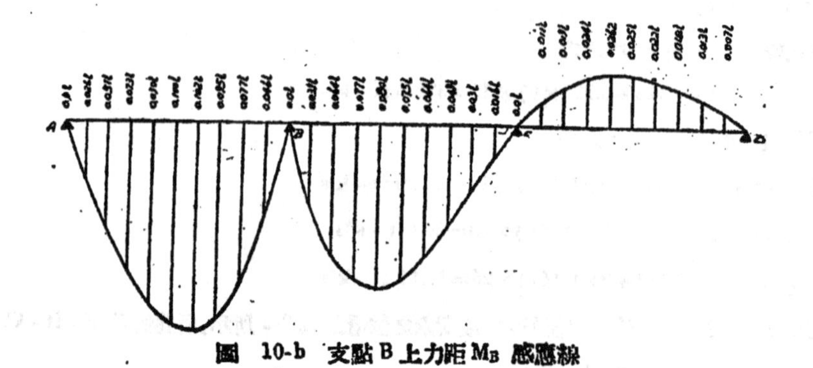

圖 10-b 支點 B 上力距 M_B 感應線

由圖 10－a, R_B ＝ 50 ×面積 AEB－1000×0.093

＝ 50 × 13.014－1000×0.093

＝ 557.7#

由圖 10－b, M_B ＝ 50 ×面積 AFB－1000×0.31

＝ 50 × 27.15－1000×0.31

＝ 1047.5 ft－lbs.

若用三力距方程式 (Three Moment Equation) 以求 R_B 及 M_B ，可得以下二式

$$4 M_B l + M_C l = -10000,000 \quad (1)$$

$$M_B l + M_C l = -93,750 \quad (2)$$

解上二式 得 $M_B = -1,020$ ft－lbs

$R_B = 516$ lbs

錯誤百份率:——

反作用力 $\frac{557.7-516}{516} \times 100\% = 8.7\%$

力 距 $\frac{1020-1047.5}{1020} \times 100\% = 2.69\%$

由此可知用感應線求得之數值甚爲準確，且在計算上甚爲方便，尤適用於連續樑架 (Continuous Truss) 方面

例 3: 設有一不對稱之剛架 (Unsymmetrical Rigid Frame) 其跨度及荷重如圖 11. EI爲一常數。

試求 a. C點水平力H_c之感應線。

b. 利用H_c感應線以求現有荷重所產生H_c之值。

c. 用其他方法求H_c以比較之。

a. 解放鉸鏈A，移去所有荷重，沿H_A之作用方向，施以1K之力：剛架用C'A'B'D扭曲成CABD之形(圖11-a)

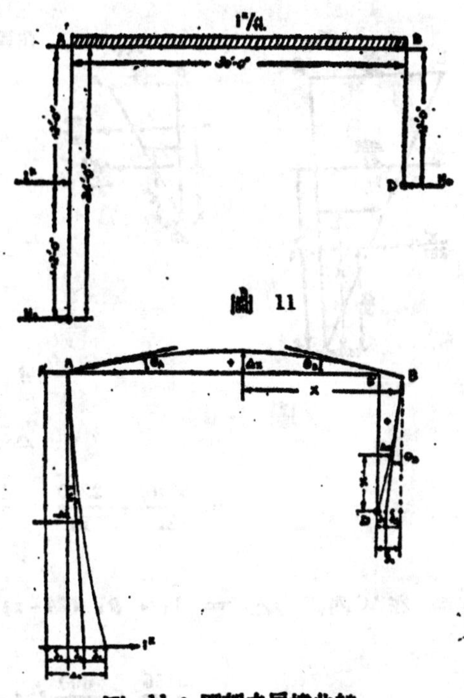

圖 11

圖 11-a 剛架之屈撓曲線

假定(Assumption)：

1. A'A = B'B

2. 接口(Joint)在屈撓後仍爲90°角。

圖 11-q 爲圖 11-a 之共軛樑。

由圖 11-c

$$\theta_A = \frac{1}{3}\triangle ABH + \frac{2}{3}\triangle AGH$$

$$= \frac{1}{3} \times \frac{180}{EI} + \frac{2}{3} \times \frac{360}{EI}$$

$$= \frac{360}{EI}$$

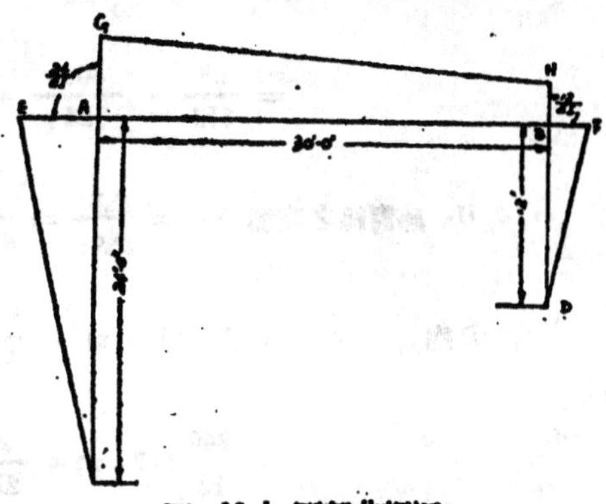

圖 11-b 剛架共軛樑

$$\theta_B = \frac{1}{3}\triangle AGH + \frac{2}{3}\triangle ABH$$

$$= \frac{1}{3} \times \frac{360}{EI} + \frac{2}{3} \times \frac{180}{EI}$$

$$= \frac{240}{EI}$$

由圖 11-a

$$\delta_4 = 12\ \theta_B = \frac{12 \times 240}{EI} = \frac{2880}{EI}$$

$$\delta_2 = 24\ \theta_A = \frac{24 \times 300}{EI} = \frac{7200}{EI}$$

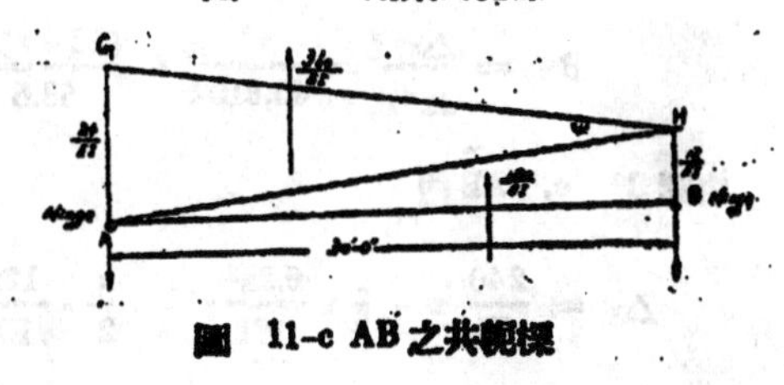

圖 11-c AB之共軛樑

25891

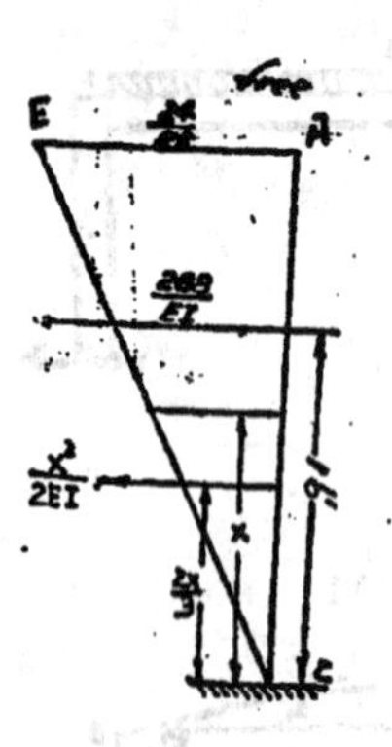
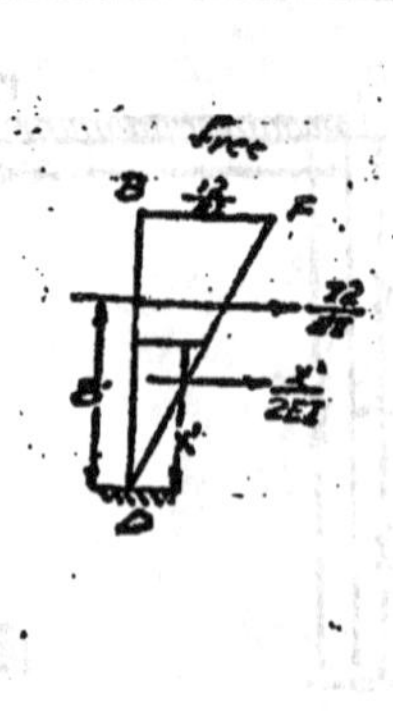

圖 11-d

由圖 11-d $\delta_5 = \frac{2}{3} \times \overline{BD} \times \triangle BDF$

$$= 8 \times \frac{72}{EI} = \frac{576}{EI}$$

$$\delta_3 = \frac{2}{3} \times \overline{AC} \times \triangle ACE$$

$$= 16 \times \frac{285}{EI} = \frac{4610}{EI}$$

$$\delta_1 = \delta_5 + \delta_4 = \frac{3356}{EI}$$

$$\triangle c_1 = \delta_1 + \delta_2 + \delta_3$$

$$= \frac{3356}{EI} + \frac{2200}{EI} + \frac{4610}{EI} = \frac{15270}{EI}$$

在AC內， $\triangle x = \delta_1 + \theta_A (24-x) + \frac{2}{3} x \frac{x^2}{2EI}$

$$= \frac{3356}{EI} + \frac{300}{EI}(24-x) + \frac{x^3}{3EI}$$

$$= \frac{x^2}{3EI} - \frac{300x}{EI} + \frac{10,656}{EI}$$

H_A 感應線之座標 $\delta_x = \frac{\triangle x}{\triangle c_1} = \frac{x^3}{45,810} - \frac{x}{50.9} + 0.698$

在BD內， $\triangle x = \theta_B (12-x) + \frac{2}{3} x \cdot \frac{x^2}{2EI}$

$$= \frac{240}{EI}(12-x) + \frac{x^3}{2EI}$$

$$\delta_x = \frac{\triangle x}{\triangle c_1} = \frac{x^3}{45,810} + \frac{(12-x)}{63.6}$$

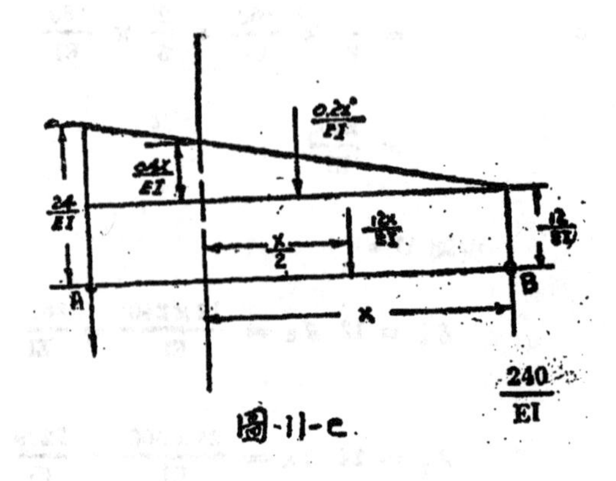

圖 11-e

由圖 11-e， AB內

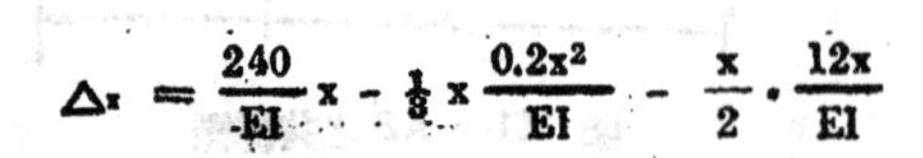

$$\triangle x = \frac{240}{EI} x - \frac{1}{3} x \frac{0.2x^2}{EI} - \frac{x}{2} \cdot \frac{12x}{EI}$$

$$= \frac{x}{EI}\left(240 - \frac{x^2}{15} - 6x\right)$$

$$\delta_x = \frac{\triangle x}{\triangle c_1} = \frac{x}{15270}\left(240 - 6x - \frac{x^2}{15}\right)$$

b. 利用感應線以求 Hc 之值.

Hc = lk × P 點之感應線座標 + lk/ft × AB 感應線之面積

$$= 1 \times \left[-\left(\frac{12^3}{45810} - \frac{12}{50.9} + 0.698\right)\right] + 1 \times \int_0^{30} \frac{x}{15270}\left(240 - 6x - \frac{x^2}{15}\right) dx$$

$$= -0.5 + \frac{1}{15,270} \times \left[120x^2 - 2x^3 - \frac{x^4}{60}\right]_0^{30}$$

$= 2.15$ k

c. 利用力距分配方法 (Method of Moment Distribntion) 求 Hc.

EI ＝常數

$$K_{AB} = \frac{1}{l_{AB}} = \frac{1}{30}$$

$$K'_{AC} = \frac{3}{4} \times \frac{1}{l_{AC}} = \frac{3}{4} \times \frac{1}{24} = \frac{1}{32}$$

$$K'_{BD} = \frac{3}{4} \times \frac{1}{l_{DB}} = \frac{3}{4} \times \frac{1}{12} = \frac{1}{16}$$

$K_{AB} : K'_{AC} : K'_{BD} = 1.06 : 1 : 2$

固定邊力距 (Fixed end moment) $M_{AB} = +\frac{wl^2}{12} = +\frac{1 \times 30^2}{12} = 75'^{K}$

固定邊力距　$M_{BA} = -\frac{wl^2}{12} = -\frac{1 \times 30^2}{12} = -75'^{K}$

固定邊力距　$M_{AC} = -\frac{3}{16}Pl = -\frac{3}{16} \times 1 \times 24 = -4.5'^{K}$

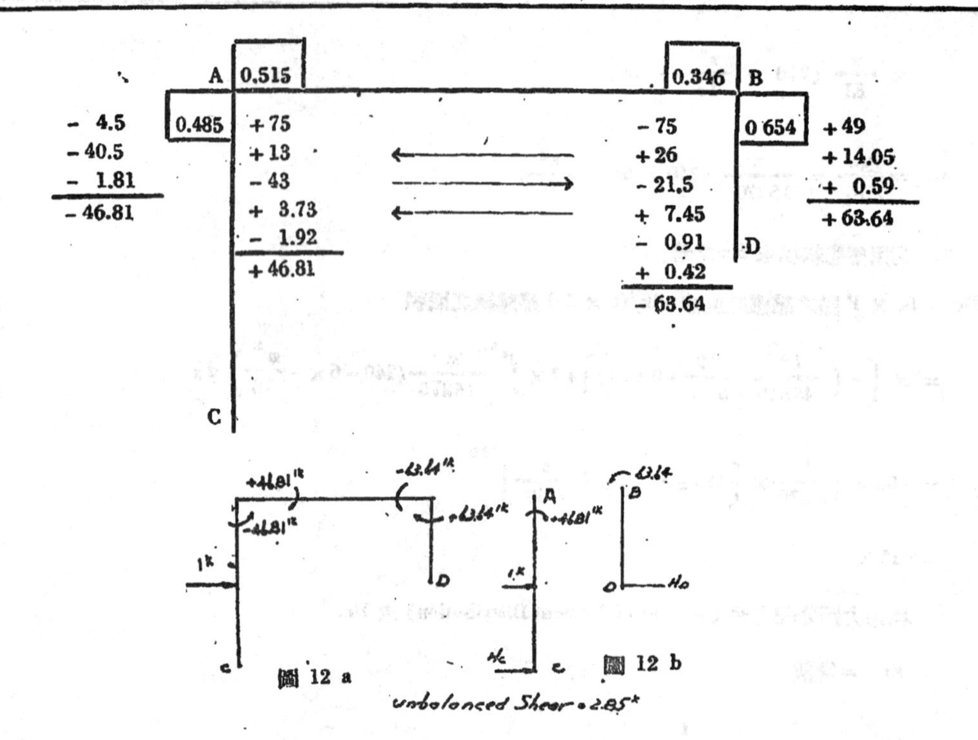

圖 12 a　　圖 12 b

根據求得之力距，吾人可繪圖 12-a 及圖 12-b 之 free body diagrams.

$$H_C = \frac{46.81}{24} + \frac{12}{24} \times 1 = 1.45K$$

$$H_D \ \frac{63.64}{12} = 5.3K$$

H_C 及 H_D 之作用方向如圖所示

不平衡之剪力 $= H_D - H_C = 5.3 - 1.45 = 2.85K$

此不平衡剪力所以發生之原因，乃基於剛架發生旁側作用（Side Sway）．而由此作用所產生之力距尚未算在內。故圖 12-a 之力距必需更正。

由於旁側作用所生之力距與 I 成正比而與長度 ℓ 之平方成反比。設 $M_{\triangle AC}$ 及 $M_{\triangle BD}$ 各爲 AC 及 BD 旁側作用所生之力距。

$$M_{\triangle AC} \propto \frac{I_{AB}}{\ell^2_{AB}} \propto \frac{1}{(24)^2}$$

$$M_{\triangle BD} \propto \frac{I_{BD}}{\ell^2_{BD}} \propto \frac{1}{(12)^2}$$

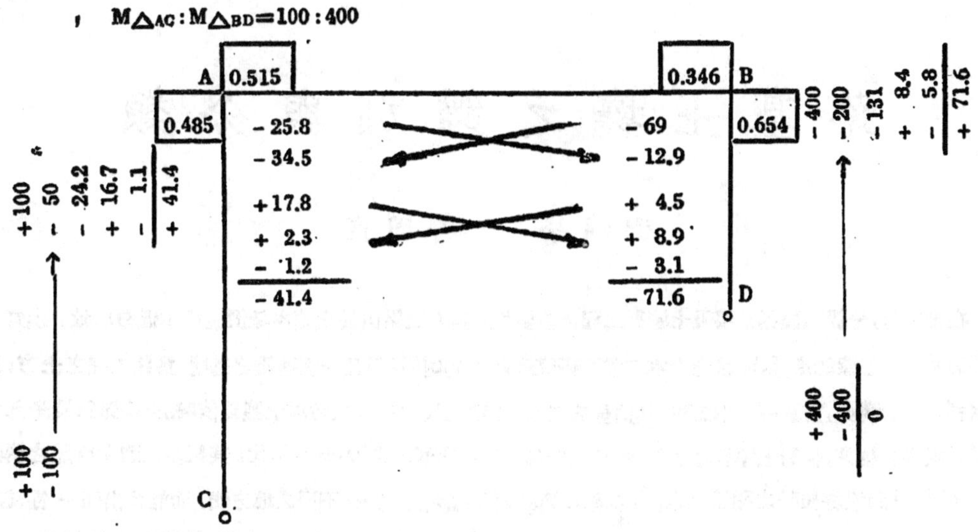

由此所求得之力距，吾人可繪圖 12-c 及圖 12-d

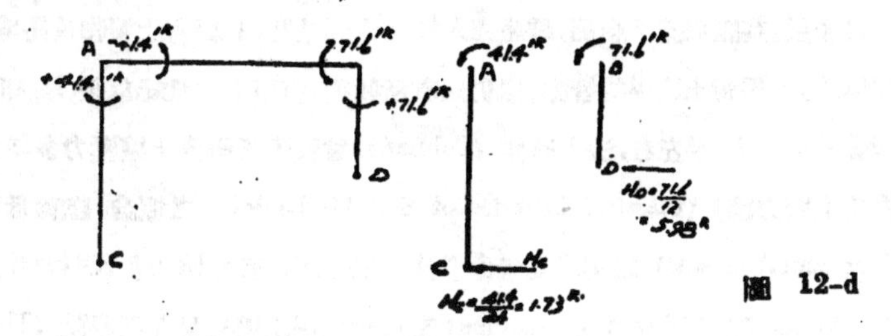

圖 12-d

Unbalanced Shear = 7.71K

圖 12-c

不平衡剪力 $= H_C + H_D = 1.73 + 5.98 = 7.71$

調整後之 $M_{AC} = -46.81 - \dfrac{2.85}{7.71} \times 41.4 = -62.11'K$

調整後之 $M_{BD} = 63.64 - \dfrac{2.85}{7.71} \times 71.6 = +37.24'K$

故 $H_C = \dfrac{62.11}{24} - 0.5 = 2.1K$

此與前用 H_C 感應線求得之值 2.15 甚為接近。

Reference:

STATICALLY INDETERMINATE STRESSES — PARCEL and MANEY
THEORY OF MODERN STEEL STRUCTURES — GRINTER
CONTINOUS FRAMES of REINFORCED CONCRETE — CROSS and MORGAN
CONCRETE PLAIN and REINFORCED — T.T.S.
THE ELASTIC ENERGY THEORY — J.A.Van den Brock

路基土壤之鑑別與分類

鄧錦榮　　鄺國良

在卅年前，一般工程界皆漠視土壤對工程之重要性，其大概係由於土壤本身價值根本低微所致，須知土壤不但是人類生產的泉源，而且是土木工程司研究最基本的對像，舉凡一切建築之基礎，鐵路工程之土方，以及水利，隧道，溝渠，市政等土木工程，其佔全部工欵，工期之最大百分數皆以土壤爲依歸。土壤雖爲最平凡而又最易獲得之物質，但對土壤智識之尋求，至爲微妙，蓋其內容之奧妙，却無倫比，其組織係踰大自然之偶然變化而成，其間錯綜複雜之關係，實毫無規律可循。天下雖大，但在兩不同之地方間，而能找出同一性質的土壤，幾成爲一絕不可能的事實，土壤學處於科學進步之今日，足引起國內外大多數工程界探討的興趣。

土壤研究之發展，其歷史殊不悠遠，雖遠在人類文化開始初期，經有土壤物質構成之簡單工程結構面世，但在十六世紀以前，人類對土壤學的智識，似仍未有所收獲，直至十六世紀以後，法人開始研究利用土壤之可壓性而建築堡壘及至1773年左右，法人哥倫 (Column) 曾發表其研究土壤壓力多年之心得，其實法人會將幾種基本『古典土壤力學』(Classical Theories of Soil Mechanics.) 之概論，獻給世界，其將『理想碎片體』(Cldealized Fragmental masses) 之解釋爲包含似沙之顆粒而佔有摩擦力及牽引力者，與種學說，可說是古典土壤力學創立之淵源。到1856年英人 (Rantine) 又曾創立一土壤質量平衡學說 (Theory of Equilibrium of Earth Masses) 藉以解決幾種基礎工程學之主要問題。約在十八世紀時，人類對土壤關乎工程結構基礎之多種原理及性能上的種種關係，似有黯淡之見解，後到1934年，奧國土壤權威太沙基氏 (Terzaghi) 講學於美國麻省理工學院 (M.I.T.，遂刺激起美人對土壤研究之興趣，兩年後，國際土壤學會 (Intemational Confevence of Soil)．及基礎工程學會 (Fouudation Engg Society) 在哈佛大學 (Harvard Vniversity)，舉行首次大會，由是土壤研究被列爲重要部門，各國內升學者均紛紛參加研究，直至最近廿年間，土壤學的研究始有所結晶，又中國土壤工程學會經於民卅二年間籌備成立，則我國工程界人士應及時努力研究邁進爲切。

土壤學識在公路路基工程上應用甚久，在1931年六七月間 Hogentogbr 在 Public Load 雜誌上發表一文 "Subgrade Soil Coustants, Their Siguificance and Their Application in Practice." 同年九月 Watkins 發表 "The Soil Profile and the Subgrade Survey." Public Roads Vol. 12 No 7，又 Wintermyer 發表 "Procedures for the Determination of the Subgrade Soil Constants." Public Roads Vol. 12 No 8. 此後，美國公路總局及試驗總局遂訂定標準試驗及分類法，採用業經十餘年，當有相當之效力。土壤分類既有標準，研究之

25896

目標遂轉移於土壤穩定 (Soil Stabilization) 利用壓實之土壤，或大小沙石顆粒之分佈，或化學藥物或油類對於土壤之處治，使成為一個便堅固而廉價之路面，是則公路之進展，更踏進另一世紀之途也。

本文謹就路基土壤之試驗及分類方法加以介紹，至土壤之穩定及利用，則俟他日有機會再作介紹：

土壤之性質

一、黏性——土壤之黏性為其不受外部壓力影响時，土壤本身之團結力，普通所謂黏性，實具有二部，一部為眞實黏性，(true cohesion)，係由於土壤本身分子吸引力所做成；另一部為外表黏性(apparent cohesion)，係由於土壤所含水之表面張力所做成。顆粒較大之土壤，其分子間吸力遠不及其質量，同時其空隙間水分之表面張力甚微，故此種土壤之黏性甚小。反之，顆粒細小之土壤，其分子間引力較其分子質量為大，空隙小，表面張力大，故黏性甚强。至土壤含水過少，則無表面張力，太多，則分子間離愈遠，表面張力亦愈小，至成液狀時，則黏性全失矣。

二、內阻力——內阻力可分為二部：一部為顆粒間之摩擦力，其值視土壤顆粒表面之光滑度而定。一部為顆粒與顆粒間之連鎖力，其值則視土壤之壓實度而定，土壤顆粒大小均匀，則連鎖力愈大。

內阻力之程度可用內阻力角 θ 表之，內阻力角之簡便測定法可將土壤堆成一錐形堆，則堆與平面所成之靜止角 (angle of repose) 即等於 θ，或可盛土壤於一盒中，抬高盒之一端，至盒中土壤開始滑動時，盒與平面所成之角即等於土壤之內阻力角。

各種土壤 θ 之值不同，碎石與沙土混合料之值最高可至 34°，最低為軟黏土，其值約為 2°。

三、毛細作用——毛細作用為土壤之性能，將水自附近水源，吸入並向四方傳達，並不因地心吸力之大小與方向而影响。毛細作用之大小與毛細管之直徑成反比例。黏土之顆粒至微，毛細管直徑頗小，故其作用亦甚强，惟其蔓延率則甚緩，沙土之顆粒較大，其毛細作用亦較弱，但其蔓延率則甚大。

毛細作用對於土壤有下列三種直接影响：

(1)土壤膨脹現象。

(2)土壤收縮現象。

(3)霜凍後路基之霜脹現象 (Heaving Phenomena)——土壤中毛細管被水分充滿後。若氣溫突降，則凝結為冰，體積增大，其力足以掀起路面。

四、可壓縮性——泥土之壓縮性為當受外壓力時。(其容積之變遷)乾土之能壓縮而不發生旁面流動 (Lateral flow) 者，蓋其內空隙中之空氣逸出。濕土之能壓縮而不生旁流者，蓋水分自空隙中被壓出也。

如加壓於含水之沙土中，水份甚易被擠出，而容積跌減甚速，蓋沙土之滲透性甚大，反之加壓於濕泥土中，容積跌減甚緩，故凝實速度與滲透率有關也。

又加壓於乾沙土上，其壓縮之容積，遠比黏土為小，蓋沙土中空隙少而黏土內空隙多也，故土壤之可壓性之大小應視其空隙之多寡而定。

五、彈性——彈性爲土壤受壓力後變形，而待壓力除去後，能返至原來狀態之能力，無一種土壤爲具有完全之彈性者，且各種土壤之彈性相差甚遠，據研究結果，土壤之所以具有彈性，全因內部含有薄片狀之土壤顆粒與有機膠體之故。

土壤樣品之採取

採取土壤樣品之先，須有全路之平面圖與從斷面圖，然後鑽探之工作，沿線而行，採取樣品之椿號，爲橫坐標，所鑽深度或土層深度爲從坐標，均應紀錄，以備他日繪成土壤縱斷面圖之用。

普通沙泥或含有碎石之鬆軟土壤，可用螺旋鑽（Soil auger）鑽取，螺旋鑽爲一T形鐵桿，上端橫桿爲柄，直桿之端爲螺旋形，其直徑爲 $1\frac{1}{2}$ 吋。鑽之總長爲三尺，如需要時，可以拆下橫桿，將直桿加長三尺。如土質堅硬，或需要深度過深，或需要採取未經驚動之土壤樣品（Undisturbed samples）則採用水冲法（Wash borings）攪碎法（churn drilling）轉割法（rotary drilling）及特製之未經驚動土壤樣品儀。

鑽探之間隔因地形與從斷面而變，大抵在巒峻之山地，地質變化急劇，距離應短。在地形平坦，或地質均勻之地，距離可達一百尺至百五尺。

至鑽探之深度，亦視地形與地面情形而定，在挖土之地，應在設計線（grade line）下六呎。在填挖土甚少地方，應達地面下六呎，在發現有水滲透之地，應鑽至蔽水地層之下，如發現地下水，其高度應紀錄。在填土段，所填土材料係取自借土坑，則應借土處加以鑽探。

每次鑽探至地層發生變化後，將鑽得之泥，選出五磅重左右，盛於緊密之布袋中，註明採取之地點椿號，深度，及其他應紀錄事項。然後置於空氣流通乾爽之處，待其晾乾，以備實驗之用，至未經動土壤之樣品，應用蠟封固，以防水分蒸發，運管送至試驗室，儲於潮溼室中備用。

土壤之鑑別

土壤之鑑別，全在其物理性，物理性則由機械分析與土壤常數而定，土壤常數包括。

塑性 Plarticity
- 液性限度 Liquiel limit
- 塑性限度 Pastic limit
- 塑性指數 Pasticity index

體積變遷 Volume Change
- 收縮限度 Shrinkage limit
- 收縮比率 Shrinkage ratio
- 直線收縮 Lineal Shrinkage

土壤含水能力 Moisture Capacity ot Soils——野外含水當量 Field Moisture eguivalent

水流阻力 Resistance to flow of water——離心含水當量（Field moisture Eguivalent）。

一、機械分析（Mechanical analysis）一機械分析爲指示出顆粒之大小與分布，顆粒直徑在 0.074 公厘（200

號篩)以上可篩分析法 (Sieve Analysis) 求其大小，在0.074 公厘以下者則用連續沉澱法 (Successive sedimentation) 或比重計法 (Hyodrometer analysis) 求之。

篩分析法將土壤分爲七種：

礫石 (Gravels) —— 在2mm以上者

粗沙 (Coarse sand) —— 在0.5至2mm之間者；

沙 (Sand) —— 在0.5——0.25mm之間者

細沙 (Find sand) —— 在0.25——0.05之間者

瘀土 (Silt) —— 在0.05至0.005mm之間者

泥土 (Clay) —— 在0·005mm以下者

膠體 (Colloids) —— 直徑在0.001mm以下者

普通所用之篩列爲 #10#20#40#60#140 及 #200 至於通過#200篩之微小顆粒則繼續用沉澱法求之。

連續沉澱法係根據斯閩克定律 (Stokers law) 而來，即『小圓球體在液體中之沉澱速度與圓球直徑之平方成正比。』

比重計法係根據土壤溶液密度隨沉澱時間而變之原理，而土壤溶液密度變化可用比重計求得之。

至以下之土壤常數測定所用之土壤均爲經過 # 40篩者。

二、塑性之測定——由前述可知土壤黏性之由來爲分子間之吸力及含水之表面張力，兩者中當以後者影响爲大，當乾燥之泥土，漸加水份至某定量則成可塑性，但更加水至另一定量則成半流體狀，塑形全失，故塑性有上下兩極限，而兩極限間則爲塑性指數也。

(1)液性限度——當土壤對於剪力，毫無抵抗，以輕微之力即能令其流動，此時之土壤稱爲在液體狀態中，而液性限度乃使土壤變成此種液體最少之水量也。液性限度指出土壤所需之水分用以潤滑顆粒之表面以抵消其內阻力及用以分離開顆粒間之距離，而令其消失互相吸力也。

液性限度之手測法——取通過 #40 篩之土壤約30公分，置於直徑約4½吋之蒸發皿中，加水用小刀拌和，然後將皿內土壤攤平，中間厚度約爲½吋用標準大小之梯形分溝器 (Grooving tool) 將土壤之中部劃開，然後一手緊握蒸發皿，輕輕向他手手心擂擊10次，若拍擊十次後，皿中土壤梯形底適巳接合，則此時土壤之含水量適達至液性限度，如未滿十次或十次以後，梯形底邊始連接，則表示水份過多或過少，應加土或水，調和後重複以上手續待至液性限度時，將溝底附近土壤少許，置於錶面玻璃，白鐵盒或鉗鍋中，稱其重量後移入溫度 110°C 之電爐中，烘乾至重量不變時爲止，移入氯化鈣乾燥器中，俟其冷後再稱其重量。液性限度之值可由下式計算之：

$$液性限度(L.L.)=\frac{土壤含水重量}{烘乾土之重量}\times 100$$

液性限度之機械測法——機械測法與手測法試驗步驟大致相同，惟改由機械碰擊杯底引起震動，使溝旁土壤滑下而相連接，俟連接時，紀錄碰擊次數，用刀取出一小部分土壤，測其含水面分數如上述，由數次所得之結果，用含水成份爲從坐標，碰擊次數爲橫坐標，繪圖在半對數圖解紙（Semi-log paper）上，在敲擊25次時與此曲線交之點卽爲該土壤之液性限度。

（2）塑性限度——土壤由乾而濕，漸呈黏性，其起始發現黏性時所含最少之水量，曰塑性限度。以試驗方法解釋之，則爲土壤加水，在玻璃上滾成圓條，俟其直徑至 $\frac{1}{8}$ 吋適生破碎現象。卽將該土壤所含水量測定如上述，塑性限度之值可用下式計算之：

$$塑性限度(P.L)=\frac{土壤含水重量}{烘乾土之重量}\times 100$$

（3）塑性指數——液性限度減去塑性限度所得之差，謂之爲塑性指數，若塑性限度等於或大於液性限度時，塑性指數爲零。若塑性限度不能獲得，則塑性指數可稱「無塑性」NP（Nonplastic）蓋此種土壤完全缺乏塑性也。

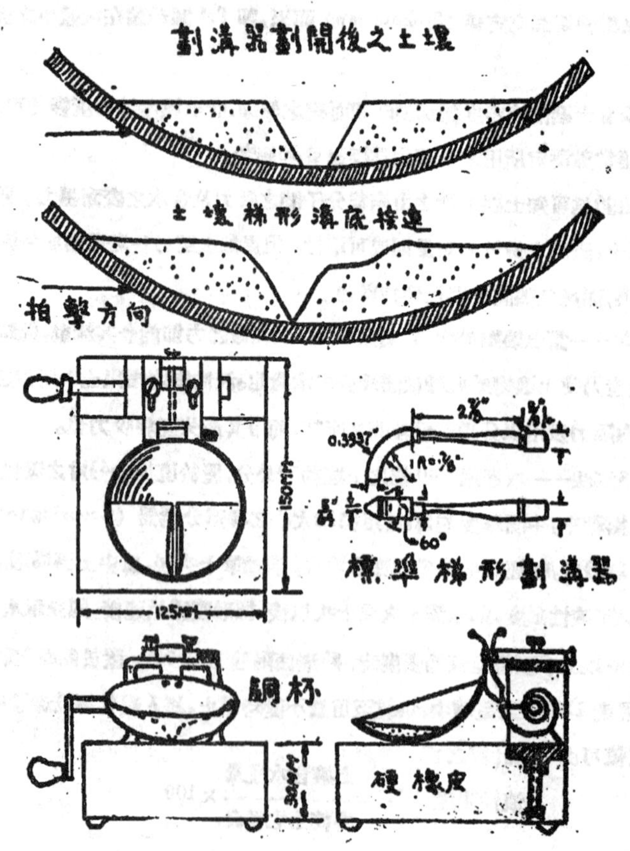

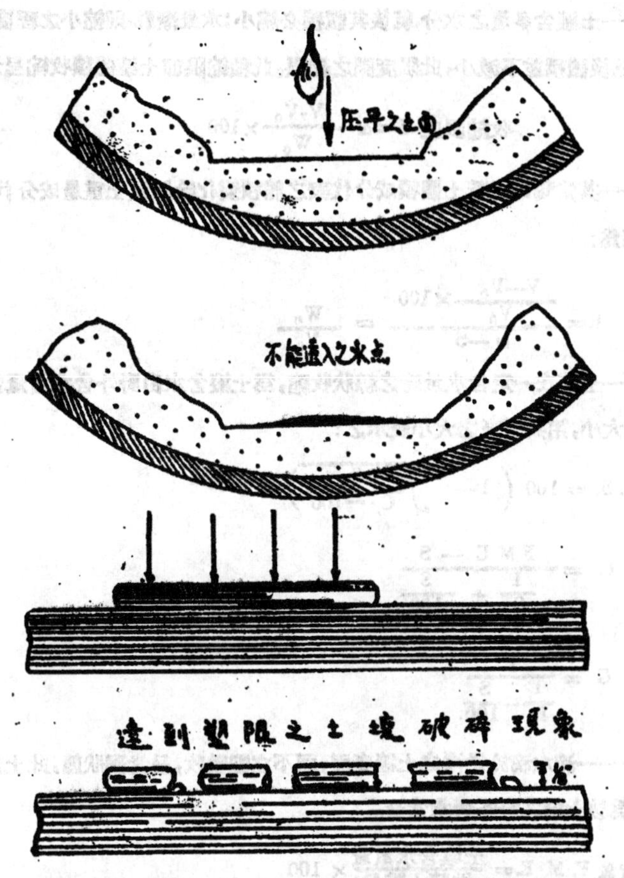

三、體積變遷——取通過#40篩之土壤樣品 30 公分，置於蒸發皿中，徐徐加水使成漿狀，移入塗有凡士林之圓形平底牛乳皿（milk dish）內。以底在厚書上敲擊，並使多餘土壤由皿之四周溢出，用鋼界尺將皿邊將餘土括去，務使土中絕對沒有空隙。隨即稱其重量，在空氣中晾乾約 24 小時，再烘於 110° C 之電爐中，至其重量不變時為止。濕土之體積即皿之容積，可用水銀注入皿中，用玻璃片緊壓，擠出多餘水銀，然後用量筒量之，至乾土之體積，可將乾土浸入水銀皿中，排出之水銀即乾土之體積。

設 V ＝濕土之體積。

V_0 ＝乾土之體積。

W ＝濕土之重量。

W_0 ＝乾土之重量。

$$\omega = \text{含水量} = \frac{W-W_0}{W_0} \times 100$$

25901

（1）收縮限度——土壤含多量之水分，乾後其體積必縮小；水量漸乾，則縮小之程度漸大；但有一限制，卽在此限制之內，水量乾後體積並不減小，此限度謂之縮限。此種縮限卽土壤體積收縮呈最小時所含之水量也。

$$收縮限度\ S=\omega-\frac{V-V_0}{W_0}\times 100$$

（2）收縮比率——其定義爲以乾土體積成分代表之體積變化除以乾土重量成分代表之收縮限度所失之含水，以公式表之，則爲：

$$R=\frac{\frac{V-V_0}{V_0}\times 100}{\omega-S}=\frac{W_0}{V_0}$$

（3）直線收縮——土壤在一定含水量時之線狀收縮，爲土壤含水由野外含水當量減至收縮限度時，所減少之一方向大小，用其原來之大小表示之：

$$直線收縮\ L.S.=100\left(1-\sqrt[3]{\frac{100}{C_f-100}}\right)$$

$$式中\ C=\frac{FME-S}{\frac{1}{G}+\frac{S}{100}}$$

$$G=\frac{1}{\frac{1}{R}-\frac{S}{100}}$$

四、野外含水當量——一滴水滴於光滑之土壤表面，而不立卽吸收，呈光潤狀態，此土之含水，以其烘乾土壤百分率表之，卽爲該土壤之野外含水當量。

$$野外含水當量\ F.M.E.=\frac{土壤含水重量}{烘乾土重量}\times 100$$

五、離心含水當量——將已飽和之土壤樣品，置於離心力大於地心引力一千倍之離心器中，一小時後土壤之含水量，以其烘乾土壤百分率表之，卽爲該土壤之離心含水當量。

離心含水當量可表示下列各種現象。

（1）可鑑別土壤是否具有滲透性。

（2）可斷定滲透性土壤所具毛細張力之大小。

（3）可鑑別具膨脹性之滲透質土壤與不具膨脹性之滲透質土壤。

土壤之分類

在若干年前，每一工程司均感到其所設計之工程構造，雖受硬性之公式控制外，其餘關及該工程結構所依附的基礎設計，確遭遇許多麻煩和困難與問題在公路工程上，更較爲常見，例如某公路能在任何氣候狀況下維持 light traffic，而又有某公路能在乾燥氣候中較爲穩定，但不能在潮濕氣候中維持穩定狀況，更有等公路之路基甚爲穩固，相反地，有等公路時會發生膨脹作用以致舖路材料分裂；有等公路可受壓力而縮至一高度性的密度，而另有則爲高度彈性而不受非天然的壓力影响者，此種種現像之發生，完全係基於土壤物理性質互相不同的緣故，今爲便於硏究起見，應將土壤分爲有系統之組別。在公路工程上欲畫分路基土壤則較爲容易，蓋路面上所承載之壓力不及房屋橋樑基礎所承載者之大，且路面之靜重較小，土壤之凝實，度亦較微，故預測路基土壤之行爲較預測建築物基礎下土壤之行爲爲易。

分組除以實驗得出之土壤常數及物理性質爲根據外，尙須參照以實地情形，方可得完善之效果，玆將均一路基土壤 (Uniform subgrade soil) 依最顯著之特性序列先後分爲八組如下：

A－1 組，　此組土壤含有粗細材料，級配均勻，並具有極佳之結合料 (binder)，內阻力及牽引力均高，缺乏有害的收縮性，膨脹作用及毛細管作用或彈性，故無論潮濕情形如何，載重時極爲穩定，爲最佳之路面，路基及填土材料。

A－2 組，　此組土壤與 A－1 組畧同，惟級配不甚均勻，結合料之品質較 A—1組畧遜，此組土壤可分爲脆性(含過份的沙)及塑性 (含過份黏土) 二種；脆性組之塑性較低，塑性組之塑性則較 A—1組爲高，脆性組需要毛細壓力供給其黏性，故潮溼時甚爲穩定，若雨水過多，路基存水不能蒸發，則甚易軟化。

A－3 組，　此組土壤僅含有粗料，而無結合料，載重時不穩定，但不受潮溼影响，不致發生霜凍作用，亦不致發生大量收縮或膨脹，對於柔性路面 (Flexiable Pavement) 或較薄之剛性路面 (Vigid Pavement) 之承載情形極佳。

A－4 組，　此組土壤係粉沙質，缺粗料及黏土，乾燥或微溼時極爲堅固，荷重除去後回彈甚微，在潮濕天氣極易吸收水分，隨卽失去其穩定性，此組土壤因霜凍作用，易使路面破裂，其承載力亦甚微。

A－5 組，　此組土壤與 A－4 組相似，但載重時極易發生形變，荷重除去後回彈亦甚大，彈性作用使路面在建築時不能壓實，建築後亦不易保持黏結狀況。

A－6 組，　此組土壤係黏土質，無粗料，但黏性甚强，乾燥時極爲穩定，載重時形變甚微，荷重除去後回彈亦甚微，惟極易吸收水份而發生膨脹，甚而至入液性狀態，浸入路面罅隙中，使路面破裂，或使塡土崩塌。

A－7 組，　此組土壤與 A－6 組相似，惟含黏土之量較多，載重時極易發生形變，荷重除去後，回彈性亦甚大，與 A－5 組相同，潮溼時所起之體積變化較 A—6 組尤甚。

A－8 組，　此組土壤含有多量有機質之極軟土壤，若不預先加以壓實，或更換以其他土壤，不能支承斷面。

(注意) 以上所述之 1，組土壤均屬於均一支承 (Uniform Support)。

各組均一路基土壤與物理性之關係可用下表簡單表明之

物理性 組別	內阻力	黏性	收縮與膨脹	毛細作用	彈性
A－1	高	高	無損害性	無損害性	無損害性
A－2	在某種情況下高	在某種情況下高	有時具損害性	有時具損害性	有時具損害性
A－3	高	無		無損害性	無損害性
A－4	高低不定	甚微		作用重要	無
A－5	高低不定	甚微		作用重要	甚強
A－6	低	含水十時甚高	恆具損害性		無
A－7	低	含水小時甚高	恆具損害性		有
A－8	低	低		具有損害性	具有損害性

各組土壤鑑別，乃可備助於土壤鑑別圖 (Soil Vdentification Chart)，此圖係表示液性限度與塑性限度，收縮限度，野外含水當量，及離心含水當量等四種物理性之關係，故倘已知路基土壤之行爲，則不難可用圖鑑別出土壤應屬於何組，或已知路基土壤係屬於某組，則亦不難可用圖預測其行爲，此鑑別圖更可用爲研究加物理或化學混料於土壤中所生之變化，所獲效果如何!? (鑑別圖畧)

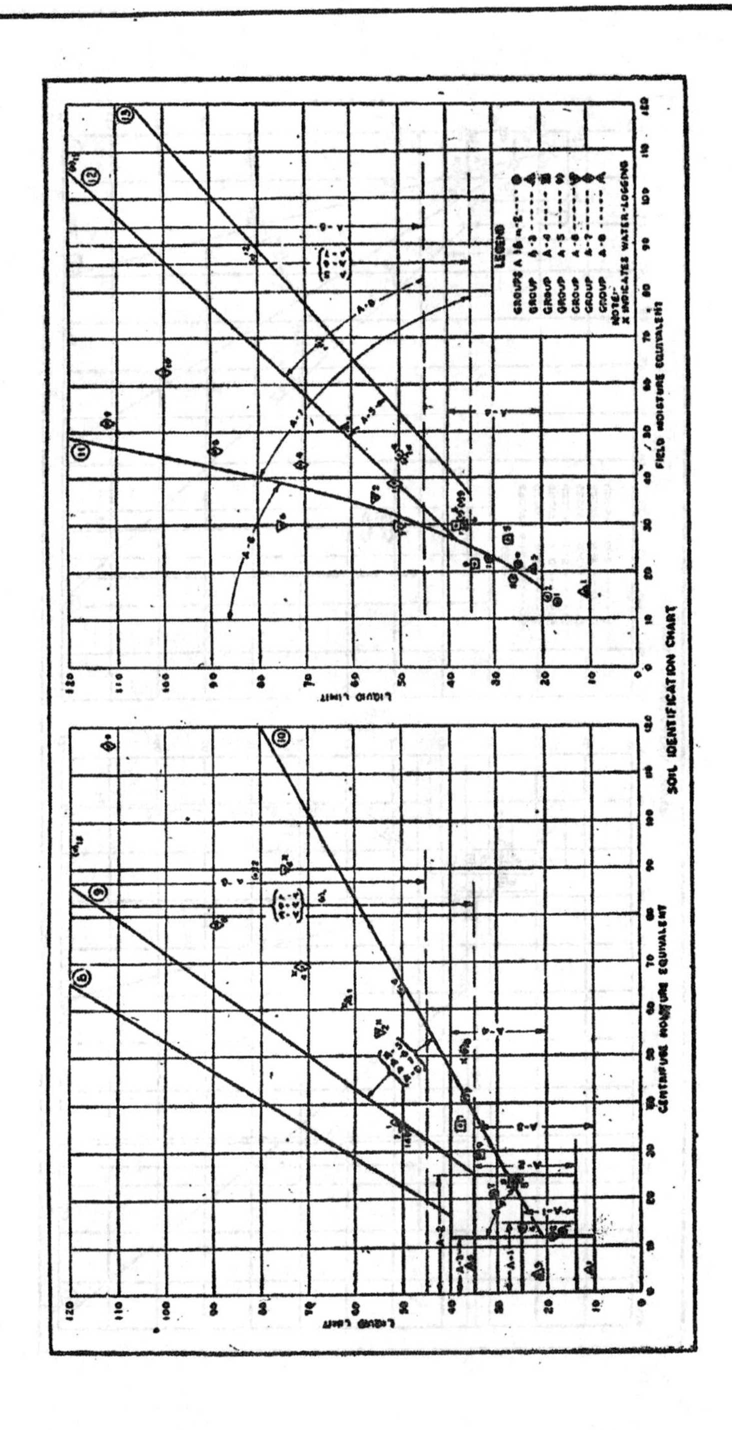
SOIL IDENTIFICATION CHART
LIQUID LIMIT
CENTRIFUGE MOISTURE EQUIVALENT
FIELD MOISTURE EQUIVALENT
LEGEND

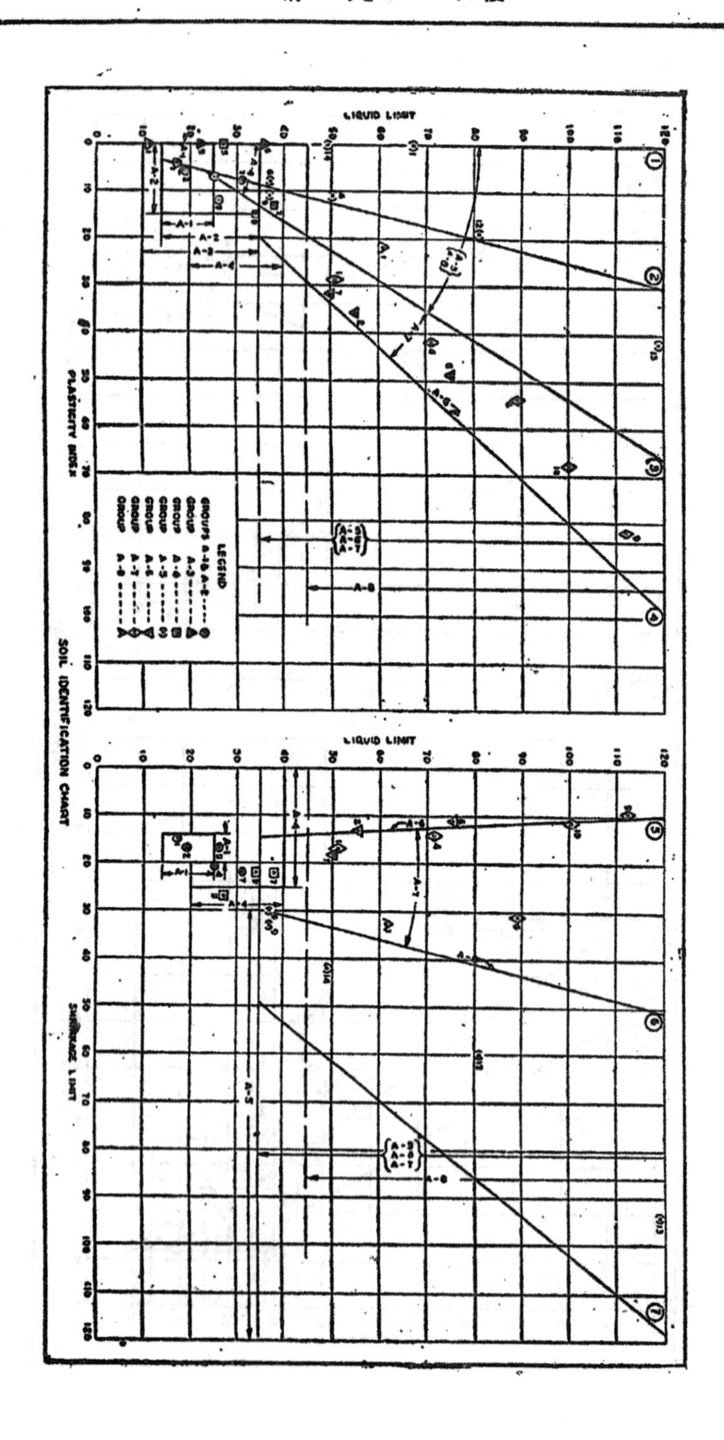
LIQUID LIMIT
PLASTICITY INDEX
SOIL IDENTIFICATION CHART
SHRINKAGE LIMIT
LEGEND

茅竹與茶桿竹之抗張試驗

——劉益信——

I 緒言

本試驗之目的爲欲知竹能否代替鋼根在建築三合土上應用，此點爲本文作者之論文工作，現尚未完成暫未能發表，不過在此工作之先必要得知竹性質及强度等等。

數千年前，東方人已開始利用竹做紙、筷、籮、筆、弓、杖等物 以後更利用它特優之彈力，負扭力，抗抵壓力、張力、柔韌、色澤、身軀、節、空洞等之特性做笛、烟管、傢俬、竹繩。美國更研究做人做絹絲之材料。

1912 年有留美學生周厚坤氏，於麻省理工學院研究竹之强度爲其論文工作，取材於浙江省出產之各種竹材，1915 年已見日人楯山淺次郎有竹筋試驗報告之發表，1934 年我國政府於第七次國際道路會議報告書中對竹筋三合土路面與竹之各種特性均有具體報告；其後戰事爆發，繼起乏人，獨日人努力研究在1937—1939年間發表研究成績甚多。

竹之可取者爲其抗張强度與韌性，抗張强度遠駕乎普通鑄鐵之上，對上等鑄鐵亦可比擬，對建築之軟鋼及半軟鋼則約及一半比木材則有 2 至 4 倍以上强度。總之竹爲亞洲之特產，在我國產量尤豐，以其能有如此强大張力，且價格低廉，長度相當，實爲一最有希望之建築材料也，希吾國學者努力研究之。

II 種類及產地

竹之種類甚繁，約三十餘屬二百餘種，盛產於熱帶亞熱帶及溫帶，尤以亞洲著名。我國產竹多在珠江長江流域一帶。以廣東、浙江、四川、台灣著名。世界竹之分類甚少專書記載。故只能介紹廣東主要者十五種如下：

名稱	學名	形態			
		高 (公尺)	第4節壁厚 (公分)	第4節直徑 (公分)	最長節間 (公分)
I 簕竹屬	Bambusa Schreber				
1.石竹	B. lapedis McClure	7.5	2.0	15	37
2.鶴南竹	B. Sinospinosa McClure	8.0	1.7	14	37
3.坭竹	B. eutuldoides Var basistriata McClure	5.5	0.9	12	41
4.水竹	B. sp.	5.5	0.9	13	47

5.臺　竹	B. Vulgaris var. striats Gamble	8.0	1.5	13	34
6.撐篙竹	B. perrariabilis McClure	5.5	0.9	11	33
7.青皮竹	B. textilis var. maculata McClure	3.5	0.5	8	35
8.木　竹	B. rutila McClure	7.0	2.2	13	43
9.觀音竹	B. sp.	5.0	0.3	13	57
II 慈竹屬	Sinocalamus McClure				
10.大頭甜竹	S. Beecheyana (Munro) McClure	9.0	2.2	12	40
11.吊絲球竹	S. Beecheyana var. pubescens Li	5.5	0.8	12	43
12.大葉甜竹	S. latiflorus (Munro) McClure	8.0	0.9	10	33
III 單竹屬	Lingnaina McClure				
13.單　竹	L. Chungii McClure var. cerosissima McClure	5.5	0.7	11	90
14.白粉單竹	L. Chungii McClure	5.5	0.5	8	70
IV 莎簕竹屬	Schizostachyum Nees				
15.羅　竹	Schizostachyum Funghomii McClure	6.0	0.4	10	70

以上爲4屬11種，4變種。

作者所試茅竹與茶桿竹：

名稱	學名	形態					產地
		高	直徑		壁厚		
		(公尺)	最大(公分)	最小(公分)	最大(公分)	最小(公分)	
茅竹	Phyllostachys adulis, Aetc, Riv.	7	15.0	4.0	1.5	0.3	廣西，桂林，東江一帶。
茶桿竹	Asundinasia amabilis, McClure	4	4.0	1.2	0.7	0.3	廣東省，廣寧，懷集，北江一帶。

III　抗張強度

1. 試材之部位及取材方法

從余仲奎氏所著「川產楠竹性質之研究」一書，內關於竹材抗張强度與部位之關係得以下結論：

a. 部位——同一節內越近外表皮者越大

b. 高向——離地越遠者越大

由此兩點作者爲欲得一平均實用數字，故取材根據以下兩點原則：

a. 部位——竹材之外表爲最大抗張强度之部份，然有等竹之壁厚甚薄，約 3 公厘，則甚難取其最外 10 %或 20%，雖然大概已知外表 40 %變化較大近內部之 60 % 則無甚分別，爲求實用與將來施工容易宜取 50 %。

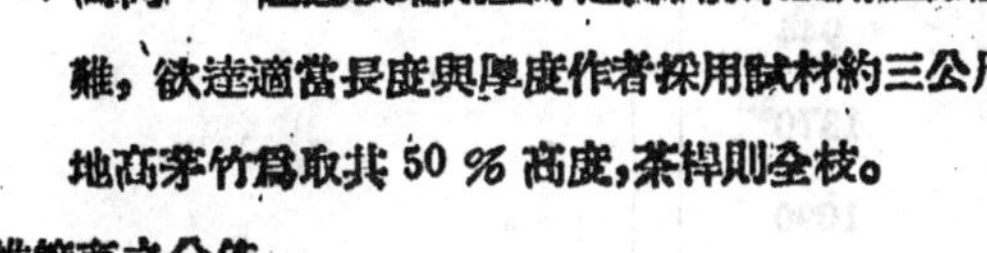

b. 高向——越近頂端則壁厚越薄，將來實用上亦越困難，欲達適當長度與厚度作者採用試材約三公尺離地高茅竹爲取其 50 % 高度，茶桿則全枝。

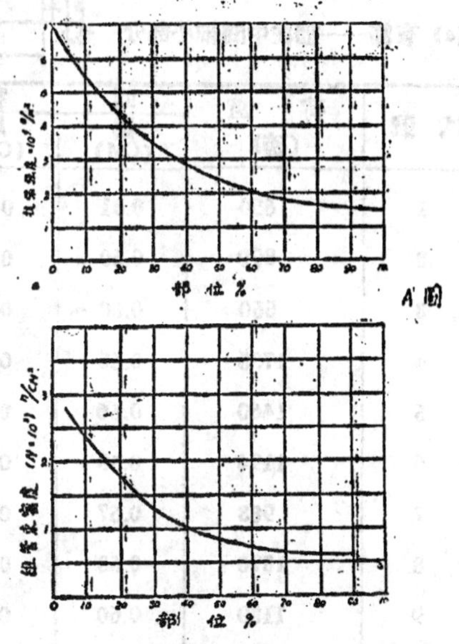

2. 維管束之分佈

余氏實驗楠竹結果：

$$N_{0-45\%}t=330\ (10^{-1.2t})$$

$$N_{45\%}t_{-100\%}t=60\ t^{-0.6}$$

N：爲單位面積內之維管束數

t：壁厚

從上圖得知 維管束之密度與抗張强度成正比。

3. 試驗儀器

用美國費城製之 25 噸螺旋力重複桿試力機 Screw-Power compound-lever Testing Machine (Universal Testing Machive) Made by Riehle. Bros. Testing Machine Co. U. S. A.

試驗之附帶零件如圖 1.

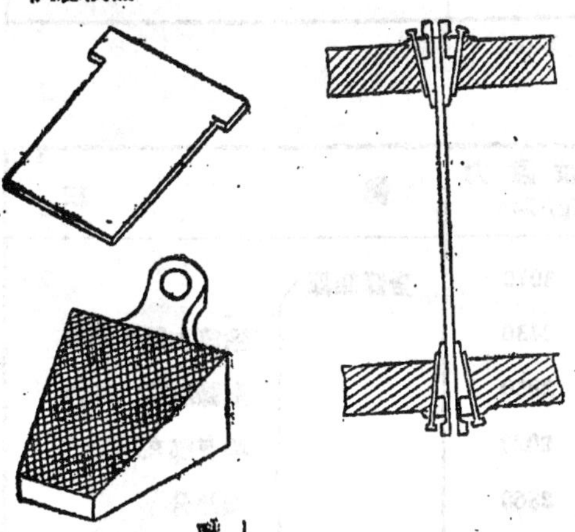

圖 1

VI. 試驗成績

茅竹

A. 試驗日期：2 月 28 日 1949

風乾日期：4 日

試材形狀:如圖 3

安置方式:用夾鉗內外表皮

(a) 有節——節在中間細小部分

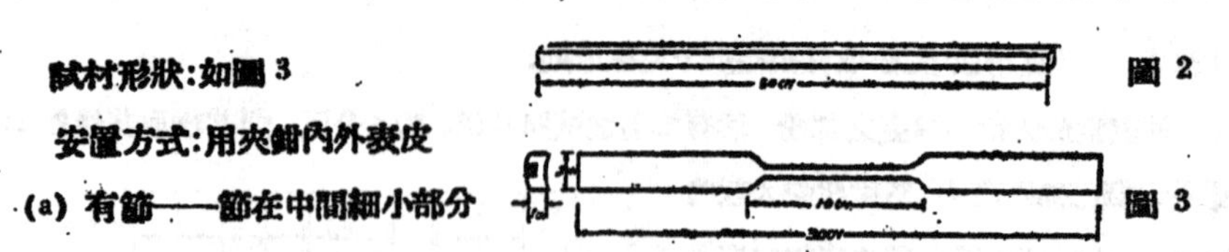

圖 2

圖 3

試 號	荷 重 (磅)	面 積		單 位 應 力 Kg/CM²	備 註
		濶 (CM)	厚 (CM)		
1	650	0.61	0.51	944	
2	890	0.60	0.50	1370	破壞在節
3	560	0.60	0.61	1090	破壞在節
4	1735	0.65	0.70	1730	破壞在節
5	1460	0.66	0.60	1670	破壞在節
6	1194	0.58	0 60	1550	破壞在節
7	983	0.67	0.67	990	破壞在節
8	1570	0.60	0.64	1850	破壞在節
9	1130	0.60	0.60	1420	破壞在節
平 均				1402	

(b) 無節——中間部份無節,兩邊或有節

試 號	荷 重 (磅)	面 積		單 位 應 力 Kg/CM²	備 註
		濶 (CM)	厚 (CM)		
1	1800	0.62	0.43	3070	全條無節
2	1605	0.65	0.46	2430	除第1第8兩號全條無節之外餘均有節在兩邊大頭部分破壞在節不論該節離任何一末端2至8 CM.
3	1870	0.61	0.47	2950	
4	1200	0.63	0.43	2020	
5	1630	0.61	0.47	2560	
6	1525	0.60	0.45	2560	
7	1235	0.61	0.41	2240	
8	1640	0.61	0.54	2260	全條無節
9	1830	0.61	0.45	2750	
平 均				2540	

B. 試驗日期：3月3日1949

風乾日期：7日

試材形狀：全長 30 CM. 等橫截面面積如圖

安置方式：用夾鉗內外表皮

(a) 有節——節在中間

試號	荷重(磅)	面積 闊(CM)	面積 厚(CM)	單位應力 Kg/CM²	備註
1	1380	0.60	0.73	1425	其餘在節口斷
2	1700	0.70	0.65	1925	
3	1675	0.67	0.60	1915	鉗斷
4	1685	0.70	0.70	1560	
5	1700	0.77	0.55	1810	鉗斷
6	875	0.73	0.67	810	
7	580	0.70	0.48	784	
8	1600	0.70	0.74	1400	
9	1320	0.68	0.56	1575	
平均				1467	

(b) 無節——全條無節

試號	荷重(磅)	面積 闊(CM)	面積 厚(CM)	單位應力 Kg/CM²	備註
1	1355	0.67	0.44	2080	鉗斷
2	1955	0.75	0.58	2040	
3	2023	0.71	0.53	2440	
4	1920	0.66	0.50	2640	鉗斷
5	1300	0.47	0.63	1990	
6	1975	0.68	0.55	2390	鉗斷
7	1720	0.70	0.63	1770	
平均				2193	

C. 試驗日期：3 月 25 日 1949

風乾日期：29 日

試材形狀：如圖 3

安置方式：夾鉗內外表皮

(a) 有節——節在中間

試號	荷重 (磅)	面積 闊 (CM)	面積 厚 (CM)	單位應力 Kg/CM²	備註
1	1335	0.65	0.60	1550	
2	1610	0.67	0.77	1415	
3	1110	0.63	0.59	1350	
4	1990	0.65	0.65	2130	
5	2060	0.68	0.64	2150	
6	1260	0.64	0.56	1590	
7	1330	0.71	0.53	1600	
8	1510	0.64	0.62	1730	
9	1445	0.62	0.61	1730	
10	1605	0.67	0.68	1590	鉗斷
11	1460	0.67	0.63	1570	
12	1645	0.65	0.51	2250	其餘在節口斷
13	1775	0.62	0.53	2440	
14	1470	0.63	0.55	1930	
15	1915	0.70	0.57	2170	
16	1460	0.60	0.53	2080	鉗斷
17	1820	0 69	0.44	2720	鉗斷
18	1100	0.57	0.40	2190	
19	1435	0.67	0.46	2100	鉗斷
20	2005	0.69	0.57	2300	
21	1135	0.62	0.51	1630	
22	1355	0.65	0.58	1640	
平均				1903	

(b) 無節——全條無節

安置方式:與 (a) 不同,只鉗兩側面,不損壞外表。

試號	荷重 (磅)	面積 闊 (CM)	面積 厚 (CM)	單位應力 Kg/CM²	備註
1	1980	0.65	0.52	2660	全部斷在中間細少部分
2	1925	0.66	0.46	2870	
3	1780	0.61	0.55	2400	
4	1630	0.60	0.66	1860	
5	1320	0.62	0.42	2300	
6	1430	0.64	0.49	2070	
7	1675	0.60	0.51	2480	
平均				2377	

茶　桿　竹

A. 試驗日期:2 月 28 日 1949、

風乾日期:4 日

試材形狀:圖 3

安置方式:夾鉗內外表皮

(a) 有節——在中間

試號	荷重 (磅)	面積 闊 (CM)	面積 厚 (CM)	單位應力 Kg/CM²	備註
1	970	0.54	0.36	2250	斷在節
2	780	0.56	0.40	1580	
3	895	0.57	0.34	2090	
平均				1973	

(b) 無節——只中間細小部分無節,兩邊大頭部分或有節。

試號	荷重(磅)	面積 闊(CM)	厚(CM)	單位應力 Kg/CM²	備註	
1	1035	0.60	0.33	2040		其餘斷在節口,節在大頭部分上。
2	1405	0.54	0.34	3310		
3	1705	0.52	0.35	4250	鉗斷	
4	1210	0.53	0.37	2800	鉗斷	
5	960	0.47	0.32	2890	全條無節	
6	1515	0.55	0.35	3560		
7	1045	0.57	0.35	3440		
8	1390	0.56	0.32	3530	鉗斷(全條無節)	
9	1290	0.54	0.40	2420		
10	1270	0.55	0.30	3420		
平均				3196		

B. 試驗日期:3月5日1949

風乾日期:9日

試材形狀:如圖2

安置方式:鉗內外表皮,與鉗兩側兩種方式。

(a) 有節

試號	荷重(磅)	面積 闊(CM)	厚(CM)	單位應力 KgC/M²	備註
	鉗內外表皮				
1	830	0.44	0.37	2300	
2	815	0.42	0.40	2200	
3	570	0.38	0.30	2270	
4	690	0.50	0.34	1840	斷鉗
5	725	0.45	0.30	2440	鉗斷

6	1025	0.48	0.27	3560		
7	815	0.55	0.30	2240		
8	530	0.40	0.32	1870	鉗斷	
9	520	0.38	0.50	1980	鉗斷	其餘均斷在節部分
10	890	0.37	0.30	3640		
11	800	0.48	0.40	1890	鉗斷	
12	610	0.45	0.30	2050	鉗斷	
13	550	0.37	0.32	2100		
14	670	0.47	0.30	2160		
15	795	0.43	0.32	2610		
16	765	0.50	0.41	1690		
17	875	0.35	0.30	3770		

平均　2389

鉗兩側面

1	705	0.38	0.55	1530		
2	735	0.50	0.23	2890		
3	720	0.45	0.28	2590		
4	560	0.40	0.28	2270		
5	585	0.36	0.57	1180	鉗斷	
6	610	0.32	0.53	1625		其餘均在節斷
7	740	0.42	0.48	1660		
8	500	0.30	0.46	1640	鉗斷	
9	510	0.27	0.32	2670		
10	770	0.40	0.34	2560		
11	305	0.30	0.30	1540		

平均　2140

兩種鉗法總平均 2242 Kg/CM²

25915

(b) 無節——全條無節

安置方式：鉗內外表皮

試號	荷重 (磅)	面積 濶 (CM)	面積 厚 (CM)	單位應力 Kg/CM²	備註
1	545	0.38	0.23	2830	鉗斷
2	700	0.39	0.22	3660	
3	730	0.37	0.28	3180	
4	615	0.35	0.23	3460	
5	620	0.34	0.25	3300	
6	770	0.30	0.34	3420	
7	930	0.53	0.26	3250	
8	550	0.38	0.27	2420	
9	550	0.26	0.35	2740	
10	890	0.52	0.27	2880	
11	625	0.36	0.25	3150	
12	760	0.36	0.40	2390	
13	690	0.34	0.23	4000	
14	650	0.37	0.26	3060	
15	445	0.20	0.24	4170	試材形狀如圖 3
16	325	0.17	0.24	3620	
17	243	0.26	0.25	1690	
18	530	0.23	0.26	4020	
平均				3180	

C. 試驗日期：3 月 25 日 1949

風乾日期：29日

試材形狀：如圖 3

安置方式：鉗內外兩面

25916

(a) 有節——節在中間

試號	荷重 (磅)	面積 闊 (CM)	面積 厚 (CM)	單位應力 Kg/CM²	備註
1	775	0.56	0.43	1485	不斷在節
2	960	0.49	0.48	1850	其餘斷在節
3	1080	0.59	0.60	1380	
4	1220	0.67	0.51	1620	
5	1350	0.64	0.50	1910	
6	1160	0.60	0.50	1750	
7	1700	0.67	0.52	2210	
8	855	0.53	0.51	1430	
9	930	0.47	0.44	2030	
10	1620	0.63	0.60	1940	
11	775	0.56	0.44	1425	
12	785	0.66	0.49	1100	
平均				1686	

(b) 無節

試號	荷重 (磅)	面積 闊 (CM)	面積 厚 (CM)	單位應力 Kg/CM²	備註
1	1615	0.63	0.39	2940	
2	1400	0.64	0.41	2420	鉗斷
3	1395	0.55	0.37	3090	鉗斷
4	1905	0.62	0.44	3180	鉗斷
5	1485	0.63	0.41	2610	
6	1705	0.62	0.47	2640	鉗斷
7	1550	0.64	0.40	2740	
8	1665	0.64	0.34	3440	
9	1900	0.63	0.37	3690	鉗斷
10	960	0.53	0.40	2010	鉗斷
平均				2876	

簡列以下結果如下：

茅竹

風乾日期	4	7	29	平均	
有節	1402(9)	1467(9)	1903(22)	1591	1980 Kg/CM²
無節	2540(9)	2193(7)	2377(7)	2370	

茶桿竹

風乾日期	4	9	29	平均	
有節	1970(3)	2242(28)	1686(12)	1967	2525 Kg/CM²
無節	3196(9)	3180(18)	2876(10)	3084	

註釋：1. 表上數字後面括號內之數目爲試驗次數以表示該數字爲從若干數目內所取之平均值。

2. 鉗斷：表示試品因鐵夾中之齒，鉗斷外表皮以致破壞。

V 試驗結果之比較

	茅竹			茶桿竹		
	有節	無節	平均	有節	無節	平均
作者	1591(40)	2370(23)	1980	1967(43)	3084(73)	2525
周傚正	1843(4)	2143(4)	1993	2108(4)	3160(4)	2634

VI 粵產各種竹之強度試驗

竹之使人最感興趣者爲其抗張强度，故研究者特多，現介紹其他學者之結果以作研究與參攷之用。

	周傚正 Kg/CM²			沈蘭根 Kg/CM²		
	有節	無節	平均	有節	無節	平均
桹竹	2972	3269	3120			
茶桿竹	2108	3160	2634			
水竹	2359	2534	2447	2060	2820	2440

撑篙竹	2110	2346	2229	1220	2460	1840
茅竹	1843	2143	1993			
箣南竹	1613	2233	1924	1980	2090	2035
篠竹	1641	2026	1834	770	2290	1550
大頭竹	1279	1889	1584			
青皮竹	1316	1066	1491	1690	2710	2200
黄甘竹	1163	1665	1414			
石竹				1900	3130	2515
坭竹				2080	3580	2830
耄竹				1320	2680	2000
木竹				800	2920	1860
觀音竹				1240	2630	1935
大頭甜竹				900	1930	1415
吊絲球竹				1990	2190	2090
大葉甜竹				370	1100	735
白粉單竹				650	2560	1605
羅竹				2110	3220	2665

IIV　竹與鐵材及木材之比較

材種	名稱	應引力 Kg/Cm²	
		外表皮 10 %	平均
竹材	觀音竹	4115	1935
	坭竹	4110	2830
	青皮竹	3875	2200
	茶桿竹		2634
	榥竹		3120

25919

竹材	川產楠竹 Phyllostachys edulis (Carr) H. de Lehaie		2620(無節)
	川產慈竹 Sinocalamus Affinis (Reudle) MacClure		3390(無節)
	川產斑竹		4340(無節)
鐵材 (土木工程計算圖表)	鑄鐵		1406
	鍛鐵		3515
	中級鋼		4218
	鎳鋼(35%鎳)		5976
木材 (土木工程計算圖表)	木構		1606
	楠木		1557
	麻櫟		1276
	野櫻桃		1144

VIII 結論

1. 節爲竹類之最弱部分。

2. 無節試材之抗張强度必大於有節者，茅竹與茶桿兩種約爲 1.6 與 1 之比。

3. 用有齒夾鉗竹應加改良，因試品多爲鉗斷並非受强大牽引力而破壞，佘氏試楠竹時用瑞士 Alfred J. Amsler Co. 出品之四公噸手搖式試驗機，而特設一附件以鉗竹以免滑下，該附件爲璽鉗形中加兩鐵塊，不拘任何形狀求與該鉗密合，且有平滑面以夾竹，兩旁更以兩口以上螺釘，通過該鉗以推鐵塊用以緊壓試品，据佘氏言用此法亦無鉗斷之弊，因兩端特大中間試驗部分細，可免滑脫之虞，不過每次最小要轉緊四口以上螺釘，未免廢時耳。

4. 試材形狀以圖 3 爲佳，如兩端大頭部份濶比厚有 2 比 1，則仍可用現在之有齒鉗設備鉗兩側面，如過濶則防扭轉。

5. 無節試材之抗張强度大以川產斑竹其次以坭竹與慈竹。

6. 有節無節平均最大以根竹其次以坭竹與茶桿竹。

參 攷 書

1. 沈蘭根:國防部科學研究發展報告第 31—1 號"粤產竹材十五種性質之研究"
2. 余仲奎:沈蘭根:航空委員會研究院第十號報告"川產楠竹性質之研究"
3. 趙仰夫譯:竹林培養法
4. 周儀正:廣東大學論文"竹筋三土合之研究"
5. 楊文淵:土木工程計算圖表

藤筋三合土

(Native cane Reinforcement in Concrete)

——吳秉俠——

譯自 Vol. 13, No 8. August 1943, civil Engineeing.

"混凝土之筋材，目前除鋼條被廣泛採用外（有等國家亦選用鑄鐵〔Wrought Iron〕），究否另有其他物質能替代鋼筋？現日尚未有一確切之解答，唯近三四十年來，各國對竹筋曾作概括之研究，尤以日本最有成就。至我國對是項研究，亦遠於三四十年前。然仍未有大規模應用於工程建築上，

除竹筋對替代鋼筋較有希望外，六年前美國劍臣學院之工程研究站（Clemson College Engineering Experiment Station）曾作藤筋三合土之試驗。其報告雖發表於六年前，然其足資參考之歷程實可達數十年以後，加之譯者有鑒於本系現正以竹筋三合土爲研究工作中之一主題，故特不嫌其發表過久而轉譯之，以供與於此項問題之同學參考也。………譯者誌"

藤之物理特性

抗張强度	=25,000磅 每平方吋
彈性係數	= 2,000,000 至 2,5000,000
抗滑强度	
當藤之較大直徑一端藏於三合土內	= 140 磅每平方吋
當藤之較小直徑一端藏於三合土內	= 60 磅每平方吋
平均抗滑强度	= 100 磅每平方吋
體積收縮率	= 10%
直徑烘乾時之收縮率	= 3,2%

劍臣學院之工程研究站對於南加老拉娜洲之藤（South Carlina river cane）曾作準確之試驗，以求其在混凝土方面之效能。其試驗之目的及綱要如下：

（一）藤之物理特性。

（二）藤筋混凝土樑及三合土板之荷重限度。

（三）尋求有關設計藤筋混凝土之紀錄報告。

對於藤筋擴張力之大小，共作以下三種不同方法之試驗，結果各方法所得之論斷，並無過大差別，其試驗方法爲：

（一）將各藤條自表皮下加以分層（用表皮部份——譯者）每標本之長八英吋。標本之尾部則保持完整狀態，以備鉗緊於重力機上。平均之試驗斷面積爲十六分三吋。

（二）將各薄片分層膠合一起而成一二分一吋直徑之試材，用以鉗緊於試機上之尾部，則爲四分三吋。

（三）此法所紀錄之藤筋擴張力乃得自因欲決定藤與混凝土間之抗滑力時而施諸藤筋之拉力，於此法中有等試材，因半徑過小及埋藏於混凝土中之部份過長，故而每先抗滑力之失敗而斷於張力作用。

抗張力試驗結果

由上述各種方法與不同標本中所求得之抗張力，各有效面積之平均值約爲 25,000 磅每平方吋，各個別試材之所得值，最小爲 13,000 磅每平方吋，最大可達 30,000 磅每平方吋，然大多數之結果皆甚接近。各試材中有新近採擇者，亦有經長久之風乾至彼此之重量相等者。試驗結果，顯示風乾對藤之擴張力無大作用。

於試驗後，得知藤之節環乃最弱之擴張部份。任何之有節藤筋標本，一經拉力作用，即每於節環部份斷裂。而應注意者，即其斷拆非發生於節之正中，而乃於節之頭尾部位，其破壞特徵爲附近之表皮受拉斷而成鋸齒狀也。

於擴張試驗時，對藤之伸長（Elongation）亦作精密之記錄，以爲計算其彈性係數之用。結果求得藤之彈性係數介於 2,000,000 至 2,500,000 之間。而青藤與乾藤應無特殊顯要之差別。

對於抗滑力之試驗，乃將一長二呎之試材藏於 6×12 吋之圓柱形混凝土內。試材有青藤及乾藤之分。其試法爲將突出之藤筋之尾部緊封於試機上，然後紀錄抗滑之重量。結果各試材間甚爲符合，其平均值爲：

	抗滑力
（甲）青藤	
（一）當藤之較大直經一端藏於混凝土中	140 磅/平方吋
（二）當藤之較小直經一端藏於混凝土中	95 磅/平方吋
（乙）乾藤	
（一）當藤之較大直經一端藏於混凝土中	190 磅/平方吋
（二）當藤之較小直經一端藏於混凝土中	150 磅/平方吋
（丙）青藤藏於混凝土中直至藤筋乾後而失去水份時試驗之	
（一）當藤之較大直經一端藏於混凝土中	110 磅/平方吋

(二)當藤之較小直經一端藏於混凝土中 70 磅/平方吋

共需注意者即遇有藤節埋藏於三合土內時,則抗滑强度可增加。

至於藤因風乾後(抑其他方法以促其乾枯——譯者注),其直經與體積之收縮率之研究,一部份試材乃受110°C之高熱烘乾,及至彼此之重量相等而後已,由十二條試材所得之結果,其平均體積收縮約爲10%,而直經收縮率則爲3.2%左近。對於上述之體積收縮率,似感嚴重,然吾人若査悉其體積之收縮部份乃爲藤條表皮下之疏空維管束區域,(抗張作用極弱部份——譯者注)即感其嚴重性下降矣。又藤之內圓周有一薄層之細胞髓存在,一遇烘乾時,此層細胞髓即行收縮,因而形成其直徑之3.2%收縮率也。然此項直徑變率現象,對本試驗無嚴重之影响。

其次對藤筋混凝土樑亦作多次試驗,樑之大小爲長四又二分一呎寬四吋,深七吋。於此等樑中,其筋材分爲靑藤,乾藤,及層藤。結果除因藤筋之安置方法不同而畧有差別外,大致上成績,皆相吻合。相等斷面積之藤筋於同様荷重下試驗結果不論藤筋爲靑藤乾藤抑層藤,破壞現象皆一致。遇有藤筋與淨三合土之比例小於3%至4%時,則此等樑每每發生偏差(Deflection),通常情形下,樑之荷重量遂減少。

試驗結果,所有樑皆爲剪力裂縫(Shear Crack)所破壞,須有抗剪設備亦然。有等情形爲剪力裂縫發生於藤筋之上部,繼卽迅速以水平方向發展而及於整條樑長,他種情形爲斜向裂縫(Diagonal crack)之發生,其特徵適與鋼筋混凝土樑同。

通常此等樑之受破壞,全因受集中荷重壓力之混凝土部份斷裂所致,而此等情況每發生於斜向裂隙出現而及於整條樑深之後也。試驗中全未發覺有等試材因拉力關係而至樑之筋材下部之混凝土先行裂開,有若鋼筋樑者,此故因樑中之藤筋比率較高時而鞏固下層之三合土之結果也。

藤筋混凝土樑之試驗

當藤筋以一吋之淨空滿佈於樑中(寬與深)時,則其荷重値減低。樑之破壞乃由於平行軸(Neutral axis)左近發生水平剪力裂縫所致,此固乃因混凝土本身在平行軸左近之有效抗壓面積爲藤筋所減少,而水平剪力適於該處爲最大,故有此破壞特徵焉。

如用乾藤作筋材,則對用水以促成混凝土之凝固應特別小心,若樑模成彼而以水浸淋之以求其達於最高强度時,則此等樑每因內中之藤筋吸收水份膨脹而引起裂隙,換言之乾藤筋必需另有防水設備,否則實不利於作筋材應用也。若於乾藤上漆以一薄層之臘膏,則上述之吸收水份問題當迎刃而解。至用靑藤作筋材,則不論混凝土於何等情形下進行凝結,皆無吸收水份至膨脹之弊。

另一頗堪注意之問題,卽藤筋部份突出之小枝節,如不加以削剝整理而用作筋材時,則此等小枝節對抗剪方面有所增效。對於藤筋之安置,應設法曳張之,以儘量放置於樑之底部爲佳。

三合土板層之試驗

為求完整之板層之實際荷重情形，乃製成兩塊一以完整之藤條為筋材及另一以徑向一半分層之藤條為筋材之三合土板，其淨距為 $11\frac{1}{2}$ 呎，寬度 4 呎，深則為 5 吋，藤筋之平均直徑為 $\frac{1}{2}$ 吋。筋材緣寬距以一吋之淨空排列，底部與筋材之淨空亦為一吋。故板層之有效深度為 4 吋。

第一塊三合土板（以完整之藤條作筋材者）內之藤筋斷面積，每尺板寬為 1.44 平方吋所有筋材皆新鮮者，其採擇期間僅先於製橫一日而已。所用之水泥之七日試驗强度為 2,750 磅每平方吋。

此一板層於 4320 磅之集中荷重試驗下未能顯明發現因抗滑力弱而起之破壞蹟象，惟有三條裂縫出現於三合土板上，最初之裂隙生於正中，其他二條則發現於中點兩側約 $1\frac{1}{2}$ 呎左近，板層之所以破壞，乃因正中之裂隙積漸發展所促成。至試驗所生之偏差（Deflection）乃因筋藤受張力而伸長所致，絕非因抗滑失敗而滑動做成之現象也。

另第二塊三合土除所用之筋材為完整之藤條之一半外（徑向分為二等分層——譯者註），其他製作方法皆與第一塊板層同。故筋材之面積，只及第一塊筋材之一半。至製造技術方面，則較第一塊板層為完密，各藤節皆為混凝土緊密包藏。試驗結果 4000 磅之集中荷重方促成藤筋受張力而破壞。

上述兩三合土板之破壞，皆因屈撓（Bending）過大所致。第二塊板層於試驗時，其最大偏差達 14 吋此項偏差發生而荷重未能增加時，則似為抗滑力失敗而起之現象。惟板層破壞後檢查結果，又發覺此項猜測並無明顯蹟象足以根據。又於試驗期間，在板層之支點附近，並無斜向裂縫產生，而板層底部之混凝土亦無裂隙。再者雖荷重結果計算到混凝土所受之壓力超過其本身之容許限度，然板層終無因壓力而呈破壞之現象發生。

其足以注意者為板層於荷重增加而未破壞之前，通常可有 6 至 8 吋之偏差，其情況適與藤筋樑相彷。

對於抗滑强度問題，在藤筋混凝土中實不足重視之，蓋藤筋混凝土之破壞每因三合土本身未能承受高壓所致而已。而施於藤筋與混凝土間之接觸面積之重力實甚低小。通常在板層荷重達 4000 磅每平方吋時，抗滑力多未能超出 20 至 20 磅每平方吋。於所有試驗中，全未發覺藤筋三合土因抗滑失敗而破壞先於因剪力或壓力對混凝土所促成之破壞者。

於多數樑之試驗中，發覺剪力乃促成破壞之一大因素。當單位水平剪力超過 70 至 80 磅每方吋時，則剪力裂隙（Shear crack）卽行發生，如有抗剪設備，則抗剪力可增至 100—125 磅每平方吋。

對於上述各種試驗之結果，未必能全然作為此等藤筋混凝土之結論，惟由上列之紀錄吾人可値此得知一藤筋混凝土樑可能設計，至當其筋材之斷面積較平行設計（Balance Design）之鋼筋樑內之鋼筋面積大十倍時，則此兩樑之荷重可相等。至藤作筋材能否保持耐久之問題，雖以前曾有學者研究，結果認不為難於堅持久遠，然本實驗仍未能將此作一決定論也。

橋墩受冲擦之模型試驗

鄒堅柏

譯自 Feb. 1949 Civil Engineering C. J. Posey 原作

"橋樑倒塌之原因,除因建造之不當外,橋墩及橋基受水流之冲蝕及挖深有以致之,然欲自混濁之各層透視河床,企可得實際冲擦情形以爲設計防範之資料殆不可得,故非有模型作研究之助不爲功。本文作者美國愛歐華大學教授普斯於去年夏曾邀集多人在洛機山水力實驗室作關于橋墩冲擦之模型實驗,本文爲其實驗經過及所得之資料,普氏實驗在橋樑界尙屬創舉,故特允爲介紹,企引起讀者對此實驗之興趣而作更廣泛更多及更加深之研究,此譯者之意也——譯者誌。"

橋墩及橋基周圍之受冲擦與挖深,爲橋樑倒塌之最大原因,故橋樑工程師必須具有豐富之渠道及水流冲擦智識。更多之實驗工作以得更多之設計資料殆爲必要。用作實驗之橋墩模型必須透明及中空,俾得在實驗進行中隨時得悉冲擦之深度,因如在截斷水流後始讀其度,則其冲擦狀況必因水流停止及含沙沉澱而改變也,用作實驗之水不以結晶體者爲佳,此種裝置不特可便利量出冲擦深度,而水床之剝蝕及含沙之走動亦可瞭如指掌也。附圖所示之橋墩模型爲一有刻度之圓玻璃筒,其容量爲一公升,現察水內情況可自一懸于銅線下端與水平傾斜45°之小鏡爲之助,該鏡可自由牽引銅線使之上下以觀察任意深度,墩橋四周之冲擦深度(即圓筒刻度)可自鏡之反射明確讀出,而橋墩附近河床之平水亦可自視線與圓筒直位置而決定,圖示之實驗其流量爲1.39立方尺/每秒,流沙量爲20磅/分,用作實驗之沙爲經過濾之幼沙,百分五殘留于八號篩,百分之七十殘留于四十八號篩共分析如下表:—

實驗用之含沙分析

篩之號數	殘存之百分比 %
8	5
14	12
28	32
48	70
65	78
100	88

圖所示之冲擦洞于實驗開始不久後卽迅速生成，如用沙將之塡滿，二三分鐘後又可恢復原狀。撤斷水流後所量之河床高度較在實驗進行中所量得者爲高，蓋因浮懸于漩渦之含沙沉澱故也。

在實驗進行中，漩渦之含沙多向下集中于水流方向下方之墩脚，橋墩附近水床面之沙土綠床面拖動，但絕不浮懸水中，漩渦在橋墩四周迅速携帶床沙，冲擦洞之形成乃由于漩渦含沙沿何止角（Angle of repose）滑下至離橋墩不遠之點停止。另一漩渦離水面不遠形成，但與另一漩渦成相反方向旋轉，且不携沙。如玻璃筒代表一六呎徑之筒狀橋墩，則此實驗所代表之水深爲四呎六吋，而最大冲擦深度爲六尺三寸，故可知水流冲擦之力驚人也。附圖如下：——

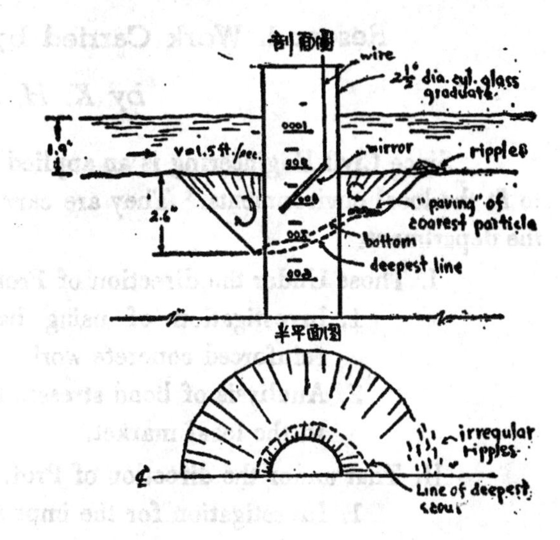

系 訊

Research Work Carried by the C.E. Department.

by K. H. Leung

Since Civil Enginecring is an applied sciences, research works are emphasized to fit the local environments. They are carried under the direction of the staff of the department.

I. Those Under the direction of Prof. C. Y. Yu.
   1. Investigation of using bamboos as material of construction in reinforced concrete work.
   2. Analiysis of bond stresses for different tensile materials, available in the local market.

II. That under the direction of Prof. C. W. Lau.
   1. Investigation for the improvement of different airfields in Canton City.

III. Those under the direction of Prof. N. Y. Luk.
   1. Design of a girls Dormitory for Lingnan University.
   2. Design of a gynasium for Lingnan University.
   3. Design of a Market Center for Lingnan University.

IV. Under Prof. K. H. Leung
   1. Investigations on the compressive strength of conerete using materials available in the local market.
   2. Investigation on the subgrade of the high way between Lingnan University and Canton.
   3. The sewage disposal and drainage problems for the campus of Lingnan.

V. Under Prof. K. H. Leung and Father J. A. Hahm.
   1. Design of different building for the new Middle School Compound of the Lingnan University.

VI. Uuder Mr. J. C. Wang.
   1. Design of a Library for Lingnan University.
   2. Design of a Student center for Lingnan University.

List of the new major equipments for the year of 1948-1949.

I. Material Testing laboratory
   1. 1-Impact testing machine.
   2. 1-Brinell hardness machine.
   3. 1-Fatigue testing machine.
   4. 1-Electrical stress and strain analyser.
   5. 1-Mechanical Stress and strain analyser.
   6. 1-Soil compaction out-fit.
   7. 12 concrete molds—4"
   8. 12 concrete molds—8"
   9. 1-Set of Precison balance.
   10. 3-Sets of sieves.
   11. Miscelleneous.

II. Workshop.
   1. 1-Grinder.
   2. 1-Precision lathe.
   3. Miscelleneous.

III. Surveying Instruments.
   1. 5-Dumpy level.
   2. 4-Transit level.
   3. 2-Theodelite.
   4. 1-Plane table.
   5. 1-Blue print machine.
   6. Miscelleneous.

IV. Electrical Eng'g. Lab.
   Ammeter, Voltmeter, Rheostats, and miscelleneous.

本系教授資歷表

桂銘敬教授兼系主任在假：交通大學工學士，美國康奈爾大學土木工程科碩士，歷任廣東大學講師，中山大學教授，廣東省建設廳公路處技正，兼工務課課長，粵漢鐵路株韶段正工程師，湘桂路天成路副總工程師，本校教授兼系主任。

梁儉卿教授兼代理系主任：嶺南大學工學士，美國麻省理工大學工程碩士，歷任美國公路總局實習工程師，交通部公路總管理處幫工程師，軍委會滇緬公路監理委員會專員，軍委會滇緬運輸局駐緬副工程師，軍委會運統局公路總處副工程師，美國軍部正工程師，現任本系代理系主任。

陳永齡教授：國立交通大學 B. Sc. 英國倫敦大學帝國學院 D. I. C. 德國柏林工業大學工學博士，國立西南聯合大學教授，中國地理研究所研究員兼大地測量組主任，交通部鐵路測量總處處長兼總工程司，國立同濟大學，國立中央大學兼任教授。

黄郁文教授：美國亞麻工程專門學校工學士，歷任咪吔洋行，富新機器製造廠，誠信洋行等工程師，廣州工業專門學校教授，本校副教授。

劉載和教授：嶺南大學工學士，歷任交通部第三區公路局副工程師，正工程師，鐵路工程師，設計組組長，本校副教授，廣州大學教授。

陸能源教授：交通大學唐山工程學院土木工程學士，美國普渡（Purdue）土木工程學碩士，程記隧道公司，中國工程公司工程師，美國普渡大學工程研究所研究員，美國土木工程師學會會員及美國 Society of Sigmaxi 之 Associate Member，嶺南土木工程系副教授。

余仲奎教授：Mass Inst. of Tech. S. B. in Elec. Engg. & S. M. in Aero. Engg 空軍司令部及航空學校檢驗員教官及學科主任，德國 Heinhel Flugzeugwerke Airplane inspector 空軍第二飛機廠設計課課長，國立中央大學航空工程系教授，空軍航空研究院研究員主任副院長，國立中山大學工學院教授，嶺南土木系兼任教授。

韓約瑟副教授：美國麻省理工大學工學士，曾任美國瑪麗諾大學教授（G. A. Hahn）

王銳鈞講師：嶺南大學工學士，歷任軍委會技師研究室研究員，技佐，技正，交通部西北公路局幫工程師，廣州市工務局技士。

凌鐵錚講師：湖南大學電機工程學士，交通大學電信工程碩士，曾任上海廣播台工程師。

江開礦講師：美國加州大學工學士。

老瑞麟助教：國立交通大學畢業航空工程學士

張興耀講師：國立西南聯合大學工學士，資源委員會中央機器廠工務員，國立南開大學助教，教員。

呂惠炎助教：嶺南大學工學士，現任土木工程系助教兼技佐。

本系歷屆教授畢業同學姓名表

1934

崔兆鼎
許寶漢
黎樹仁
劉戟和
李文泰
吳慶鳴
彭震東
黃錦袭
潘祖芳
劉　登

1935

陳景洪
陳國柱
陳錦珊
鄭漢新
高永譽
黎均霖
林榮棟
李文遠
李廷汾
梁卓芹
梁寶琨
文鏡堯
韋金信
容永樂

1936

陳秉准
陳睿馨
陳元力
夏進興
李卓傑
龍寶鎏
莫佐基
黃樹邦
黃惠光
王宏猷
陳守勤
崔世泰
趙宗武
趙長有

1937

陳士敦
鄭煥新
彭奐原
沈錫琨
曾德甫

1938

陳振泰
陳棣濂
陳自仁
林家就
梁健卿
吳潤嵩
黃文海
溫兆明
王銳鈞
余廣才

1939

陳樹彬
馮葆鎏
馮佑增
夏傑榮
黎辛才
林文贊
李卓平
李梁材
伍瑞明
黃寬昌

1940

區錫齡
陳德泰
黎廣杰
楊啓讓
余槐欽

1941

陳尙武
朱錦池
李世能
盧炳煌

1942

陳乃鼎
馮漢邦
何耀波
何育民
賀普定
林漢中
王啓群

1943

程肇興
許銓錦
劉柏江
梁挺燊
阮樹鈞
張炯翰
甘光儀
陳世柏
符和豐
黃炳禮

1944

關啟明
黎思圻
林永泉
梁百衡
馬錦源
徐承焕
龎寶光
余進長
余伯長
張錦華
趙礪艮
朱士賓
李偉廉
楊克剛
梁光能
冼補勳
林顯夫
陳賡鏘
陳光明
李湛深
黃華照
陳榮士
劉文漢
伍慶常
蕭漢輝
卓華耀
文耀芳
黃匡原
梁廷俊
麥立賢
阮其江
王霖威
譚　啟
歐允文
朱慕澄

1945

無

1946

鄭觀志

1947

無

1948

呂志炎
趙浩然
陳潤初
羅明亨
歐陽讓
黃漢基
何思源
李克勤
雷永庶
黃煥林

25931

會務報告

——余文傑——

本學期土木系同學突然增加，連轉學借讀同學，共達一百九十一人。這個數目可算是本系有史以來最高的紀錄。居本校人數最多的第三系。本會很能够增加這樣多的新血輪並希望在各同學協助下本會工作得以順利展開。玆將本學期工作分述如下：

三月四日下午七時半假工學院繪圖室舉行新舊員生聯歡大會，到會者有馮教務長，梁健卿，黃郁文，王銳鈞等教授，同學有一百七十餘人，席中馮教務長講話，畧介紹本系新教職員，聚餐後，由生活組担任遊藝，秩序豐富，笑話百出，至深夜盡歡而散。

三月五日假怡樂村舉行音樂晚會，赴會者計四十餘人，是夕各同學心情特別輕鬆，情興揚溢感情極爲融合。

三月十四日晚假俱樂部召開第二次職員會。推選工學會日籌備委員會會員。預算是日各實驗室開放。老大哥爲事。上午開紀念會，中午聚餐，下午並假鐘亭設鷄尾茶會另於小學球場賽球晚上有音樂欣賞會。

此外智育組鑒於有增加同學間之課外常識之必要，擬於四月卅日舉行參觀西村士敏土廠，由高開績先生領隊。相信同學對於此行不但極感興趣，且必獲益不少。

至於羣育組方面，亦加緊工作，最近擬於工學會日後舉辦集體旅行，藉以聯絡各同學友誼，到時必定有精彩節目演出。

至於體育組方面，亦極力進行。曾於本學期初舉辦足球比賽——「新疆」對「中原」。又於連日內與各外系作友誼足球比賽。戰局異常緊張，本會各球員，異常賣力。其運動精神殊堪可嘉。

本學期土木系畢業同學，約二十餘，人數之多。亦爲本系有史以來之罕見。故本會擬於本學期來特別提早舉行歡送畢業同學大會希各同學參加。

貶眼間一學期又將過去，而同學間能够集合一起的機會又不多所以對同學間之意見不能完全互相解，况鄙人深知才疏學淺，不能在各方面使各同學絕對滿意，惟希各同學不吝賜敎，則鄙人幸甚，工學會幸甚。

草於一九四九年四月廿五日

25933

25934

電油發動：Concrete Mixer, Aircompressor,
Piling Machine, Stone Breaker,
Rock Driller; Concrete Vibrator,
Electric Motor, Centrifugal Pump, Plastic Cables

建築工程材料，英國漆油，化學工業原料

總 代 理

新 志 利 洋 行

香港 法國銀行大厦四樓 電話22765
廣州 沙面敦睦路四號 電話13024

25935

三十七年度下學期本會職員表

會　　長：余　文　傑
副 會 長：姚　保　照
智　　育：林　崇　義
總　　務：黄　兆　歡
羣　　育：鄧　翰　儀
文　　書：尹　行　賢
財　　政：王　　　嗣
體　　育：鄧　國　樑

南大工程

康樂再刊　第五期

嶺南大學工程學會出版

主編：吳秉俠

編輯：林崇義　潘世英　潘演強　何逢康　姚保照　余文傑

廣州教育路十六號
蔚興印刷場
電話一一六一七

市政工程年刊

民國三十二年度（第一次）

市政工程年刊

吳鼎昌題

中國市政工程學會編印

民國三十三年出版

工程師信條

（中國工程師學會第十屆年會通過）

一．遵從國家之國防經濟建設政策，實現國父之實業計劃。

二．認識國家民族之利益高於一切，願犧牲自由貢獻能力。

三．促進國家工業化，力謀主要物資之自給。

四．推行工業標準化，配合國防民生之需要。

五．不慕虛名，不爲物誘，維持職業尊嚴，遵守服務道德。

六．實事求是，精益求精，努力獨立創造，注重集體成就。

七．勇於任事，忠於職守；更須有互助互磋親愛精誠之合作精神。

八．嚴以律己，恕以待人；併養成整潔樸素迅速確實之生活習慣。

中國市政工程學會

第一屆（三十二年至三十三年）職員名單

理事長：凌鴻勛

常務理事：鄭肇經　譚炳訓

理　　事：朱泰信　李榮夢　余藉傳
沈　怡　吳華甫　林逸民
哈雄文　袁夢鴻　張劍鳴
過守正　薛次莘　蕭慶雲

候補理事：朱希尙　段毓靈　胡樹楫
陶葆楷　嚴宏溎

常務監事：茅以昇

監　　事：李書田　趙祖康　裘益祥
關頌聲

候補監事：袁向華　蔡　臚

總幹事：譚炳訓

市政工程年刊「三十二年度（第一次）」目錄

目錄

轉載

附錄

發刊詞

凌鴻勛

我國新興市政事業，正在萌芽茁長之時，而敵人挾其暴力任意摧殘，沿海各城市之淪陷者無論已，卽腹地都市，亦歷經敵人之空中蹂躪，此六七年來之變遷，固市政之浩刦，然亦新市政勃興之時機。現勝利在望，今後各地城市應如何復興改造除舊佈新，並依據抗戰之經驗，建國之需要，以建設新時代之市政，配合國家之一切進步，任務之艱鉅，實不可思議。同人旣組市政工程學會，復徵集各項資料，與同人研究所得，彙爲年刊，以供於世；倘有助於市政建設之前途，與市政當局及學者研究之參考，則爲幸甚矣。

論著

戰後我國之都市建設

譚炳訓

（一）

戰後建國工作中，工業與農業之輕重先後，和城市與鄉村建設之輕重先後，都是時論所爭的焦點，其實皆爲不必爭辯的問題。工業原料仰給於農業，而農業本身又漸趨於工業化，工農業之界限日泯，沒有什麽輕重先後之分，只有如何合理配合以促其加速發展這一問題。同樣，代表工業的都市與代表農業的鄉村，也是互相依賴，都市是經濟文化的核心，近人所提倡的田園市與星形市的城市的鄉村化，而鄉村中公用設備完善，也可以有城市生活上的享受，所以城鄉建設不可有輕重先後之分，而應該並行不悖。將來城鄉的界限逐漸泯除，城鄉建設卽融爲一個——人民公共生活所需要的建設。

歐美工業發達的國家，城鄉人口的分佈，從城鄉各半，到城佔百分之八十鄉佔百分之二十。我國欲迎頭趕上歐美，完成國家的工業化及農業之自足自給，人口分佈，也將達城鄉各半的比例。

我國市組織法所定十萬二十萬或百萬人口以上爲省轄市與院轄市，這是立法上的規定。英國村鎮人口在四百以上者卽有市的公共工程與衛生的設施。時人有以一千人口爲小型市的基本數者，作者以爲我國大體上可以「鎮」（卽北方之集店，西南之墟場。）爲舉辦市政的最小單位。我國千人以上之鎮市數目約有五萬個左右，佔全國總人口的百分之四十以上，所以從「市政工程的觀點看，目前我國已有半數人民的居住飲食行路衛生與安適，需要市政工程師爲之處理。

在戰後隨工業化而發生的新都市之建立與舊都市之改造，工作更繁鉅，市政工程師的任務更艱偉，全國人口需要市公用設施之比例數也必更加增高，所以市政工程不是少數人享受的一種奢侈，而是大多數人民日常生活上的一種必需。不是粉刷門面的一種裝飾，而是利用現代科學工程的成果，實際增進人類福利，代表人類文明的一種事設。

〔二〕

市政工程之範疇至廣，從狹義講，包括道路，溝渠，建築給水及衛生工程等項；從廣義講，包括供給市光的電氣工程，供給市熱的煤氣暖氣工程，解決民食的冷藏倉儲工程，以及通訊運輸等皆為市政上不可缺少之事業。而都市計劃又為一市市政上之基本設計。

我們舊城市大都是汚穢湫隘，實為文明古國之玷。惟有北平一市，有整個都市計劃，宮殿園林城郭皆有完美的配置，有系統的寬廣街道與排洩雨水的溝渠，有點綴風景的引水工程，市民也無形中分區居住，堪稱為代表中國文明的世界都市。新興都市中，首都南京市建設計劃與大上海市建設計劃皆未得全部完成，租界的市政富有殖民地的風味，缺少根本遠大計劃，只顧到少數人的享受，置大多數窮苦市民康適於不顧。惟有一青島市俱備現代都市的條件，而非由國人所建設。至於抗戰以後，大後方新興的大小城市，除少數例外，最顯著的特點有三：第一為疎散，這是空襲的教訓，惟疎散無計劃，未作星式的細胞發展，大半沿公路排列到十餘公里以外，如此轟炸目標仍在，居民交往不便，且使幹道交通擁塞，汽車肇事增多。第二為審美觀念的缺乏，戰時講美是奢侈，但審美代表一國的文化程度，同等的材料人力，為何不求其配置之美。本來人民已經是彼此摹倣，只顧建造同式的簡單房屋，又加上地方機關規定一種不中不西的門面圖案，強迫人民照樣改造，使全城房屋成為一個樣式，這種單調的卑醜，表現了民族的文化水準的低落。第三是工程常識的缺乏，例如街道上舖的石板道折毀了，新修的碎石路無排水設備，雨後盡為泥塘，建設成為破壞。這三種特徵所生的結果，是市民的損失，市政的浪費，只要稍開智慧，加以指導，這些缺點都是極容易避免的，雖然這是政府的責任，也是我們從事市政的工程師，未能盡其應匡助的義務，以普及市政工程的常識所致。

〔三〕

我國之市政建設，過去與現在，除少數外，無全盤與遠大的計劃，自然擴展的結果，不知造成了多少災難與罪惡，摧毀了多少市民的性命，浪費了多少寶貴的人力物力。租界畸形發展的市政，就是一個顯著的例子。今後在偉大建國工作中，首先要對市政建設的觀念，及其在整個建國工作中的地位，作正確之認識。市政建設成裝飾品，是過去所走的歧途，它是實際為多數人民謀福利的民生建設。要與工業農業及國防等建設相配合而並進，不僅建造能抵抗空襲的城市，減少轟炸的損害，還要使國防據點的城市要塞化，成為打擊侵略者的堡壘，我們要充分運用「遠見」與「計劃」，以免蹈以往之覆轍，我們在勝利之前，就應及早準備以下各事：

第一為「都市計劃法」之修正及附屬規章之草擬。市組織法是市的政治基本法，市計劃法是市的技術基本法，不但規定市政建設的規範，並且標明都市發展方向及都市發展理想，如都市人口的經濟限度，都市發展的型式（集中，分散，衛星式，環形式，帶形等），各區帶（園林住宅工商業政治交通等區帶）人口的密度各區道路及建築面積所佔之比例，各級市（從都市到鎮市）公用設備之標準，都市國防上最低限度之設備，及其他市政建設上的基本原則，皆規定於市計劃法內，以為以後市政發展的指針。二十八年內政部公布的「都市計劃法」，必須加以修正與充實，始能滿足戰後都市發展的要求。至於建築法規，給水與溝渠法規，其他公用事業之法規，皆要及早編定，以應戰後建設之需。

第二項工作是市政工程常識之宣傳。國人對市政工程常識的缺乏，尤以縣城及鎮市，無聘市政工程師之能力，我國亦無如許工程師可供小市鎮之用，所以宣傳的對象，應以縣鎮的市為主，文字以外，着重於各種簡易的工程圖案，最好將工程圖案作成通俗的圖畫，使縣政人員及鄉鎮長能够一看就懂，能夠參照着因地制宜的摹倣。這種效果，費省效宏，快而普及，可以補救訓練技術人員緩不濟急之弊。其他各市政工程叢書之編纂，市政工程模型之展覽，皆為普及市政工程常識的有效方法，也應積極倡辦。

第三項工作為市財政制度與土地政策之確定。美國的市經理制，其機構形式，我國不必一定要摹倣，但經理制的精神——專家政治及權責合一，值得我們取法。市經理一定為市政專家，他負行政全責，也有行政全權，尤以市預算之「統一集權行政負責制」，可以使市經理制發揮最大的效能。一個現代化的市政府，其支出的大部為事業費而非行政費，市政事業費隨人口增減及人民需要而隨時變動，尤以戰後要激進工業化的我國，一市的人口在一年以內增加一倍兩倍並非意外之事，在財政統收統支的系統下，一年前編的預算既不適用，而追加預算又緩不濟急，市政上的設施，必不能追及需要，發展則更不可期。市財政制度關係市政興替者至鉅，不可不早日參考歐美市制及我國戰後情況，確立一種能促進市政發展的市財政制度，都市土地之增值，為政府建設之力所造成，土地收益為都市最大之財源，應歸之政府，民生主義已明確言之，應根據平均地權的理想，制定都市土地政策實施綱領或土地法及其實行細則，早日公布，以杜戰後市區土地投機之風，為市財政確立一個最基本的來源，同時可以免除將來整理都市土地的困難。

第四項工作為淪陷區及內地都市復興計劃之研究。此種研究應先確定戰後國都之所在，繼之國防工業據點城市，國際貿易港埠，國內交通核心城市，文化及風景遊覽城市等之選定，根據此基本分類，再爲各都市之個別研究，以確定一個初步復興建設計劃，否則戰後百廢待舉之時，將無所措手足，全國都市，恐仍不免於自由放任之發展，不能與整個建國工作結成一環。

〔四〕

中國市政工程學會經過半年的籌備，本月二十一日卽在陪都成立。這是全國市政工程專家和工程師所組織的學術集團。希望其努力於工程學術探討之外，對於我國戰後都市建設許多實際問題，聯絡其他市政學術團體，擬製各項方案，貢獻國家。前節所列的四項工作，不過是一個舉例。

展望戰後的建國大業，全國成千成萬的市鎮，有悠久的歷史，俱備天時地利的優越條件，居住着純樸敦厚的人民，他們將爲建設現代化國家的中堅份子，他們的日常生活——起居，飲食，健康與工作效率，全靠市鎮的道路溝渠飲水燈光和一切市政設備與市政管理的良好與否，我們中華民族給予國際人士的第一個印象，也就是全國市政建設與市政管理的成績，這是民族生活的代表，文明程度的尺標，應列爲建國的基本重要工作之一，我們要配合其他建設事業，並肩齊進，萬勿落後，落後就是失策與災難。盼中國市政工程師努力，祝中國市政工程學會成功！

實業計劃上之城市建設

朱泰信

一 實業計劃輪廓作爲我國城市建設之背景

國父實業計劃，並無專章討論我國城市之建設。然按諸六大計劃之內容，則城市之建設，隨在可見。如海港計劃，頭二三等及漁業港共計三十七個。鐵路計劃中之起訖點城市，約有二百，水運商埠碼頭所代表之城市，約在一千二百。至於一百六十萬公里之公路建築，設以平均每一十字交點（相隔一百公里，或以四條路線五十公里所及之範圍計算）可爲一中型或小型之內地城市者，爲數將達八千。其餘可引起城市之興造者，如「居室工業」「水力發展」「移民殖邊」等項計劃之實現，則新興城市，不知凡幾，此外如農鑛工各種生產事業之創立，則直接或間接影響於現存城市之改進，與未來城市之佈置者，尤未易衡量。換言之，一部實業計劃，實可作爲我國城市建設之背景，宜其不必立專章討論之。茲僅就管見所及，實業計劃上之城市建設問題，可歸納爲下列六端：

（一）城市集中；

（二）城市分散；

（三）維持平衡；

（四）改造與復興；

（五）新建城市；

（六）城市防衞。

（一）「城市集中」云者，乃謂鄉村人口趨向于城市，由村鎮而變爲小城市，由小城市而變爲大城市，此種人口集中現象，原爲工業化之自然趨勢。尤以在十九世紀，以煤水爲蒸汽動力之源者，爲不可避免之事態。實業計劃原爲我國工業化之方案，一朝實現，城市之集中趨勢，自然大增，因之所有城市集中化之弊病，亦連帶而至，此其所以成爲問題也歟。

（二）「城市分散」云者，乃指近代都市之另一趨勢，即龐大城市（百萬人口以上者）之人口，疏散至其郊外區域，形成所謂「衞星城市」或「城市系統」是也。此在吾國抗戰前，五個百萬以上人口之城市，均有必要，亦且可能。必要云者，乃爲吻合國防之需要，所以疏散物資，減少空襲威脅，是也。「可能」云者，乃以近代電力與內燃機之發展，雖疏散至廣闊空間之內，而可聯繫於縮短時間之中，實則近代測量城市距離之遠近者，原不僅以「里程」，且須以「分秒」計矣。

（三）「維持平衡」云者，乃指城市人口已達「最佳密度」，如在十萬左右，每因交通與實業之開發，而有激長驟增之勢，此則有待於種種之限制措施，方可以維持其平衡。此外尙有繁榮之城市，每因他處有劃時代之種種發展，而呈就衰之情形，如我國之揚州即爲一例。則求所以維持其平衡者，亦爲今後城市建設分內之事也。

（四）「改造」或「局部改造」，或「全部改造」而達到「城市復興」之理想者，實爲我國今後城市建設之必要手術。此種「手術」如過去「拆城放寬馬路」所代表者，多不免遭外科損壞肢體之譏。良以我國現有城市，率爲歷史上之勝地，斯其生長與我民族生命息息相關。故欲爲城市施近代「手術」者，應注意於「城市生命」之生理與心理需要，萬不可魯莽從事也。

（五）「新建城市」，如上述（一）（二）（三）項城市建設之實現，業已包括在內。此處所須指明者，則爲一種技術問題——即「新建城市」之手術，如選擇地點，實爲一切城市建設工作之基本是也。時賢當有以「新建城市」可解決我國一切城市問題者，實犯「單純論」者之謬誤。實則，非任何新地點均可供建築新城市之用，一也。「新建城市」，非如新造機器，可一蹴而躋，二也。城市以歷史爲其生命，新建城市，倘完全忽視此點，將爲無靈性之城市，而不合於吾人安居樂業求進步之條件，三也。

（六）「城市防衞」云者，謂今後城市建設，應針對世界戰略思想之新發展，有以自衞與抗敵也。尤以第一、二兩次大戰所予吾人之教訓，似近代城市不僅須堡壘化——如古代城市所必備之條件——尤須「戰場化」，如最近「斯達林格勒」所表見者。「自衞」云者應以減去空襲威脅爲其主要設施之對象。「抗敵」云者則在準備有利條件之下，作巷戰是也。

上述六項，實爲近代城市規劃學術之主要題材。而此項學術之泰斗，英人亞當姆斯，嘗謂過去之城市規劃，由一城市之內部着想，而推及於一「自然區域」乃至全國者，實爲一種不合理之程序。合理之程序應在先有「國家規劃」為「自然區域規劃」之背景，而以城市建設充實其子目與線格。準此以談，將見　國父實業計劃之所示之輪廓，實為一種國家規劃。我輩研究城市建設者，正可藉之為背景，以求國防與民生二種建設，如何配合聯繫，形成集體生活，如大小城市村鎮所能代表者，斯為合於理想也已。

二　我國城市概況及其分類

「一切科學，起始於分類」。故城市建設，作為一種科學研究，必有待於分類。此項分類，計可分為兩種：一種係以城市大小分類者，一種係以城市性質分類者。先論第一種之城市分類。

城市之大小，率以人口數目為準，此種分類之詳細理論，頗值玩味，然非本篇所宜涉及。姑就編者一得之見，將城市之大小分為四類如下：

（一）龐大城市　人口在百萬以上者；

（二）大城市　人口在十萬以上百萬以下者；

（三）中型城市　人口在一萬以上十萬以下者；

（四）小型城市　人口在一千以上一萬以下者；

據傑佛生一九三一年所發表「世界城市人口之分佈」文中，估計全世界大都市，（人口在十萬以上者）共計五百七十三個，其中一百十二個為我國城市。但在抗戰前夕，我國「龐大城市」計有五個。「大城市」計有一百一十個。據孫本文教授估計我國「中型城市」共計四千五百個。「小型城市」（自一萬人口以下至二千五百以上者），共計二千九百一十個。似嫌估計太低。（據孫本文教授估計五萬至十萬人口者，為六百四十八個，一萬至五萬人口者，為三千八百八十個，五千至一萬人口者，計一千九百四十個，二千五百至五千人口者九百七十個，二千五百人口以下者，即列入鄉村）。按就民國二十年至二十一年內政部所發表之「全國縣城鎮市村落人口分配表」而論，一千人以下者，稱為村落，計佔我國人口數百分之五七、一九。一千至五千人口之「鎮市」人口佔百分之三〇、五三。五千人以上之「鎮市」人口，佔總數百分之六、五五。縣市人口共佔總數百分之五、七三。而一萬人以下者則為其成分百分之七十二，即約佔人口總數百分之四。準此估計，設以全國總人口數為四萬萬者，則一千至五千人口鎮市人口總數，應為一萬萬二千萬。設以其平均數為每一單位鎮市之人口（三千人口）則可得「小型城市」四萬個。而五千人口以上之城市人口總數，約為五千萬人，試以一萬人口為一平均單位數，則可得五千個城市，故小型城市之總數應在四萬以上。（按設「市」之法定人口數目，與劃分城鄉界線之人口標準，並非一事，後者乃就社會學立場，以城市為人口集居之單位，可謂「城市生活」者立言。）

編者試將一千人口之鎮市列入「小型城市」者，理由厥有兩端：即（一）此類集居單位爲數至多，實構成我國城市佈置網之基礎。（二）一千人口以上之鎮市，固可以近代市政衛生工程新技術，使之城市化，而作爲近代工業之疏散據點。（按英國在抗戰前，凡有四百人口以上之集居，即無不有給水，溝渠，馬路與電燈之設備。故此次應付德國空襲，即以疏散大城市人口，爲其主要設施。至於日本鄉村與城市人口之分類，如以二千人爲界線者，都市人口竟佔百分之九十四，作視之，雖覺至不合理，而日本正恃此小型城市普遍化，可利用農閒及家庭工藝以與歐美城市集中化之國家，作商業戰爭之製造根據地也）。換言之，即此種分類，可令吾人發深省。我輩似不必震炫於世界大城市之發展，而忽略我國在一千人口以上之城市，其人口總數實佔百分之四二、八三，若吾人能注意於應用近代市政衛生工程技術於此種小城市者，則有一萬萬七千萬人口，可列入城市人口範圍，作爲將來生產戰爭中之「工業化部隊」，亦足以雄視全世界矣。

第二種城市分類，係按照性質劃分，似應以「國防科學」立場爲歸宿。「國防科學」之思想，如近世所週知者，係在研究「全力戰爭」因而實現之。所謂「全力戰爭」者，據編者於年前分析之結果，可分爲五類力量，（見時事月報「中英美三國互救論」一文）即除一般人所習知者之人力物力財力而外，仍應加上「心力」與「史力」。後二者實構成一國之「精神力量」與軍隊之「風紀」「士氣」，外國人士統稱之爲"Morale"者是也。準此以談，則我國之城市可分爲五類：即（一）歷史名城（二）國際商埠（三）內地工業城市（四）內地商業城市（五）農業集鎮。

（一）「歷史名城」——如北平、南京、西安、洛陽等城，原爲吾國歷史上之都城，與我民族生命有不可分割之關係者。至於武漢廣州，則因辛亥革命前後流血起義之結果，其地位亦已「神聖化」至上述四大名城之列。更因此次抗戰，「閘北」「台兒莊」與「長沙」等處，爲我將士浴血犧牲成功成仁之地，在勝利之後，其地位亦將提高與前者相等。此類名城將爲我國「史力」所繫，對其古跡遺物，雖一木一石之微，亦應加以珍護愛惜，永垂紀念。不僅有紀念性質之建築，如城牆宮室，須加意修葺；即戰爭中所遺破壞之殘蹟，凡可以表揚前烈之勳績，起後人之景仰者，均應保存，而加以必要之裝璜佈置，如此，「史力」方可附金石以永垂不朽，較之照耀簡冊者爲更有效也。

（二）「國際商埠」——此類又可分爲三種，第一爲過去有「租界」之城市。第二爲過去訂有國際條約而無租界之「通商口岸」。第三爲一般對外貿易城市。第一第二兩款城市，雖因不平等條約之取消，而可同納入於第三類，但畢竟有其特徵，如上海天津將來歸還我國，其管理行政方面，即對爲吾人政治能力之試金石，尤以上海五方雜處，國際性質極端重要。編者於抗戰前，即主張應早組織一種「接收研究委員會」，一方面研究如何接收，一方面訓練接收人員，庶可按照既定步驟，從容將事。此外，此類城市既當爲「外力」之據點，又爲國恥之陳跡，自可爲我國「心力」（即報仇雪恥與不甘屈服之決心）發揮之場所，至因對外貿易之關係，此類城市又爲「財力」之源泉，故其關係於我國前途之命運，固不因抗戰前後而有所變更也。實則保佛生所論我國十萬人口以上之城市，一百十五個，均屬此類。

(三)內地工業城市，係指接近於工業資源之場所，因工業發達之結果，而形成近代工業之城市。如唐山可爲此類城市之典型。然此類城市，過去任其自然發展，以致凌亂汚穢不堪，影響居民壽命與生產效率至鉅，此外尙有我國固有「歷史工業」之城市，如景德鎭將來受機械化之洗禮，自亦有「近代化」之必要，按此類城市實爲我國「物力」之源泉，正以其能將原料在國內製造，發展精進之技術，將爲我國「機械化」之據點故也。

(四)內地商業城市，自與前一類城市不可分割，但此類城市之存在與繁榮，率以地理交通條件爲其主要原因，及其發展至相當程度，自又可爲工業化之場所，故可視爲「財力」所繫，國家金錢因賴之以圓滑周轉者也。

(五)農業集鎭，大概均爲一千至五千人口之小型城市，正因其爲數至夥，其積聚影響，乃至不可忽視。此類城市實爲「人力」之源泉，良以鄉村居民，不慣於集團生活，無論平時工廠，或戰時軍隊之生活，對之均不免係一種威脅與痛苦。而農業集鎭，乃爲集居生活之初步訓練，倘能因勢利導，將見此類城市可爲發動「人力」之最有效機器。同時在近代工業疏散之趨勢下，此類小型城市又可視爲各種生產事業之搖籃與據點也。

三　城市建設之三大問題

我國城市建設，厥有三大問題，即：

（一）城市集中化之程度問題；

（二）城市之大小限制問題；

（三）城市之近代化問題。

「城市集中化」之趨勢，爲十九世紀至二十世紀初葉一大特徵。如英國城市人口在一八五一年，佔全國人口總數百分之四十五，至一九二一年此比率乃增至百分之八十。德國在一八七一年，城市人口僅佔百分之十八，至一九二五年，乃增為百分之六十二。美國在一八八〇年，城市人口僅佔百分之十五，五十年間，乃增至百分之五一・四。日本在一八九八年，城市人口佔百分之十五，至一九二五年乃為百分之五十六（此尚係以五千人口為城鄉之分界者），此僅就比率之增加而言，已覺可觀。如按照各國人口總數之與年俱加，而計算城市人口激增之總數，尤為驚人！美國某作家，嘗描寫近代城市人口之增加情況，略謂：鄉村初在睡夢之中，乃因機器之聲響噪醒，於是人口一倍兩倍，按照幾何級數增加，今日尚為鄉村，不須十幾年光陰，即如暴發戶成為十萬至百萬之「巨富」都會矣。蘇聯在進行工業化中自一九二六至一九三六之十年中，城市數目自一千九百二十六，增至三千一百一十，即十年之中增加一千二百城市，尤可見「城市集中化」即在廿世紀新進工業化國家，猶方興未已。此種「城市集中化」之勢力，顯現於一國城鄉人口之分佈者，尤不若人口十萬以上之都市數目，及其與總人口比率為更清晰，良以近代城市生活之充分發揮，有待於人口超過十萬以上故也。茲將傑佛生於一九三一年所編之表，節錄如下：

國別	城市數目	佔全國人口之比率（百分數）
中國	〔一一二〕	六、四
英國	四二	四四、二
美國	九三	二九、六
德國	四六	二六、六

故我國城市建設之第一問題可分為兩項子題即：

（一）我國城鄉人口之比率將如英國者乎？抑如美國者乎？

（二）我國城市集中之勢力將如英國者乎？抑如其他工業國家者乎？編者對於此兩項問題試解答如下：

（一）我國城鄉人口之比率，應以英國為戒。而以美國為標準，實則，倘將一千人口以上之鎮市劃入城市範圍之內，將見城鄉人口各佔半數之理想，在我國工業化二十年左右，應可達到。

（二）我國十萬人口以上之城市人口比率，顯見太小，正為我國尚未工業化之象徵，將來達到百分之三十五可矣。

總之我國城市集中化之程度問題，應以城鄉人口各半為歸，而以十萬人口以上之城市，佔全人口四分之一，如此方足為「工業化」之人力基礎，以與世界列強相抗衡耳。

據一九四〇年國際勞工局所發表之統計，就業之人數，稱為「動性人口」"Active Population"約佔各該國總人口百分之四五十之間，至農林漁鑛部門之就業人數對「動性人口」之百分比率如下：

美國	（一九三〇）	二四、五
英國	（一九三一）	二[illegible]、九
法國	（一九三一）	四八、〇
比國	（一九三〇）	三三、〇
德國	（一九三三）	二九、〇
日本	（一九三〇）	五〇、〇
意大利	（一九三六）	五〇、〇

此種比率尤可表見城鄉人口分配之內在意義，如英美德比四國均爲高度工業化國家，而英國之比率太小，正表示鄉村人口之低落，乃有「過分工業化」之危險，如最近英國國會訪華團某氏所稱，英國在現在需要農業，正如中國需要工業，同樣迫切。並謂英國一朝「農業化」，將維持其農業於永久。而美德比三國之比率，在百分二三十之間或謂自四分之一至三分之一爲鄉村工作者，至於日本與法國原爲輕工業發達之國家，同時其農工配合，無畸輕畸重之弊，實可作爲我國城市建設之珍貴例證也。

其次論及「城市之大小限制問題」乃爲近代一般城市社會學者與規劃專家所最注意之點。

插圖及說明

人口集中

P.I.

A

B

C

時間（以年計）

城市人口在一定之面積內，在某一定時期中，（統稱之爲「一定生活空間」）其與年俱增之情形，可示如一種長S式之曲線，謂之爲「合理曲線」（"Logistic Curve"）。按此種曲線普遍適用於一切「集體生物學」現象，（"Group Biology"）A段代表生長之起始遲緩，漸漸加快，達及B段，生長激增，至相當限度，而入於C段飽和狀態，終至不再增加，或甚至減少，P.I.爲轉捩點。余以爲城市人口之最佳密度，當在B段之內，P.I.點之上下，如此規劃城市大小，方爲合理。

試一按問城市人口集中之現象，究依據何項法則乎？曰：增加人類之生活效率故也。良以一百人在一百處各營其「魯濱遜」式生活者，祇可以「飄流」二字形容之，較之於一百人在一處集聚生活者，相形之下誠有天壤之別矣。準此推論，城市之生活效率，高於鄉村，而大城市之生活效率，又高於小城市者。特以集居生活，究爲生命現象之一種，乃逃不出「合理曲線」之大律（見插圖）即謂此種推論不能作直線式無限延長耳，一萬人口之城市，其生活效率，固可高於一千人口之村鎮，十萬人口之城市，則高於一萬人口者，而一百萬人口乃至一千萬人口之大城市，其生活效率，是否不低於十萬人口之城市，且有問題矣。實則，近代城市規劃學家幾盡同意於城市，作「人口集居單位」存在，以十萬數目為「最佳密度」(Optimum density)余嘗就城市人口增加之合理曲綫分析之，計可得三種階段。第一階段，在城市初起時，人口增加速度甚為和緩，殆不易覺察。第二階段，在城市繁盛時，人口增加既達一定數目，如三萬至五萬者，便似獲得增加之「運動量」，而增加甚快。第三階段，在城市穩定或甚至就衰時，人口數目漸進於飽和狀態，不特不再增加，且有減少之現象，如大城市中之中心區域人口之變動，最足代表此種現象，實則就大城市中之各市區，作此種研究，最為合理，以其界限既定，「生活空間」乃為有限者故也。余意「最佳密度」當在第二階段中之轉捩點上下，以其「生活空間」富有餘隙，正可維持其居民之生活「位能」，使不致受災害或失業之影響，而可保繁榮於永久耳。如英國社會學家布斯研究倫敦，克德堡研究約克之城市居民及工作，結果認為英國大城市居民有三分之一在貧窮困苦中，度其「非人生活」此則為「合理曲綫」第三階段中所不能表示者也。

余在此處，應加聲明者，即所謂「生活空間」乃指集居人口生活需要影響所及之區域，此在古代閉關自守時期，內地城市之距離率為一日來往路程之所限，換言之，即其「生活空間」不過建立在半徑五十里之圓圈面積上。因之此類城市人口若能在一、二萬左右，佔其區域內總人口十分之一者，已達其極限矣。(國家都城，當然為例外)迨乎輪船火車既通，一日路程之距離，乃增至二三十倍，則其生活所需，可以影響之區域面積且五六百倍於昔時，此從簡單幾何學之原理，已可說明近代龐大城市之人口何以可達五六百萬以上者也。當然近代城市之龐大，原不祇此一項原因，而此項原因乃為其最重要者無疑。蓋無近代長距離之運輸交通，即無近代之龐大城市也。

雖然，此一原因僅足說明龐大城市之可能性，尚不足以說明其必然性，必然性之原因，則在如蓋德斯所稱「舊機械時代」(Paleo-technical era)所恃之力源，如水與煤，均為體積甚大，價格甚微之物品，雖以近代之長距離運輸工具，亦嫌搬運不合經濟，於是工廠就煤水便利之地而建立，人民乃依工廠之左近而集居，漸至市廛，商場興起，工商居民遞為因果，由村鎮而成小城市，由多數鄰接之小城市，生長混合而成龐大城市。如今日之倫敦，實為十九世紀城市最著之一例。但廿世紀開始，此必然之原因，已隨機械上之發明，而有所改變，即動力之源，有以電與汽油代替煤水之趨勢。運輸方面則以汽車代與，道路四通八達，頓恢復其機動性，因之大城市乃有疏散之可能，而小城市無增加其人口之必要。於是，限制城市之大小，至「最佳密度」，乃從理想而至有實現之可能。如在此次大戰前，英國即早訂有計劃建立一百五十小城市，以疏散其龐大城市(如倫敦，格拉斯哥之類）而世界各國之城市規劃技術，亦均趨於限制過大之城市，如莫斯科柏林等之新計劃，均擬有「綠色地帶」（或農業地帶不准建造街巷連接式之房屋）以環繞而隔絕之。抑更有進者，具有「龐大城市」之國家，經此次大戰中空襲之教訓，已習得「疏散」乃爲最安全之防空方法，則廿世紀之後半期或將爲中小型城市之時代矣。

復次論及城市之近代化問題，應首先研究「近代化」名詞之內容及其條件。「近代化」之特徵，厥有二端，卽「分工」制度之繼續精進與「機械化」之勢力普遍是也。試就「分工」立場，略論城市中之「近代化」條件。

一般城市之興起在能先滿足其基本之需要，卽飲食與交通是也。地球表面上乃先以河道之四岸，供給此項合宜之地點，「飲水」「用水」與「交通」俱能滿足其需要。時至近代，由此三種原始需要，乃漸發展有其各別之技術，卽「給水工程」「陰溝工程」及「市街工程」是也。此三項工程，乃爲「市政衛生工程」之宗支，英國作家合稱之爲「文事工程」（Civil Engineerings），卽謂某一地址，須經過此種工程手術，始可由鄉野進而爲文明是也，故編者以爲近代城市之第一資格，應在先具有此三項工程建設，此三項建設具備，卽可見近代城市之每一市街，殆如昔日城市相依爲命之河道然，不過「供給飲水」「排洩用水」，（汚水）與夫「交通運輸」等三項功能，絕對分開，而各不相擾，（如巴黎市街之截面上，爲街道之舖整，下有排洩大量汚水之「運河」，更於此「地下運河」之兩岸，則緊以高架之給水管道，乃成爲世界一大奇觀，可爲此種「分工」成就之極致）故此種分工可稱之城市基本需要之技術分工。

其次爲「工作」與「休憩」設備之分工，此類分工，雖略表見於近代城市之「分區」辦法（Zoning）但分區辦法，原係爲救濟已存城市之紛亂而設，故當遷就事實而未必盡合於城市規劃之原則。故眞正「工作」與「休憩」設施之分工，應以「園林城市」之發展方式爲歸。良以此類城市之發展方式，限制工商業地帶面積（代表「工作」生活者）甚小，而其餘則爲住宅地帶與休憩遊樂所必需之花園公園及運動場所等空曠地帶。如英國之「第一園林城市」辣他渥斯總面積爲三千八百十八英畝，僅以三分之一充作街市建築之用（其中尙留有六十英畝作爲公園敞地）而工廠基地僅佔一百一十英畝，尙不及建築街市基地十分之一。此項面積分配比率，已爲城市規劃大家安文爵士引用，至成爲衡量近代城市建築「空」「實」分配之標準。良以近代機械生產，生活緊張，固必有悠閒休憩之空間，以涵養調節之也。至於「分工」趨勢之最近發展，則爲城市「居住」與「行動」兩項功能之劃分，此在大城市規劃方面，有所謂「限制飄帶式」發展者，卽在取消自古以來沿街道建造房屋之習慣辦法，而代以「運輸孔道」與「住宅街道」之劃分。良以街道之設，在昔一方面作為交通要道，一方面作為建築門面或居住基地之用。近則以此兩種功能，動靜異趣，乃有互相干涉之弊。故近代城市之近郊運輸孔道，應專為行車之用，而另以支道引入較幽僻之地址，從事建築所謂「囊底式」之居住區域，（Cul-de-sac Development）其中街道不必甚闊，舖墊路面可容單行車足矣，此種分工趨勢，及其限制辦法，即包含有限制城市大小之意義在內。何則？十九世紀龐大城市之形成，率由鄰接較小城市作飄帶式之建築發展彼此混合吸收，殆如鯨吞魚然，乃如此「龐大」雜亂耳。至於不准沿孔道建築房屋，即謂在既存市區之外，須留不建築之地帶，或稱農業地帶，或稱綠色地帶以為近代城市之界限，所以容其自成單位，昔之混亂龐大城市，乃將為近代有秩序之「城市系」System of cities所代替也已。」

至就機械化而言，近代城市殆依之以為表見其近代生活之特徵，尤以龐大城市之運輸系統，機械化程度最高亦最複雜，而形成為近代有機組織意識者之幻想對象。但試一按問此種「機械化」是否必要？將見其不為龐大城市錦上添花者幾希！換言之即此類過度之機械化種種措施，常為救濟龐大城市之缺陷而設，殊不知「龐大城市」即因其龐大，而有本身缺陷，非可以任何機械彌補之者。故疏散大城市既成為近代趨勢，則「機械化」之方向，應在促進此種趨勢，如電訊汽車之類，均應為近代城市必備之條件。更因近代機器製造之精確程度日增，兼之為避免空襲威脅之故，「閉窗工作」與「地下工廠」乃為經常事件，因之通風取暖及調節空氣工程所謂「工業環境衛生技術」者，（Industrial sanitation）亦為近代城市機械化之一端。總之，我國城市之近代化，一方面應以利用最新機械，使中小型城市可作為工業之搖籃，另一方面則須循「分工」之趨勢，使各種城市可以實其近代化正常生活也。

四、從國防科學立場論我國城市之佈置

城市之佈置問題，可從兩方面尋求其解答，即（一）個別城市之內部佈置（二）其在全國之分佈情形是也。

（一）個別城市之內部佈置又可從下列三觀點討論之，即：

（1）防空觀點（2）巷戰觀點（3）行軍觀點。

從防空觀點，近代城市既為空襲之目標，必應構成為消極防空之有效單位，「有效」云者，在此處至少應具有下述之四大條件，即「空曠」，「偽裝」，「抗火」，與「抗震」是也。前二者，以近代園林城市之佈置，頗能吻合要求。而後二者一方面有待於合宜建築材料之引用，（如鋼骨混凝土）一方面有待於「動態的平衡」構造工程之設計，如水管網，有為防空立場而規劃者，則將其主要管道與抽水機，以鋼骨混凝土掩護周到，使其不畏轟炸，同時其他管道附有自動截

水之浩瀚，俾破壞時水不至損失過多；此外更以充足之水量與水頭，可以迅保消滅同時多處起火之火災；其尤要者，則在利用地下水源，可保受空襲之威脅，凡此種種設施，苟有賴於「機械化」與精密規劃也已。

從「巷戰觀點」上，將見近代城市不僅須如古代之堡壘，且將為全民戰爭之戰場。實則，巷戰之重要性，似隨近代歷史開端而起始，良以巴黎大革命原為一種巷戰，巴黎市民則藉之以推翻法國之王朝，其後拿破崙第三稱帝令郝斯曼改造巴黎，其動機實在鎮壓巴黎市民之暴動耳。而自第一次大戰以後，城市作為戰場，似益加肯定，其最著者如西班牙內戰中之馬德里，如我國抗戰中之上海、閘北，最近乃如斯達林格勒之戰。實則關於近代城市作為戰場之嚴重性，戰略家如德國之邦士教授，早已指明第一次大戰開始時德國攻入比國利日及其工業區所受之損失，而加以警告，以為不可輕於嘗試，邱吉爾在此次出任首相之初，對於德國軍隊攻入英倫三島之可能性，未加否認，但謂如德國果攻入倫敦者，以倫敦之大，即將以巷戰消耗乃至殲滅之耳。邱吉爾當然非一般空言之政治家可比，其對於近代戰略，尤有特殊貢獻，則其所言自有其內在意義，意者以倫敦之大，各種建築物之複雜幽奧，加以地道車，電線網，機械等等設備，則在英國科學家工程家以及軍事學家指揮組織之下，改造為種種之「抗戰機構」與「殺人陷阱」者，自為意中事也；「與城共存亡」至近代戰爭，始由難能可貴之理想，變而為可能與必要之事實，此在我國富有國際性之城市，同時為我國門戶與咽喉之地，如上海、如天津，以及青島、武漢、廣州，等城市，均當視為可能之戰場，以求可在我方最有利條件之下，進行「巷戰」也。

從「行軍觀點」，城市一方面供給食宿，以為軍隊休息遊樂之資，一方面在疏散軍隊，可得較安全之掩護，德國軍事學家林德樂爾估計城鄉容納駐軍之能量如下：

一般區域供給一星期食宿，可容至三倍其人之軍隊（僅供住宿，食由軍隊自購），在富有之農村地帶，每一人口之容納量為士兵十名，（即謂一千人口之村鎮，可供一萬士兵之住宿。）

同上，在城市及工業地帶，每一人口之容納量為士兵五名。

此種估計原在假設正常狀態時，其有難民麕集者，自應比照減少其容納駐軍量。反之，如居民撤退之地帶，則可以比照容納較多之駐軍，編者特引此數字者，乃在指明我國中小型城市及其民居問題之重要性，同時亦足以表示國防與民生之一致，尤以在我國「瘴癘」地帶，城市不發達，原為「瘴氣」之直接原因（近代研究熱帶病學者，常稱瘧疾為鄉野之病，有城市文明，即可掃除「瘴氣」矣），而行軍於此類瘴癘地帶，其死亡疾病之嚴重，以及其傳播疾症之可能，如雲南思芽一例所示者，尤足令吾人不寒而慄。總而言之，編者嘗主張我國市政應置放於衛生工程之上，而衛生工程之技術，尤當以「雪中送炭」之精神，改良我國中小型之城市，並為吾國城鄉民衆先謀居住問題之解決者，非專為「民生」，實亦為「國防」打算耳。

（二）就城市在全國分佈之情形而論，將尤見其與國防思想不可劃分，試溯城市分佈情形之由來，當可發見其演進秩序如下：

（1）「首都重心式」——即以「首都」為重心之城市分佈，此可代表最原始之分佈式樣，良以在君權時代，「朕即國家」之思想，乃需以首都為君主之甲殼，君主乃藉此「甲殼」，以伸其勢力於全國各地。

（2）「要塞外圍式」——「首都重心」，在國土尚小時，不見其弱點，及區域既廣，外敵較之內亂，尤為可憂，於是依據天然險要，作為邊塞，大河流最為合乎理想，重要城市乃沿之興起，作為外圍。（如我國沿長江黃河之城市，多在兩岸，乃為顯明之例，所以我國歷史所示，嚴重之外侮，均係由北而南。）

（1）（2）兩式之合併，乃形成蛛網式之城市分佈，即以城市為中心，以主幹道路為網絲，而普及其軍政勢力於全國各處，此種佈置，似深中吾人之心，故雖在民主思潮澎湃之十九世紀，似亦無人過問其國防價值究竟如何者，一般人常以其外表之有系統，認為天經地義不敢置疑，但試衡之以近代全民戰爭之思想，將見此種佈置，原有其先天之缺陷，即首都集中，目標顯著，一也。首都龐大，禍生肘腋，二也。外圍要塞城市列為陣線，陣線一處突破，沿線駐防軍隊，不易集中，敵人即可長驅進迫首都，作城下之盟，三也。首都，要塞城市之外，以狃於「強幹弱枝」之見解，概不設防，人民不習兵事，可以傳檄而定，四也。有此四弊，內亂外患，率同時並起，欲求國防之鞏固，不可得已！余意近代城市規劃，既已破一城市之界綫，而作為其所在「自然區域」之一單位，乃有所謂「區域規劃」，而「區域規劃」又必須超越其界線，作為國家之一單位，而有所謂「國家規劃」，城市規劃之能事，在使一城市內之各種單位，如街道，建築，工廠，居宅等等，作和諧之配合，同理「區域規劃」乃在調整其包含之城市單位與鄉野背景，形成「城市系統」。再進一步，國家規劃乃在形成一種「城市網」，以地理為背景，以歷史為綱領，以各種交通運輸為線索，使國家作一「全力戰爭」之整體，而永遠存在。此種「城市網」將有類於漁網，結結自緊，線線有力。平時則藉之以使全國民衆休戚相關，戰時則外力侵入，有如自投羅網。美國作家孫德安氏嘗盛稱德國戰略思想，此亦乃為「立體或三度者」，實則，一切戰爭均有「時空」相合之「四度性」。過去之戰略，在先佔有「空間」，而後漸漸克復「時間」，即以種種經濟手段與名利心理，毀滅被征服者之歷史觀念，因而取消其復仇雪恥之思想耳。至於此次大戰初期，德國所發動之閃電戰，實為一種「四度戰爭」，即在利用其一切有形武器，佔領空間以外，更用第五縱隊，發動所謂「心理戰爭」，如法國名作家莫洛亞氏所描寫者，實為在短期內征服「時間」，或謂之為先期克服對方「時間」阻力——此其所以為「閃電」者歟。換言之，即德國以其「四度戰略」擊破歐洲一般蛛網陣式之國家，設如余所言之「漁網陣式」，則外力一入，各「線」各「結」，均可分受其影響，因而減輕其能力，所謂「人人敵愾，步步為營」之理想，始易實現，使略者之技亦窮如黔驢而已！

五　結論

國父實業計劃原在利用外資，開發我國實業，故特別注重於沿海港埠之建築。但試一細心研究其六大計劃之先後次序，當不難發見其苦心孤詣之處，同時目光如炬，不特對於國際形勢瞭如指掌，且將資之以為我國建立國防之根據，如再與國防十年計劃書之目錄對照讀之，更可有所發明，其尤有關於我城市建設，除主要者列入於「社會」部門外，其他如(一)(丙)(30)「各地軍港要塞砲台航空之新建設計劃」，(三)「政治」(41)「收回我國一切喪失疆土及租借地，租界劃讓地之計劃，以及(四)經濟部門內所列各款，均具有「國家規劃」之意義，而以城市建設為其對象，故編者，首即說明此點。機乃根據此項「國家計劃」，討論我國城市建設之細節。茲試作結論如下：

(一)發展中小型城市，疏散龐大城市；

(二)城鄉和諧，工農兼顧；

(三)疏散與聯繫，並行不悖，以形成一種「抗侵略性」之國家城市網；

(四)城市成為人民安居樂業求進步之場所，務須工作條件與生活條件一致——抵抗外來侵略時之戰爭，亦應視為平民正常工作。

本此結論，作為我國城市建設之基本假設，將見其工作繁雜，非同小可，所需各種人才，其數尤多，試就英國之經驗為根據，每三十人口，即需地方行政（當然以有關城市建設為主）人員一名，準此估計，欲將我國五千人口以上之城市近代化者，即須一百六十萬人員，（最近美國之統計數字，其各邦市地方行政人員達三百萬名，約合每四十人口一名。）此處「地方行政人員」一名詞，當然包括各種人色，自行政主管人員，醫工專家，乃至小學教員警官巡士等，均計算在內，但此類人員，必須有相當訓練為其基礎，方可勝任愉快，從此亦可見國父國防十年計劃書內所列「教育」門內（55）「訓練國防基本人才一千萬計劃」，估計原不算高，因城市建設一項所需人才，即按美國標準，亦應在一千萬以上也。

民國三十二年七月四日重慶寓園

城鎮改進設計須知

胡樹楫

一、弁言

依國民政府頒佈之都市計劃法，凡已成立之市，已闢之商埠，省會，聚居人口在十萬以上及其他經國民政府指定之地方，應儘先擬定都市計劃，由內政部會同關係機關核定，轉呈行政院備案，交由地方政府公佈施行。至於一般較小城市（人口十萬以下），如縣治，鎮，集等之設計，於法雖無明文規定，然因事勢之需要，此等城鎮，或早已着手改進，或今後亟待整理，尤以戰時曾遭兵燹者之除舊佈新為最迫切。為謀建設之合理化，則適當之計劃與方針自亦不可少。

商埠省會等之都市計劃所涉之範圍既廣，研討所資自有專籍，主持其事應由專家，殊無代為借箸之必要。反之，一般城鎮之改進大都毋需大刀闊斧之手段，亦不必懸最高至上之目標，則若干重要原則之貢獻或已敷實際上之採擇。本文之撰擬即本此旨趣，不尚多談理論，但求便於實行，藉供謀地方新建設者之參考。

二、總綱

一、在謀城鎮改進之先，應就地方情形詳加研究，如交通（水道，鐵路，公路）與物產（包括尚未開發之富源如礦山等）是否將使工商業更趨發達，抑僅足維持現狀，改進所需經費如何籌措及為數可達若干，何處宜於建築（如平坦乾爽之地），何處宜留為原野（如陡坡及低窪之地），庶成竹在胸，發為計劃，無不切實際或眼光過短之弊。

二、擬製城鎮改進計劃之先，應備有當地及附近一帶之地圖，如無現成地圖（如各省陸地測量局所印行者）可得，最好從新測繪（必要時可借調測量，或工程機關人員担任此項工作）。此項地圖之比例尺，最好在二千五百分之一以上，最小亦不宜在一萬分以下。一切改進設施，如街道拓寬及開築之界線，公園菜場設置之地位等等，均於圖內標示，附以文字說明，呈報主管行政機關備案並公佈。

三、萬一無現成地圖可得，又限於人力與財力，不能自行測製，所有改進計劃及其實施程序，至少應以書面說明，附以草圖，呈報主管行政機關備案並公佈。

四、城鎮改進計劃應就今後工商業與交通上及居住上之實際需要及環境衛生着眼，并兼顧一般人民之經濟能力，不可好大喜功，過於理想與誇張。在交通發達（或將發達）及富源充足之處，又須具遠矚最近三十年以後之眼光，勿過於遷就現狀。

五、計劃所包括之各部分，何者應一氣改造或建設，何者宜逐步實行，須有妥善適當之籌劃。例如拓寬街道，可令全街舖戶同時拆讓，一氣實行，亦可隨沿街舖戶之翻建房屋或改造門面逐步整理，應視財方之豐嗇，需要之緩急，及市面之榮枯等情形決定之。

六、計劃之效力，兼含積極消極兩種性質。例如規定某處爲未來道路所經過，是爲積極性。因闢路計劃一時未能實現，而根據規定阻止該地面上永久性房屋之建築，以避免將來實施計劃時之困難，則爲消極性。徒有積極之規定，而無消極之設施，計劃往往成爲紙上空談，不可不注意。

七、城鎮改進如築路拓路等所需之土地，屬於人民私有者，如係無償或低價收用，應用協商或名譽獎勵（如政府機關謝函，路旁立碑勒名等）方式，激發其急公好義之心，以徵得同意。因築路拓路而無條件拆讓之建築物亦準此辦理。勿用強迫手段以免糾紛。收用之土地，如原有稅捐之負担，應請由主管機關比照劃除之。

八、城鎮改進所需之費用，如不能由政府機關撥給，或由地方自治機關籌措，宜向地方熱心公益人士勸募（並予以名譽獎勵）或向享受利益者徵集，勿用附捐等名義強向一般人民收取。

九、城鎮改進所有經費收支情形，應呈報主管行政機關備案並公佈，尤以經費由捐募籌集時爲然。招工購料等手續亦應公開。

三　水陸交通

一、主幹水道之疏浚與改良，自有專管水利機關主持其事，但支流之疏浚改良，碼頭堤岸之修築，應由地方商明水利機關辦理之。

二、水道之過往商旅貨物有在當地與鐵路聯運之需要時，宜及早與鐵路機關商酌鐵路與碼頭之銜接辦法，以便預留設站舖軌之地位。

三、鐵路與公路之車站及水道碼頭與商業中心之間，應有汽車路（在將來工商業發達時或並有電車路）之聯絡。

四、交通殷繁之公路線最好不經由熱鬧市街，而於城鎮外通過。此於商旅雖稍有不便，然可免市街塵囂擁擠之弊。

四　建築

一、一般城鎮之商業（包括手工業）建築與住宅建築，大致無劃分地區之必要（可聽其自然發展，將來各自成區）。但機器工廠，例如碾米廠、麵粉廠等，以及一切以蒸汽爲原動力之工廠，足使空氣惡濁，環境囂噪，應另指定地點設置。工業區之專容納小規模機器工廠（其出產品以供給當地及附近需要爲主，如碾米廠、地方電廠、水廠等）者，最好離城鎮邊一公里以上，並對城鎮居逆風之方向（以一年間最多之風爲準）（圖一）。其專容或兼容大規模工廠（其出產品以遠運爲主，例如鋼鐵廠、兵工廠、大電力廠、較大紡織廠等）者，則應離城鎮邊十公里以上，以防戰時空襲危險。其對城鎮之方向，則可毋需深究，但應遠離風景區及戰時人民防空疏散之適宜地點。

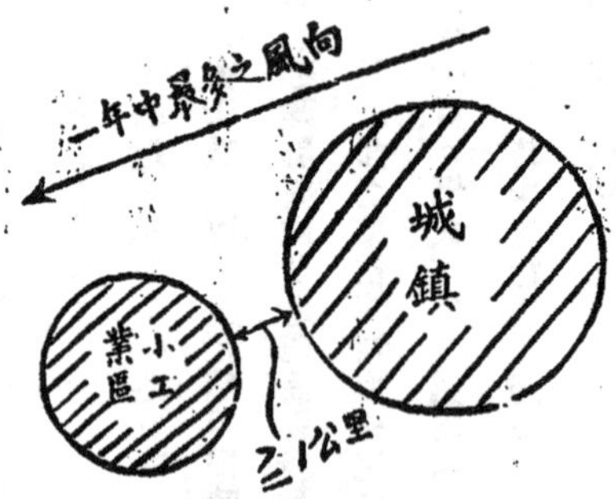

一年中最多之風向

城鎮

小工業區

一公里

圖（一）

二、城鎮爲謀人民居住之安全與衞生，及促進交通之便利與市容之觀瞻起見，應訂立簡要建築規章或公約，監督勸導人民遵行。

三、建築規章或公約應包括下列各款：

(一)沿街建築物前面應讓出規定之街道線。（理由：促成街道之逐漸拓寬）

(二)毗連之建築物，尤其深二進以上者，後面（或側門面）應另有出向公巷。現無公巷時，應由各家於後面（或側面）各劃出寬一公尺半以上之地段，以便闢通公巷。（理由：防火災危險）

(三)毗連或左右相距不滿三公尺之房屋間，每家應於左右界址各建封火牆。同屬一戶之房屋有多間並列時，並應至少每隔若干間（例如六間）或每隔若干公尺（例如二十五公尺）各建封火牆（封火牆應高出屋面，至少用磚二道砌成，木柱架不得砌入牆內）。（理由：防火災蔓延）

(四)每家房屋（尤其住宅）應有充分庭院空地；庭院（天井）之寬深，至少應與其旁屋簷或圍牆之高度相等。（理由：使光線充足，空氣流通，以利居住衞生）

(五)沿街道邊房屋（不由街道邊退後起造者）之高度（即屋簷或裝飾面牆之高度），至多應以與街道寬度相等爲限。（如退後起造，得將退後尺寸加入街寬計算）。（理由：同(四)）住屋之高應以二層樓爲限（即連地面共三層；下層屋簷上頂屋瓦之部分不以樓論）。（理由：高過二層樓時，住戶上下不便，火災時尤危險）

（說明）關於(四)(五)兩款房屋之高與其距離或街寬之最大比率，理論上應以冬日陽光能射入屋內爲標準，故與地方所在之緯度有關，即距赤道愈遠應愈小，距赤道愈近則愈大；又與房屋之方向有關，即東向或西向者應較南向者為小。如就吾國住屋通常南向之習慣而論，上文所擬之比率（一比一），理論上在華南似稍涉嚴格，而在華北又嫌寬大。但實際上此種比率之選擇，須兼顧地方氣候，地勢等情形。華南各地或夏日酷熱，或多雨潮濕，或位於山嶺間，為求爽豁通風計，殊以街寬屋低為宜。華北各地氣候亢爽，而冬日寒風逼人，則又不妨街窄屋高。故一比一之比率，似為適用於全國之折衷數。

（六）河溪湖沼公井等之旁至少在十公尺之範圍內不得建築。（理由：維持水流清潔及水濱交通）

（七）建築物（尤其公共建築如會場戲院旅館茶樓等）之有特殊情形，如梁桁特長（開間特大），柱架特高，或應用鋼鐵或鋼筋混凝土為柱梁樓板屋頂等者，其設計監工應由有工程學識經驗之技師辦理。（理由：保障建築之堅固，以防發生意外）

建築規章或公約以適用於新建築為限，不追溯以往。

四、建築發展之趨勢，宜使由城鎮中心向郊外伸張，而不團簇擁擠於固有城鎮範圍之內，故如舊有城垣城濠無保存之必要，應及早拆除填濠（惟如城濠水流清潔宜保留為風景之點綴品），並多闢出郊街道。市內原有菜圃空地等，宜獎勵人民盡量保留，不作建築之用，必要時或收買為公園林場等，以資建築之疏散，而利居住衛生與戰時防空。

五　道路

一、一般城鎮之舊街道，大都狹隘曲折，不便交通，且因空氣不易流通，陽光難以照射，潮濕污穢，有礙衛生，故須予以拓寬與改良。

又城鎮為適應交通與建築之發展，每有添闢新街道之必要。

在謀改造及新闢街道之先，應就各街道對中心市區（原有熱鬧街道如「正街」之類）與碼頭車站及四鄉交通上之形勢，加以研究，以決定孰將為車馬行人往來之通衢（下文簡稱「交通路」），孰為有事於商場與市廛者所麕集（下文簡稱「商業路」），孰僅供居住者及其訪候者之出入（下文簡稱「居住路」），庶可視交通之繁簡，分別選定寬度，以收經濟之效。最好根據研究結果，于地圖上將所有街路之路線概行繪入，而成「道路系統圖」，為整個城鎮改進計劃之重要部分。

二、選擇街道之寬度不宜過小，過小則不足以應將來交通上之需要，又不宜過大，過大則路面修築保養之費用無謂加大，在舊街道改道時困難亦較多。

茲假定城鎮各種街道之寬度如下表，以供謀城鎮改進者之參考：

路別	車馬道寬度（公尺）	每邊人行道寬度（公尺）	總寬度（公尺）	備考
一等交通路	一四、五〇	三、二五	二一、〇〇	通行電車之交通路城鎮幅員將在五公里以上者應有之
二等交通路	一三、〇〇	三、二五	一九、五〇	車馬道兩邊停放車輛僅較大城鎮有之
三等交通路	一一、五〇	三、二五	一八、〇〇	車馬道之一邊或中央停放車輛
四等交通路	九、〇〇	三、〇〇	一五、〇〇	汽車交通繁密時不得停放車輛
甲種商業路	九、〇〇	三、七五	一六、五〇	行駛汽車但以不停放車輛爲原則行人交通甚繁
乙種商業路	九、〇〇	二、二五	一三、五〇	同上但行人較少路旁不植樹
丙種商業路	六、〇〇	二、二五	一〇、五〇	同上同時汽車交通甚稀
丁種商業路	—	—	九、〇〇	不行駛汽車無車馬道人行道之分
甲種居住路	九、〇〇	（三、〇〇）二、二五	（一五、〇〇）一三、五〇	直接通入交通路之居住路
乙種居住路	六、〇〇	（三、〇〇）一、五〇	（一二、〇〇）九、〇〇	僻靜之居住路
丙種居住路	—	—	六、〇〇	里巷性質無車馬道人行道之分二層樓房屋應由路邊後退
房屋之後面出路（後巷）	—	—	至少三、〇〇	房屋宜由路邊後退

附註）（一）上表中各項總寬度係以三公尺（及其半數）之倍數爲標準，取其便於記憶，且免各級之間相差懸殊（例如按五公尺遞進）。

（二）車馬道寬度之計算，係假定雙軌電車及其站台佔五公尺半，行駛之汽車每列各佔三公尺，停放之汽車（及其他車輛）每列各佔二公尺半，行駛之慢車（人力車脚踏車等）每列各佔一公尺半。例如寬一四、五〇公尺之車馬道，即由雙軌電車道五、五〇公尺，雙列汽車行駛線六、〇〇公尺，及雙列慢車行駛線三、〇〇公尺所組成；寬一三、〇〇公尺之車馬道，即由雙列汽車行駛線六、〇〇公尺，雙列慢車行駛線三、〇〇公尺，及雙列汽車停放線五、〇〇公尺所組成；餘仿此。

（三）前此一般城市之人行道多嫌過窄；行人往往擁擠滯塞，或侵入車馬道，殊非所宜。在商業繁盛之街道，其人行道尤應從寬設置，例如三、七五公尺，則無電桿行道樹等之障礙時，可容五人並行，有樹桿時亦可容四人並行（每人以佔七十五公分計算）。如因舊路拓寬不能過多或其他困難，亦不可減至二、二五公尺以下，俾無樹桿等阻礙時，至少可容三人並行。交通路之人行道亦宜寬三公尺以上。居住路之人行道不妨寬一、五〇或二、二五公尺，但如植樹最好增至三公尺，以利樹冠發展（參閱表中括弧內尺寸）。

(四)純粹商業路最好不通汽車，以便購物者從容徘徊瀏覽於兩邊商店之間，如表中所列丁種商業路是。

(五)較小城鎮(人口五萬以下)大致採用三等以下之交通路，乙種以次之商業路與居住路，已足敷需要。

三、街道(尤其交通路)曲折（即路綫改變方向）之處，應插入圓形曲綫其半徑(以路中線爲準)最好在一百公尺以上，俾凸出處視綫爽煞，而汽車可以每小時五十公里以上之速度行駛。惟曲線半徑（R）愈大，曲折角度(B)愈小，則凸角處房地退讓之尺寸(K)愈多，故在街道由舊路改善，或有其他特殊情形（例如附近有名勝建築物或古樹等必須保存）時，往往有將曲綫半徑，減至最小之必要。茲就十二公尺至十八公尺寬道路，假定最小行車速度爲每小時十五公里，凸角退讓尺寸在六公尺以下，並計及車馬行人之安全，計算最小曲綫半徑如下表：（參閱圖二）

道路寬度(公尺)	行車速度(公里/小時)	直視線長度(公尺)	曲折角	最小曲線半徑(公尺)
一八以上	五〇	七〇	一五〇—一八〇度	一〇〇
	三〇	五五	一二〇—一五〇度	四五
	一五	三〇	九〇—一二〇度	二〇
一二以上	五〇	七〇	一五〇—一八〇度	一〇〇
	一五	三〇	九〇—一五〇度	二〇

圖（二）

(附註)(一)汽車在灣道上以一定之速度行駛，除應有相當長之直視線，以免與其他車輛衝突外，又須所發生之離心力不致影響安全，上表即參酌此種情形擬定。(二)在二十公尺之曲綫半徑（在窄於

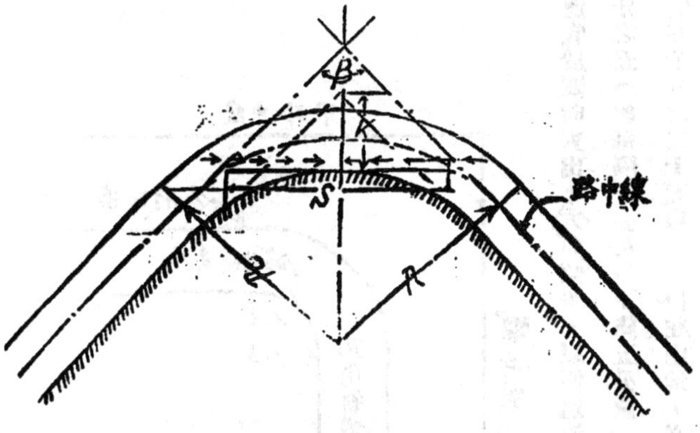

十二公尺之街道宜增至二十五公尺〕之灣道上，因直視線較短，而行人又每喜斜越車馬道以取捷徑（如圖二中小箭號所示），故最好更將行車速度限制在每小時十五公里以下以更求安全。

方向相反之兩曲線間應介以長十公尺以上之直線圖（三）。

兩路交叉處路角應設半徑三公尺左右之圓曲線或與其弦相當之斜截直線（圖四）。

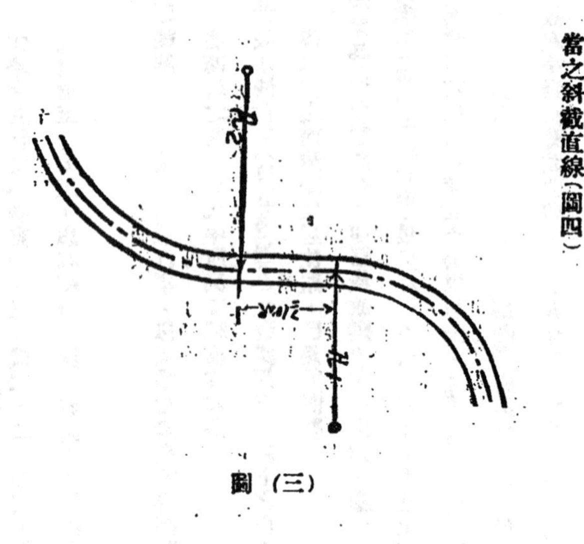

圖（三）

四、街道宜於縱向對出水方向傾斜，以便迅速排洩雨水，其斜度應在千分之五（即每隔一公尺長下降五公釐）以上，為便利交通計，又宜在千分之二十五以下，但在山地，必要時可增至千分之八十左右。

甲路中線
人行道邊
路角圓曲線
路角斜截直線
乙路中線

圖（四）

五、計劃道路系統時，首宜注重交通路之設置，次則為有商業路與居住路之改良及擴充。舊有城基不妨設置環城路。沿河流鐵路宜闢貫通之街道，以免背河流鐵路建築之不雅觀。又城鎮舊有建築稠密部分，往往前街房屋與後街房屋毗連，致建築物僅有前門出路，發生火患時，影響住戶之安全甚大，宜於前後街間添闢寬三公尺以上之通巷以爲救濟（參閱四—三—(二)）。

六、實施築路之程序，應先從接通碼頭車站及商業繁盛之街道着手。在市面勃興之處，出郊道路亦宜及早興築，環城路則應以緩開闢，以促進建築之向四郊發展（參閱四—四）。其一般舊有街道與巷，可隨沿街巷房屋之翻造修理，逐步拓寬改良。

七、路面（車馬道及人行道）之舖築，在一般城鎮，可暫用下列各種材料：

(一)石板或石塊（整齊砌舖）。

(二)碎石（粗細分層，或混合壓舖）；（在石料缺乏之處，往往用沙礫或碎磚代替）最好於其上加舖水泥三和土面層，則晴日揚塵雨天泥濘之弊較少。

(三)石片鵝卵石（不規則砌舖）。

(四)煤渣或沙（鐵舖）。

除熱鬧街道外，車馬道與人行道必要時可僅舖建一部分，（例如車馬道三公尺寬，人行道一公尺寬，）其餘部份挖塡平成爲土路，或略舖沙或煤渣，以防泥濘。又車馬道與人行道亦可暫不劃分。

車馬道路面應由中央向兩邊傾斜，人行道路面應自路邊向車馬道傾斜以便洩水入溝（明溝或暗溝之納水口在車馬道與人行道之間）視路面之平糙，縱向斜度約爲千分之二十至五十。

八、人行道之寬度在三公尺以上者宜植樹以添風景。樹與樹之距離約六至十二公尺，其地位勿對向沿路建築物之門戶。但人行道之在路南及灣道凸邊者，因缺乏陽光照射或妨礙行車視線，均不宜植樹。

九、跨人行道之騎樓妨礙植樹，在灣道之凸邊又遮蔽行車視線，建築物內部之納光通風亦蒙不良影響，除特殊情形外，以不設置爲宜。

六　公園及其他公共用地

一、在人口較多，幅員較大之城鎮，尤其市內建築稠密，而四郊無天然風景者，宜設置公園，以供人民之游憩，兒童之嬉戲，而陶養愉快活潑之精神。

二、公園之面積不宜過小（大抵至少應佔地十市畝以上），否則效用甚微。與其設置小公園數處，不若擁有較大公園一所。其佈置宜委由富有美術眼光與園藝知識者設計。而市內無充分地面可得，不妨設於郊外（最好利用山邊或水濱之荒廢地）。

三、公園面積較大時，除兒童游戲場外，可附設體育場及民衆教育設備（如民衆閱覽室，動物園等）。

四、舊時之城垣城濠基址，宜佈置為公園帶，以供人民游憩，而環城道路之開闢則待諸將來（參閱五—六）。

五、市內舊有叢樹溪池，及一切可資點綴風景者，應盡量保存，最好並與公私園囿及其他空曠地（參閱四—四）聯成一片，與郊外田野山水相接，成為整個風景地帶。

六、郊外童山荒地，宜收用或獎勵人民植林，既可點綴風景，復有調節氣候防範洪水之效；至於材木果實之利用，猶其餘事。

七、名勝古蹟應予保存。一切公地應留供公共用途，勿貪近利，售予私人。

八、人烟稠密交通較繁之城鎮，宜設置菜場，其有市集者，並設置市集場，以維街道之秩序。

七　公共設備

一、按照一般城市之情形，電燈之裝置，比自來水較易普及。故小城鎮，尤其有水力，或低廉煤料可利用者，亦宜從早開辦電廠（公營或商營），以杜煤油輸入之漏巵，並以代替一般小工業（如碾米廠等）所用之舶來液體燃料。

二、至少重要街道（交通路及商業路）應有路燈之裝置，以利夜間交通。

三、城鎮宜設取費低廉之公共浴所，以重平民衛生（附以少數較優設備，以應較富裕階層之要求）。如初時風氣未開，就浴者少，不妨定期開放（例如每隔三日或一星期或以市集為期）。

四、廣告標語宜指定地位粘貼，或另立廣告標語牌於路旁，以維觀瞻。

八　環境衛生

一、一切城鎮應注意環境衛生，如飲水之清潔，雨水之排洩，污水與糞便垃圾之適當處置，街道之掃除，蚊蠅鼠蚤之撲滅，以保公眾之健康。

二、城鎮最好有自來水之設備（由專家主持設計）。如限於人民經濟能力，不能辦到，至少應維持水源之清潔。如鑿井，或引泉，或開蓄水池塘（宜養魚以除孑孓藻菌），應在人烟稀疏，環境清潔之處。如取用溪河之水，應在溝渠尾閭之上游。飲水井塘及溪河汲水處之上游，應禁止洗濯泳浴（必要時另指定洗濯泳浴處所）。水源附近不得有廁所等之設置。

三、城鎮所有雨水與污水，最好設置有系統之新式陰溝網排洩之；並由專家主持設計。如限於經濟能力，不能辦到，至少在建築稠密之處應有明溝或暗渠，以洩雨水於適當尾閭（最好為較大溪河之下游，不宜爲蓄水池塘）。溝渠底對於尾閭須有一貫之斜度（最好在千分之十以上，即每隔一公尺至少降低一公分），並時時予以清除，以防積泥淤塞。至於盥漱之餘水，廚間之棄汁，及其他汚水，則應勸導人民用以澆園，洗廁或飼畜，而勿隨地傾潑。

四、水冲廁所在未有自來水及新式陰溝系統（與公共淨化汚水設備）以前，雖期普及，而糞便向爲一般農民利用爲肥料，故一般城鎮處置糞便之方法，要不外容以坑桶，隨時外運。（在未有新式陰溝及公共淨化汚水設備以前，如偶有水冲廁所之設置，亦應附築化糞池以容納糞便，不得放入雨水溝渠內。）惟民間廁所之建造率欠完善，應示以範式，提倡改良：尤注意於坑內底壁之粉刷防濕（防滲汚地下水源），穢氣之設管高洩，容納與儲留汲取部分之分離，蠅類出入之阻止，（最好更時洒石灰乳以消毒）其地位應遠離居室廚房水井等。公廁尤應儘先改良，並設於隱僻之處，（勿臨通衢，最好在空地上，叢樹間）。又糞便之外運，應限用有蓋器物，並以泥土塞縫，最好並在清晨行之。

五、民間之垃圾廢物，勿令隨時傾棄。首宜勸導人民維持舊時民間分別處置之良好習慣，（廢棄食料用以餵飼雞豚，灰土用以肥園，字紙收集焚化，破銅爛鐵售予收買舊貨者），可以利用者盡量利用，必須傾棄者（尤其廢棄食料等易腐物質）最好由公衆集貲雇工逐日按戶收集，運往郊外偏僻之處（至少離城鎮三五百公尺，最好居下風方向），或用以肥田或填塞窪地，或設灶焚化。舊時一般城鎮傾棄垃圾於水濱之惡習，必須廢除禁止。

六、街道應由公衆或集資雇工逐日（清晨或深夜打掃）清潔。亢旱時並時時洒水，以防揚塵，而利衞生。

七、蚊蠅鼠蚤之撲滅，宜從積極方面着手。如地無積水（良好溝渠設備），室無廢食（適當垃圾處置），穢物隱蔽，（良好廁所設備及適當垃圾處置），沐浴便利（公共浴所設備），則蚊蠅鼠蚤無從滋生，自可漸趨消滅。消極撲殺方法，僅宜資爲輔助。

九　結論

城市爲人口集中之地，其榮枯攸關國計民生，欲順應世界經濟文化進步之潮流，其改進發展自不可少。惟就國防方面而言，農村爲立國之基礎，全國多數人口集中於城市，殊非所宜。必不得已而思其次，則少數大都會之畸形膨脹宜及時加以阻遏；而全國工商業及文化設備應盡量分散於多數小城市。故一般城鎮今後所負之使命尤覺重大。將謀完成此項使命，必拓闢道路，改良建築，增置公共設備，促進環境衞生，使行便而居安，近悅而遠來。然亦宜顧及當地人民經濟能力建設無取過事舖張，面目不妨姑從簡樸，此本文撰擬所本之微意也。

工廠設計須知

胡樹楫

一·弁言

將謀民族復興，必先發展工業，以應人民生活上之需要，而供給國防上之資源。前此我國天然富藏多未開發，重工業固尚無規模，即人民日常服用物品亦未達到供求相應之地步。此次抗戰已激發國人發展民族工業之自覺心，將來隨交通事業建設之進展，各處工廠之勃興，勢必如雨後春筍。顧工廠成立之條件，除應備之資本外，有原料，人力，運輸等種種因素，而建築與設備本身對意外災禍（如火患等）之預防亦不容忽視。尤以重要工業，或為國防上直接資源所倚賴，（如鍊鋼廠，鍊鎢廠，鍊銅廠，兵工廠，飛機汽車火車船舶製造廠，石油鍊製廠等，）或為大宗民族經濟集中之企業（例如大規模之機器工廠，紡織廠及化學工業）戰時之敵方勢必謀加以破壞，如猶於平時環境，漫無對策，一旦燬於敵機炸彈之下，國家將蒙莫大損失，國防將受意外威脅，關係之鉅，不言而喻。故下文特注重安全之一點，就工廠地址之選擇，建築之方式，及運輸問題等方面，列舉原則若干條，以供設立工廠者之參考。

二·地址之選擇

一、工廠地址之選擇，應注意下列各點：（一）原料貨品運輸之便利，（二）防空上之安全，（三）地價之低廉，（四）人力之易得，（五）給水排水之便利，（六）其他一般有關建築方面之問題。（註一）

（註一）以上各端，有時實際上不可得兼。例如就（一）（四）（五）等項而言，設廠地址普通宜莫如城市，而以大都會為尤佳，而（二）（三）等項之要求則趨向鄉間。然為整個民族工業與國防之關係上着想，工廠地址應以遠離城市（尤其大都會）為原則，而因此所致之缺憾並非無法加以補救。（參閱下文）

二、有兩地於此，一則便于原料之取得，一則利於製成品之運銷，宜儘先擇用便於取得原料之地；但原料由國外輸入時則反是。（註二）

（註二）原料價值自較製成品為低，加以製造時不免有一部分廢棄，其量亦較製成品為多，故原料之長途運輸為不經濟。反之，由零散之原料製成整齊之貨件後，運輸上自較便利。故如冶鐵廠鍊鋼廠之設於鐵鑛附近，鋸木廠之與森林鄰接，紡織廠之在產棉地區，殆爲當然之事。至原料之由國外輸入，究爲例外情形，或屬過渡期間不得已之舉。如在各消費區分別設廠，將貨品就近銷出，則原料方面運費之增加，亦可藉貨品方面運費之減少抵消之。（必要時可由政府就各產銷區對外來同樣貨品課稅，以保護當地工業。）採用本原則，可避免前此工業集中沿海一帶之流弊，而促進民族工業在國內之普及化與疏散化。

三、運輸問題應與防空問題兼顧並籌，故工廠雖應有水道鐵路或汽車路之聯絡，而不宜逕傍水道或鐵路公路幹線，尤不宜位于城市（尤其大都會）附近，（註三）應擇距上列各項較遠（至少十公里）之處設置之。（參閱四、一—二）

（註三）城市（尤其大都會）及延長之河流鐵路公路線均爲由空中易尋覓之目標，且城市（尤其大都會）每爲敵機轟炸之目的，交通幹線之上空每爲敵機往來所經，故應遠避。

四、爲求防空上之安全起見，工廠地址並應符合下列條件：

（甲）遠離國防線。

（乙）勿與其他工廠團簇一處，最好各自分立，必不得已亦祇可魚貫羅列，切忌前後疊置。

（丙）地位在較大城市（尤其大都會）之後方（對國防線之形勢而言）。

（丁）地勢上有隱蔽之資，如：

（1）綿延之森林，

（2）山谷或山塢平原，

（3）山陰（山嶺之北面）及在敵機可能空襲時間陽光所不直接照射之處。（註四）

（戊）與已有鄉村隔離（至少相距半公里）。（註五）

（註四）例如距國防線較近時，敵機空襲時間多在上午，則廠址宜在山嶺之西面，距國防線較遠時，敵機空襲時間多在下午，則廠址宜在山嶺之東面。

（註五）防萬一遭受空襲時波及鄉村。

五、平房式與高棚式廠屋之建築（參閱三、一）需地較多，加以將來之擴充不可不預爲計及，故廠址應擇地價低廉之處，以減輕開辦費用。（註六）

（註六）就此點而論，工廠亦應遠離城市與交通綫。

六、人工招募之難易，與各地區人民之生活習慣有關。大抵沿海區域人民伶俐活潑，北方各省人民勤苦耐勞，均較適於工廠生活。惟廠址所在，應以原料運輸與防空爲先決問題，則在內地設廠初時對於雇工方面雖稍有困難，亦不必過予顧慮。補救之法，可由他處（尤其沿江海大城市之附近）招募所需之工人內運，並添招若干藝徒，為必要時之補充。（註七）

（註七）沿江海區域人口較密，且因農村及一般經濟最先感受列強侵略之影響，失業人民亦較多，故由沿江海區域移民至內地各省，有裨於國計民生者至巨，政府方面宜設法獎勵其實行，如減免移民舟車費等。反之，內地正病荒地多而人口少，其農村狀態最好勿使為工業發展所牽動。

七、工廠為取水便利起見，最好設於山旁可引泉下注之處，否則井水或溪流河流亦可資利用。

為排水之便利起見，工廠（尤其有大宗汚水排出者）所在應有溪流或河流，以四時水流而流急者為最佳，庶廠內汚水放入後可為所滲淡而淨化，沿溪河民居之飲用水源，須維持其清潔。

八、其他一般關係建築方面之問題大抵甚易解決，祇須注意下列各點：

（甲）無洪水為災（可詢附近居民而知之）。

（乙）地勢宜平坦，而稍向一面傾斜，但原料由高處來貨品往低處運時，則較峻山坡亦無妨礙。

（丙）地質相當堅實，無沉陷滑塌之虞。

（丁）地面宜乾爽，以免建築上防潮設施之煩。

九、地址宜成長條形，以便平房式（或高棚式）廠屋向兩端延展及擴充。

十、地址之大小可約略按每工人一名需廠地三五—四〇平方公尺，或廠地面積合廠屋面積之二倍計算。此外並酌加將來擴充時所需之面積。（註八）

（註八　參閱 Blum, Stadtebau, 2, Auflage Hutte III, 26. Auf, 1 S.384

十一、如為員工等設備住宅，其地點宜距廠址稍遠（約半公里至三公里），並對廠址居風向（一年間最頻者）之上方（註九）。所需面積可約略按每二百人至三百人（包括員工眷屬）佔地一公頃（一萬平方公尺）計算。

（註九）謀員工居住上之衛生及空襲時之安全。

十二、廠址如與鐵路接軌，其長邊宜與接軌線大致平行。（註十）

（註十）如此，接軌線或其岔道可直通廠屋之各部，而無假助轉轍臺之不便。

三・建築之方式

一、工廠建築應盡量採用平房式，而於必要時採用高棚式。（註十一）

（註十一）按工廠建築大抵不外三種形式：卽樓房式、平房式、及高棚式。樓房式廠屋係由西人樓居之習慣，遞邅而來，除可節省地面與較便取暖設施外，並無特殊優點。如在地價低廉之鄉間建廠，節省地面既可不必，而我國各地冬日大都無嚴寒，取暖設備亦屬次要問題。就樓房式工廠之劣點而言，則笨重貨料之傳遞，龐大機器之裝置，房屋本身之擴充，均屬不便，而因機器設備及貨料等層疊於較小之地面上，一旦發生火患，或為敵機投彈命中，損失之巨，將不堪設想。故今後工廠建築，普通應採用平房式，使機器與貨料分散於較大之地面上，以免上述之弊。此外平房式廠屋因墻身較低，炸彈斜着之機會既少，碎片之防禦亦易，（可於屋外堆置沙袋以保護墻垣薄弱部分及窗戶，）故宜盡量採用。如製品龐巨（如鐵路機車及車輛等），或機器高大，則必要之部分可採用高棚式建築。

二．工廠建築最好為單間式，成長條形延展，（註十二）並藉窗戶納光，而避免天窗之設置。如必不得已，剖面最多勿過三間，天窗之設置亦以光線從旁面（氣樓式與鋸齒式天窗）而非從上空（屋面一部分蓋玻璃瓦）輸入為宜。鋸齒式天窗應對向北方。（註十三）

（註十二）建築物愈狹長，則敵機空襲時投彈愈難命中，故工廠橫向開間以少為宜。

（註十三）完全暴露而承受陽光之玻璃天窗，向天空反射強烈光線，易惹敵機注意，故應避免。（參閱鄒恩泳譯「房屋建築及城市設計對於防空之趨勢」新工程第三期四九頁）

三．工廠建築之牆柱等應盡量採用防災材料（磚，混凝土及鋼筋混凝土），但萬一被燬時急需修復之部分則宜採用木料。（註十四）

（註十四）因木質建築為火焚燬後，火場最易清除，重建亦較速（參閱 Hutte III.26. Aufl.S.401）。關於建築上防火問題，並參閱胡樹楫譯「建築物之防火效能」新工程第五期

四．工廠建築應至多每隔三十公尺設自地面至屋面以上之封火牆至少自半高至屋面之阻火籬。（註十五）

（註十五）參閱Hutte III.26 Aufl. S.402

五．工廠之屋面宜採用易燬散之材料（木椽蓋瓦）。（註十六）

（註十六）屋面所以宜用易燬散之材料，而不宜用木板或鋼筋混凝土等滿鋪者，因發生火患時，可任其焚燬散落，使烟及熱氣迅速發散，而於消防方面較為便利（參閱Hutte III.26. Aufl, S.403），又遭受空襲時，如炸彈落入屋內，爆發之氣體亦易外洩，而免醞釀較大之損失。

六．工廠建築除應作長條形延展，並多設封火牆或阻火籬外，最好並

將業務上聯絡較少之部分相當隔離。

七、戰時最爲敵方注目，如兵工廠等，最好設於山洞內，以防空襲。其出入口宜在山腰，而不宜在山脚。

四、運輸問題

一、工廠輸入之原料，或輸出之貨品，爲笨重性質，或特別繁多者，最好與水道幹線有支河之聯絡，或與鐵路接軌。

二、在下列情形之下，工廠宜以汽車路與水道碼頭，鐵路車站或公路綫相聯絡：

甲、輸入原料與輸出貨品爲輕小性質，而不甚繁多者。

乙、附近無支河可利用，與鐵路可接軌者。

丙、廠址在山嶺間者。（註十七）

丁、防空上須特別注意者。（註十八）

（註十七）汽車路受地形之束縛較少。

（註十八）汽車路比水道鐵路較不惹敵機注意，如舖以煤屑等晦黯物料尤佳。

三、如與鐵路接軌，宜與路局商定，由最近之車站起舖設專線，逕達廠內收貨交貨之處，並供給運輸車輛，以免轉駁週折。（註十九）

（註十九）參閱 Hutte III.26. Aufl, S.427—429

建都之工程觀

譚炳訓

〔1〕前言

中國工程師學會十月在桂林舉行的第十二屆年會，關於戰後建都問題，在市政衞生工程小組中，曾經就工程的觀點，作熱烈的討論。建都的實際工作就是市政工程。因爲我們是工程師，對於國都的選擇，應從國防建設與市政工程方面，來加以研究，以爲國都最後取捨的標準之一。

近來建都問題的討論，可謂盛極一時，就地理歷史以立論者有之，就國防經濟與交通以立論者有之，就氣候與民族康健以立論者亦有之，觀點不同，主張也就各異。

關於建都的論文及其主張，在報章雜誌上所發表的，列如下表：

作者	文名	何時何處發表	主張
張其昀	論建都	卅年十二月思想與時代第五期	就地理觀點及海陸兼顧主張建都南京
錢穆	戰後新首都問題	三十一年十二月思想與時代第十七期	就歷史的觀點主張以西安爲首都北平爲陪都
近眞任	論建都	三十二年九月五日大公報	建都西安

大公報最近刊載的兩篇星期論文皆論建都之作。第一篇是傅孟眞先生的「中國之中樞區域與首都」，第二篇是沙學浚先生的「戰後建都問題」，都是主張建都北平。這兩篇文章立論的精神是一致的，「戰後建都問題」一文，除說明北平建設的理由外，同時將武漢、西安、南京等主張及歷史說地理中心說，也一一加以評論。沙君的結論是「都南京是守成，都北平是進取」。這兩篇文章都有歸納各方意見，導國論於一是的意義。如再由工程方面加以論列，研究建都的實際問題，這一場論戰，似乎就可以告一個段落。

〔2〕各種建都的主張

張君俊	戰後首都問題	九月七日大公報	以民族生物學之立場主張建都北方或卽西安
陳爾壽	國都位置與地理中心	九月十六日大公報	建都武漢
龔德柏	武漢與西安孰適於建都	九月廿日大公報	西安第一北平次之
柯璜	定都之我見	九月廿四日大公報	建都西安
紀文達	戰後國都問題比較	九月廿五日大公報	第一北平第二南京第三西安第四重慶
大公報社評	戰後國都宜在北方	九月二十五日	建都在北方以北平爲首選
黃霖覺	勝利不容有折扣戰後應建都北平	九月二十六日大公報	建都北平
胡秋原	長春建都論	同前	飛機時代地理中心說已不適用爲防日本之再起應建都長春以「迎敵」因避敵爲下策
陳學公	論國都	三十二年十月四日重慶掃蕩報	在襄陽南陽間山岳地帶建造新都
柴眞（附）	戰後首都位置的檢討	三十二年十月二十五日大公報	建都北平北平對各方面皆可作攻勢根據地
谷鳳	論建都	十月二十八至三十一日東南日報	主張不確定南京武漢或汴洛一帶
現代農民（月刊）	本刊的建都意見	十一月十日六卷十一期	建都北平
谷春帆	遷都商兌	十一月十八日大公報	建都東北之松遼平原或卽遼寧
傅孟眞	戰後建都問題	十一月廿九日大公報	北平爲資源最富的北十省之中心及國防之前衛不以安樂而以憂患建國則應建都北平
沙學浚	中國之中樞區域與首都	十二月十九日大公報	北平可兼顧東北與西北海與陸並均衡南北之發展建國之八項重要工作皆在北方

上表所列十八篇論文中，主張北平者七，主張東北或北平者二，主張西安者五，主張南京者一，主張武漢者一，主張未確定者二。武漢有南京之弊而無南京之利，建都東北爲矯枉過正之議，皆可置而不論，現僅就北平南京西安三城，比較其建都工程條件的優劣。

〔3〕建都之工程觀

一個城市之適否建爲國都，就市政工程的觀點而論，須具備十個條件：（一）「交通」，（二）「地形」，（三）「氣候」，（四）「舊市區之利用與新市區之闢建」，（五）「近郊之風景名勝」，（六）「工程建設」，（七）「公用設備」，（八）「公共建築及建築材料之供應」，（九）「食料燃料與人力之供應」，（十）「代表國家儀容與民衆精神」。現將北平南京西安三市，按照這十個條件，分別比較如下：

一、交通：南京海陸交通皆極便利，惟其控制力僅及長江及珠江流域，此其缺陷。西安不通水運，在鐵路交通上又處在盲腸位置。隴海路尚未築至天水，必須天水至成都，天水經蘭州至迪化，蘭州經寧夏包頭三條鐵路築成後，西安才有內線交通，始够首都陸上交通應備的條件。北平有水運而無海運，北運河可通航至通州，通州距北平二十公里，通航亦無問題，傳說敵人已完成疏浚工作，將來平津間通航數百噸之輪船，工程上無困難。鐵路則北平原已四通八達，現除平包、平漢、北寧及津浦等四線外，敵人又增築數線。第一線爲北平經古北口至錦州之鐵路，並有支線通至熱河之赤峯。第二線爲北平經門頭溝涿鹿至大同之線，此綫可減少平包路青龍橋段運輸之擁塞。第三線爲北平經通州至唐山之線。現時北平，計有三條鐵路分經熱河唐山及天津通至東北。有兩條鐵路通至西北。（從平綏路終點之包頭，沿綏新公路至哈密爲一千六百公里；從寶天鐵路終點之天水，沿甘新公路至哈密爲一千七百五十公里，所以到新疆最便捷的路，是由北平經綏遠之一線。）一條經武漢而廣州。一條至南京，再延伸至皖浙贛。全國鐵路交通之便，當以北平爲第一。其控制力可及東北西北東南及西南的大部。平津間公路最近已全部改爲水泥路面。熱察晉冀魯豫等省之公路網，亦已完成。就交通而論，陸運以北平爲第一，距海在二百公里以內，且有運河之水運。南京海陸運皆備。西安則無水運，陸運恐戰後十年以內，亦無法解決。至於就空運而論，北平位置，亦近理想。以北平爲中心，劃一兩千公里半徑的圓周，則香港、廣州、昆明、昌都、玉樹、哈密、赤哈、伯利、東京、台灣，皆在此圓周以內或附近，即新式運輸機五小時航程之內。所以擴大眼光，就亞洲全局而論，北平的位置，更顯得重要。

二、地形：南京有起伏之地形，建花園都市及築下水道皆爲理想之地，湖沼大江及邱陵爲南京之特點。西安與北平皆位於平原上，市區地形無起伏之可言，惟近郊皆有山嶺，北平地層爲河流冲積而成，地下水甚好，全市皆呈向東南方向之徵坡，對於溝渠之建設，很多方便。

三、氣候：北平之氣候，有海洋大陸氣候之優點，而無其缺陷，夏季不必避暑，冬季亦不致妨礙戶外活動，四季氣象分明而嚴肅，人有爽朗靈敏之感。全年雨量六百餘公厘，乾濕適度，日夜溫度變化不速，有裨於健康，對於花木菜蔬水果之生長，亦甚適宜。西安爲純大陸氣候，春季乾燥，夏季日中甚熱，晝夜溫度之變化較大。南京濕熱，夏季不能使人淸醒深思，尙不及受有海風影響之上海，亦不及西安，更不及海陸氣候兼備之北平。南京實處於海與陸，南與北之臨界氣候中，似南似北，非海非陸，這是南京的最大缺陷。

四、舊市區之利用與新市區之闢建：南京建都雖有十年，新市區除住宅區稍有建設外，政治區商業區及水陸運輸總站皆尙未着手，舊市區可利用之價値不大。至新市區之闢建，南京城內城外，皆有發展餘地，惟以湖沼及長江關係，稍受限制。西安城內面積約十平方公里，現在人口約三十萬，將來可容納五六十萬人。現在市區雖可利用，但去容納首都之量太遠，必另闢新市區，且其規模必大於原有之城區，西安南郊可供闢建新市區之首選，東西北三郊亦可利用，但建一大於舊市之新市區，財力以外，「羅馬不是一日造成的」，時間上也非數十年不爲功。北平之內外城有六十平方公里，現有人口一百餘萬，將來可容納至二百萬，舊城區可全部利用。至於北平新市區之闢建，四郊皆宜，北郊恢復元代之大城，可闢爲政治及駐軍區，西郊可闢爲風景文化及住宅區，南郊可闢爲商業區及陸空交通總站，東郊連至通州可闢爲工業區及水運站。在還都北平之初，政府如欲集中辦公，則東交民巷之使館區可全部收用，另在西南郊跑馬場一帶，擇地爲各友邦建新館舍。如此旣可收中樞集中辦公之效，復可將滿淸遺下之國恥特別區，予以根本取消。

五、近郊之風景名勝：都市近郊之風景名勝，在調劑市民生活，陶冶市民情緒上，具極大價値。南京近郊除陵園爲後起之秀，尙在經營中外，有燕子磯、采石磯、湯山及棲霞山諸勝，其淸幽偉大與人工佈置，皆不足稱。西安近郊有翠華山、灞橋、驪山諸勝，華淸池則離城稍遠。其容納遊衆之量及人工之布置亦有欠缺。北平近郊之風景名勝，近者有頤和園、玉泉山、淸華園、香山、西山、湯山、溫泉。稍遠者有明陵、八達嶺、長城。城內壇廟園林及池沼等風景名勝更不可勝數。其容納量及人工佈置皆達上乘，從飛機上俯瞰，北平是一個林樹蒼萃，和諧而美麗的城市。

六、工程建設：北平全市街道網已配置的很勻稱，只須修整，不必大事拆改，主要街道已全築成柏油路面及水泥人行道。全城點綴風景的引水工程與雨水溝渠，雖為舊式設計，但皆為有計劃的系統建設，稍加改良，仍可為流通全城血脈之用。污水現時流入雨水溝渠，將來人口增加一倍後，再另建污水溝渠系。南京街道尚在闢建中，有柏油路面者僅中山中華等數路。溝渠僅完成設計工作，尚未開工。西安街道的配置甚好，有系統的溝渠工程則未着手。

七、公用設備：南京之首都電廠規模甚大。自來水廠已開辦而未完工，市區與下關有鐵路而無交通上之價值。娛樂及市內遊憩場所甚少。西安電廠太小，現時供電已感不敷，自來水未辦，而井水多帶鹹味。北平除電廠自來水廠俱備外，電車路已貫通南北東西四個城區，自來水廠應加擴充，北平地層富於「地下水」，掘井皆得甘泉，舊日之井稍加改良，仍可利用。娛樂及遊憩場所菜場食店滿佈城區。

八、公共建築及建築材料之供應：西安現時之公共建築物存者不多，勉可敷省會之需，無容納國都之可能。建築材料除磚瓦可就地燒製外，木材須伐之於秦嶺，運輸困難，其品質適否建築之用，尚待研究。石灰須取之於豫省。水泥鋼鐵玻璃皆須遠道陸運。南京可以利用之公共建築，除十六年建都後之少數房舍外，幾一無所有。建築材料之供應則甚為便利，磚瓦就地燒製，龍潭之水泥近在咫尺，湘贛之木材及鋼鐵五金玻璃皆可利用水道運輸。北平之公共建築，為全國財賦積近千年之所經營，可以容納中樞全部官署而有餘，此為北平建都最便利而獨有之條件。北平公共建築物的價值，等於全部國富或且過之，此非誇大之辭，僅頤和園一處即浪費了李鴻章氏建設北洋海軍的全部經費，戰後我國首要之建設為國防與工業，遷都北平後，將建設新都及官署宿舍的經費，建設渤海海防與平津陸防，當有餘裕。北平之建築材料，有唐山之水泥，龍烟之鋼鐵，秦皇島之玻璃，當地之磚瓦與琉璃瓦，美洲南洋及東北之木材，也可水運至通州或北平近郊。

九、食料燃料與人力之供應：食料如糧食菜蔬肉類水菓等之供應，西安現在供給量不敷首都之需。南京附近之食物與燃料，產量豐，水道運輸便。北平燃料產自近郊，糧食取自河北平原，菜蔬肉類水菓之質與量俱佳。南京與北平皆可得淡水及鹹水之水產食品。人力之供應分勞心者與勞力者兩種，西安勞心者之供應須仰賴他處，勞力者也需要豫省供給。南京處東南人文會萃之區，勞心者之供給自無問題，勞力者則靠江北。北平為舊日京華，現仍為文化中心，勞心者之供應自無問題，勞力者則冀魯豫三省皆標準壯丁之產地，更是取之不盡了。

十、代表國家儀容與民族精神：一國的首都，必須能代表國家的儀容與民族的精神。儀容與精神徵象於國都者有二，一為歷史上民族之遺跡，一為市容的氣象。西安在民族遺跡上佔第一位，黃帝的衣冠塚及周秦漢唐四朝帝王之陵墓，皆在西安附近，這是歷史論者主張建都西安的最大理由。南京有明太祖陵及國父陵墓。北平則有明清兩朝的宮殿，明成祖以下十三代皇帝陵寢亦在西北郊，國父的衣冠塚在近郊之西山。至於市容的氣象，為街道與建築二者所形成，西安的街市很宏偉，南京的首都氣象尚未建設起來，北平有長達十餘華里的寬闊通衢，各種樓閣牌坊壇廟的美術建築，佈滿全城。北平所代表的儀容，是「偉大」、「莊嚴」、「肅穆」與「和平」，不但為國內之第一城，也是世界的名都。法人「明日之城市」著者戈必意氏，就北平市之平面圖論，卽譽之為「以此地圖與巴黎城地圖比較，則吾人須襲取中國，因有必要襲取其文明也」。

〔4〕結論

建都之實際工作為市政工程，以上十項標準，對於市政工程，有直接或間接之影響與關係，皆為國都所必須俱備之市政工程上的條件。南京、北平、西安三市，其符合於十種條件之程度，自以北平為第一，南京次之，西安又次之。

（三十二年十二月二十日渝）

計劃

大青島市建設計劃

過守正

吾國對於市政建設向不注意，所有城市，一沿舊習，爲不合理擴展，甚少改革。除北平、上海、南京、青島等設市區域以及各省省會等重要都市，曾作有計劃之建設外，其餘較大城鎮之所謂革新，亦不過加寬馬路，修葺新式店面之一種消極改善而已。

市政工程建設，關係全市市民之健康與生活，至爲重大，目前抗戰勝利在望，一切建設均待付諸實施。市政工程建設實爲主要建國工作之一。考歐美工業先進國家，人口之集中，已由城鄉各半增至城八十鄉二十之比例，我國復興工作，以完成國家工業化以及農業之自給自足爲目標，將來人口之分佈，亦可達到城鄉各半之比例。是則市政建設之良否，不獨有關工商業之發展，而半數以上同胞，日常生活亦受其影響。惟建設改造，必須先有完善之工程標準，可資參考，並須有比較完備之建設先例，以資倣效，方能有所準繩。查青島市區，地位優越，環境優美，歷經三十餘年之經營，所有現代都市應有之設備，均稱完善，在過去全國市政建設之中，有特出之地位，雖大部份爲外人所創置。但德管時代範圍過小，計劃局促，日管時代範圍擴大，發展欠缺計劃。我國接收以後，力矯前弊，曾擬具擴大計劃，因勢利導，逐漸改革，惜因戰事發生，未克全部實施，不無遺憾。現雖巳隔日黃花，但其中一切規模特點，頗足爲將來計劃重建淪陷區內都市之參考，並可爲目前改造後方城市之效法。作者對於該項計劃，忝曾主持，因就所憶，分述於后。

（一）青島之優點

（1）環境地位之優美：青島市區三面沿海，地臨膠洲灣，水深而闊，可泊巨艦，北接嶗山山脈，屏障天然，市內有萬年太平諸山，層巒疊嶂，秀麗異常，前面小島臚列，點綴海空，市區南向，氣候冬暖夏涼，尤爲避暑勝地，全境地質爲花崗岩層，塵土絕少，故市內清潔異常，況建築基礎既佳，工程亦易，取用建築材料，尤稱便利。

（二）土地制度之建立：確定土地制度，爲市政計劃最重要之步驟，亦卽爲解決都市建設一切問題之根基。青島市區對於土地之整理，至爲完善，凡市內土地，均劃爲公有，按照建設計劃及使用目的，重行分配，編定地號，嚴訂建築高度及面積，限制法規，所有私人建築，均應按照規定向政府登記，領用修建，較之其他都市之自由買賣，自由建築，以致一切改革計劃不能實施者，不可同日而語。

（三）衞生工程之完備：青島市之衞生工程，藉其天然條件之優越，堪稱國內城市之最完善者，其自來水工程，取給於白沙、李村兩河，該兩河係發源於嶗山，經長距離之流濾，而入引水井內，再佈流於市區，故飲料至爲清潔，溝渠工程，對於雨水汚水均有完備之排洩系統，此外垃圾之處置，亦極妥善。青島市之見稱於世者，爲「潔」與「美」，而衞生工程之完備，實爲達到「潔」之主因也。

（四）森林管理之嚴密：青島原係漁村，歷經三十餘年之經營，將造林計劃全部完成，故青市森林園地幾佔全市面積之半。尤其對於森林管理之嚴密，非其他各市所能及，青市之所以能得到美麗都市之稱者，亦得力於此。

（二）青島市擴大市區計劃

一）青市全盛時代之推測：青島係一國際通商之港埠，凡港埠之發展限度，須視港口貨物吞吐之數量而定，青島港之地位，係介乎東方大港與北方大港之間，爲膠濟鐵路之終點，將來膠濟線延經彰德，展長至寧夏甘肅新疆而達蘇聯時，該港卽爲大西北與黃河淮河流域經濟動脈之尾閭，前途之發展，未可限量，當時爲便於計劃起見，姑就最低限度，暫照二十四年度之人口調查，增加五倍，以一百萬爲目標，計劃市區面積，北至滄口、李村，東至麥島一帶，計一三七〇公頃。本計劃一切市政設計，卽以該項擴大市區範圍爲根據。

（二）計劃原則：

（甲）實用與美觀並重：青島市之特性以居住遊覽爲主，而以工商爲副，提及青島，莫不有美麗的印象存於腦際，爲保持此優點，並發揚而光大之，則市政計劃之目標須實用與美觀並重，但有時二者不可得兼，在工業港埠兩區採實用主義，在商業區則以實用爲主，美觀爲附，在住宅區則以美觀爲主，實用爲附，所以適應環境也。

（乙）新舊區域連成一氣併盡量避免變動舊區：青島舊市區，係計劃於三十年前，中經間斷之發展，無論形式與實際，局部與整個，均欠連貫，故一小部份舊市區之改造，殆爲不可避免之事實，惟力求各方面減少無謂之損失起見，務在可能範圍盡量避免之。

（丙）適合於將來擴充計劃之連綴：近代都會計劃之趨勢，公認人口必須分散，欲達到此目的，惟有在原有市區之週圍，建設若干附屬之小市區，必須與中心市區有適當之交通聯絡，如此則人口之增加至無限，而中心市區之過分擁塞繁榮可以避免，邊區則逐步繁榮與中心市區保持均衡。

(三)全市分區計劃：分區計劃係將全市面積，按其使用之性質劃分區域，而對於市內一切建築加以地區之限制是也。如地勢平坦，放出煤烟，不擾及市民者，宜設為工業區。交易繁盛，交通便利，位置適中者，宜設為商業區。僻靜之地，風景幽美，空氣新鮮，合於衛生者，宜設為住宅區。其餘近於港埠鐵路之地，合於運輸堆置貨物者，宜設為港埠區。一市行政總機關，與市民接觸最密，宜擇適中地點闢為行政區，商業區與行政區為一市精華所在，又為各大道之集中點，普通稱為市中心區域。以上雖按建築物之用度，分為工商等置，雜亂之現象固能免除，但對於人口居住過密，交通過繁，光線空氣不良，火害容易蔓延鄰舍等弊，仍不能加以取締。故除用度分區之外，尚有高度分區及面積分區。高度分區，即規定在某種地帶，建築房屋之高度為若干公尺。面積分區，即規定在某種地帶，其房屋面積與空地面積不得超過若干成。青市用度分區規定如下：

(甲)港埠區：青市現有大小兩港，小港已發達至極限，擬規定於四方大港間另建工業小港一處，原有小港則專供本地商業運輸之用，該新築小港毗連陸地，即規定為港埠區，在該區內容許建築倉庫、堆棧以及其他運輸上必需之建築物。大港內水面遼闊，除原有碼頭外，尚可容納新碼頭五座，當該項碼頭告成時，亟需大宗土地以為堆積及運輸貨物之用，故規定大港周圍約五百公尺之地，均劃作港埠區。

(乙)工業區：工業有大工業與小工業之分，小工業之不衛生程度較弱，在特許情形下，可計入商業與住宅區內，大工業因妨害衛生與安寧，應另行規定工業區域。茲將工業區選擇標準列後：

(子)水陸運輸便利，且以鐵路能通至工廠之內或附近者為最佳。

(丑)地勢平坦而廣闊，將來足資發展。

(寅)遠隔住宅區，且必在最頻數風向之下方，庶煤烟不致紛飛入於清幽之住宅區。

(卯)地價低廉，不致累及工廠之成本。

查青市四方東面及沿鐵路直至滄口一帶土地，約與以上四項條件相合，作為工業區最為相宜，且地勢平坦，鐵路均可直達工廠之旁，運輸非常便利。

(丙)商業區：查大港之南，南至中山路一帶，北至遼寧路一帶，事實上已形成商業地帶，因勢利導，故應規定為商業區，同時，為便利各較遠住宅區及工業區內人民購買用品起見，在住宅及工業區內，擇適當地點設置日用市場及商場，範圍及數目酌量當地情形而定。

(丁)住宅區：市民生活於住宅之時間為最多，其一生之幸福與健康全繫於此，故計劃住宅區最宜審慎，舉凡空氣、日光、風向、林木等要素，均須經充分之考慮。茲規定凡風景優美，無工廠煤烟吹入可能之地，及隣近商業區處所，均一律作爲住宅區，按環境之優劣，又分特種及甲、乙、丙各等，茲分列如下：

(子)特種住宅區：湛山浮山一帶風景優美之處。

(丑)甲種住宅區：築城路東一帶。

(寅)乙種住宅區：齊東路，萊蕪路等處，

(卯)丙種住宅區：西嶺、四方、舊嶺莊南，文昌閣，桑子村東等一帶及不屬於特種暨甲、乙兩種住宅之地。

總計本計劃所規定各區面積之分配，計行政區占百分之〇・三，商業區占百分之五・四，工業區占百分之一〇・九，住宅區占百分之四三，港埠區占百分之四，園林區占百分之三六・四。

(四)全市街路計劃：計劃市政建設有兩大要點，一爲區域之劃分，一爲交通設施之配置。交通設施之最重要者，莫如街路之佈置，苟街道配置不得其當，則一切市區建築如園場公私建築，電車綫路衛生工程計劃，均將隨之而蒙不良之影響，是以街路之計劃，宜經詳細之研究，與通盤之籌劃，在舊式街路，其系統有不規則式，棋盤式，放射式，圓形式等，其實街路既爲交通之需要而設，則其主要之因素，當以交通情形及地勢限制而定，且近代之市政建設，着重於分散化，尤須視各區域實在情形而定，絕不可固執一定形式。青市在市中心區部份，因減少改造之損失起見，大都依據舊路組織，故仍採用棋盤式，但在較遠部份，如吳家村一帶，則採用較新之蛛網式，其在更遠處如浮山滄口等處，則大部採用細胞式。各區域街路所採方式，雖有不同，但其相互間之聯絡仍極密切，茲將本計劃之大概分述如下：

(甲)幹路組織：幹線爲聯絡各重要部份，使其交通便利，易於呼應，故各分區之間，以及車站碼頭各重要住所，均應貫以幹道，俾車輛來往毫無阻礙，因市內交通之方向，多以市中心區爲目的，爲適應此種趨勢之要求，故幹線組織在市中心區所需之幹路必多於郊外，而近市中心幹道之寬度，必大於郊外之所需。

(乙)支路組織：支路之設原爲便利本區內局部交通，與供給各建築物以充分光綫空氣之用，故其寬度不妨略小，其佈置固無一定形式，視幹路間所隔成面積之形狀與地勢之是否平坦而定。

至街道寬度應視全盛時代實際需要而定，以免拆屋讓路之困難，及收買高昂地價之損失，茲將路寬標準規定如下：

(甲)林蔭大道：自三〇公尺至五〇公尺。

(乙)交通幹道：

(子)鄉市交通：自三五公尺至四五公尺。

(丑)本市交通：自二五公尺至四〇公尺。

(丙)次要幹道：自一六公尺至二二公尺。

(丁)支路：自一〇公尺至一六公尺。

(戊)里街：自三公尺至五公尺。

(五)全市園林及空地計劃：市區之有園林空地，猶人身之有肺部。現代歐美都市計劃學者規定園林空地，應佔全市面積至少半數以上。世界各都市，經過此次戰爭空襲轟炸所受損失之教訓，以後市政計劃當以防空爲主要目的，而採取疏散式，全市空地面積百分數更需增加。青市於分區計劃內，規定房屋高度及密度，故房屋建築不致過於繁密擁擠。且市內多山林起伏之處，不宜建築，最適宜於培植森林，使成爲天然之園地，茲規定凡山地山谷不宜建築房屋者，一律規定爲園林空地，茲分別說明如下：

(甲)森林：凡山地高度在海面以上六十公尺者，一律規定爲森林保留地。

(乙)公園：規定凡建築區域面積每半平方公里內必須有小規模公園一所，其位置酌量情形規定之。至於太規模公園，除現有第一公園外，併擬於砲台山，吳家山，四方西及滄口飛機場之東等地，分別建築大規模之公園。大公園與小公園之間用帶形園林互相聯系，形成市內園林系統。

(丙)運動場：規定每一公里內有小運動場一所，爲便利市民鍛鍊身體及兒童遊戲之用。至於大運動場，除現有之匯泉體育場外，擬加建浮山，四方北，李村等體育場，建築不必宏麗，但規模必須偉大。

(丁)海水浴場：青島爲避暑勝地，故海水浴場應盡量擴充，以增加市民正當之遊樂場所。凡沙灘純潔平坦之海岸，應一律加以整理建爲浴場，計有(子)四川路浴場(丑)太平路浴場(寅)匯泉浴場(卯)山海關路浴場，(辰)湛山浴場，(巳)浮山所浴場，(午)李村河浴場，(未)滄口浴場，(申)船渠港浴場等九處。

(戊)廣場：凡街道特別寬大地點，通稱之爲廣場。廣場之設置爲市政計劃中重要問題之一，佈置得宜，不但合於應用，且可增加市容之偉大與美麗，凡街道之集中點與交叉點可稱爲街道廣場。可就街道交叉情形，設置圓形廣場或三角地。此種設備，青島市內原有甚多，整潔美麗增加市容不少。

(己)公墓：青島現在公墓爲公園式，既可予市民以精神上之安慰，復可增加城市之風景。惟將來墓地逐漸增加，爲土地經濟着想，可採用深埋式，俾墓地表層可供耕種之用。至公墓地點應根據(子)不礙市區之發展，(丑)不礙水源之清潔，(寅)土質宜深厚等原則選擇之。

(六)全市一般交通計劃：市政計劃對於飛機起落之場，輪船停泊之處，以及鐵路入境之途，均應有適當之規定，青島將來繁榮之程度，至少有人口一百萬，其需要充分之交通設備乃屬必然之事實。茲分別路述如下：

(甲)陸地交通可分爲(子)鐵路交通，(丑)市內交通兩種。

(子)鐵路交通：青市現有鐵路僅有膠濟一綫，按照將來內地之需要，尚有由山東東南部來會之膠徐鐵路，(自徐州經臨沂諸城等縣至膠州)及由山東東北來會之青烟鐵路(由烟台至青島)二線。惟此三線入青市之途徑，因地勢關係，仍衹有沿膠濟線之一法，不過此數鐵路會齊入青市時，必須有一綜合車站，以支配數路車輛之按時入市。此車站之位置以膠州東部爲宜，該平原位於大沽河之下游，地極平坦，面積極大，可敷設多數軌道，與佈置大塊堆貨場地，且有大沽河水運之便。

(丑)市內交通：市內交通繁密以後，必須採用高架式或地下式交通，本市在市中心部份，以地價高昂，街面交通擁擠，擬採用地下式，路線之組織，擬採用自市中心向四處放射，並互相環繞而成8式。至原動力擬完全採用電力，以保持一市之淸潔。爲補高速交通之不足，尚須有電車及公共汽車之設備，其路線酌量實際情形再行規定之。

(乙)水面交通港埠：青島市之發展，完全繫乎水陸交通能否聯繫。現在大港有碼頭五處，以備外洋船艦停卸之用，小港爲近航船舶停卸之用。前海棧橋爲軍艦渡船停泊之用。將來海外航運發展，大港碼頭之增加實爲解決水運問題之主要目標。查第五碼頭卽將完成，但五個碼頭壁岸總延長亦不過四、一〇〇公尺，與全盛時代停船噸位相差尚遠，除在四方附近添建一工業港外，尚須就大港內增建碼頭二處，壁岸總延長可達八、〇〇〇公尺至一〇、〇〇〇公尺。關於港灣之深度，規定大港不得少於十一公尺，工業港不得少於八公尺，小港不得少於六公尺。

(丙)空中交通：青島多山，平原又少，除現在滄口機場一處尚使用外，實難覓第二場地。查塔埠頭東南一帶海灘寬闊平坦，將來可築大飛機場，以應全盛時代之需要。至水上機場擬設於團島附近，因該處水面廣闊，沙灘平坦也。

(七)全市衛生問題之解決：據一般調查與統計，人類之死亡率與虛弱症，城市大於鄉村，其所以致此者乃因都市之居住過密，一切空氣、日光、空地，以及飲食排洩等問題無圓滿解決之方法，故城市之衛生工程問題其重要與衣食住行相等。有關市民衛生之範圍甚廣，除公園運動場及森林等已詳前節，房屋建築問題另詳於後外，本節所及爲衛生工程如自來水溝渠，及垃圾等之處置問題。

(甲)自來水：本市自來水關於給水方面與香港情形相同。全市全係石質山地，既不能鑿井吸水，附近又無巨大河流可資利用。市內給水，全憑自遠處河流吸引供應，送水系統工程浩大。現有水源計三處，卽海泊水源地，李村水源地，及白沙河水源地是也。

海泊水源地開闢最早，設備方面有吸水井十五個，每日送水量爲二、七〇〇立方公尺。李村水源地距市內約十一公里。設備有吸水井計五十餘個，送水總管長一一、五〇〇公尺，每日送水量約七、三〇〇立方公尺。白沙河水源地距市內二十二公里。分東西二廠，東廠有井四十餘個，西廠有特製大井十二個，送水總管經滄口，四方而達市內。計長二二、八〇〇公尺，每日送水量爲七、〇〇〇立方公尺。

以上三處水源每日送水量不過一萬七千立方公尺，勉強足供今日之用。將來人口增加，市面繁榮，非增闢水源不足以解決水荒根本問題。郊外河流，可作自來水源者，惟白沙河上游之月子口，及膠州境內之大沽河二處而已。月子口係白沙河上游之盆地，茲擬築壩以截取嶗山境內來匯之水。計該水源地完成後，每日送水量不過二萬立方公尺，連同現有總計送水量不過三萬七千立方公尺，尙不過足供六十萬人口之用。將來人口增加，決非六十萬所可限制，故最後永久水源計劃，擬在大沽河成立大範圍給水總廠，大沽河水終年不涸，其量之大決非白沙河所可比擬，將來該廠引水，不特可供青島市區，卽塔埠頭港所需水源亦可仰供於此。

(乙)溝渠：青島市之溝渠係採取雨水污水分流及合流二系統，設備之完善爲我國各都市之冠。現有雨水管共長九萬餘公尺，污水管八萬餘公尺，混合管二萬餘公尺，雨水溝一萬餘公尺，合計溝全長約二十萬餘公尺。雨水管之設置，係依地面自然傾斜之坡度自高而下，以入於海，雨水管不排洩穢物，故入海處可卽在市區沿海偏僻之處，因其於衛生及觀瞻上均無妨礙也。至於污水排洩，不若雨水之簡單，本市現有設備係根據地勢情形分全市爲四個排洩區，各據其最低處設沉澱池及淸理廠，污水經淸濾廠處置後輸送入海，所有糞渣另行儲運，惟現有溝渠設備，僅限於市區南部，四方滄口等處尙未普遍安設，於擴大市區計劃內，自應統盤籌設，關於污水集合區域按照實地情形分全市爲四大集水區，將原有第一二三排洩區及西嶺淸理廠之範圍劃爲第一區；與湛山一帶住宅區劃爲第二區；四方東鎮一帶工商業區劃爲第三區；滄口一帶劃爲第四區；每一集水區域設總污水淸理廠一，分淸理廠若干，其多少視地形而定。

(丙)廢物之處理：城中之廢物，包括動植礦物，經毀棄不堪再用者而言，每日所出數量甚大，茲擬定處理方法如下：

(子)灰燼之處理：灰燼係由工廠及住宅火爐內所產生，用之塡築窪地最稱相宜。

(丑)垃圾之處理：垃圾係由清除街道所得之雜物，用之填築窪地亦稱相宜。

(寅)廚渣之處理：廚房內產生之廢物，大都爲有機體，易腐爛，有礙衛生。處置方法擬分(1)煉脂法，(2)飼料法(3)藥餵法三種。

關於廢物收集辦法，必須注意於節省經費，便利住戶，及保持清潔三原則。茲規定收集廢物之辦法如下：

(甲)居戶盛放廢物之方法：以廢物之性質及處理方法之不同，規定灰燼垃圾與廚渣二類，應分別盛放於二桶，不得混合於一桶。

(乙)運送廢物之辦法：收集之次數無論冬夏應每日一次，其時間以早晨爲宜。運送車輛應有蓋，併應不漏水以保持清潔。

(八)公共建築物之配置及私有建築物之管理：

(甲)公共建築物之配置：公共建築有關市容之莊嚴偉大，以及市民使用之便利至鉅。如政府機關，地方公共團體之辦公室，圖書館，博物院，市禮堂，紀念堂，戲院及音樂院等均與民衆發生密切之關係，尤宜選擇相當位置，建築於市中心區域。但距市中心較遠之地，爲便利民衆之要求起見，亦當有此項設備，惟規模可小。至於學校及醫院，應選擇鄰近園林空地及交通稀少之地。惟小學應分佈於住宅區內，醫院除於市中心區設立總醫院外，併於各區設立分院。

(乙)私有建築物之管理：青島市私有建築物之實施整理，已有三十餘年之歷史。對於建築物之用度，高度及面積之分區均有條例可循，建築工程之規則亦有章程可據。制度最爲嚴密，實施最爲澈底，爲吾國各都市之冠。故擴展計劃，應就原有制度，發揚光大。但爲增進市容美觀起見，對於建築圖樣之審核，擬組織房屋審查委員會，聘請專家爲委員辦理之。優美者則予以獎勵，如此則建築事業得以進步，一市之觀瞻得以改良。

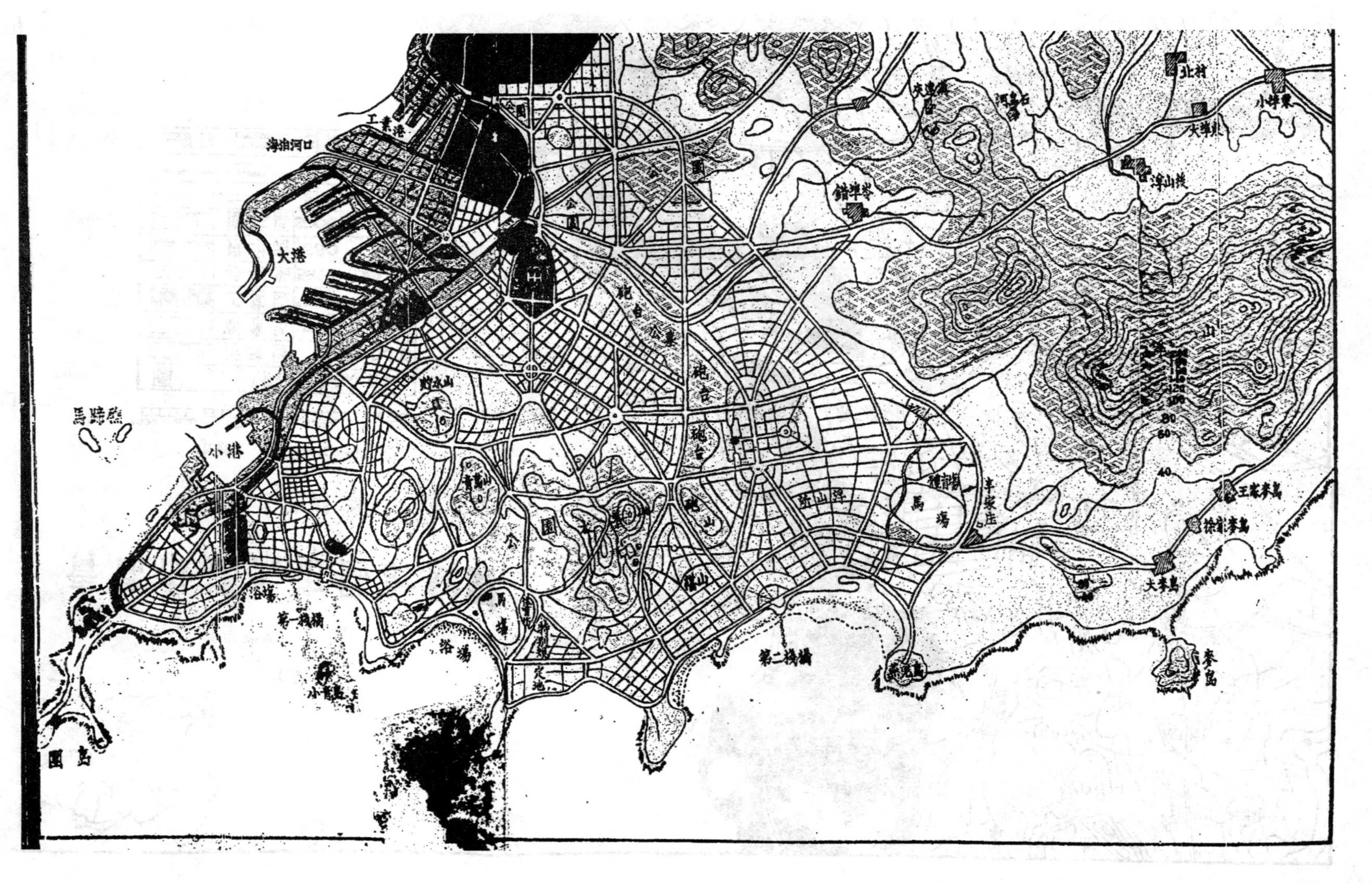

海泊河口
工業港
大港
小港
馬蹄礁
團島
第一棧橋
第二棧橋
浴場
北村
淨山後
大麥島
辛家庄

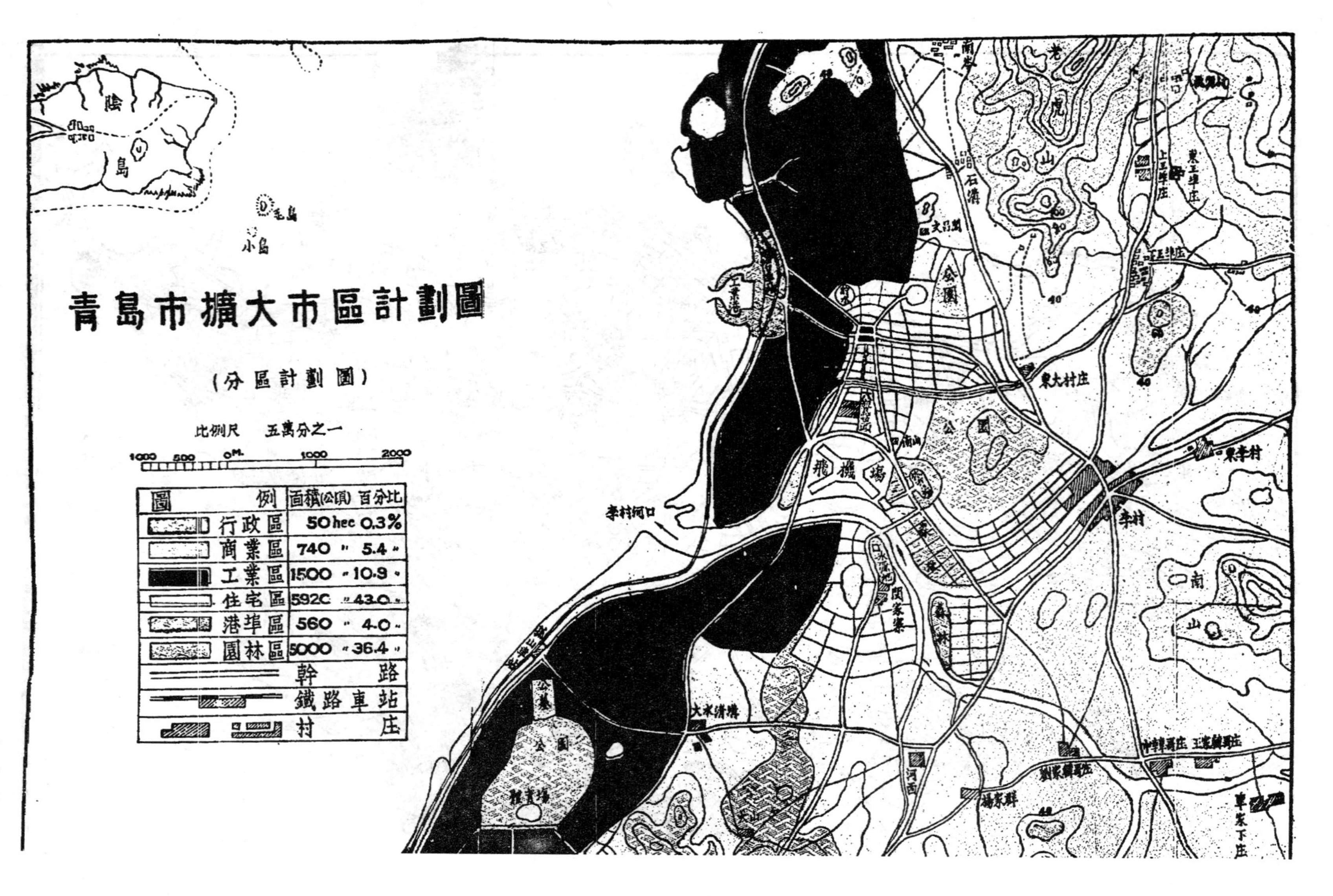

青島市擴大市區計劃圖
(分區計劃圖)
比例尺　五萬分之一
1000 500 0M. 1000 2000
圖例 面積(公頃) 百分比
行政區 50 hec 0.3%
商業區 740 " 5.4 "
工業區 1500 " 10.9 "
住宅區 5920 " 43.0 "
港埠區 560 " 4.0 "
園林區 5000 " 36.4 "
幹路
鐵路車站
村庄
陰島
毛島
小島
老虎山
石溝
東大村庄
東李村
李村
李村河口
飛機場
公園
南山
閻家寨
大水清溝
河西
楊家群
劉家韓哥庄
王家韓哥庄
車家下庄
體育場
上王埠庄
東王埠庄
文昌閣

關於今後大後方及收復區舊城市之改造及新城市之重建，亟宜於斯時集合全國市政專家研討市政工程基本原則，並規定實施時應有之步驟，以免都市畸形之發展，而達成略合現代化市政之要求，茲根據青島市之市政計劃草擬市政工程設計原則，以供研討時之參考。

（一）規定市區單位：市政工程之設施，非爲少數人之奢侈享受，而是爲大多數人民求生活之舒適，故在多數人民集合之處，卽應有市政建設之施行，英國村鎭人口在四百以上，卽有市政設備，我國則不妨以一千人口以上之村鎭爲舉辦市政建設之最小單位，並可稱爲鎭市，在十萬至二十萬人口爲城市，百萬人口爲都市，以市之等級而確定市政工程實施之標準。

（二）確定市區性質及範圍：市之性質，係指該市將來發展之趨勢及使用之目的，如工商業中心，或文化中心，抑專供遊覽之用等等，市之範圍，係指將來可能發展之限度，以及全盛時代人口之估計，該兩項之確定實爲市政計劃之先決條件，因根據市區發展之趨勢，方可計劃各種分區面積之大小。根據全盛時代人口之估計，方能確定市區將來發展後需用之面積。能於計劃市區之先，將未來之性質與人口爲設計對象，則將來市區建設，可期其在計劃範圍以內進行，不致有畸形發展之弊。

（三）實施土地政策：我國城市建設，最感困難者，厥爲土地糾紛之不易解決，因而阻礙計劃之推行，影響整個市政建設之發展，青島市之所以有蔚然市政建設之成績者，實賴有完善之土地市有政策。查我國土地政策，中央早已明令實施，並提前辦理城市之土地測量及土地陳報，城市之土地，政府可隨時收用，於實施市政建設之時，所有土地糾紛之困難，自可獲得解決。該項土地政策，盼能早日完成，以利將來市政建設之推行。

（四）規定園林空地：我國舊有城市，隨其自由發展，往往建築櫛比，缺少公園空地，是項密集之組織，對於市民之健康，交通之安全，以及整個城市「美」、「潔」，影響至鉅，每遇空襲，損害尤大，故園林空地面積之規定，實爲市區計劃切要之圖。大都市方面人口衆多，交通頻繁，其林園空地面積，至少須在百分之四十以上，普通城鎭範圍較小，離郊外較近，空地面積可酌予減少，鎭市除建築物應互相隔離外，如無其他原因，則可酌留空地。總之，市區愈繁榮，人口愈衆多，則園林空地面積必須愈擴大。

（五）普遍衞生設備：在普通鎭市因範圍狹小，可設雨水污水合流之簡單溝渠系統，以資排洩，垃圾穢物爲一般農作物之肥料，應予嚴格管理，以期搬運便利掃除清潔，惟廁所之建造，應嚴格依照規定建築及取締，至於城市都市之衞生設備：（甲）必須設置完善之溝渠系統，視其實際情況，採用雨水污水合流制或分流制。（乙）污水之處置應酌量實際情形，予以完善之設備。(丙)垃圾之儲運及消毀，應予以嚴厲之管理。(丁)自來水工程應有清潔之水源，完備之配水系統，以及普遍之消防裝置。總之，衛生設備之要求，在於完成市內之環境清潔，市民之供應清潔，促成市民之生活清潔，使一般身體漸入康樂之途。

(六)實施建築管理：建築爲市政之骨幹，關係全市之美觀與安全，吾國舊有城鎮之建築，式樣簡單，配置失當，並且互相毗連，缺乏空地，經此次抗戰以後，前後方各鎮市，多被烽火破壞無遺，將來必需分別重行建設，自應把握時機，改革以往缺點，詳訂管理及取締建築辦法，嚴厲執行，使所有建築，配置合理化，式樣美觀化。

(七)舊有市之改造與新闢區之聯繫：我國舊有城市，缺乏全盤遠大計劃，任其自然發展，造成畸形狀態，着手改造，殊感困難，如予徹底改革，則犧牲過大，勢不可能，如局部設計，則趨向粉刷門面，不切實用，是以担任此項計劃時，須詳加研究，捨重就輕，盡量避免無謂犧牲。關於區域之劃分，尤應就原有情勢而予以決定，其交通系統，亦不必拘泥於一定形式，以期因勢利導，漸入合理之境，至於新闢市區計劃固可自如，但須注意新舊市區之密切聯繫，使之呵成一氣，俾能達成平均發展之進度。

市政工程，係利用現代科學方法來增進人類福利，代表人類文明之專門建設，市政計劃之完善與否，爲一市盛衰所系，歐美先進諸國，莫不設立專司機關，廣羅市政工程專家，主持全國市政建設之計劃與實施，此項專家政治精神，頗值得吾人倣法。我國已確定戰後發展工業爲建國中心工作，而市政建設，實爲工業建設之先導，在我國市政未有基礎之階段中，欲於短時間內舉辦各大都市城市鎮市之市政建設，其單位數量之多，計劃工作之繁，實屬千頭萬緒，欲解決是項嚴重問題，似應在中央專設全國市政局辦理一切市政法規之草擬及裁決全國市政計劃方案，並督導各大都市及各城市之市政實施，省設省市政局，負責督導各城鎮市之市政實施，都市則另設都市市政局，直屬全國市政局，縣設城市政局，管理城鎮之市政實施，如此分層負責，職責分明，吾國市政建設之前途，庶能收到偉大之成果。

陪都市政建設

吳華甫

一

重慶自國府西遷，明定陪都以後，成爲軍事政治經濟之樞紐，戰後更將爲西南建設之中心。我國抗戰以後，國際地位日益增高，吾人今日之戰時首都，已爲舉世所矚目。質言之，重慶不僅爲我國抗戰司令台，亦且爲世界政治中心之一。

重慶自「五三」「五四」以後，迭遭敵機狂炸，物質損失，自不待言，然亦未始非重建重慶之無上機會。吾人應一本不屈不撓之精神，以堅定之決心與毅力，於斷垣殘壁中迅速產生更完善更優越之都市。本市建設，經二十八年闢建太平巷，及二十九年規定道路網後，始基乃得稍具，從此依照計劃，慘淡經營，五年於茲。時至今日，差幸尚有若干成績。惟以經費所限，所定計劃，就整個市區言，僅及局部，其能實施者，又祇爲全部計劃中之一部而已。

計劃爲推行事業之張本，亦即行政三聯制中最要之一環。欲冀事業有成，首賴計劃周詳妥善，庶可循序而行，達到預期目標，而建設都市之計劃，其重要與迫切，可以想見。惟都市建設，經緯萬端，大而至於分區之擬定，小而至於一學舍之規劃，無不需要精密審慎之考量。就現時客觀條件而言，固難立即大舉興建，但爲建設陪都，較要諸部門之設計，似應事前籌劃，俾能逐步推進。筆者願就從事陪都建設五年經驗之所得，對較要各點，略抒管見，尚希海內賢達有以正之。

二

計劃都市，應先注意交通。重慶地位之重要，既如上述，則市內外之交通，自應有適宜之配備。市外交通如飛機場公路鐵路及碼頭之佈置與聯繫，以及市內交通道路網之分佈與交通工具之設備，均應妥爲規劃。

飛機場爲現代都市必有之設備。本市原有珊瑚壩機場，地點雖稱適宜，但以面積過狹，不敷應用，且每值洪水時間，輒被淹沒。爲謀航空發展，應盡量在接近市區及飛機起落安全等原則下，另覓適宜地點，重新建築。以鐵路交通言，成渝路車站已設總站於九龍舖，分站設菜園壩，便利客運，頗稱合宜，將來更可在九龍舖附近架設橋樑，以與南來之鐵路線相啣接。北來鐵路，似以唐家沱附近設置總站爲宜，再於江北附近設一分站，以利行旅。

本市襟帶嘉陵揚子兩江，南北兩岸，各居一隅，而山坡起伏，爲道路網發展之大礙。爲期兩岸聯繫，繁榮市面，必須興建兩江大橋，近有一般人士，以爲興建大橋，在戰時易爲敵機轟炸之目標，如能改建隧道，既得溝通之便，戰時復可移作市民躲避空襲之用，此在理論上固不無理由，惟事實上則困難甚多，江底水深，在低水位時，約十五公尺，隧道深度至少應在河底下二十五公尺。洪水時期，江水陡漲，較低水位高約二十餘公尺。故洪水時期壓力水頭常在六十公尺以上，且河底石層大部砂石，漏水甚烈，加以水深關係，壓力增大，防水滲漏，殊匪易事，此其困難者一也。依照實際地勢，水流深度及隧道深度，兩岸隧道引道長度，至少在一公里半以上，其與地面啣接之處遠在熱鬧地區以外，對於交通方面，亦頗不宜，此其二也。基於上述理由，當以興建橋樑爲佳。惟橋樑式樣，應注意戰時情形而加以設計。爲配合戰時狀態，似以上承式鋼架桁橋爲最宜。每橋之跨度，應儘量相同，其橋面應用鋼骨水泥建造，以資維護。平時可多備若干桁架，一遇轟炸損壞即可將準備之桁架提出應用，趕工修理，庶於交通無礙。至若鋼索懸橋，在經費上固未必經濟，而於轟炸時期，又少保障，殊不適用。

市內交通尤爲重要。設無妥善佈置，縱有良好商港與鐵路亦難收指臂之效。是以過江大橋與路網之聯絡，關係至大。跨越嘉陵江之橋樑應有二座，以在沙坪壩附近之石門及大溪溝附近爲宜。至長江大橋，一應在望龍門附近，俾與南岸龍門浩相溝通，一應在九龍舖附近。如是兩江之上，四橋相連，交通自更便利，南北兩岸，自可日趨繁榮。至市區道路網，應注意幹路之佈置，務使各交通據點能互相聯絡，而各區中心，亦有完善之聯繫。庶幾將來人口增加，不致有阻塞之虞。

今日談路網者，多注意圖案式之佈置，如棋盤形或放射形等等。重慶地形崎嶇，如徒重形式，不無削足適履之嫌。結果必使工程經費增大。而兩旁土地亦難於經濟使用。此種教訓，在國外已數見不鮮，可爲吾人前車之鑑。故陪都道路網之設計，必須顧及實際情形，凡道路之濶度坡度，均應在交通安全原則下，儘量遷就地形，分別設計之。

都市應有寬暢大道，已爲識者所公認。例如巴黎「香榭里赴」大道寬達八十公尺，紐約市「皇后大道」寬亦六十公尺，整齊莊麗，舉世聞名。本市因受地形限制，現在最寬幹線，僅定爲廿二公尺，計中部車道十四公尺，兩旁人行道各四公尺。實際上此項規定已有施工不易之感。總之吾人建設重慶，不應徒重裝璜，而以適合需用爲原則。

三

城市臨江建設，最爲適宜。沿江一帶，如將土地善爲利用，多建築碼頭倉庫及堆棧等，則於發展城市經濟，大有裨益。如再闢以廣闊沿江道路，綴以花木，作爲簡單公園，市民游行其間，尤可曠怡心神。試觀近代都市，凡臨近水道交通地區，莫不儘量建設，旅客一履其地，頓覺井然有序。而重慶沿江一帶，走道狹小，棚屋雜居，對於市容與實用兩不適合。若干部份，雖已闢爲輪埠碼頭，但以設備簡陋，仍不足以應需要。爲整頓市容及發展社會經濟起見，應建設沿江隄路及碼頭。此在抗戰勝利以後，應即立時舉辦。

建築隄路，工程浩大。本市舊城區江北及南岸三處，俱臨大江，同時興修，恐亦不易。似宜先從舊城區着手，分期舉行。餘者可俟過江大橋完成以後，再行逐一舉辦。

欲論重慶隄路與碼頭之建築，則應明瞭川江水位情形。蓋川江漲落均較其他都市之河流爲甚。其每年一度之普通洪水位，大致高出低水位約二十公尺之譜。高洪水位較低水位約高二十八公尺。每十年之中，或見一二次。至最高洪水位，三十八年前曾有一次，高出低水位約三十六公尺。惟以上三種高水位，一年內僅或一二天有之。至建築隄路及碼頭高度，則與水位有密切關係。取之過低，則一方面雖易遭受水侵損失，但工程較易，而平時效用亦可較高。反之，建築過高，則工程費大增，而實用亦已降低。筆者認爲其適當高度，應較普通洪水位略高，其附近之建築物，則用耐久材料，期於江水侵襲時，不致發生危險。隄路以上，迄舊城區城墻間之土地，尚稱寬裕，必須妥爲利用。其較要工作，莫若土地重劃，使每一土地，均能有最合宜之使用。隄路完成後，地價自必增漲，原住該地之平民，勢難繼續居住。關於遷移地區，亦應事前擬定，庶建設與平民生活得兼顧並籌也。

隄路爲市道路網之一部，應與其他道路啣接，自無疑問。惟市區道路高出隄路，恆在二十五公尺以上，聯接殊感不易。臨江門有與大溪溝連接之北區幹路可通，儲奇門已有聯絡道路規範，朝天門在舊城區之尖端，地勢亦並不高，均可設法與幹線相接。而此三處，平時交通已極頻繁，地位又極適中，如能計劃完成，則其有助於水陸交通之聯繫，當更非淺鮮。

四

分區爲都市計劃工作中之較要工作之一。分區云者，卽將市內主要部份之面積，按其使用性質，劃分若干區域，規定其使用，而對其建築及設備，加以地域限制。例如城市中心，大率爲商業薈萃之處，需有廣闊之街道與堅實之房屋，應規定爲商業區。水陸交通便利，便於貨物運輸，而有充分動力供給之地，宜劃為工業區。至若僻靜之地，名勝之處，則宜劃為住宅區。此乃分區之大概標準。但近來對分區制度，頗有不同之見解，有主張廢除者，以為各種建築聚集一處，一旦發生戰事，有被集中轟炸之虞，易於招致大量損失。亦有主張不宜廢除分區者，以為惟有分區，乃能便市政之管理，而利人民之生活；設使住宅、貨棧、學校、工廠等，雜處一地，任意興辦，凌錯雜亂，匪特居住其間者，深感不安，市政發展，亦受莫大影響。即以重慶而言，數年來經長期猛烈之轟炸，深知爆炸之損失，未若燃燒損失之為烈；故分區後，房屋建築如有相當防火設備，及將區內之建築物，作疏散之佈置，則損失自可大為減少。且分區使用，非即集中建築。集中建築，戰時固屬有害，平時亦為不利。分區之主要目的，乃有系統之支配，就建築物使用之性質，而分類集合之。觀乎重慶以往自由發展之結果，舊城區內競相聚集，各類房屋密集，而××及×××一帶因水陸運輸及動力供給之便，工廠林立，雜處學校建築之間，實屬畸形現象。為便利管理及顧及市民健康起見，確有詳加研究，重行計劃支配之必要。

本市依照實際情形，應分為商業區、工業區、風景區、住宅區及市中心區，茲將各區範圍分述如下：

(一)商業區：重慶舊城區，新市區之一部，江北城區及南岸區之一部，已具商業雛形。若於交通要道，架以大橋，橫跨兩江，聯成一氣，商業必可益趨繁盛。故仍以舊城區新市區江北城區及南岸沿江一帶為商業區，較為適宜。

(二)工業區：本市揚子江下游，江北縣城以東之雞冠石唐家沱一帶，江寬水深，可泊巨舟。將來此處鐵路築成，可設法與成渝鐵路啣接，水陸運輸，均極便利。且以位在全市之東，相距不遠，兩面崗嶺高峻，可免煤烟之污濁，且在兩江下游，污水不致妨及飲水水源，劃為重工業區，尤稱合宜。嘉陵江兩岸地勢頗佳，交通亦便，可為輕工業區。

(三)住宅區：住宅區為市民安息之所，選擇地點，不宜煩囂。且為便利市民往返，應與作業地點較為接近。約言之，可分數部：中三路以西至李子壩一帶，緣該區開拓未久，新式住宅建築已多，且尚幽靜，適合商業市民之居住，化龍橋沿成渝公路至小龍坎一帶，交通極便，各界居住，均尚適宜。嘉陵江北岸一帶地勢頗佳，位在工商業之間，無工商之煩擾，工人居住最為適宜，南岸商業區附近，可供商民居住，往返亦極便利。

(四)風景區，都市之中，園林至為重要。園林廣地可以陶冶性情，調節空氣，非徒足資游樂已也。重慶之自然環境甚佳，背山面水，風景甚美，如再予以適宜之佈置，則所謂都市田園化，可成為優美之天然大公園。故本市建築規則內，對於風景區之建築，已經加以規定。本市東面黃山、汪山，西面山洞、歌樂山等地，林巒秀麗，風景幽雅，公路交通，亦甚便利。且東西遙對，分佈適宜，兩者均應劃爲風景區。

（五）中心區：中心區爲全市行政中心之所在，近來研究市政建築學者，莫不注意及此，當於下節申述。

五

市政發達以後，行政事務，及管理人員，勢必大增。其工作地點，若仍如今日重慶市屬各機關之分散各處，則未免過於散漫。且公文往返接洽，亦殊不便。爲增進行政效率計，殊有集中辦公之必要。且員司既經集中，眷屬當亦隨之居於附近，羣相聚居，自然形成另一區域，故事前必須作有計劃之佈置。舉凡公共建築，除辦公房屋外，其他較大建築，如大禮堂、圖書館、博物館等，凡能會集一處者，均應設置一區內。務使全區建築之佈置，整齊合理，使能代表全市精神之所在，而予一般人士以深刻之印象。

市中心區既爲全市行政樞紐，其所在地自應與各區交通，脈脈相連，本市復九路與復新路交叉之間，地勢尙稱平坦，村落稀少，可收平地建設自由佈置之功，而免改造舊市，拆卸搬移之煩，費省而效宏，殊値吾人考量。

六

市區建設工作，至爲艱巨，如港務碼頭之經營，下水道之設備，公園之佈置，及公共建築計劃等，均須妥爲規劃。本文僅就較要各點，作原則上之商討，至於設計細則，非本文所及。

都市計劃之擬定，應具遠大眼光。計劃之執行，尤須切實努力。各項重要建設，均應以「國防第一」爲依歸。且都市計劃，非一朝一夕之事，具有繼續性，絕不可因人事之變遷，任意更改，故在計劃未定之先，必須鄭重考慮，既經決定以後，則非因技術上發生困難，或因其他重大原因，不應輕率變更。即以道路系統爲例，已經規定之道路，如果隨意更改，則市民將受無窮之損害，而負責行事者，亦易招致物議。是以中央對於都市計劃之核定，無不十分愼重，並明令規定，非經呈准不得變更也。

都市建設之實施，應遠在需要之前。如此則工程費用輕，施工時之糾紛少，進展乃井然有序。觀乎改善舊城市之難於進行，而發展新闢地區之易於舉事，其理即明。以言建設，最要者莫若經費之籌措。考都市建設用費之來源，不外下列數種：一·在市稅總收入內支付；二·徵收受益費；三·借款或發行建設公債；四·中央撥款補助。以市稅收入支付建設費用，則建設進度必甚遲緩。蓋市行政費用所需至多，其能用於建設者，必不充裕。徵收受益費，較為公平合理，既可減輕市庫負担，而其來源亦無窮。依照中央規定，凡建設一工程，其工程一切費用，得徵擬百份之六十於受益地區，餘由市庫負担。惟推行不易，必須人民與政府互相深切了解，通力合作，始克有濟。但為迅速推進建設，似仍應舉行借款或發行公債，以建設後之生產能力償還之。重慶既已定為陪都，則中外觀瞻所繫，中央亦必能視工程情形，酌予補助。凡此四種辦法，均應兼籌並顧，庶使陪都建設，得以蒸蒸日上。

都市計劃之完成，固有賴於積極方面之工程建設，亦宜注意消極方面之機績維護。此可以個人強健體格之養成為譬喻。在積極方面個人應注意飲食營養，在消極方面亦應講求清潔衛生。維護都市之工作固不勝枚舉，而其最重要者，厥惟營造管理。按營造管理較要之目的有二：一為安全，一為符合都市之計劃。如在計劃馬路之處，建築房屋，必須退讓，再如工程建築等，在某種地段不得興建時，即應予以取締。再重慶以往之建築物，多係易燃材料構成，徒重外表粉飾，忽視構造安全，以致塲尾慘劇，時有所聞，一遇火警，輒成燎原，似此情形，必須改正，惟此種管理工作，實行時事務既煩，而每一部門之手續，概與技術有密切關係。欲達完善之境，即細微之處亦須注意。惟市民對此頗多漠視，以致推行之際，頗費周折，必須以堅強之毅力，儘量開導解釋，方能收效。至於各機關之互助合作，亦為達到建設目的之重要條件。

南京下水道

馬育騏

引言

我國城市設有下水道者，寥寥無幾，其間管道具有系統，依照計劃而設置者，尤少如晨星，南京下水道可謂此少數中之一也。先事調查，繼以設計，舉凡全市下水管之大小坡度，俱有規定，各處均可按照安置，逐漸推廣，彼此啣接，以至於全部之完成，其於設計之際，引起工程問題頗多，足資討論，而爲他市之借鏡，故敢冒昧，重提舊題，願同行者其矯正之。

本文擬略述南京下水道設置之起因與南京市之地勢及下水制之採擇，再就與該工程有關之測量、製圖、調查、研究與設計之經過，工程實施之情況，摘要報告，以供我工程界之批評糾正，俾將來從事下水道者，多得一正確之參考材料。

南京下水道設置之起因

南京下水，向以秦淮河爲歸宿，南水污水沆瀣一氣，以致昔日盛名之秦淮，今卽變爲一泓臭水矣，自建都以來，開闢馬路，多有安置下水管，但爲數有限，缺乏系統，下水問題之解決，僅屬局部，非及於全城也，加以建築興始，處處需地，往日之池塘，漸次填塞爲房屋之基地，而舊式之溝渠，時常淤積，未予疏通，雨水污水之宣洩更感困難。綜上數因，南京需要一下水道，日切一日，惟以經費無着，組織缺如，無能應事。迨至民國二十二年，荷蘭庚款，退還我國，劃百分之六十五爲南京市水利工程之費用，並於南京市政府下設下水道工程處，以主其事，作者忝任主辦工程師，承長官之提示，荷籍顧問之指教，以及同仁等之協助，經二載之努力，完成大南京之防水建設，而城南下水道之設計，亦隨復告竣，繼以建築，時逾二年，七七事變，工程遂告中止矣。

南京城之地勢及下水制之採擇

南京城之周約三十公里，城內面積約四十平方公里，形非正方，地勢以鼓樓一帶爲高，劃分南北二區，城南大致偏東地低，人煙稠密，城北大致偏西地高，而多曠野，城外則東北多山，南面山與平原參半，西臨揚子江，地勢特低，所設下關，則在城北偏西瀕江之低處，玄武在城之東北，處於城山之間，莫愁湖在城之西，昔可瀦水，今已淤塞，惟秦淮河位於城內部，縱橫交貫，通於城外護城河，而達於江，向爲宣洩城南之雨水與污水也。

據上地勢，並依照建築之發展，城北與城南之排水，不能採取同一辦法，城北適用分流制，而城南則以合流制爲宜，因城南需要下水道之迫切，故先予設計，而實施建築矣。

測量及製圖

南京秦淮河與城區地面之高低，均與貯水及排水有關，故自下水道工程處成立以後，則着手測量，先爲秦淮河橫斷面，俾便求算各期水位，淺水面積及其容量，（附圖一）繼爲城南之水準與水準標椿高度之確定，（附表一）爲製等高線圖及安置下水管之用，椿數五十，分別埋置於城南各重要地點，有備蓄之生鐵箱保護之，椿身爲鋼筋混凝土，椿頂爲銅叙，與椿身同時倒成，而露出其上端。

製圖方法，先就馬路中心之測點，記載於財政局二千五百分之一之地圖上，算出二十公分等高距之點，移於一百分之一之陸軍測量總局之航空攝影地形圖上，而繪出等高線，再加繪已公佈之道路，以定下水管之路線。

調查與研究

調查之對象，爲雨量與人口，因其影響下水量，而爲設計下水管所必需也。時以南京氣象台自動雨量記載器之紀錄，僅及五年，時間短促，不足爲設計之用，遂採取上海徐家匯氣象台自民國八年至二十三年之紀錄，（附表二）以其期間較長，較爲適當，而京滬相距不遠，雨量相差亦無幾也。

由以上紀錄而計算其自五分鐘、十分鐘，至三百六十分鐘之雨量，及此不同時間每小時所降雨量之公厘數，即所謂五分鐘、十分鐘，至三百六十分鐘之雨量密度是也，（附表三）此密度又分爲一年，一年半、二年、二年半、五年、七年半，及十五年之常率，而繪成曲線圖，（附圖二）可以表示在一定時間內，每分鐘所降之雨量。

人口密度表係根據首都警察廳所載民國二十三年各區人口數依照地點分別統計製成也。（附表四）

設計

南京下水道之設計係根據假定與規定之二種原則，茲分述於左：

假定原則

（一）人口密度：南京城南之人口密度，並非一致，爲適合實情及便利估計起見，將之劃分爲二區：

第一區：東北以秦淮河為界，西南之城牆為止。

第二區：第一區以外之城南部份。

第一區之人口密度，當時平均每公畝三、四〇八，即每英畝一三八人，該區房屋密集，殊少發展之餘地，故以每公畝三、四六八（每英畝一四〇人）假定。

第二區之人口密度，當時平均每公畝〇、六八人（每英畝二七、四八），該區地多空曠，殊有發展之餘地，然其建築，將來多係住宅，勢必留出道路與庭園，空地亟多，故其人口無論如何增加，決不致如第一區之稠密，縱以每公畝一、二四八假定之（每英畝五〇人）較之事實，相去無幾。

（二）污水量：污水為淨水之變態，其出量可以「用水量」比擬之。第二區既係未來之住宅區，居戶之生活程度當高於第一區，故其每人每日之用水量及污水量亦應較多，茲估計之如左：

第一區：每人每日三七、八五公升（即十加倫）

第二區：每人每日七五、七〇公升（即二十加倫）

（三）入管滲水量：下水道管之接縫，頗難絕對嚴密，地下水管滲入，勢所難免，惟其滲水量之估計方法不一，或根據人口，或以下水管之長度計算，或依照接縫之長度而估計，然所估皆係約數，本設計之假定，乃仿照歐洲之方法，以「滲水量等於污水量」較為簡單。

（四）雨量：雨水量之估計，係根據於上述之雨量曲線圖，參照南京情況，而採取一年半之常率，其最大雨量為每小時二、七四吋，雨水集中時間最少則有二十分鐘，倘其雨量超過此項數目時，當地即有氾濫之虞，然經計算，為時極暫。

（五）徑流因數：雨水降落後，因滲漏蒸發之損耗，實際徑流入管者，僅佔所降雨量百分之幾而已，此項百分數，名曰「徑流因數」，其與地質之堅鬆，地面之疏密，地勢之高下，氣候之寒燠等均有密切之關係，南京路面及住戶庭院之舖砌，絕不能如歐美之嚴密，故其徑流因數之假定，可因較少數目，例如左列：

人煙稠密處：百分之三十五。

住　宅　區：百分之二十。

空地較多處：百分之十。

（六）雨水集中時間：此時間如上假定為二十分，如以雨水流速除其經過地域之長度（即集中時間）結果不及二十分鐘者，亦以二十分鐘計算之。

規定原則

（一）管內流速：為免污物沉澱計，管理最低流速規定每秒鐘〇、七八公尺（二呎半），間因地勢較陡，須用較大流速者，亦不能超過每秒鐘二、一四公尺（七英尺）以防下水管之冲蝕，而免污物之落後。

（二）折合徑流面積：城南部份，根據其等高線，而劃作若干排洩區域，各該區域之面積，與其徑流因數相乘，而曰所謂「折合徑流面積」。

（三）下水管橫斷面：下水管之橫斷面，以圓形爲最普通，歐美各國其於合流制，亦用蛋形或橢圓形者，藉以增加污水之流速，按南京情況，此種下水管，增加污水流速甚少，而其價格遠過於圓管，故不採納，而用圓管。

（四）下水管設計及糙率「n」：下水管直徑，橫斷面積，坡度，容量及流率等之設計，皆以麥克及益蒂氏之圖表爲根據，管內糙率「n」概用〇·〇一五。

（五）下水管埋置之最大深度：爲防水及通航計，秦淮河之水位則決定保持於京滬路零點以上五一、六〇公尺與五二、〇〇公尺之間，是故合流管之管內低高，不能低於五二、〇〇公尺，以求水流暢行無阻也。

（六）下水管中心線及道數：設計之管，其中心線概須連續，除過狹之街道及事實上或經濟上所不許者外，每一街道概埋下水管兩道，左右各一，以便利住戶接管，而免除時常挖路舖路之損失。

（七）溢口：尋常污水流入截水管，一至暴雨，水量超過污水量四倍者，則由「溢口」徑流入秦淮河，溢口下檻之高度，須在最高水位以上，即五二、〇〇公尺是也。

（八）陰井距離：爲便於清除下水管起見，陰井距離，原定管徑在九十公分以下者，用三十公尺，其大於九十公分工人能入內冲洗者，則增至五十公尺，嗣爲經濟計，管徑在九十公分以下者，陰井之距離改爲五十公尺，其以上者改爲一百公尺，灣曲管線，以直線成之，而在轉灣處各設一陰井。

（九）抽水站：下水管底降至高度四九、〇〇公尺時，該處則應有抽水站設備，以提高污水使之下流。

（十）小陰井：小陰井取簡單形式，無封吸作用，其距離爲三十公尺。

（十一）叉管：下水管每間十五公尺，須設叉管，以便居戶接通溝管之用。

規定細則

爲求下水管耐久管基堅固計，特規定左列細則，爲製下水管與設置管基之規範。

（一）製造下水管細則

下水管採用兩種，一種在管徑五十公分（二十英吋）以下者，得酌用釉面陶管，一在五十公分以上者，概用鋼筋混凝土管，管內面須光滑，形須一律，接縫處尤須不漏水，而於管之載重力，結實量，吸水量，均須依照製造，此種工程材料普通之規定細則，予以嚴格試驗，而其及格者方准應用。

(二)管基設計細則

先舖二十公分厚之碎磚，夯打堅實，上設鋼筋混凝土一層，厚度依照設計，設計方法，與橫梁同，跨度爲二節溝管之總長，荷重則包括土重，動荷重及兩節滿流清水下水管之重量，繼則安置下水管於鋼筋混凝土上，下部填滿水泥混凝土，至圓週四分之一處，藉以均佈載重至較大之面積，倘地質優良，土重及動荷重，可以減半或至四份之一，如在硬黃土上，則僅舖設水泥混凝土一層，厚度不得超過八公分足矣。

汚水之處理

關於南京城南汚水之處理，凡其適用者，如砂濾，表面接觸，噴水過濾，養化泥，養魚澄清及稀釋等方法，均予考慮，作初步之估價，彼此比較，結果以稀釋法爲最適當。

依照稀釋法，城南之汚水，藉常經馬路之下水管，導流入埋置於秦淮河邊之截水管，引至水西門，與漢西門間之抽水站，以抽水機提高而使之流至下關三汊河口外，以達於江，深入水中，爲二一、七〇〇倍江水之稀釋，借大長江之水量，當可養化該汚水，至於無害也。

一遇大雨，汚水與雨水同流入管，循截水管而進江，迨雨水量四倍於汚水量時，下水管內之水早已稀釋並冲洗淨潔，則由截水管邊之溢口，徑流入秦淮河，如此秦淮河可以保淨潔，且不至於乾涸，衞生以外，風景與航船，兩得裨益也。

工程估價

依照以上辦法，全部工程需款六百餘萬（當時之幣價）然其最要部份，不過百九十萬元耳，此項最要工程一經築竣，則新舊溝管出口及處理汚水之問題，即可解決矣。至其餘四百一十萬，係爲建築馬路下水管之費用，固儘可度量財力，分別緩急，循諸築路程序，而次第籌備也。

工程實施程度

城南下水道，一經設計完竣，不久則循開闢馬路之程序逐漸設置，其重要者如中央醫院前，鼓樓一帶及大中橋附近等之幹管，及至招標埋置截水管與與築漢西門與水西門間之抽水站時，則七七事變，工程遂告中止，誠可惜也。

（附表一）

水準樁位置及高度表(一)

TABLE OF B. M.

樁號	高程	路名	所在地	備註
1	54.565		市政府大禮堂左邊	
2	55.088	中山東路	逸仙橋江南汽車公司六號樁對過	
3	54.608	東廠街	第一公園出口左邊靠東廠街牆脚下	
4	56.954	白下路	建康路白下路交叉口白下路416號門牌牆下	
5	54.767		通濟門外東水關旁	
6	54.852	東冶路	東關頭東關閘西南角水泥欄杆旁	
7	56.167	長樂路	長樂路武定門西面城洞脚街石旁	
8	55.106	中山東路	中山東路與上元路交叉口(漢府街)花圃內	
9	54.732	常府街	常府街38號門牌地下(四條巷附近)	
10	55.177	白下路	白下路239門牌西牆下(中正街小火車站之東面)	
11	54.576	長樂路	長樂路江甯地方法院西面圍牆角下	
12	54.246	長江路	成賢街與四牌樓交叉口西北角電桿下	
13	54.875	珠江路	浮橋南與宗老爺巷交叉口東北角	
14	55.820	國府路	碑亭巷與國府路交叉口東南角電桿下	
15	56.088	中山東路	大行宮太平路中山東路交叉口花圃內	
16	55.262	太平路	太平路淮海路交叉口上海酒樓牆下	
17	54.572	太平路	太平路太平巷世界飯店牆下行人道內	
18	54.915	太平路	白下路太平路交叉口中南銀行門右側	
19	54.581	建康路	建康路朱雀橋交叉口上海銀行旁行人道內	
20	55.759	黃河路	成賢街與九眼井交叉路口東南角	
21	54.146	長江路	大石橋東中大實校西東牆角下	
22		珠江路	蓮花橋洪武街蓮花橋東南角	因築珠江路業已遺失
23	55.136	國府路	國府路香舖營南角	
24	55.821	中山東路	中山東路與洪武路交叉口西北花圃內	
25	55.041	洪武路	淮海路洪武路交叉口洪武路98號門口旁	

(附表一)

水準樁位置及高度表(二)

TABLE OF B. M.

樁號	高程	路名	所在地	備註
26	55.279	白下路	內橋中國銀行東南角奠基石下	
27	55.876	中華路	中華路建康路交叉口中華路198號門口旁	
28	56 051	中華路	中華路長樂路交叉口中華路417號牆下	
29	55.958	中華路	中華路鎭淮橋北祥豐號牆下(596號)	
30	57.654	中華路	中華門外雨花路門牌13號協泰和牆下	
31	56.873	北門橋	北門橋西鷄鵝巷匯文里交叉口均昌泰南牆角	
32	62.061	中山路	鼓樓中山路交叉口西南角中山路495號門牌附近	
33	56.369	中山路	乾河沿金陵中學外中山路花園內	
34	56.028	中山路	中山路華僑路交作口路旁花園內	
35	55.741	中山東路	新街口交通銀行西南角	
36	55.083	中正路	中正路與甪閣老巷交叉口東南角汽車間旁	
37	57.203	中正路	中正路與昇州路交叉口西北角牆下	
38	56.788	中正路	中正路與長樂路交叉口長樂路1號牆下	
39	56.048	秣陵路	豐富路與秣陵路交叉口	
40	55.488	建鄴路	建鄴路179號門牌前牆下(與豐富路交叉口相近)	
41	54.997	昇州路	昇州路221號東首牆下(光華路之西)	
42	54.472	光華路	柳葉街33號門前牆下	
43	58.000	集慶路	集慶路136號(鳴羊街之東)西首牆角下	
44	57.226	長干路	陳家牌坊2號對面牆角下(長干路)	
45		漢中路	漢中路與莫愁路交叉口	
46	56.016	莫愁路	莫愁路與秣陵路交叉處	
47	56.549	莫愁路	朝天宮西街莫愁路口(西區憲兵隊門口)	
48	55.313	莫愁路	水西門大街與莫愁路交叉口	
49	54.688		婁角市實業部全國度量衡局門前牆下	
50	59.142	漢中路	漢中路漢西門大街交叉口	

計劃　南京下水道

（附表二，三）

26012

雨量密度表

INTENSITY OF RAINFALL

根據徐家匯天文台民國八年至二十三年之紀載

連降時間	5分鐘	10分鐘	15分鐘	20分鐘	25分鐘	30分鐘	45分鐘	60分鐘	80分鐘	100分鐘	120分鐘	180分鐘	240分鐘	300分鐘	360分鐘
雨量密度以每小時公厘為單位	228.0	180.0	159.0	136.8	127.0	117.8	90.5	69.5	56.6	46.2	39.1	30.0	31.9	28.3	26.6
	204.0	160.2	144.0	23.0	108.7	117.6	87.5	68.6	37.3	33.1	36.7	25.6	13.5	13.1	12.8
	140.4	160.0	120.4	113.7	105.0	112.4	82.6	49.2	34.4	22.8	28.4	13.9	12.7	12.0	8.0
	128.4	134.0	118.0	103.5	103.0	102.4	69.5	44.1	33.0	28.0	27.3	13.8	12.6	11.7	7.8
	127.2	120.0	15.2	94.5	84.7	93.4	61.0	42.3	32.3	20.4	17.8	13.7	11.0	10.5	6.9
	126.0	119.0	114.8	91.5	82.8	89.8	49.0	38.9	28.8	19.0	17.2	12.7	10.6	9.2	6.8
	122.4	116.5	99.2	85.0	76.8	75.2	48.0	38.4	20.8	17.4	19.9	12.4	9.4	8.1	6.5
	120.0	116.0	91.2	73.8	72.0	71.4	46.9	36.6	19.2	15.3	16.5	12.0	9.0	7.6	6.4
	119.0	114.0	86.0	72.6	67.4	66.4	46.7	35.8	17.3	15.2	16.2	9.9	8.9	7.5	5.4
	118.8	104.0	85.6	66.0	65.5	63.0	45.8	35.2	15.5	14.2	13.6	9.5	8.5	5.9	4.3
	117.6	102.0	83.6	65.7	62.9	58.4	38.8	31.9	15.1	13.5	13.4	9.3	7.8	4.7	
	99.6	96.0	75.6	64.5	62.4	58.2	38.3	26.6	14.6	13.0	13.3	9.1	7.6		
	96.0	86.4	75.2	63.6	59.0	57.4	34.5	25.7	14.0	12.8	11.8	8.1	6.4		
	94.8	83.4	72.4	63.0	58.8	56.0	32.3	25.5	13.8	12.7	11.6	7 9	5.6		
	93.6	82.8	72.0	61.8	57.2	55.6	31.8	22.6	12.8	11.5	11.5	6.5	5.3		

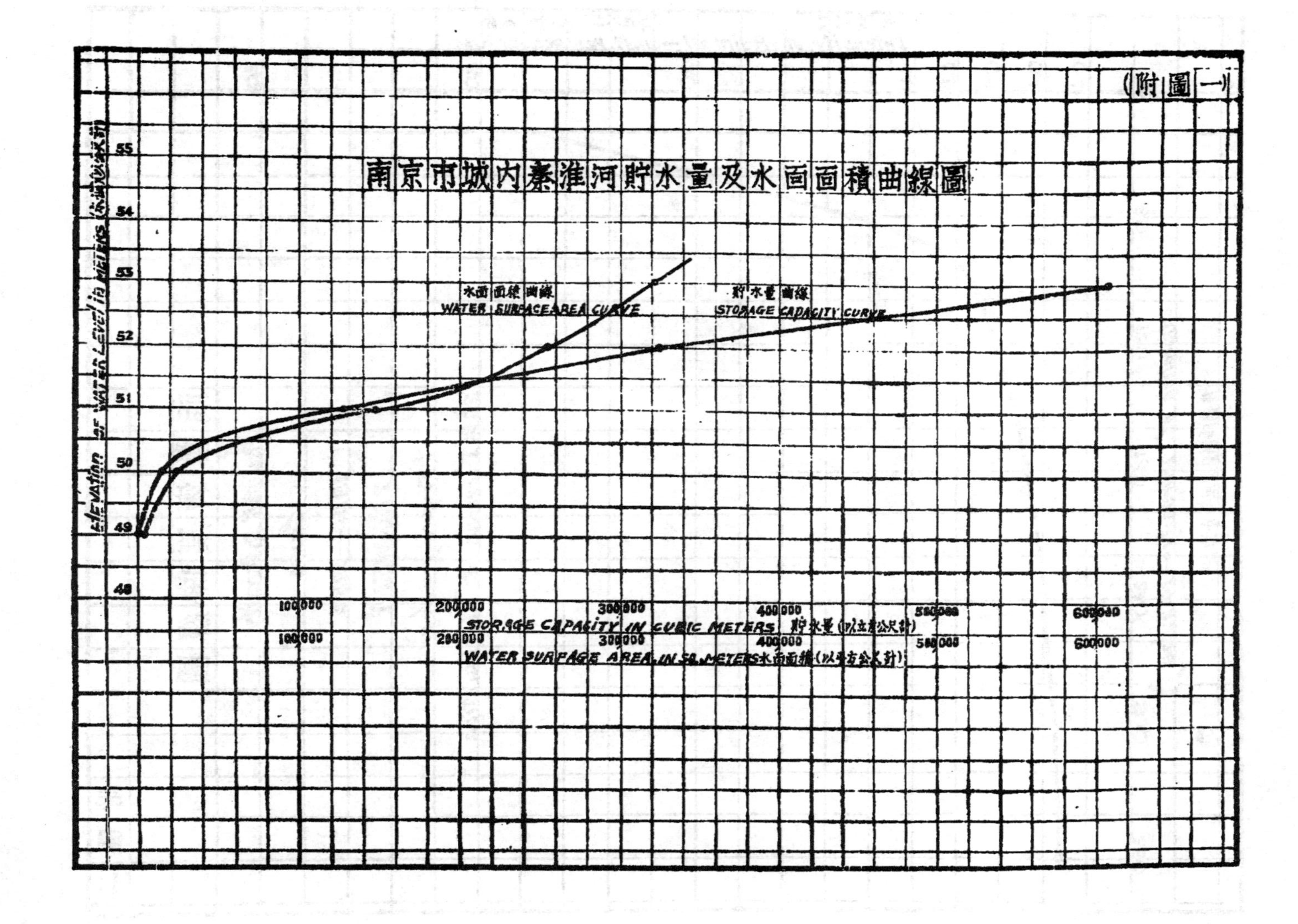

(附圖一)
南京市城內秦淮河貯水量及水面面積曲線圖
ELEVATION OF WATER LEVEL IN METERS
48
49
50
51
52
53
54
55
水面面積曲線
WATER SURFACE AREA CURVE
貯水量曲線
STORAGE CAPACITY CURVE
100,000
200,000
300,000
400,000
500,000
600,000
STORAGE CAPACITY IN CUBIC METERS 貯水量(以立方公尺計)
100,000
200,000
300,000
400,000
500,000
600,000
WATER SURFACE AREA IN SQ. METERS 水面面積(以平方公尺計)

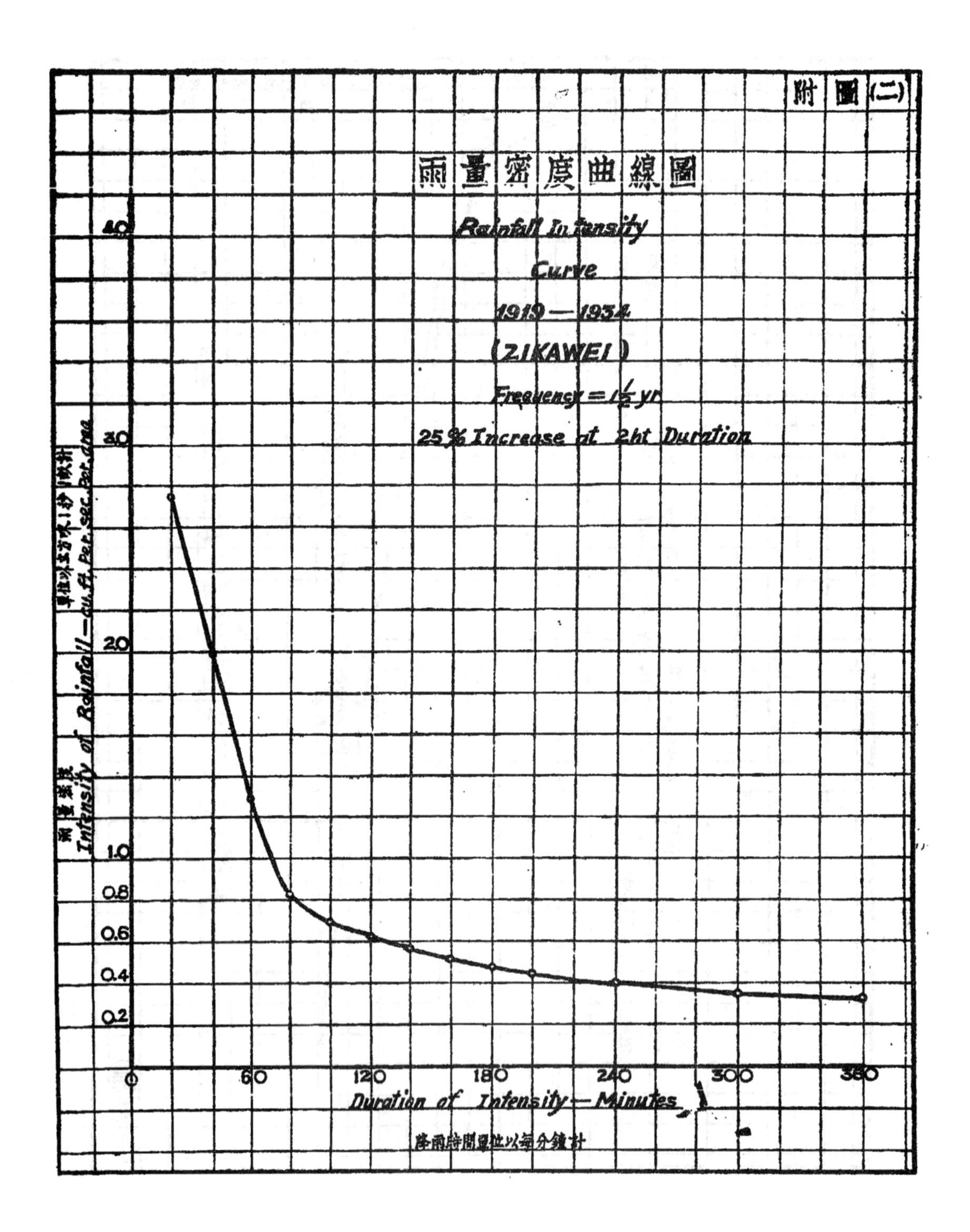

附圖(二)
雨量密度曲線圖
Rainfall Intensity
Curve
1919—1934
(ZIKAWEI)
Frequency = 1½ yr
25% Increase at 2hr Duration
雨量密度 單位以立方呎1秒1畝計
Intensity of Rainfall—cu. ft. Per. sec. Per. acre
4.0
3.0
2.0
1.0
0.8
0.6
0.4
0.2
0
60
120
180
240
300
360
Duration of Intensity—Minutes
降雨時間單位以每分鐘計

(附表四)

南京市人口密度表(民國二十三年二月)(一)

POPULATION DISTRIBUTION OF NANKING

局名	分駐所所在地	人口數	面積		人口密度	
			英畝數	公畝數	每英畝內人數	每公畝內人數
第一局	大行宮	9,662	57.3	2,318	168.5	4.16
	通賢橋	9,585	57.6	2,328	166.8	4.12
	將軍巷	18,151	96.6	3,913	188.3	4.65
	韓家巷	12,261	85.1	3,443	44.1	3.56
	國府路	16,074	77.8	3,150	206.5	5.10
	羊皮巷	8,067	79.5	3,218	101.2	2.50
	戶部街	10,265	71.1	2,880	144.1	3.56
	壽星橋	12,339	57.8	2,340	213.9	5.30
	大逸巷	13,238	538.0	21,893	24.7	0.61
	太平橋北	15,291	846.0	4,268	18.2	0.45
	總計	124,933	1966.8	79,751		
第二局	申家巷	6,796	101.5	4,118	66.8	1.65
	白下路東段	7,175	68.7	2,780	104.5	2.58
	釣魚巷	9,507	45.5	1,845	208.3	5.15
	膺福街	8,868	58.4	2,363	151.8	3.75
	龍王廟	13,708	108.5	4,400	125.9	3.11
	三條巷	11,886	81.5	3,308	145.3	3.59
	東廠街	5,023	183.5	7,425	27.6	0.68
	中山東路	2,467	316.0	12,825	7.7	0.19
	光華西街	8,972	582.0	21,195	17.0	0.42
	共和門	7,860	196.5	7,942	40.1	0.99
	總計	82,287	1,742.1	68,201		
第三局	貢院街	5,176	28.9	1,170	179.0	4.42
	大全福巷	13,719	85.7	3,473	160.3	3.96
	織作坊	17,554	80.5	3,263	217.9	5.38
	毛府園	19,295	94.4	3,825	204.1	5.04
	姚家巷	6,097	51.6	2,093	117.3	2.90
	鷲峯寺	16,639	136.2	5,523	121.8	3.00
	新路口	16,188	19.2	4,837	135.2	3.34
	總計	94,668	596.5	24,189		
第四局	珍珠巷	17,933	283.0	11,475	63.2	1.56
	養虎巷	16,253	255.5	10,350	63.6	1.57
	中華門	19,885	73.8	2,993	267.9	6.61
	釣魚台	15,684	86.8	3,512	181.0	4.47
	小膺巷	14,379	80.7	3,262	178.5	4.41
	五福街	17,307	191.0	7,717	90.8	2.24

(附表四)

南京市人口密度表(民國二十三年二月)(二)

POPULATION DISTRIBUTION OF NANKING

局名	分駐所所在地	人口數	面積		人口密度	
			英畝數	公畝數	每英畝內人數	每公畝內人數
第四局	張家街	18,294	80.6	3,262	226.5	56.00
	總計	119,735	1051.4	42,571		
第五局	石鼓路東口	19,233	194.0	7,852	99.2	2.45
	石榴園	19,819	108.0	4,388	182.7	4.51
	馬巷	15,361	51.2	2,070	301.2	7.41
	程善坊	17,117	84.3	3,415	202.5	5.01
	石鼓路西口	14,517	718.0	29,062	20.2	0.50
	朝天宮西街	17,059	193.8	7,830	88.3	2.18
	水西門街	15,646	109.0	4,410	143.3	3.54
	鳳凰街	8,819	78.0	7,200	49.5	1.22
	瓦廠街	10,657	234.2	9,495	15.4	1.12
	總計	138,219	1,870.5	75,722		
第六局	鼓樓北	6,589	252.0	10,194	25.9	0.64
	保泰街	2,455	104.0	6,210	159.9	3.95
	虎踞關	1,561	522.0	21,105	3.0	0.07
	門樓上	4,495	345.0	13,837	13.0	0.30
	馬台街	8,595	384.0	15,570	22.2	0.55
	靜界寺	6,370	613.0	4,861	10.4	0.26
	妙香	6,092	215.5	8,730	28.3	0.70
	和平門	4,006	658.0	25,662	6.1	0.15
	五洲公園	1,144	124.5	5,045	9.2	0.23
	丹鳳街	13,420	250.0	10,125	53.6	1.33
	總計	54,727	3,468.0	142,337		
第七局	靜海寺	11,935	49.9	2,025	238.0	3.87
	復興街	11,897	59.2	2,390	201.0	4.97
	美孚街	7,662	153.2	6,200	19.8	1.23
	天保路	14,426	63.8	2,587	226.0	5.53
	二馬路	8,504	44.5	1,800	191.6	4.73
	虹門口	6,550	33.3	1,350	196.3	4.85
	寶塔橋	7,303	139.0	5,625	52.6	1.30
	北祖師菴	6,841	33.5	5,400	51.1	1.26
	總計	75,120	676.4	27,377		
第八局	小河南	13,753				
	下碼頭	13,350				
	六股道	5,362				
	總計	32,465				

計劃　南京下水道

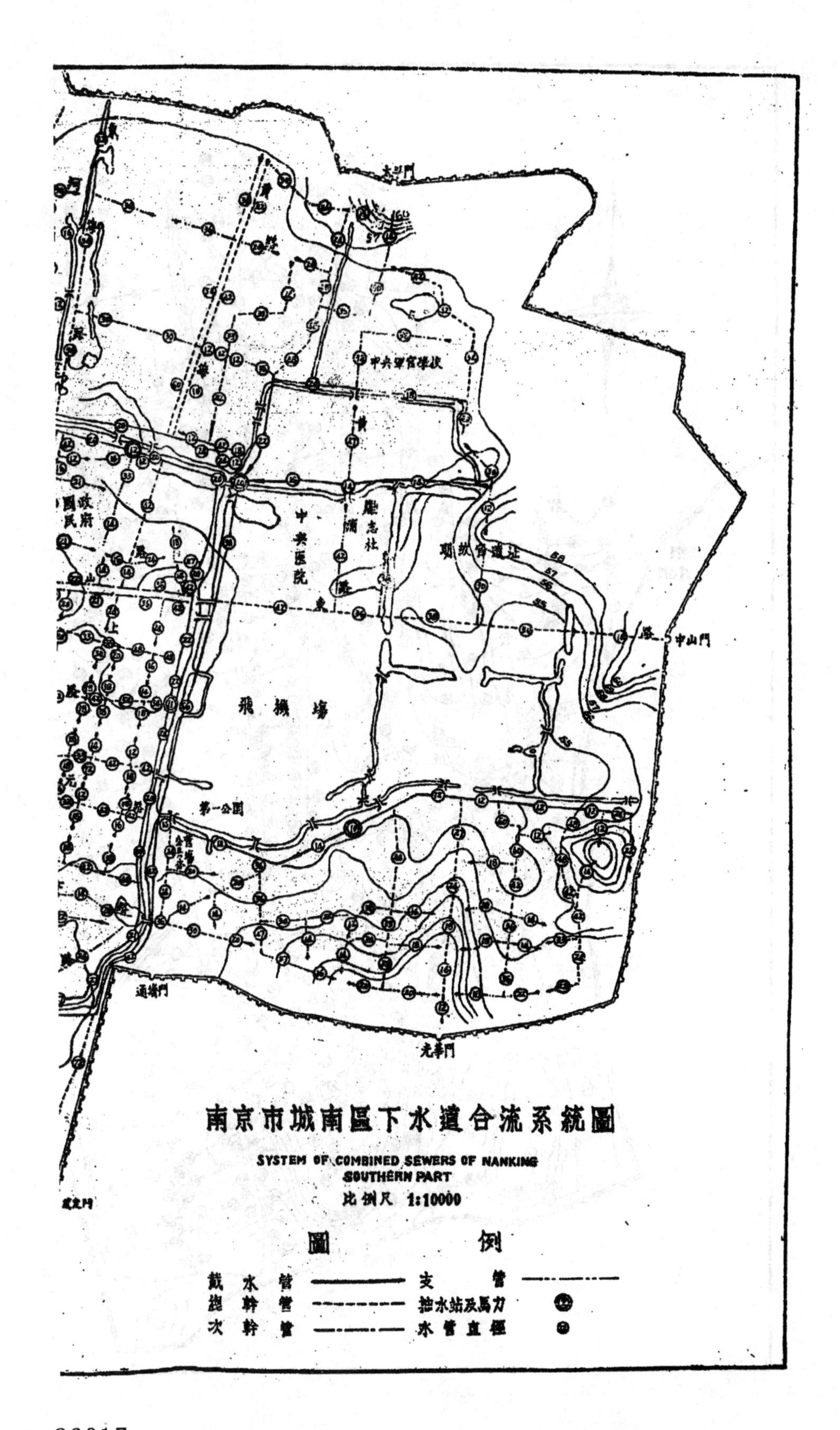
南京市城南區下水道合流系統圖
SYSTEM OF COMBINED SEWERS OF NANKING
SOUTHERN PART
比例尺 1:10000
圖 例
截水管
支管
總幹管
抽水站及馬力
次幹管
水管直徑
中山門
光華門
通濟門
飛機場
第一公園
中央醫院
國民政府

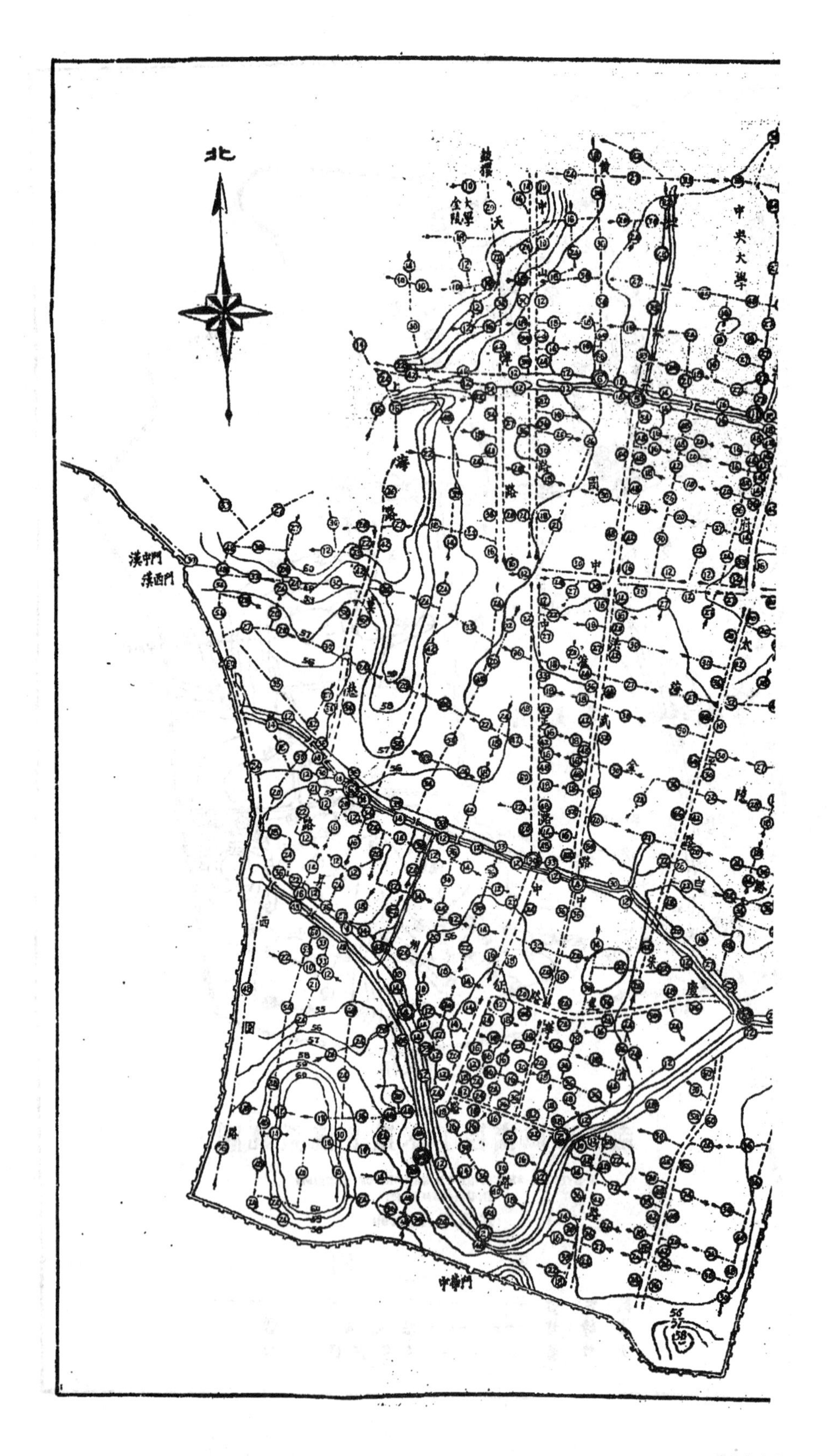
北
漢中門
漢西門
中華門
中央大學

北平市溝渠建設計劃

譚炳訓

溝渠建設設計綱要

一　序言

溝渠爲市政建設之基幹，爲市民新陳代謝之脈絡。道路藉溝渠之排水，路基始得穩固，路面始免冲毀，是以道路與溝渠，爲市政上不可分離之建設，須相輔而行者也。惟道路與溝渠在設計上有截然不同之點在，即道路可就目前需要之程度以定其路面之寬度，他日交通增繁，可隨時就路傍預置之空地展寬，昔日所修之路面仍可完全利用。溝渠則反是，若僅就目前建築狀況及一區域之水量而建造，則將來本區建築增多或鄰區安設溝渠須假道此區以排水時，則原設溝管必難容納，另改較大溝管，則昔日所埋設者，發工挖出，大半拆毀，不能再用，投資化爲烏有，此誠市政建設上之一種浪費，爲市政工程上所應竭力避免者也。故溝渠創辦之初，雖可就市民需要及財力所及，舉辦局部之小規模建設，但設計時必須高瞻遠矚，作統全市之整個計劃，以適應市內溝渠全部完成後之情況。如此則脫胎於整個計劃中之局部建設，雖爲局部小工程，亦可永爲市產之一部，永爲市民所利用，不致中途廢棄，使建設投資變爲一種無益之消耗也。

根據以上所論，雖在市財政尚未充裕之時，亦應先行草擬全市溝渠建設之整個計劃，以爲局部建設之所依據。惟全市溝渠整個計劃，非有精確之測量，縝密之研究，難期完善。北平爲已臻發育成長之城市，市民習慣及已成建設，均須顧及，而籌適應利用之策，故尤須有詳確之調查及各項預備工作，方克着手設計。但設計所依據之基本原則，基本數字及基本公式，須先行討論研究，經專家之審定後，以爲擬定整個計劃之所根據。此本設計綱要之所由起草也。

二　舊溝渠現狀

北平市舊溝渠之建築時期，已不可考，傳稱完成於明代，迄今已有數百年之歷史。內城分五大幹渠，由北南流，皆以前三門護城河為總匯。內城幹渠之最大者為大明濠（即南北溝沿）與御河（即沿東皇城根之河道）。現除御河北段外，均改為暗溝。什剎海匯集內城北部之水遊入三海。西四南北大街與東四南北大街各有暗溝二道，惟出口淤塞。外城之水大部匯集於龍鬚溝，流入城南護城河，龍鬚溝西段現亦改成暗渠。據北平市內外城溝渠形勢圖而論，幹支各溝，脈絡貫通，似甚完密，實則溝渠宛延曲折，甚少與路線平行；且曩時市民築房，漫無限制，致多數溝渠壓置於市民房屋之下，難於尋覓。溝之構造均為磚砌，作長方形，上覆石板溝蓋，無人孔，滲漏性甚大，小量之水洩入溝中，常不抵出口，即已滲盡，掏挖時須刨掘地面，揭起石蓋，方可工作。現大部溝渠或淤塞不通，或供少數住戶之傾洩污水，雨水則多由路面或邊溝流入幹渠。此項舊溝渠，昔時完全用以宣洩雨水，迨後生活提高，市民污水漸感有設法排除之必要，遂多自動安設污水管接通街溝，近年本市工務局為便利市民並資限制計，代為住戶安裝溝管，而舊溝渠之為用遂成污水雨水合流矣。舊溝渠淤塞之病，由來已久，歷年雖有溝工隊專司掏挖，然人數既少（僅百餘人）又無整個計劃，此通彼塞，無濟於事，蓋積病已深，非支節掏挖之所能收效也。

三　溝渠系統

溝渠之為用，可大別為二：（一）排除工廠及市民家屋（廚房浴室廁所）中之污水。（二）宣洩路面房頂及宅院中之雨水。污水含有多量之污穢物及微菌，終年流洩，不稍間斷，故須設管導引至市外遠處，經過清理手續，再洩入江湖或海洋中。此種污水流量無多，且無陡增陡減之現象，故清理之工作雖繁，所需以導引之管徑則小。雨水之流量當數百倍於污水，設管導引，所需之管徑極大，但污穢較少，不必清理，即可洩入市內之河流池沼。污水與雨水之質與量，既有如上所述之不同，溝渠之系統遂有「分流制」與「合流制」之區別；分流制為一街之中分設雨水污水兩種溝渠，各成系統，不相混亂，適宜於舊城市已設雨水溝渠設備之區域，及雨水易於排出之處，無須設大規模之雨水溝渠系統以導引者。此法多數市街可單設污水管，雨水則藉明溝或短距離之暗渠以流洩至雨水幹渠或逕達消納雨水之處。此種分流制度之優點，以污水流量較少，所需導引之管徑亦小，用機器排除，亦較簡易，大雨時因與雨水分管而流，無雨水過多，倒灌室內之弊。故各國城市多採用之。合流制為雨水與污水同在一混合管內流出，適宜於新闢市街，雨水不易排洩之區，污水雨水當同時設管導引或同需機器抽送者。此法一街之中僅設一道混合管，開辦既較同時安設兩管爲省，而平時維持費亦較分流制爲低也。

北平市舊溝渠現況，污水與雨水混流，似爲合流制，然此種現象之造成，實因本市人口增加，生活提高，新式浴室廁所日多，無污水溝渠以消納污水，於是洩入舊溝，而演成今日無溝不臭之現狀，不可據以認爲污水混流於雨水溝內爲合理，而斷定本市溝渠爲合流制也。新式雨水溝渠不適於宣洩污水爲盡人皆知之事實，而本市之舊式雨水溝渠，不適於污水之流洩，其理由更爲顯著，因溝底不平，坡度過小，污水入內，幾不流動，與其名之溝渠，勿寧視作滲坑，附近井水，莫不被溝內滲下之污水所濁，因而病菌繁殖，侵害市民，此本市舊溝渠不適於合流制之最大理由也。

本市溝渠系統應採何制？舊溝之不能用以合流雨水污水，已如上述，即本市將來建設新式溝渠，亦不能用合流制而應採分流制，其理由有五：(一)分流制之水管可用圓形管，合流制之水管則多用蛋形(下窄之橢圓)管，因圓管水滿時流速大，水淺時流速小，故量少之污水流過時發生沉澱；蛋形管則無論流水之多寡，水深與水面寬度，常爲一不變之比，流速無忽大忽小之弊，故流大量之雨水或少量之污水，皆不致發生沉澱。惟此種蛋形管，管身既高，且下端又窄，所需以埋設之必深，而本市地勢平坦，不易得一適當之坡度，且土質鬆軟，安設此種溝管，不特所費過鉅，且工事進行亦困難，此本市不能用合流制而應採分流制之理由一也。(二)什刹海三海及內外城之護城河皆可用以宣洩雨水，而不能任污水流入，以臭化全市，此污水雨水應分道宣洩而採用分流制之理由二也。(三)本市舊溝渠雖不適於運除污水，若加以改良疏濬，大部分尚可用以宣洩雨水，利用舊時建設，排除今日積潦，爲最經濟之市政計劃，此本市溝渠應採分流制之理由三也。(四)市民生活程度漸高，衛生設備日增，現據自來水公司之報告，新裝專用水管者，每月有百戶之多，近年來全市水量消費亦日增，因而時感不敷供給。全市之穢水池皆苦宣洩不及，穢水洋溢於外。前三門護城河中污水奔流而下，爲量可驚。由以上三點而論，本市之污水排洩問題已日趨嚴重，自應另行籌設全市之新式污水溝渠，自成系統，與雨水溝不相混亂，專用以排除全市污穢，此不僅現時市民之衛生狀況因以改善，且可一勞永逸，樹市政建設之百年大計，此本市溝渠應採分流制之理由四也。(五)污水量少，所需之溝管直徑亦小，約自二百公厘(八吋)至六百公厘(二十四吋)，故建設費所需較少，估計第一期工程費約爲一百四十萬元(詳見北平市污水溝渠初期建設計劃)，初期建設完成後，雖不能逐戶安設專用之污水管，但利用新式穢水池消納多數住戶污水之效力，則今日污濁橫流，穢水漫街之現象，當可免除。此就建設經濟言，本市應採分流制之理由五也。

根據以上之討論，本市之溝渠系統問題可得一合理之解決，即改良舊溝以宣洩雨水，建設新渠以排除汚水，即所謂分流制者是也。此不僅為理論上探討之結論，亦本市實際情況所需要，且為比較經濟之市政建設計劃也。

四　舊溝渠之整理

欲整理本市之舊溝渠，須先明瞭舊溝渠弊端之所在。本市之舊溝渠，其弊有五：(一)全市排水均以環繞內外兩城之護城河為總匯，而以二閘為洩出之尾閭。但二閘以上，多年未加疏濬，河身淤淺，且有較溝底為高之處，以致水流不暢。大雨時洩水過緩，遂有路面積水，溝渠淤塞之病。(二)支渠斷面過小，不足容納路面及宅院排出之水。(三)溝渠坡度太小，多數均不及千分之一，直如一水平之槽溝。非雨水注滿，水不流動，即或流動，速度不足，不能攜泥沙以同流，易致沉澱，故溝常淤塞。(四)本市柏油路不多，土路及石渣路面上之雨水，常攜多量之泥沙，沖積溝內。(五)汚水藉舊溝流洩，不獨有第三節所述之各種弊害，且溝中常存積汚水，大雨時則洋溢於外。(四)(五)兩項可藉道路之舖修，汚水暗渠之建設，以免除之。(一)(二)(三)三項乃溝渠本身之已成事實，設局部支節挖濬，一年之後，又復淤塞，非一勞永逸之計，徒耗財力。為謀澈底改善，茲擬定整理之大綱如下：

(一)護城河之疏濬　護城河源始於玉泉山麓，至城西北之高梁橋以東，分為二道，環繞內外兩城，終復匯流於二閘，成為通惠河之上流。二閘以下是否淤塞，尚待調查，但以二閘上下流高度之差觀之，(二閘上下流河之底差為三·四公尺，設上流挖深二公尺，相差尚有一·四公尺)即二閘以下不加導治，於上流之洩水，亦無妨礙。高梁橋以上，雖亦淤塞，但於本市溝渠之整理，關係尚小，茲不備論，故亟需疏濬者，為環繞內外兩城及貫通三海之一段，而尤以前三門護城河為最要，以其為內外兩城洩水之惟一幹渠也。疏濬之次序應斟酌緩急，分為五期進行。

第一期　前三門護城河(自西便門至二閘一段)

第二期　西城護城河(自高梁橋至西便門一段)

第三期　什刹海及三海水道

第四期　外城護城河(自西便門環繞外城至東便門一段)

第五期　東城北城護城河(自高梁橋至東便門一段)

(二)舊溝渠之整理　本市各街溝渠淤塞已久，位置在市民房屋之下者有之，湮沒無從尋覓者亦有之，皆淤積過甚，非支節挖濬之所能濟事。設一區之幹溝挖通，支渠未治，大雨後各支渠淤積之汚穢泥沙順流而下，已挖濬者有重被堵塞之虞。反是若支渠疏通，幹渠不治，則水流遲緩，仍難免巨量之沉澱。故擬採分區分期疏濬辦法，就全市地勢高低之所趨及各幹溝分布之情形，分為若干排洩區，每區之疏濬整理必須於一個時期內完成之。舊溝渠之斷面過小坡度過平者，則設法縮短其洩水路程，以期於可能範圍內充其量以利用之。其實不堪應用者，則另行籌設新式雨水暗渠，如此分期進行，市庫不致担負過重，且可一勞永逸，不數年間，全市溝渠可望無淤塞或排洩不暢之弊矣。

按本市舊溝渠系統迄今尚無詳確調查，工務局雖有一萬七千五百分之一之溝渠形勢圖，及十八年份工務特刊中之內外城暗溝一覽表，但此項圖表僅表示流水方向，溝渠寬深及長度。溝渠之位置及坡度則未詳載，故可供參攷之價值甚微。且圖中所示已疏濬之一部，有壓佔於市民房之下者，（如燈市口等處）有僅有檅水袖洩水之用，雨水則另藉明溝或路邊以排洩者（如西安門大街等處）。故全市溝渠中究有若干尚可利用，若干須另設新溝，以及疏濬舊溝渠與另設新溝渠經濟上之比較，均無由着手，整理工程之概算，亦無從估計。故詳確之測量調查，實為整理舊溝之基本工作，而須首先着手進行者也。

（三）雨水溝渠流量計劃法 （即整理舊溝渠用作根據者）

雨水流量之計算擬採用準理推算法（Rational method），此法較用其他各種實驗公式（Empirical Formulas）為宜，因後者係就歐美各城市之經驗而定，各國各市之情況且各不同，本市強予採用，有削足適履之弊也，準理推算法之公式如下：

$$Q=ciA$$

Q為每秒鐘流量之立方呎數，c為洩水係數(Coef.of Rain-off)，i為降雨率(Intencity of Rain-fall)每小時之吋數，A為集水區域之英畝數。其中之ci可規定如下：

（1）降雨率（i）降雨率在本公式中須視雨量大小，降雨時間（Duration of Rainfall）之久暫，及降雨集水時間（Time of Concentration）之長短而定。本市降雨量無長久精確記載，北平研究院雖有自民國三年至二十一年之最大雨量表，但其記錄中最大降雨率（民國三年）每小時僅三七・二公厘，清華大學本年（二十二年）之雨量記錄，最大為每小時四四・五公厘，清華所用者為新式之自動雨量計，北平研究院所用者為普通標準雨量計，由所用方法上比較，則前者所得結果自較後者為準確，且本年「二十二年」本市雨量不為過大，而清華之記錄即達每小時四四・五公厘，由此可證明研究院之每小時三七・二公厘之記錄，不足深信。清華大學亦僅有二年「民國二十一年及二十二年」之雨量記錄，亦難用為設計之標準。華北各城市之雨量記錄可供參考者甚少，青島之雨量記錄年限稍久，但僅有每小時之最大雨量測驗表，降雨時間，亦無記載。青島溝渠設計所用之最大降雨量為六二．六公釐（二．四六吋），本市為大陸氣候，全年降雨總量雖不甚大，每小時之雨量則不能斷定其小於青島（據翁丁二氏合著之中國分省新圖中之全年平均等雨量區域圖，北平與青島之全年平均雨量均為六〇〇公釐至八〇〇公釐），因夏季多驟雨故也。據此暫假定本市之最大雨量為每小時六十五公釐（二．五吋），似較為合理。再本市多舊式瓦房，路面坡度又甚小，降雨集水時間（t）當稍長，設計一切溝渠均按照假定之最大降雨率設計，殊不經濟。然本市之降雨率及降雨時間既無記載，上海市雖有五年十年之降雨率循環方程式，因與北平氣候懸殊，不能採用，華北各地亦無可供參考者。茲就美國各城市之雨量統計加以比較，擬採用梅耶氏（Meyer）公式第三組之「降雨率五年循環方程式」：

$$i=\frac{122}{t+18}$$

為本市設計之標準，而最大以每小時六十五公釐為限，即降雨集水時間在三十分鐘（t等於三十分鐘時，i等於六十五公釐。）以上者用方程式，在三十分鐘以下者用假定之最大降雨率。

（一）進水時間（Inlet time）按十五分鐘計算。

（2）洩水係數（c）與地質，地形，房屋之疏密，街道之構造，及地上之植物均有關係。此等係數有從平日實驗而得者，有根據情況相近之城市已有之記載而定者，茲限於時日，採用第二法。

本市繁盛區域，多為四合房（即四面建房，中留空地），房頂與房地全面積之比為四：五，道路面積與房地面積之比約為一：五，（本市街寬至無規律，同為繁盛區，王府井大街及西單牌樓寬二十餘公尺，大柵欄鮮魚口等處街寬不過八公尺，計算時須按照各街實況斟酌變更），道路假定為瀝青路面，屋頂為普通中國瓦舖成，院地為磚砌或土地，準此情形，列為下表：

洩水係數表（甲）

承雨面積種類	道路	屋頂	院地	統計
面積百分數 R	二〇	六四	一六	一〇〇
洩水係數 C	〇．八五	〇．九	〇．五	
R X C	一七．〇	五七．六	八．〇	八二．六

住宅區域，道路多為石渣路或土路，院地較大，空間處且種植草木，洩水自少，如次表：

洩水係數表（乙）

承雨面積種類	道路	屋頂	院地	統計
面積百分數（R）	一五	三五	五〇	一〇〇
洩水係數（C）	〇．五	〇．九	〇．二	
R X C	七．五	一．五	一．〇	四九．〇

由甲乙兩表可定繁盛區洩水係數為〇．八三，住宅區洩水係數為〇．四九，此項數字係一假定之例，設計時須就市內各處實際情況斟酌為變更。

五 污水溝渠之建設

本市原無污水溝渠，設計時得因地勢之宜，作統盤之計劃，而無所遷就顧忌。惟污水溝渠之運用，有需於自來水之輔助，本市自來水設備，尚未普遍，市民大多數取用井水，取之不易，用之惟儉，恐污穢難得充量之水以冲刷溶解，溝渠內難免有過量之沉澱。然此爲初設時之現象，將來市政進展，自來水飲用普及，此弊自免。

（一）污水出口之選擇 選擇之標準，須（1）地勢低下（2）旁近湖泊或河流，（3）須距市區稍遠。本市地勢，西北凸起，東南趨下，最大水流爲護城河匯集而東之通惠河，該河二閘附近，遠在郊外，人煙不密，地勢較低，以作污水出口，尚稱合宜。惟河狹流細，恐不足冲淡氧化巨量之污水，故須於總出口附近，設總清理廠，於污水洩出前清理之。

（二）污水排洩之劃分 本市地勢平坦，土質鬆軟，設路面掘槽過深，不獨工勞費鉅，滯礙交通，且恐損及兩旁房屋。冬季氣候嚴寒，污水管敷設過淺，則有凍結之虞。故假定管頂距路面之深度最少以一公尺爲限，最多以四公尺爲限，幹管坡度最小千分之一，支管坡度最小千分之三，準此劃分污水排洩區如下：

第一區 內城東部，鐵獅子胡同以南三海以東之區域，幹管設南北小街，遥達於二閘之總清理廠。

第二區 內城西部，即三海以西之區域，幹管設西四北大街，南至宣武門附近設污水清理分廠以送水至總清理廠。

第三區 內城北部，即北皇城根街以北之區域，幹管起於護國寺街西端沿北皇城根街以至鐵獅子胡同以東，擇地設污水清理分廠以送水至第一區之污水幹管。

第四區 外城全區，幹管有二：一起於宣武門外大街，繞西河沿經正陽門大街以趨於天壇。一設於廣安門大街，至西珠市口東端與由北來之幹管會合於一處，即於天壇東北設污水清理分廠以送水至二閘總廠。

（三）污水之清理 擬採以下二法：

第一步 篩濾法：於各清理分廠及總清理廠內各設篩濾池一座，以除去固體物質及渣滓，池底每日清一次或兩次，沉澱物取出後，積存於積糞池，以備售予農戶，用作肥料。

第二步 沉凝法：於總清理廠設伊氏池（Imhoff Tank）污水經此池後，能變易其性質，使污穢沉凝，微菌減少，洩出之水色淡無臭，然後注入河中，即河水不甚充足，亦無大害。

（四）污水之最後處置 污水經清理後，雖污穢大減，若河水量小，或冬季無水，污水洩入後，仍難免有停滯凍固之弊。茲就其可能採用者，擬定以下三種辦法，但採用何法，須待詳爲測量調查後決定。

(1)汚水經清理後，卽於二閘下流，洩入河中。此法簡易經濟，惟河水須終年不斷，並於冬季結冰時，須保有足量之潛流以資沖淡汚水。

(2)沿河設汚水水導管直引通至通州之北運河中。此法須設長二十餘公里之導管，所費稍鉅。

(3)汚水經清理後，導流至附近農地，以利灌漑，此法驅無用爲有用，可使不毛之地，變爲良田，法之至善。惟建設費過昂耳。

(五)汚水流量計算法　汚水流量須視市街之人口密度，每人每日最大用水量，及地下水滲透量而定。平市人口密度，據二十二年公安局之調查統計，人口最密之外一區，每公頃(Hectare)四〇七人，普通住宅區，如內一區，每公頃僅一五〇人。惟現值國難時期，百業衰微，將來市政發展，人口密度，當不止此，且市街兩旁，迄今尤多平房，此等平房將來均有改造樓房之可能，人口密度亦必有大量之增加。我國城市之繁盛者，人口最多區域之密度多在每公頃六〇〇人上下，故假定商業繁盛區之人口最大密度爲每公頃六〇〇人(約合每英畝二五〇人)，普通住宅區每公頃三〇〇人(約每英畝一二〇人)，每人每日用水量參照青島市與上海市之統計，假定爲七〇公升(約合一五英加侖)，再每人每日用水時間，各隨其習慣而不同，茲據歐際頓氏(Ogden)之假定，每人每日用水時間爲八小時，每公頃每日總流量之半於八小時內流盡，準此可求得每公頃一秒鐘最大之汚水流量(Q)如下：

$$\text{繁盛區}\ Q=\frac{600\times70}{2\times8\times3600}=0.73\text{公升}$$

$$\text{住宅區}\ Q=\frac{300\times70}{2\times8\times3600}=0.365\text{公升}$$

地下水滲透量，須視所用水管材料之品質，接裝方法，安設之良否，及地下水位之高低而定，普通每公里長管線之滲透量最少一一八〇〇公升，最多九四〇〇〇公升。本市地下水位甚高，水管擬用唐山產之缸管，滲透量假定爲每公里管綫二五〇〇〇公升，將來再按各處實際情況，酌爲損益。由上所述，可規定汚水總流量之計算如下：

任在何處，汚水總流量等於該處以上流域公頃數與每公頃一秒鐘流量相乘之積，再加該處以上管綫之地下汚水滲透總量。

汚水管計算之標準，假定如下：

(1)庫氏公式(Kutter's Formula)中之N爲〇・〇一五。

(2)全滿時每秒鐘最小流速爲六公寸(約二呎)。

(3)幹管坡度最小爲〇・〇〇一，支管坡度最小爲〇・〇〇三。

(4)計算汚水管徑以水流半滿爲度。

(5)街巷公用污水管徑最小不得小於二百公厘。

污水溝渠初期建設計劃

北平市溝渠之建設，宜採用分流制，即整理舊溝渠以利宣洩雨水，另設新溝渠以排洩污水，於溝渠設計綱要第三節中，曾反復申述，原有溝渠僅可用爲雨水道，不應兼事宣洩污水，故爲溝渠之澈底整理計，除舊溝渠須逐漸爲有計劃之疏濬外，污水溝渠之建設，亦應及時籌劃。

全市污水溝渠之建設，非數百萬元莫辦，需款過多，非本市目前財力之所能及，且値此國難當前，公私交困之際，卽設有大規範之污水溝渠，因市民接用，亦須耗相當財力，恐難望使用之普及，且自來水之飲用，尙未普及，亦不必街街設管。玆擬具初步污水溝渠建設計劃，卽先於繁盛街市(如王府井大街前門大街等)及稠密之住宅區域安設污水幹管及支管，以便用水較多之住戶接用，而於各街內或路口另設新式之穢水池，以備一般市民之傾倒穢水，如斯費少效宏，可期全市污水之大半，得科學之排除方法，於本市之公共衛生，實大有所裨也。

初期建設之污水溝渠爲全部污水溝渠之基幹，雖經費務期其節省，但規模宜求其宏偉，俾可籠罩全局，謀永恆之發展。故須由通盤整個之設計中，定爲分期實施之建設計劃。玆擬具設計步驟如左：

(一)測量及製圖 設計污水溝渠系統必備之圖樣有二：(1)全市地形圖，比例五百分之一或千分之一，(2)各市街水平圖，長度比例須與地形圖一致，高度比例爲百分之一。

(二)污水總出口之勘定 污水總出口以二閘(慶豐閘)爲較宜，於設計綱要第五節第一段中曾加申述。但確定之前，須爲如下之考查：

1.該處河道水位及流量之測驗。

2.流往二閘各河水源之考查。

3.上下流人民利用河水情況，如飲用，漁業，灌溉等及約需水量，此項調查須上自水源，下抵通州。

4.自二閘至通州地勢之草測，以爲設管引至通州北運河時估計之根據。

5.二閘附近之土壤及農產調查。

(三)市區繁盛之調查，各街市及住宅區域房屋之疏密，人口之多寡，並推測其將來發展之程度，以定何處爲繁盛區(工商業區)，何處爲疏散區(住宅區，名勝區)。

(四)溝管之設計 可就全市地形圖，確定管線之位置，幹渠支管分佈之系統。同時按照市街水平圖，確定管線之高度。至於管徑之大小，可根據污水流量計算法與(三)項調查之結果，及庫氏公式圖解(Diagram for the Solution of Kutter's Formula)計算而得。

(五)唐山缸管產量與單價之調查，及品質之檢定。

(六)排洩區及清理總廠與各分廠廠址之勘定，及伊氏池，篩濾池，積渣乾洒池，積糞池，積水池，及機器房，工人宿舍等之設計。

(七)污水最後處置方法之確定，根據(二)項勘查之結果以定何種方法為最適宜。

(八)訂立：

1.市民自動安設溝管獎勵章程

2.建設污水溝渠徵費章程

3.污水溝渠施工及用料規則

(九)製圖及造具詳細預算。

以上九條，已概括污水溝渠設計之全部，蓋全部設計定為初期建設方可擇要施行。初期建設中擬完成之五部分如下：

(一)內外城污水幹管及支管(見附圖)共長約九萬公尺。

此項幹管平均管徑約為四百公厘，概用唐山產缸管，柏油與麻絲接口。清理分廠送水管承受壓力之一段，用鑄鐵管，軟鉛與

麻絲接口。人孔為圓形，底砌石灣，人孔蓋用鑄鐵鑄成，人孔牆及人孔底用一：二：四混凝土打成。

(二)總清理廠　設於二閘附近，廠地面積需地約八公頃(約一百三十畝)，可容面積約占一百平方公尺之伊氏池(Imhoff Tank)十座，於初期建設中暫設四座，篩濾池，積糞池，積渣晒乾池，各設一座，機器房一座，須能容五十馬力電動抽水機六台，先設三台。修理廠，儲藏庫各一座，辦公室一間，工人宿舍七八間。

(三)污水出口之設備，須俟污水最後處置之方法確定後，再行計劃。

(四)清理分廠　清理分廠共三處。第一分廠，設內三區東四九條胡同與北小街交道口附近。第二分廠，設於宣武門東。第三分廠，設於天壇東北角附近。每廠設篩濾池，積糞池各一座，機器房一座，須能容二十馬力動升水機四台，暫設二台。辦公室一間，工人宿舍四五間。各清理分廠須預備空地，建小規模之材料廠。

(五)積水池　街市公用之積水池須採用缸式構造，中設鐵篦，俾臭氣不得外揚，渣滓易於清除。池牆與池底用一：三：六混凝土打成，上加鐵蓋或建造小房，以保清潔。出水管徑為二〇〇公

北平市内外城汚水溝渠初期建設系統圖

比例尺　三萬五千分之一

民國二十二年十一月

北

圖　　例

汚水管	——→——	鉄路	═══
清理廠	▨	馬路	═══
河渠	～～	門洞	▬
橋梁	)(	水平石標	■

26029

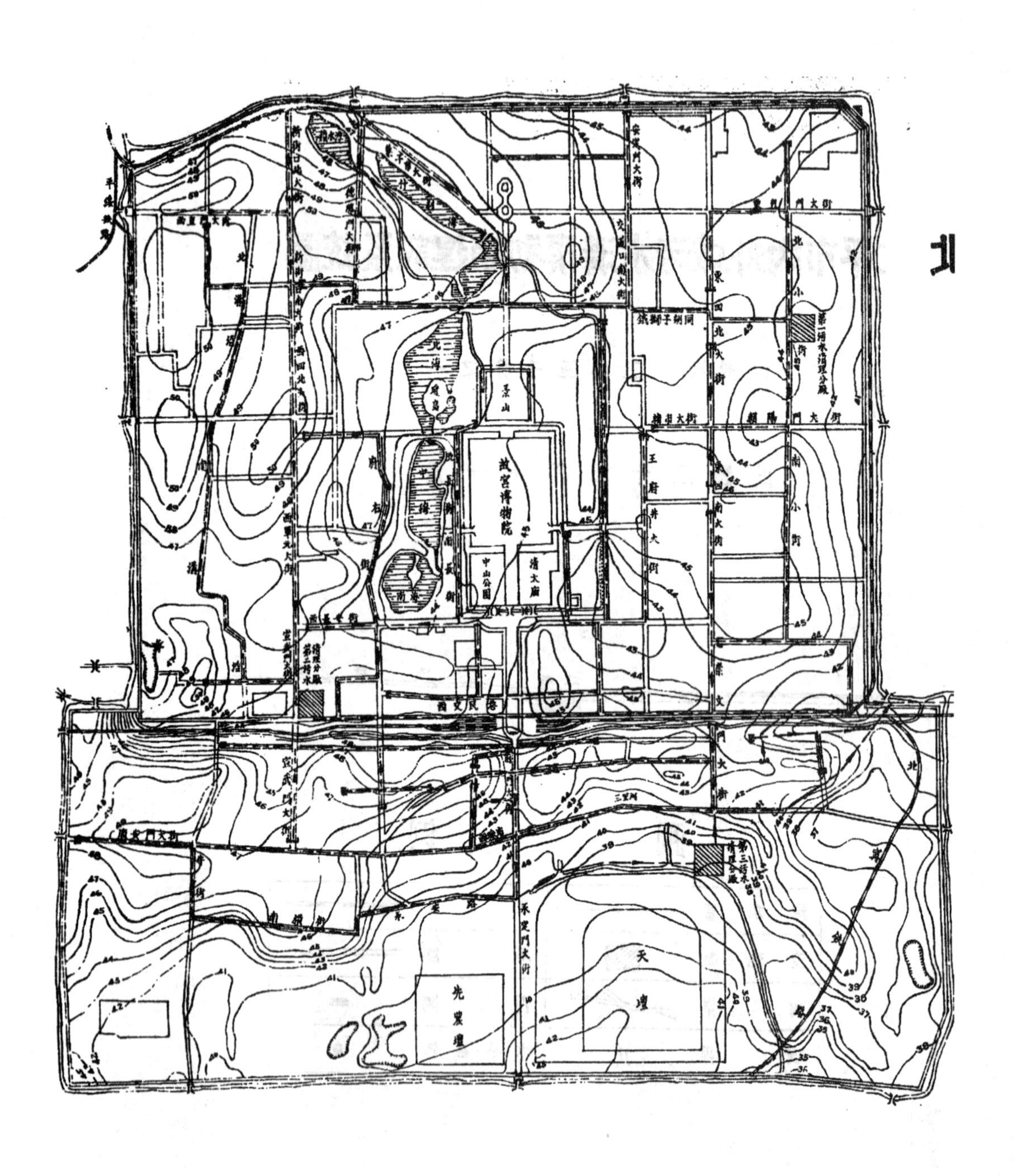

26030

厘，全市蓄水池約計須設四〇〇處。

汚水溝渠初期建設計劃概算

名稱	單位	數量	單價（元）	共價（元）	備考
管線	公尺	九〇，〇〇〇	一二	一，〇八〇，〇〇〇	平均管徑按四〇〇公厘計算土工接管工人孔等均包括在內
總清理廠	處	一	八〇，〇〇〇	八〇，〇〇〇	包括購地費全部機件及建築用費
清理分廠	處	三	二〇，〇〇〇	六〇，〇〇〇	仝上
蓄水池	處	四〇〇	二〇〇	八〇，〇〇〇	
其他				七〇，〇〇〇	汚水最後處理之設備等用費
共計				一，三七〇，〇〇〇	

註：測量及設計用費，由工務局經常費中支付，未列入本概算內。

徵求北平市溝渠計劃意見報告書

本府技術室前擬定之「北平市溝渠建設設計綱要及汚水溝渠初期建設計劃」，爲集思廣益起見，曾分寄國內工程專家徵求批評，現覆函皆已遞到，特歸納諸家高見，分爲問題七種，參以本府技術室意見，擬具報告如左：

一・溝渠制度問題

關於平市溝渠應採之制度，各家與本府之意見完全一致，即「改良舊溝以宣洩雨水，建設新渠以排除汚水，即所謂分流制者是也」。溝渠制度爲溝渠之根本問題，得各家一致之主張，本府自當引爲溝渠建設之準繩也。

二・溝渠建設之程序

溝渠之制度既定，溝渠建設之程序可隨之而決，即先整理雨水溝渠，次建設汚水溝渠是也。然溝渠設計之程序，不能同於溝渠建設之程序，雖建設可分先後，設計則須同時完成。蓋街市上雨水汚水兩種溝渠之配置，交錯時不相衝突之高度，或某處因分設兩管之特殊困難，須採局部之合流制者，必雨水汚水溝渠同時設計，始可兼籌並顧，以謀所以配合適應之道。至溝渠建設之施工，不但須分期進行，在分期之中，尚須分區工作，就工程進行上之便利及減輕經濟上之困難言，實爲溝渠建設所必採之步驟也。

三・雨水溝渠設計之基本數字

「設計綱要」中所假定之雨水溝渠設計基本數字如下：

降雨率　每小時六十五公厘（即二・五吋）

洩水係數　商業區83%　住宅區49%

降雨集水時間　三十分鐘（由梅耶氏[Mayer]降雨率五年

循環方程式求得）

進水時間　十五分鐘

北洋工學院院長李耕硯先生認爲降雨率不必假定如此之大，本府當根據今後平市之降雨率精確記錄，酌爲減低。清華大學教授陶葆楷先生認洩水係數所假定之數字稍嫌過高，本府「設計綱要」所列之二表，係舉例性質，同爲居宅區，其區內房屋疎密及道路情況，未必盡同。設計時當就各洩水區域，分別加以調查，列表備用，既可與實際情形相脗合，亦可免管大浪費之弊也。

四・污水溝渠設計之基本數字

「設計綱要」中所假定之污水溝渠設計基本數字如下：

人口密度（每公頃人數）　商業區六百人　住宅區三百人

每人每日用水量　七公升（或十五英加侖）

地下水滲透量　每公里管線二五〇〇〇公升

庫氏(Kutter)公式中N爲〇・〇一五

人口之密度，李先生仍認爲太密。上海市工務局技正胡贊予先生亦同此意見，並發表具體之主張如下：

「按二十二年平市公安局人口密度調查統計，人口最密之外一區每公頃爲四百零五人，普通住宅區每公頃一百五十人。此種情形，在最近若干年內，似不至有多大變動。即將來工商業發達，人口激增，亦宜限制建築面積與高度，及闢設新市區以調劑之，不宜聽其自然發展，致蹈吾國南方城市及歐美若干舊市區人煙過於稠密之覆轍，使文化古都，成爲空氣惡濁交通擁擠之場所，而失其向來幽雅之特色。鄙意平市商業區將來之人口密度仍宜以每公頃四百人爲限，住宅區以增至每公頃二百人爲限。」

平市人口，就民元以來二十一年之統計觀之，實有穩堅增長之總趨勢。雖六年至十五年之九年間，人口總數之變動甚微，而十五年以後之人口激增，迄今賡續前進，勢不稍衰。若根據二十一年之人口增加率，按等差級數法，推測二十五年以後之人口密度，則商業區每公頃可達五百六十人，住宅區可達二百一十人。若就最近七年來之增加率，按等差級數法推測二十五年後之人口密度，則商業區每公頃可達七百五十人，住宅區可達二百八十人。推算人口增加，以等差級數法所得之結果，最爲保守。按二十一年之平均增加率，顧得將來之人口密度，尙在五百與二百人之上，而開闢新市區以減低人口密度之法，平市以城墻關係，較之他市稍感困難，似將來人口密度之假定，商業區不能小於五百。住宅區不能小於二百。惟平市商業區與住宅區不能明確劃分，且漸有變遷移轉之勢。民元前後商業區皆集中於前門外一帶，現則東城以王府井大街爲中心之商業區發展甚速，西城以西單牌樓爲中心之商業區亦有突飛之興榮，故平市有趨於細胞發展之可能，

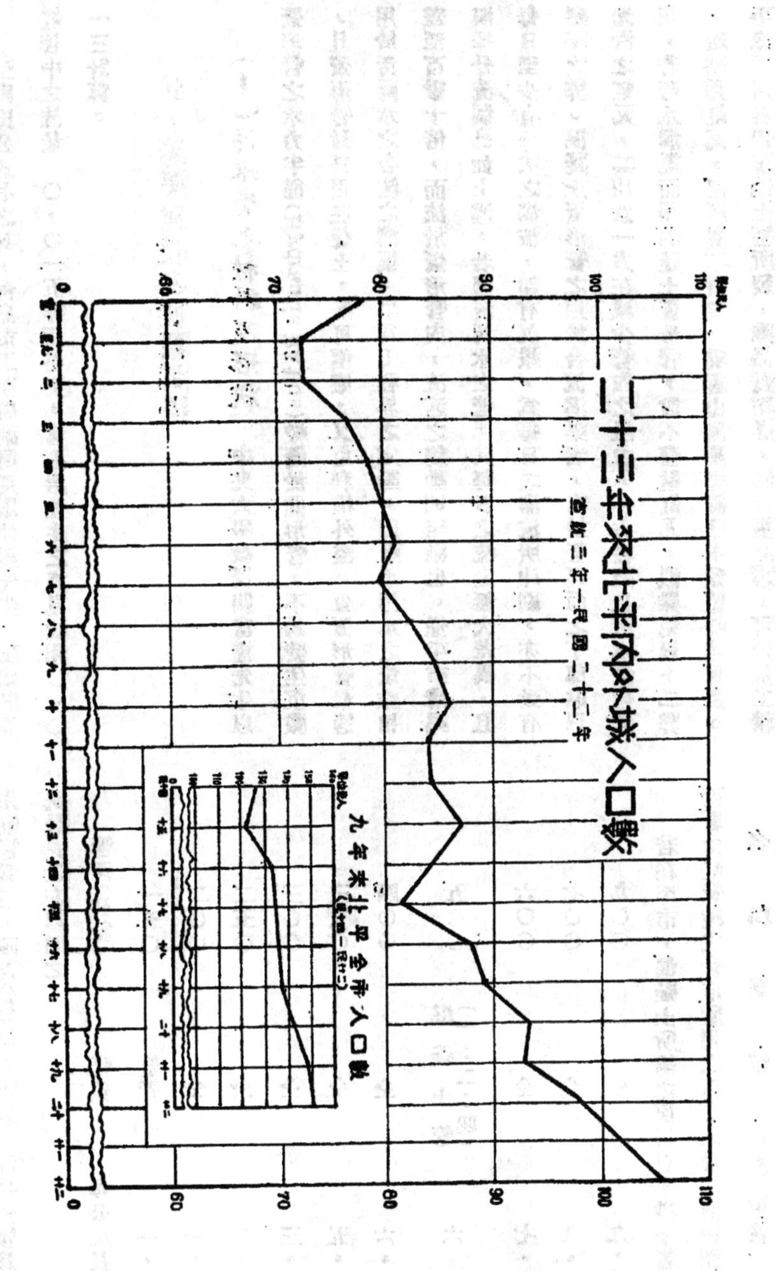

人口增加之推測，亦以分區估算爲較妥，所謂商業區及住宅區不過籠統而言，其間自應就各處特殊情形而斟酌損益也。

至庫氏公式中之N，青島市工務局副局長嚴仲絜先生，認爲計算缸管中之流量，〇•〇一五非所必要，當遵嚴先生之意見改用〇•〇一三計算。

五•溝渠建設之實際問題

(1)污水管之材料及形狀

中央大學教授關富權先生以蛋形管之水力半徑(Hydraulic Radius)較優於圓形管，不易發生沉澱，且蛋形管材料用混凝土，既可價廉，又免利權外溢。查蛋形管最適用於污雨水之合流之溝渠，早爲工程界之定論。因雨水污水之量雖相差至百數十倍，而流於蛋形管內，流速之變動則至微也。惟平市溝渠擬採分流制已如上述，若僅流污水之管，其每日之流量無大差異，且每日至少有一次之滿流，即有沉澱，爲每日之滿流所冲刷，亦不致有壅塞之弊。混凝土蛋形管之用於合流溝渠者，有於管裏面之下部貼以光滑之缸瓦，其用意一方在減少管內之阻力，一方在防止污水侵蝕洋灰，若污水溝渠而用混凝土蛋形管，設不滿貼缸瓦，似難免以上二弊，若貼用缸瓦，則所費不貲矣。現唐山開灤煤礦已不兼營缸管買易。平津所用者皆該地土窰所製，雖品質稍遜，倘大量訂購，可使加工精製也。故購用缸管，並無利潤流入外商之弊。再就經濟方面言，缸管亦較混凝土管爲省費。按青島市溝渠工程之統計，四百公釐以下者以用缸管爲省，四百公釐以上者以用混凝土管爲省。茲列青市工務局之統計表於左以明之：

管徑(公釐)	管質	每公尺長工料價共計(土工在外)
一五〇	缸管	一•一六元
二〇〇	仝	一•六六
二五〇	仝	二•〇二
三〇〇	仝	三•七二
三五〇	仝	五•二六
四〇〇	仝	六•三一
五〇〇	混凝土管(一：二：四)	六•〇〇
六〇〇	仝	七•〇〇
七〇〇	仝	八•〇〇
九〇〇	仝	九•五〇

若在平市，混凝土所需之原料石子砂子皆較青市昂至一倍左右，而唐山缸管較之青市所用之博山缸管，價尚稍廉。茲列比較表如左：

名稱	單位	青島價格	北平價格
石子	立方公尺	二•四〇元	三•九〇元
砂子	立方公尺	一•五〇元	三•七〇元

缸管　半徑四百公釐，一公尺長，　六•〇〇元　三•八四元

(此係唐山交貨最上等雙釉缸管價格，北平交貨另加運費每公尺約一元左右)

故就材料之經費而論，平市溝渠之宜用缸管，殆尤迫切於青島市也。

(2)接管用之材料　下水道水管間結合之材料，普通用者有柏油麻絲及洋灰砂漿二種，用洋灰砂漿之優點在堅實省費，其缺點在換裝支管困難，接頭處無伸縮性，若基地下陷或壓力不均，缸管有折裂之虞。用柏油麻絲之優點在換裝支管甚易，接頭處有伸縮性，缸管不致折裂，其缺點在用費稍昂，略欠堅牢。嚴先生主張用洋灰砂漿接管，在街傍用戶於建造溝渠時皆同時裝接支管，則該處以洋灰砂漿接管，尙無不便，否則以用柏油麻絲接管爲較妥。至地基之堅實情形，亦爲決定採用何種接管材料所應考慮之因素也。

(3)水管埋設之深度　原計劃假定管頂距路面之深度最小以一公尺爲限。關先生以爲〇•六公尺即足以防凍，似無須埋設一公尺之深。查北平市冬季嚴寒，〇•六公尺之深度是否足以護管，尙待考究，惟爲防止車輛震裂水管計，〇•六公尺之深度稍嫌不足。因通衢雖有柏油或石碴路面，電車之震動則甚劇，而平市電車軌之下並無鐵筋混凝土基礎以抗禦之。若水管敷設於步道或小巷中，鐵輪大車於雨後恆陷入路面〇•三公尺上下，則所餘之〇•三公尺實不足以護管。故水管埋設之深度，除防凍外，尙應斟酌交通情形而規定之也。

(4)反吸缸管之採用　污水管橫過護城河時，如該河水流橫斷面有限，不容污水管直穿時，嚴先生主張用反吸缸管，由河底穿過。查平市護城河流量多嫌不足，自以照嚴先生所言辦理，最爲妥善。

(5)消污池(Septic tank)之採用　關先生主張每胡同或數戶合建一消污池，以減污水之量，改進污水之質，並減輕未來專用污水管住戶之負担。用意實深，惟事實上則未能與理想之結果相合。公共消污池不能建於私人土地之內，必設於街衢，既置妨礙交通於不論，池之通風筒放出多量之亞摩尼亞氣，行人掩鼻而過，與今日糞車滿街情況無異，有失建設污水溝渠之意義，此其一。全市建造數百消污池，較之建一大規模之總清理廠，所費更鉅，按天津英租界工部局建造鐵筋混凝土消污池之統計，供給二十人用之池，造用約八十元，即平均每人需費四元。平市飲用自來水之人口約爲十萬，則建造消污池之所費，即有四十萬元之多，且數百消污池之清除管理，亦非易舉，此其二。消污池所減之污水量甚微，而其所剩之污渣，不能用作肥料，此其三。酸性污水，或天寒之時，池內霉腐作用，幾全停止，此其四。有此四端，故消污池不能大規模採用於平市。原計劃中有建造試

水池四百處一項，卽便於不裝置專用污水管之中下市民傾倒污水而設，故市民無論貧富，皆有使用污水溝渠之便利，市民之負担與享用，並無畸重畸輕之弊也。

六．清理分廠之地址問題

城內清理分廠之設立及其位址，完全爲地形所決定。因污水藉地面天然之坡度，由高趨下，全市污水總清理廠既設於城外之二閘，則全市各處之污水欲其皆能藉天然之坡度，匯集於二閘，爲平市地形所不容許，因城內有數處低窪之區，水流至此，若不以機器提高水位，污水卽停滯該處，而無術排除，此清理分廠之所以設及地址之所由決定也。兩先生以爲宣武門內等處，人烟稠密，設清理廠，不免有臭味，不如設於天壇地廣人稀之處。此爲北平市地形所限，不能不分設於宣武門內，東四九條胡同及天壇東北角等過於低窪之處，實爲無可奈何之事。如青島市之污水清理總廠雖僅西鎮一處，而清理分廠則有四處，其太平路之一廠，在市府之前，爲交通要衝，亦風景佳地，而因地形關係，不設廠則水不能前進，故德管時代已闢地設廠矣。日本東京市復興計劃完成後，全市有污水清理分廠八處。各廠設備有「沉沙池」及「唧筒室」等。污水經篩濾後，再提高水位，送至清理總廠。其第一排洩之區「錢瓶町唧筒場」卽在東京驛之北傍。此種污水清理分廠若設計周密管理得法，並無臭味外溢，因所清理者爲新出之污水，不同於消污池所出之霉腐污水，且僅經過篩濾一種手續，卽以抽水機送出。若清理濾池，運除污渣於夜間行之，附近居民當不致有不快之感也。嚴先生主張將東四九條之第一清理分廠移於朝陽門一帶，因該處更爲低下也。此論極是，惟朝陽門至東便門間無可供安設幹溝之街道。本市現正測製二千五百分之一地形圖，等高線之差爲半公尺，此圖測竣，各污水清理分廠之地址，當再重行通盤籌劃也。

七．污水之最後處理方法

污水之總清理廠設於東便門外之二閘，該處地價不昂，污水經清理後，卽洩入通惠河，該河之水並不充作飲料。根據以上三種情形，並爲節省財力計，故污水清理採用篩濾池及伊氏池(Imhoff tank)之法，雖此法佔用廠基稍多，然地價不昂，所費無多。清理效率雖不能十分圓滿，然全廠構造簡單，不藉機械之力以工作，且河水不作飲料，故亦無須再經他法以清潔之。採用伊氏池法，不僅建造費低，且管理易，經常費尤省也。至污水中之肥料，大部存留於篩濾池中，沉澱於伊氏池及洩入河中者爲量無多。李先生囑仿上海英租界辦法，採用「活動污泥法」(Activated Sludge Process)，以保全肥料，用意至善。惟「活動污泥法」清理污水手續皆藉機械之運轉，設備費既昂，管理亦難，經常費尤大。其優點在清理效率高，污渣之肥料價值大，而廠基佔地，在各種清理污水方法中爲最小。上海租界地價昂，或亦採用此法之一原因也。上海英租界污水清理廠之成績，本府已派員調查，以供參考；並擬選定現出污水地點數處，按時往取污水加以化驗，如每月化驗一次，則積一年以上之記錄，於污水最後處理方法之取捨，定有所助也。

污水清理廠工作系統圖

（Imhoff設計）

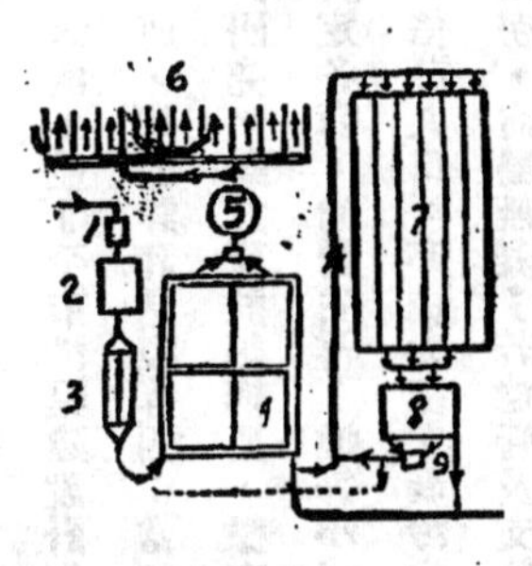

1.粗篩 Coarse racks
2.漉油池 Skimming tank
3.沉砂池 Grit chambers
4.伊氏池 Imhoff tanks
5.污泥再消化池 Secondary sludge digestion tank
6.污泥晒床 Sludge drying beds
7.吹風池 Aeration units
8.最後沈澱池 Final settling tanks
9.污泥抽送機 Sludge pumps

按伊氏（Imhoff）最近主張伊氏池與「空氣活動污泥法」，爲績之污水清理法，卽污水先經伊氏池，再入吹風池，完成「空氣活動污泥」手續後，始行排除。惟吹風池之污泥一部仍回吹風池，一部則又送至伊氏池內，助該池內污泥之消化。其工作系統如附圖所示。

圖中所示清理系統之特徵，約有兩點：

(1)伊氏池爲初步之清理，吹風池爲高度之清理，但視污水情形，吹風池可完全不用，僅經伊氏池，卽行排除，以減消耗。

(2)吹風池內之污泥必經伊氏池，與該池內之污泥混合後，始得送至污泥晒床，故吹風池與伊氏池之污泥不能分別保存。

由以上二點而論，北平市污水清理總廠，暫先設伊氏池，二閘距城稍遠，通惠河水不用作飲料，僅伊氏池已可勝清理污水之任；否則隨時加建吹風池，以前之建設仍可充分利用，固無棄置之可慮也。

此次承海內工程專家，不吝賜教，爲北平市之溝渠計劃，建一完善合理之基礎，本府實深感荷。今後在詳密計劃完成過程中，與諸位工程先進商榷之問題正多，爲百餘萬市民造福利，想諸君必樂爲助也。

茲覆按書收到之先後，附錄於後，以資觀證，而便研究。

李耕硯先生覆函

——天津國立北洋工學院院長——

前奉大函，並北平市溝渠建設計劃綱要及北平市污水溝渠初期計劃綱要一册，囑即詳評見復等因；當即與敝院衛生工程教授徐世大先生，共同研究，對於原計劃綱要，微有鄙見，茲約略述之：謹按原計劃綱要所定各節，尚屬妥善；惟每小時最大雨量，較青島爲高，似可不必。又估計人口，亦嫌太密。蓋污水溝之最小者，有一定限度。人口估計太密，尚屬無礙；若總管及支管埋設既深，人口過密，不免靡費；且北平並非工商業重要市區，人口未必增加甚多；即或某一區域有增加之時，亦可安設支管，隨時應付。

清理廠之估計似太低，但因未知其計劃，無從懸斷。又查吾國人向以人糞溺爲肥料，事不可忽視。如用因姆好夫池，即不能得肥料之用。上海所用促進污泥法 Activated Sluge Process，雖或用費稍增，而保全利益頗多，似應加以調查，以定清理之法。

以上數端，鄙見如是，是否有當，尚祈卓裁。

（二十二年十二月十五日）

嚴仲絜先生覆函

——青島市工務局局長——

頃奉大函，附北平市溝渠建設設計綱要及污水溝渠初期建設計劃，拜讀之下，具見規模宏遠，擘劃周詳，無任欽佩，猥以謭陋，辱荷垂詢，殊愧無以報命，惟念千慮一得，或有補於高深，謹將管見所及，略陳如左：

（一）污水溝渠初步計劃，採用分流制，並規定各清理廠地點，布置甚爲妥善周密，惟第一清理分廠之東南一帶，地勢仍漸趨低下，若該處人口繁盛，市面發達，有宣洩污水之必要，似可於朝陽門附近，另擇適當地點，移建第一清理分廠於該處，兼可吸收由朝陽門向面一帶之污水。（二）污水管橫過護城河或大溝，如該河溝內流水橫斷面有限不容污水管直穿，似宜用反吸虹管由河底穿過。（三）計算缸管流量應用庫氏公式時，係數（N）可減爲〇・〇一三，尚不嫌小也。（四）污水缸管接口用一比二洋灰漿，確屬堅實省費，滲水亦少，較之用柏油與麻絲爲優，以上各節，是否有當，敬乞卓裁，肅箋奉復，即希查照爲荷。

二十二年十二月六日

陶葆楷先生覆函

——淸華大學衞生工程教授——

接讀來函，並北平市溝渠建設設計綱要及汚水溝渠初期建設計劃一册，藉悉貴府計劃北平溝渠系統，有裨民生，自非淺鮮。承囑批評討論，謹就管見所及，逐條敍述如下：

（一）整理北平市溝渠系統，大體採用分流制，卽利用舊溝渠，以宣洩雨水，建築新溝管，以排除汚水，爲極合理之辦法。不過汚水溝管之安設，需款甚鉅，値此社會經濟，異常窘困之時，卽有汚水溝渠，恐市民接用者，亦屬少數。北平自來水之飲用，尙未普及，以內一區而論，僅百分之二十，接用自來水，其餘均恃井水。且糞汚均須作爲肥料，如建築汚水溝渠，導汚水至附近農田，以供灌溉，究當如何應用，亦須先作相當之研究。故鄙意今日北平欲整頓溝渠制度，宜先從雨水着手，換言之，北平市宜暫時集中財力與人力，整理並完成全市之雨水溝渠制度。倘欲建築汚水溝渠管，亦當分區進行，爲內一區需要較大，不妨先事安設，次則內二區，內三區，逐漸推廣。計劃幹渠時，當顧到將來之發展，自無待言。

（二）整理舊溝渠，宜同時疏濬護城河，高亮橋以上，如暫時不加疏濬，則宜在該處設閘。目下北平市舊溝渠淤塞者過多，故第一步在疏濬舊溝，其不能再用，或容量不足者，則須另設新溝。

（三）設計新溝，用「準理推算法」最爲合宜。惟該項計劃中所算得之洩水係數，繁盛區爲〇・八三，住宅區爲〇・四九，似嫌稍[illegible]。平市柏油路面尙少，住宅中亦多空地，卽使路政逐漸改善，但各胡同馬路之改爲瀝青面，恐爲極遠之事。繁盛區域，如因降雨量用五年循環方程式，而加高其洩水係數，尙有特別理由。至住宅區域，用降雨率五年循環方程式。已稱充裕，故洩水係數〇・四九，似可稍事減低，藉省經費。

（四）汚水溝渠，宜分區進行，已如上述。通惠河流量，能否冲淡汚水，使不致發生汚穢現象，須先作試驗，視通惠河水所含氧之成分而定。設河水之冲淡力不足，始設汚水調治廠，蓋設廠需費頗大，不可貿然決定也。計劃書有汚水淸理分廠三處，查宣武門內等處人煙稠密，汚水調治廠，不免稍有臭味，故地點宜愼貞選擇。天壇旁地廣人稀，頗稱相宜，不過本市汚水量暫時不致過多，設一淸理總廠，已足應付，如此可省經費不少也。

（五）唐山產缸管，如用口徑稍大者，最好先作試驗較爲可靠。

（六）敝校土木工程系，設有材料試驗室及衞生工程試驗室，將來貴府進行溝渠建設時，如需試驗上之幫忙，自當效勞。

（七）是項設計綱要，尙係初步研究，故鄙人所述，亦屬普通的理論。將來實際測量設計時，如須共同研究，亦所歡迎。

（二十二年十二月十八日）

胡寶予先生覆函

——上海市工務局技正——

昨由贛至滬，接奉上年十一月三十日　惠函，及附寄北平市溝渠建設設計綱要及汚水溝渠初期建設計劃一册，拜讀之餘，具仰　擘畫精當，無任欽佩，尤以「溝渠系統，採用分流制，利用什刹海三海護城河及舊溝渠之一部分宣洩雨水，另設小徑管專排汚水」一點，鄙意以爲確爲經濟合理之辦法。至關於護城河之疏濬，舊溝渠之整理，汚水溝渠之建設等計劃，自屬初步性質，須待詳細測量調查後，始能製成具體圖樣預算爲逐步實施張本，惟「溝渠建設設計綱要」第五章第五節內，（假定將來商業繁盛區之人口最大密度，爲每公頃六百人，普通住宅區每公頃三百人，揆諸現代城市設計，力趨人口分散之原則，及平市公安局人口密度調查統計，似嫌稍多。按二十二年平市公安局之調查統計，人口最密之外一區每公頃爲四百〇五人，普通住宅區每公頃一百五十人，此種情形在最近若干年內，似不至有多大變動，即將來工商業發達，人口激增，亦宜限制建築面積與高度，及闢設新市區以調劑之，不宜聽其自然發展，致蹈吾國南方城市及歐美若干舊市區人烟過於稠密之覆轍，使文化古都成爲空氣惡濁交通擁擠之場所，而失其向來幽雅之特色。鄙意平市商業區將來之人口密度仍宜以每公頃四百人爲限，住宅區以增至每公頃二百人爲限，一切市政施設與規章均以此爲目標，則不僅建設溝渠之費用可以節省已也。梓學識謭陋，猥蒙　垂詢，用抒管見，拉雜奉陳，當否仍祈　卓奪爲幸。

二十三年一月三日

盧孝侯先生覆函

——中央大學工學院院長——

前奉台函，敬悉　貴市府爲規劃平市溝渠，卓樹大計，猥蒙　垂詢芻蕘，將所擬北平市溝渠建設設計綱要及汚水溝渠初期建設計劃兩種，囑爲具列意見函復，等由；按查　貴市府所擬溝渠原計劃，備極詳盡，至堪欽佩，惟管見所及，尙有兩點，茲附陳於下：

（一）圓形缸管是否較省，其省出之經費，是否足以抵償將來沉澱淤塞之損害（按蛋形洋灰管不易沉澱淤塞）又是否能儘量購用國貨。（開灤管子雖佳然非國貨）

（二）可否在各街口設鋼筋混凝土雜爛箱，將汚水先局部清理，節省管子費用，（可用較小口徑管子）使用汚水管者負担較大，不用汚水管者減輕負担。

上述諸偷　參酌，自慚學術譾陋，無補高深，倘祈見原爲幸，耑復。

（二十三年一月十八日）

關富權先生建議書

——中央大學衛生工程教授——

1）蛋形與圓形管之比較

在水力學理論上着想，蛋形管之水力半徑 Hydraulic Radius 較優於圓形管；故水中固體物在蛋形管中不易沈澱，其沉在圓形管中，如滿或半滿時，雖亦不易沉澱，然實際不易適遇全滿或半滿之流量，故圓形管普通之水力半徑不如蛋形管之優，結果在同一圓周中，圓形管所載之流量易生沉澱。

至在費用上着想，蛋形管多用洋灰大砂子，就地用模型製造，爲監視配料及製造合宜，不特可工精價廉，且因就地製造，可免碰壞傷損及車船運費。且如用八英吋至二十四英吋口徑之大管，其質料必須極佳，如爲永久起見，目下自推開灤雙釉缸管爲最佳，然利潤流入外商，諸多不宜；至北平地勢，雖極平坦，然坡度並不致因圓形管或蛋形管有所增減，至在已舖馬路處之掘地費用，多半費在傷壞路面，（此在土瀝靑路爲尤甚）至污水管下多掘尺餘，所增費用較之全體工價相差甚微。

（2）集用戶數家合設一霉腐桶之利益

污水之設置，其經費無論出自募集公債或增加捐稅，其結果皆[illegible]凡用自來水污水管者，與不用該項設備者，同負一樣担負，此則未免使市民之担負，有畸重畸輕之弊。今爲使不用自來水之用戶減輕担負，且爲減輕穢水清理費用起見，可使數家或一胡同內之住戶聯合出資，照指定圖樣各建一鐵筋混凝土霉腐桶Rein. Conc. Septic Tank。使其污水先注入此桶內，經過微菌作用，將一部份污水變成氣體，經過照指定圖樣締成之管子放出，結果再將餘留液體，注入市設穢水支管（此際已化成極稀之液體且臭氣大減）如此有以下三利：

（甲）凡有污水管之住戶負担較多（建築霉腐桶用）而不設污水管之住戶可以大減担負，因經霉腐桶流出之液體，不特體積大減，（多半已化成氣體）且質已由濃變稀，如此管子之口徑可大減，換言之，即費用大減。

（乙）污水已經過局部清理，則最後之處理設備及經費可以大減，亦可使一般不用污水制度者減輕担負。

（丙）將來污水管用之年久必生滲漏，如已局部整理，過後即偶有滲漏亦不致大妨井水之清潔。

（3）污水管爲防凍起見，有〇．六公尺深之埋深即足，因污水管起微菌作用時，發生多量之熱，故不易受凍，且污水結冰點，亦較普通淨水爲低。

（二十二年一月二十日）

26041

一、本會成立經過

〔1〕籌備及成立

一、醞釀時期

民國三十二年二月間，在重慶工程界朋友間的談話中，都感覺都市建設，為戰後一大問題，有組織市政工程學術團體預事研究籌劃之必要。至四月二日，在一個集會中，有茅唐臣、李耕硯、林平一、李榮夢及譚炳訓諸君，會對市政工程學術團體的組織，作了一次比較具體的討論，當時談話的結果如下：

1. 市政工程團體的名稱，定為中國市政工程學會或中國市政工程師學會。
2. 市政工程應包括衛生工程，現已有衛生工程團體，名稱方面有無將「市政」與「衛生」並提，如並提時，是否即稱為中國市政衛生工程學會。
3. 將國內市政工程界知名之士，開一名單，以便分頭接洽為本會發起人。

經過這次談話以後，第一步就由譚君與在渝之市政工程界同志與華甫、薛次莘、朱泰信諸先生商談，皆承熱烈贊助，並由薛先生處借到了各工程團體組織章程。所以很快的就將本會初稿的組織章程草案擬出。第二步就將草案分寄昆明陶君葆楷、貴陽孫君壽元、桂林裘君向華，江西過君守正，請他們參加意見，並担任在當地徵求發起人及會員的工作。四月初薛次莘及譚炳訓二君將章程草案攜至蘭州，於四月十五日在蘭州西北公路局辦事處，與沈君怡、凌竹銘兩先生，開了一次談話會，這次談話的收穫很大，除修訂章程草案外，並商定：

1.本會名稱訂為「中國市政工程學會」。

2.本會會員應抱寧缺勿濫之旨，將各級會員之資格予以嚴格之規定。對於名譽會員亦僅限於對市政建設之有特殊供獻者。

3.本會成立後注意實際工作，不作表面宣傳。總幹事一職，以專任為原則，以利會務推進。

4.本會暫由當時與會之四人及吳華甫先生負籌備之責。

蘭州市工同志，對本會都熱烈贊助，西北會務就請沈先生主持。至此本會籌備工作漸具體化，而醞釀時期可謂完成。

二．籌備時期

五月十四日下午四時，在上清寺聚興新村十二號驛運總處招待所召開在渝發起人第一次會議，凌竹銘先生因公來渝，也請他參加，共計出席凌鴻勛，譚炳訓，吳華甫，朱泰信，茅以昇，楊哲明，薛次莘，蕭慶雲，張培林，張佐周，李中襄，高宗義，于均祥等十三人，會商重要結果如下：

1.修改章程草案

2.九月間開成立會

3.推定籌備員

到六月二十五日，因為薛次莘先生及譚炳訓先生皆先後又將有西北之行，在兩浮支路薛先生處與吳華甫先生商定，籌備工作請他偏勞，在七八兩月，吳先生徵求了很多會員，並向社會部辦了備案手續。

至八月十八日，薛譚二君相繼由西北歸來，此時昆明，貴陽，桂林，江西等地的發起人及會員都有回信，一致贊助，吳薛譚三君在吳先生處商定於八月二十三日正式召集在渝全體發起人開會，討論成立大會一切事宜。

八月二十三日下午三時，在國民外交協會開發起人會，到方福森，鄭肇經，高宗義，孫輔元，于均祥，薛次莘，譚炳訓，張劍鳴，尤巽照，張丹如，楊哲明，張培林，李榮夢，吳華甫，盧毓駿等十五人，重要決定如下：

1.九月二十一日召開成立大會。

2.成立「籌備」「會員資格審查」及「司選」三委員會，名單如下：

籌備委員：譚炳訓，吳華甫，薛次莘，張丹如，蕭慶雲。會員資格審查委員：鄭肇經，吳華甫，張劍鳴，方福森，李榮夢。司選委員：薛次莘，尤巽照，朱泰信，張劍鳴，鄭肇經。

3.籌備時期及成立大會所需之經費，由發起人樂捐。

4.修改會章文字。

會後，本會籌備時期的一小包檔案，又由吳華甫先生移交譚炳訓先生接管。

到九月十一日，關於成立大會的許多重要問題，亟待解決，在聚興村十二號鄭權伯先生處開「籌備」「會員資格審查」及「司選」三委員會的聯席會議，出席尤巽照，張丹如，譚炳訓，鄭肇經，張劍鳴，李榮夢，吳華甫，朱泰信等八人，並請社會部馬專員參加，重要決議如下：

1.大會會場改在社會服務處，推薛次莘，吳華甫，譚炳訓三先生爲大會主席團。

2.增加理監事候選人。

3.審定各級會員共計二百一十一名。

4.參加桂林聯合年會，並請朱泰信先生徵集年會論文。

三、成立時期

成立大會的籌備工作很繁重，除了籌備委員張丹如先生接洽會場，會員梁錫伯先生主持文書工作外，並請會外的樊琪先生辦理事務，陳憲章先生幫辦文書及收發，趙效沂先生担任招待及攝影，所有繕寫油印是由驛運總處代勞，本會皆應表示謝意。

經幾半年的籌備，成立大會照規定日期於三十二年九月二十一日舉行，就是中國市政工程學會可紀念的誕生日期，兩路口社會服務處下午秋陽酷晒中，更增高室內百餘會員來賓的熱烈情緒，先由薛次莘先生報告本會成立意義，以戰後全國各城市應如何建設，淪陷區及受敵人破壞之城市，應如何復興，一切設計規劃等工作，實爲市政工程界的重大任務，再者本會籌備時間極爲短促，會員未及廣泛徵求，希望各位分別通知未曾參加本會的市政工程同人，踴躍加入爲幸。次由譚炳訓先生報告本會籌備經過，並將本會在籌備時期所謹守的兩種精神，貢獻給大會：

(一)對於會員資格嚴格規定，重質不重量，希望每一個會員都對學會發生實際關係，有實際工作，負學會一份子應負的責任。

(二)一切皆本實事求是之旨，不作俗流的鋪張，在成立會儀式上所表現的是簡單樸實，將來在會務上希望對市工學術及市政建設有實際的供獻。

繼由社會部曾科長沛滋致詞：

(一)希望本會工作應有一詳細計劃，更要規定中心工作。

(二)戰後的都市建設，本會負有重大責任，希望能把最高深的學識和技能，以通俗的方法介紹到社會上去。

次由吳華甫先生主席討論會章，經一致通過。蒞會來賓及陪都與論界，對本會皆抱極大期望。大會中的提案，亦極扼要，都是戰後市政建設的迫切問題，更有中國工程公司，泰山實業公司爲出席會員預備茶點，表示慶賀。當場全體會員通過電將主席致敬。大會票選理事十五人，監事五人，候補理事五人，候補監事二人，會後茶點攝影。

本會呈報社會部成立籌備會情形，已經准予備案，成立大會情形亦已呈報。根據大會決議，本會第一屆年會與中國工程師學會在桂林召開第十二屆聯合年會合併舉行，承該會復函歡迎，同時進行赴桂林車輛交涉，並通知赴桂林會員先行登記。

九月二十八日召開第一屆理監事第一次聯席會，除詳細討論大會付議的決議案，並商定本會今後工作計劃，票選凌鴻勛先生爲本會理事長，鄭肇經、譚炳訓二先生爲常務理事，茅以昇先生爲常務監事。工作已逐步開展，恰巧凌理事長因公來渝，理監事會在渝同人於十月七日開會歡迎，卽席商討國都問題，及會務方針，並公推譚炳訓先生兼任本會總幹事。截至此次理事會經審查合格之會員，計基本會員一〇一人，會員一〇二人，初級會員五四人，共計二五七人。

以上爲本會籌備經過及成立前後之實錄，亦卽本會會務之第一頁也。

四．附本會成立大會紀錄

時間：三十二年九月二十一日下午三時

地點：兩路口社會服務處社交會堂

出席人：丁基實等八十三人

列席人：社會部代表曹沛滋　馬人松

主席團：薛次莘　譚炳訓　吳華甫

紀錄：陳振亞

行禮如儀

一．薛次莘報告本會成立意義：（略）

二．譚炳訓報告籌備經過：（略）

三．社會部代表曹沛滋訓詞：（略）

四．通過章程：

照原案通過，章程條文文字之修飾由大會授權主席團負責辦理。

五．選舉理監事（名單從略）

六．討論提案：

（一）籌備會前函全國工程師學會，參加聯合年會，提請追認案。

決議：追認。

（二）以市政工程之觀點，研究戰後國都應設之地點，建議政府採擇，提請公決案。

決議：通過，原案交理事會詳細討論。

（三）籌集本會基金，設立固定會所及參考室，提請公決案。

決議：原則通過，原案交理事會擬定辦法。

(四)研究戰後我國都市公私建築之風格，以表現抗戰建國之時代精神，提請公決案。

決議：通過，原案交理事會指定新建築研究小組會同建築師學會從事研究。

(五)研究戰後我國都市計劃之設計綱領，以爲今後建市之規範，提請公決案。

決議：原辦法（由理事會指定小組研究，於三月內提出報告，徵詢會員意見，俟彙齊修訂後，再酌量公諸社會討論或建議政府採納）通過，並由本會推選代表兩人參加 國父實業計劃研究會研究。

(六)建議政府添設「公共工程」部統籌城市建設提請公決案。

決議：原案交理事會組織研究小組，詳細研究後提本會年會討論。

(七)本會會員參加全國工程師學會年會應如何籌劃案。

決議：交理事會討論，參加者姑先在理事會登記，再爲設法解決交通問題。

七．臨時動議

(一)以本會全體會員名義電 國民政府 蔣主席致敬案。

決議：通過電稿。

八．茶敍

九．攝影

十．散會

〔2〕會章

一．中國市政工程學會章程

第一章 總則

第一條 本會定名爲「中國市政工程學會」

第二條 本會以聯絡市政工程同志研究市政工程學術促進市政建設之發展爲宗旨

第三條 本會設於國民政府所在地

第四條 本會得在各市縣設立分支會其組織另定之

第二章 任務

第五條 本會之任務列左

(一)編印與發行市政工程刊物

(二)接受市政機關之委託研究及解答關於市政工程上之問題

(三)舉辦市政工程之學術講演試驗研究及展覽事項

(四)徵集國內外市政建設書報圖籍資料

(五)關於市政工程改進之建議事項

（六）關於市政工程人員之登記介紹及學術進修事項

第三章　會員

第六條　本會會員分為（一）基本會員（二）會員（三）初級會員（四）團體會員（五）名譽會員

第七條　凡土木建築及公共衛生等有關市政之工程師有八年以上之工程經驗內有三年係負責辦理市政工程者由基本會員三人之介紹經本會理事會之通過得為本會基本會員

第八條　凡土木建築及公共衛生等有關市政之工程師有四年以上之工程經驗內有二年係負責辦理市政工程者由基本會員或會員三人之介紹經理事會之通過得為本會會員

第九條　凡土木建築及公共衛生等有關市政之工程師有一年以上之工程經驗者由基本會員或會員三人之介紹經理事會之通過得為本會初級會員

第十條　凡在國內外大學之市政或土木建築及公共衛生等科三年肄業期滿者作為工程經驗一年畢業者作為工程經驗二年凡在國內大學市政土木建築及公共衛生等工程專科學校或工程研究所講授或研究市政工程者以市政工程經驗論

第十一條　凡與市政工程有關係之機關學校或其他團體由基本會員或會員五人之介紹經理事會之通過得為本會團體會員

第十二條　凡對於市政工程事業或學術有特殊之貢獻者由基本會員十人之介紹經事會會之通過得為本會名譽會員

第十三條　本會會員應享權利列左

（一）基本會員及會員有選舉權及被選舉權

（二）初級會員有選舉權

（三）本會所舉辦各種事業上之利益

（四）其他會員應享受之權利

第十四條　本會會員應盡義務列左

（一）遵守本會會章及決議案

（二）担任本會所委辦之工作

（三）繳納會費

第十五條　會員及初級會員符合第七條及第八條之規定時得向理事會聲請經其通過後分別改為基本會員及會員

第十六條　凡違反本會會章及損害本會名譽者經基本會員或會員五人以上之檢舉得由理事會予以警告或除名

第四章　組織

第十七條　本會以會員大會為最高權力機關在會員大會閉會期間理

事會代行其職權

第十八條　本會置理事十五人候補理事五人監事五人候補監事二人由會員大會選舉之組織理事會與監事會理事會得互選常務理事三人監事會得互選常務監事一人

第十九條　本會得置理事長一人就常務理事中推選之

第二十條　本會得設總幹事一人副總幹事二人幹事及助理幹事若干人由理事會聘任之

第二十一條　本會理監事每年改選三分之一連選得連任

第二十二條　本會各種選舉得以通訊方式辦理之

第二十三條　本會於必要時得設置各種委員會

第五章　職權

第二十四條　本會會員大會之職權如左

(一)審議理事會監事會之會務報告

(二)通過本會章程

(三)選舉理事監事

(四)決定經費預算

(五)其他重要事項之決定

第二十五條　本會理事會之職權如左

(一)對外代表本會

(二)對內處理一切會務

(三)召集會議

(四)執行會員大會決議

(五)核准會員入會

(六)辦理監事會移付執行案件

第二十六條　本會常務理事之職權如左

(一)執行理事會決議

(二)辦理日常事務

(三)召集理事會議

第二十七條　本會監事會之職權如左

(一)監察會員履行義務事項

(二)經濟之稽核事項

(三)辦理其他有關監察事項

第二十八條　本會常務監事之職權如左

(一)執行監事會決議

(二)召集監事會議

(三)辦理日常事務

第六章 會議

第二十九條 本會會員大會每年舉行一次必要時得舉行臨時會議

第三十條 本會理事會監事會每三個月開會一次必要時均得舉行臨時會

第七章 基金與經費

第三十一條 本會之基金與經費分別以左列各款充之

(一)基金 甲・會員入會費基本會員入會費二百元會員五十元初級會員十元團體會員一千元 乙・經理事會接受之捐款 丙・基本會員之特捐(凡基本會員一次捐助基金在壹仟元以上者得免收入會費及常年費)

(二)經費 甲・會員常年會費基本會員常年會費五十元會員二十元初級會員五元團體會員二百元 乙・經理事會接受之補助費 丙・基金之孳息

第三十二條 本會基金由常務監事保管之經理事會之議決始能動用

第八章 附則

第三十三條 本會各項辦理細則另訂之

第三十四條 本章程如有未盡事宜得提會員大會決議修正後呈請社會部備案

第三十五條 本章程經會員大會之通過呈請社會部核准備案施行

中國市政工程學會第一屆職員名單

理事長 凌鴻勛

常務理事 鄭肇經 譚炳訓

理事 朱泰信 李榮夢 余籍傳 沈怡 吳華甫 林逸民 哈雄文 袁夢鴻 張劍鳴 過守正 薛次莘 齊慶雲

候補理事 宋希尚 段毓靈 胡樹楫 陶葆楷 嚴宏溎

常務監事 茅以昇

監事 李書田 趙祖康 裘發祥 關頌聲

候補監事 婁向華 蔡膺

總幹事 譚炳訓

二・中國市政工程學會分會組織通則

第一條 本通則依據中國市政工程學會章程第四條之規定訂定之

第二條 各地分會應加以所在地名稱以區別之

第三條 各地分會組織應依照非常時期人民團體組織法之規定辦理之

第四條 各地分會理監事每年改選一次連選得連任

第五條 分會會員大會每年至少舉行兩次

第六條 分會理事會每月舉行一次必要時得舉行臨時會

第七條 分會會員大會爲最高權力機關在會員大會閉會期間由理事會代行其職權

第八條 分會於必要時得設置各種委員會

第九條 分會會務推進情形每月應向總會報告並分呈當地主管官署

第十條 分會所收會費之半數得留作分會經費團體會員會費悉數交總會

第十一條 本通則如須修正時應提總會理事會議決並呈請社會部備案

第十二條 本通則經總會理事會通過並呈請社會部備案後施行

二、年會

（一）第一屆年會籌備記

譚炳訓

本會第一屆年會，參加在桂林召開之中國工程師學會暨各專門工程學會聯合年會，是三十二年九月十一日本會籌備委員會所決定的。在九月二十一日又提經成立大會通過，於是函請中國工程師學會總會，轉知桂林年會籌備會，照各專門學會例，招待本會會員，並參加各項會程。經該會董事會通過，復函歡迎，惟赴桂交通工具則須自行設法。

當即一面登記赴桂會員，一面洽借車輛，並電在桂林之理事林逸民先生、監事裘向華先生，先向籌備會接洽。登記赴桂之會員有十位，車輛亦已借妥，出發之前，得悉中國工程師學會車輛甚窘，本會會員亦有臨時因事不能成行者，承該會重慶分會之協助，本會會員得全數搭乘該會車輛，此應向該會表示感謝者也。

會員雖得解決交通上之困難，但裘監事在桂向聯合年會籌備會多次交涉，皆不得要領，最後無結果之結果，即必須有工程師學會總會之正式通知，始准本會參加年會。裘君電渝告急，炳訓立即電話與該會總幹事顧毓瑔先生洽商。請速電桂會，並由本會理事兼該會董事薛次莘先生，另致桂會主任籌備委員胡瑞祥君一電，告知董事會確有歡迎市工學會參加年會之決議。

炳訓原訂搭乘十月十二日班機赴桂，以便當面向籌備會交涉，並準備年會各項事宜。啓程前夕，因某項要公，未克成行，祇有仍託裘監事再行積極交涉。依情據理而論，學術團體之會議，爲交換智識，研究學理，不同於政治性質之會議，有權利之爭，必須依照法定手續辦理者。況本會參加年會的法定手續，業經完備，似乎不致再生枝節。殊不知炳訓於十月二十日搭赴會專機到桂之下午，同裘監事親到籌備會接洽，該會祕書仍以未奉命令，不能作主爲辭。質以董事會的決議，顧總幹事辭董事的電報有效無效，則答以這是給私人的電報，並非給本會的。不得已只好再向胡主任委員交涉，胡先生倒很痛快，他說要參加的學會太多，甚麼衛生工程，造船工程，自動車工程等等的學會都來了，現在報到的已經有一千多人，食宿會場都生了問題。炳訓當即聲明本會會員來桂非爲「吃喝玩樂」住宿，宴會，游山玩水及娛樂節目皆可犧牲，我們只要出席大會，作市工學會的會務報告，增設市工小組，討論市工提案與論文。胡先生仍然躊躇不決。好在顧毓琇先生也同機來桂了，於是同找顧先生，把他從床上拖起來，證明了董事會千眞萬確通過了本會的參加年會，然後胡先生才首肯了我們的兩點要求。我們爲求確實起見，又跑回省府內的籌備會，將這個結果告知祕書處。這時已是黃昏時分，早晨四點鐘跑珊瑚壩飛機場，空中振盪了三個鐘頭，到了桂林又經這半天的奔波，費了許多唇舌，此刻眞是筋疲力竭了。

翌日上午，聯合年會在廣西省府禮堂行開幕式，到會會員樓上樓下皆擁滿了，炳訓自渝帶來的「中國市政工程學會特刊」一千份，分贈一空。特刊內容(一)爲籌備成立經過(二)爲會章(三)爲職員名單，特刊紙張是用捐來的新聞紙，爲大會中最漂亮的印刷品。下午第一次大會中，工程師學會暨各專門學會會務報告的會程中，本會繼礦冶、化工、水利、電機、機械、土木、及紡織等工程學會之後，名列第八，由炳訓代表凌理事長出席報告，時間限五分鐘，將應報告者全部報畢，幸未被主席驅逐下台，報告辭另詳，玆不述。二十二日上下午市工與衛生的第七小組在極熱烈情形下舉行討論。至此籌備工作才算確實完成。

年會籌備，經過如許波折，焉可不向會員諸公一告。

（2）本會在聯合年會之會務報告

譚炳訓

時間：三十二年十月二十一日下午四時

地點：桂林廣西省府大禮堂中國工程師學會第十二屆聯合年會第一次大會

一、前言

市政工程學會會務報告。

本會理事長凌鴻勛先生，因事不能出席年會，特派敝人代表報告本會會務，市政工程學會爲新成立之工程學術團體，所以首先要報告的是：

二．爲什麼要成立市政工程學會

抗戰勝利在望，後方及淪陷區爲敵人所破壞及轟炸的城市，戰後亟待復興，并須有整個的具體計劃，組織市政工程學會，就是準備城市的復興與計劃。

戰後隨工業化之進展，人口必將集中，舊都市需改造擴充，新的工業城市要創建，成立市政工程學會，就是要對城市新發展之設計，有所貢獻。

都市要田園化，鄉鎮則要城市化，鄉鎮也要有新式道路，溝渠自來水等之工程設施，使全國人民能普遍享受現代化的科學成就。成立市政工程學會，就是要促成此新理想之實現。

市政建設在平時與戰時的重要性，可舉兩個例以爲證。蘇聯第一五年計劃，爲國防重工業建設計劃，工業與電氣建設的預算爲一九五萬萬盧布，都市建設預算亦有六十一萬萬盧布，爲工電預算額的百分之三十一（約三分之一）。英國內閣在戰時增設兩部，一爲飛機生產部，一爲公共工程部（Public work ministry）。所謂公共工程者，卽都市與鄉村國防與民生所需要而由政府舉辦之工程。有關民生的公共工程，大部是市政工程。

最近倫敦有一展覽會，將戰時被毀的都市，如伯明罕等，皆做了復興建設的實體模型。小如村鎮，也設計一種公共暖氣供給工程。這些計劃皆包括新的理想，可以改變今後人民的生活方式，是民生建設的主要部份。在勝利之前，英人已經準備好了復興的具體圖案了。

由上可知：（一）都市建設在工業化中之地位。（二）戰後都市復興應於此時先行準備。我國市政建設向來落後，將來都市與鄉村的公共建設，工作艱巨繁重，更必須在勝利之前，就研究方案，準備計劃，以供戰後政府採擇實施。所以全國各市政府工務局的曾任及現任的工程師，聯合組織本會，期以共同集體之努力，對戰後都市建設有所供獻。

三・市政工程學會籌備及成立經過

本會籌備及成立經過，已有書面報告，載市政工程學會特刊，於上午分送諸君，不再贅述。僅有補充報告一點，卽本會會員之人數，現已審查合格的會員，基本會員一百零一人，普通會員一百零三人，初級會員爲五十四人，共爲二百五十八人。

四・今後本會的中心工作

（1）研究工作

（甲）戰後國都問題。

（乙）都市計劃法（二十八年國府公佈三十二條，現由內政部徵求各方修正意見）。

（丙）戰後都市復興方案。

（丁）戰後公共建築的風格問題。

（戊）搜集市政工程參考資料。

（2）編輯及宣傳工作

（甲）編輯市政工程叢書及淺說，注意縣鎮一級之公共建設。

（乙）刊行市政工程年刊及專題研究之不定期刊。

（丙）籌辦市工模型及市工展覽會，以普及市政工程的常識，因我國不能於短期內，造就大批的市政工程師及工程員。

（3）組織工作

（甲）成立各地分會——桂林已成立，其他各地在籌備中。

（乙）調查全國市政工程人材。

五．尾言

本會是九月二十一日下午在渝成立，即上月的同日同時，故今日為本會之「滿月」吉日。以一個月的嬰兒，不自量的參加聯合年會，就是因為：(十)感覺勝利在望，戰後我國都市建設問題需要及早解決，不能待諸來年。(十一)藉聯合年會的機會得以向全國工程先進諸君多所領教。

(3) 本會在聯合年會之提案

建議政府添設「公共工程」部統籌城市鄉鎮建設案

理由：我民族歷史輝煌，建設都市最早，過去，東西鄰邦，以我國為「天朝」「上國」者率皆由於仰慕我國都市之繁華雄偉，徒以近百年來，西洋機械文明代興，我國承遜清專制之弊，追蹤不及，遂至國勢衰弱，備受侵凌，尤以城市建設方面，相形見絀！然北平一城，自明朝建都以後，銳意經營，規模宏偉，至今猶能雄視全球，此亦足徵我國在民族復興之際，建設都市之偉大力量，雖歷五百年而不衰，世界歷史學家，固有稱我民族為「建設都市之種族」者，建設能力，蘊藏於我民族血液之中，固不以一時歷史上之蹇運而有所損毀，亦可概見矣。方今抗戰勝利在望，不平等條約取消，最近十一中全會，政治及經濟建設決議案中，對於「復興城市」「建築民居」以及「發展工業」等項已訂定綱領，則我國今後之城市建設，頭緒萬千，自不待言，亟應遠紹我國過去之「冬官」「工部」近仿各國之組織成例，添設「公共工程」部，統籌兼顧，一方面可表示中央對於地方建設之積極控制性，一方面可集中運用此類建設專才，以發揚我民族復興之大業，抑猶有進者，我國今後之經濟建設，基於三民主義之原則，應循計劃經濟之途徑，至少對於國營工業以及國防上之種種建築物，不應聽其各自為政，必須有通盤計劃，使其標準化，並形成各種合理佈置與排列，以實現其最高價值，庶國防與民生之合一理想，由城市建設而實現，如英國為最注重實際之國家，戰後新添兩部，一為飛機生產部，一即為公共工程部，後者之主要任務，即在統籌並積極辦理英國戰事期間之公共建築與城市修建等項工作，大戰後將為英國城市復興之主管部去。至我國之「公共工程」部，似應以城市鄉鎮建設之指導及監督，爲其主要任務，是否有當？提請公決。

辦法：

(1)與中國衛生工程學會聯名向政府建議。

(2)本案作爲本會對本屆工程師學會年會提議案之一。

本會成立大會決議：原案交理事會組織研究小組，詳細研究後，提本會年會討論。

(4)第一屆年會討論會紀錄

——參加中國工程師學會十二屆年會第七小組——

地點：中國工程師學會十二屆年會第七小組會會場(桂林廣西省政府內)

時間：三十二年十月二十二日至二十四日

出席：

主席：譚炳訓　過守正(市政工程學會)

過祖源(衛生工程學會)

紀錄：袁相堯(市政工程學會)

王家診(衛生工程學會)

壹．市政工程學會

(一)各會員自我介紹。

(二)主席報告中心工作(補充二十一日大會所未報告)：

(甲)研究工作：

㈠戰後國都問題(參攷譚炳訓先生著建都之工程觀論文)。

㈡都市計劃法(國父實業計劃研究委員會交議)。

㈢戰後都市復興案

㈣戰後公共建築的風格問題(應時代之需要將有新的風格產生如中式西式或中西合抄式均非應時上選)

㈤搜集市政工程參攷材料(成立資料室)

(乙)編輯及宣傳工作：

㈠大學叢書(由大學教授會員諸君著作)——高級刊物。

㈡專刊(由總會編印不定期刊物)——高級刊物。

㈢編印工程圖案畫——初級刊物。

㈣製備陳列模型——初級刊物。

以上高級刊物爲供應學校及專家之用，初級刊物爲應社會普遍採用，一則爲貫注民衆市政工程常識，再則爲一般城市無專門人才爲主持者之寶助。

㈤編印會員一覽(總會編印)。

(丙)組織：促各地成立分會。

(三)國父實業計劃研究會都市建設小組研究之結論（國父實業計劃研究委員會交議）——討論結果：

(甲)都市計劃擬加入防空事項。

(乙)內容需另行排列。

決議：本案包括範圍甚廣，需要參攷資料切實研究，非短時間所能決定，將原案油印分發各會員帶回個別研究，擬具方案，寄由總會審查彙印後再送研究會。

(四)戰後國都問題：譚炳訓先生宣讀「建都之工程觀」論文——歸納大公報十三篇論文，並根據交通、地形、氣候、市區、風景、名勝、公共設備、公共建築、食料、燃料、與人力之供應，以及國際觀瞻等觀點，比較北平、西安、南京三都市之優劣點。

麥蘊瑜先生提：國防建設，以軍事政治爲中心，工程條件乃人爲的條件，無甚重要。

王正本先生提：國防觀點不在工程觀點範圍內，本案有提請大會審查價值。

(五)宣讀論文：

(甲)南京市下水道——馬育騏先生報告。

(乙)大青島市建設計劃——過守正先生報告。

(丙)北平市溝渠建設計劃——譚炳訓先生報告。

(六)審查論文報告：

(甲)實業計劃上之城市建設——（朱泰信先生著）。

王正本先生報告：

(一)實業計劃之輪廓作爲我國城市建設之背景。

(二)我國城市概況及其分類。

(三)城市建設之三大問題。

(四)從國防觀點論城市佈置。

(乙)大青島市建設計劃——（過守正先生著）。

麥蘊瑜先生報告：本計劃爲國內市政建設計劃中最完善最詳盡之一篇，堪爲建設市政計劃之參考珍品，「土地公用法」尤堪爲特別注意之一點。

(丙)北平市溝渠建設計劃——（譚炳訓先生著）。

馬育騏先生報告：同意分流制，數目字有待研究之點。

(丁)南京市下水道——（馬育騏先生著）。

過守正先生報告：本計劃中之各項數字與北平市，青島市下水道計劃中所採用者互相比較頗多出入，暴雨率或係因南北地域之不同，而洩水係數相差甚多，是否因柏油路面房頂院地多寡之各異，地形坡度之懸殊而然，尚待研究，此項數字在我國各都市尚有特長久時間實地考查與試驗，由會指定會員專任研究與審核，以期合於我國都市規定標準以後，各處設計下水道時可有所根據。

(七)都市計劃法：推請林逸民、麥蘊瑜、過守正、王正本四位專任討論。

決議：依照麥蘊瑜先生所擬「修正都市計劃法原則草案」原則，由總會重行起草修正意見書送請修正。

(八)市政工程學會，衛生工程學會，兩會合併或合作辦法，推請馬育騏、過祖源、林逸民、過守正四位研究適當處理辦法。

決議：由兩會另組理監事聯合委員會。

(九)交審論文：「大青島市建設計劃」為市政工程名譽提名論文，提交大會審核。

審核結果：通過。

貳．衛生工程學會

(一)戰後建國中之衛生工程，在全國技術會議中曾經討論以下各題：

(甲)訓練問題。(乙)技術標準問題。(丙)推行工作經費問題。(丁)衛生工程器材製造問題。(戊)全國衛生事業方針。

(己)會員在地有關衛生事業(工程事項)之調查。

(二)交審論文：

(甲)新橋鎮簡易給水工程——(過基同先生著)。

由黎樹仁，王家振二先生審查。

(乙)戰後建國中之衛生工程——(過祖源先生著)。

由王正本，過守正二先生審查。

(丙)高架水櫃設計——(劉景賢先生著)。

由俞微，王正本二先生審查。

(丁)快性砂濾效能之研討——(過祖源先生著)。

由俞浩鳴，馬育騏二先生審查。

(戊)熱酸酵堆肥處理糞便垃圾法之介紹——(汪德晉先生著)。

由黎樹仁，王家振二先生審查。

(三)宣讀論文：交審之衛生工程論文，由各審查人分別宣讀論文內容綱要以及審查意見。

(5)本會在聯合年會活躍之側影

——轉錄廣西日報等——

本會第一屆年會，參加桂林召開之中國工程師學會暨各專門工程學會聯合年會舉行，市政工程界駿彥咸臨，與會研討周詳，尤關戰後建都問題，論戰認眞生動，給社會人士留下深刻的側影，茲摘錄各報記載如下：

(一)戰後國都問題

——市政衛生工程學會昨熱烈討論

昨天(十月二十二日)下午各專門學會的討論中，空氣最活潑的算年紀最輕，人數最少的市政衛生工程學會。而在市政工程學會討論中，討論得最熱烈的是戰後建都問題。

這一個問題以譚炳訓先生那篇「建都之工程觀」作中心。譚先生先將「思想與時代」及重慶「大公報」、「掃蕩報」及王寒生先生在參政會關於戰後建都的提案等有關戰後建都的文章及陪都對此一問題熱烈討論的情形，作了個大概報告。他說：「他們都是就地理、歷史、交通、國防、經濟、氣候及民族健康等立論的，我們現在是就工程的觀點立論。跟着提出了十個標準：(一)交通，(二)地形，(三)氣候，(四)舊市區之利用與新市區之闢建，(五)近郊之風景名勝，(六)公用設備，(七)公共建築及建築材料之供應，(八)食料與燃料之供應，(九)人力之供應，(十)代表國家與民族之精神。結論認爲是北平第一，南京次之，西安又次之。

跟着，出席的十餘人便開始了熱烈的討論，首先替西安說話的是馬育騏先生，他說：「西安附近有個終南山，有山有水，論外表，城偉路寬，有個全國最好的洗澡堂，全國最好的車站。」說到人力，則西北人又高又大，這可以算是側重風景說」。

湖南大學教授王正本先生却着眼於國防上，他說：「北平離海岸太近；而戰後海軍建設不易於短期內有辦法，因此未免危險。」譚炳訓先生立刻說：「離敵人太遠，有時反爲消息不靈，東北事變就是一個教訓。其次，現在空軍發達，兩個尾巴的飛機一小時可飛六百多里因之距離觀念要從新確立」。

於是蔚然先楊生說：「北平不好：(一)水道交通，沒有；(二)北平太北，難於照應全國；(三)欠水；(四)太冷。譚炳訓說：「目前確沒有水道交通；說到地位，北平並不北，我們若以北平為中心，劃個一千公里或甚至二千公里的半徑，牠的地位正適中得很；而且將來不能單看全國，要放眼看全世界！至於水，不知楊先生去過北平沒有」？

「沒有。」楊先生說沒有。

於是譚說：北平是永定河的冲積層，鑿地數尺，便得甘泉，好得很！至於冷，對於衞生正有好處，據有些人的研究，因為凡是在北緯三十三度以南地帶的地域，沒有雪，一切病菌不容易死亡。而且太熱和太潮的地方，頭腦昏眼，精神不容易清爽，必須要四季分明的地方，人才易聰明，偉大」。

最後麥蘊瑜先生用廣東國語振振發言，他說：「將來建都的問題，仍是以國防及政治的問題為首要，至於工程，只要有錢，便沒有什麼問題。他說有人現在已在準備世界第三次大戰，因此國防仍是最重要，若論國防，我仍是主張建都西安！至於北平，我是最愛的，將來諸位中有誰同北平去當市長，就是科長我也願意幹。」

問題討論開去，「觀」點已經離開了「工程」，所以譚先生又說：「我們是工程師，其他我們暫時不論。」到最終並未有得到具體的結論，不過這場討論，對於戰後建都問題，已不能說是沒有一些啓示了。

(二)戰後建都北平第一

市政工程學會和衞生工程學會，因為新近成立不久，這一次出的代表人數很少，所以合併舉行。但分組討論的第一天，參加人數算是最多的也不過是四十七個人。和其他學會尤其是電機工程學會的人數比較起來，未免相形見絀。他們人數雖少，但他們討論時態度的認眞，恐怕沒有第二個學會能夠和他們相比。會場的空氣相當嚴肅，沒有其他學會討論時那樣自由而且輕鬆。在會場裏面，市政工程學會的人數較多，也較出風頭，有兩個主席，一個代表市政，是譚炳訓；一個代表衞生，是過祖源。過先生發言老成持重，譚先生則是議論風生，在討論「建議政府設立市政工程部」及譚先生的大文「建都之工程觀」一文時，譚先生精神煥發，舌戰羣倫，充分發揮了他的演說天才。會中有一位女代表，她坐在最後，沉靜寡言。會場上最冷靜的時候祇剩十多個人，第七會場又是一個長形的房子，人和空處相比，更加顯得落寞。王天俊先生聲音很大，說話急切而有力。馬育騏先生在西安住過一個時期，所以他說西安眞是一個好地方，街道寬廣，氣候又好，城外的風景也不錯，如果撇開一切問題不談，我們未來的國都，西安也是很適宜的。

譚炳訓先生辦事認真，他的大文曾經用油印印出來，發給各位出席的代表，但是，當記者們問他領取一份時，他却十二分愼重地要我們寫收條，並聲明不在報上發表。

建廳潘超科長，也是市政工程代表之一。但他的事情頗忙，第一天出席之後，便很少再見他的蹤跡。

麥蘊瑜先生發言扼要，他幾次都是在討論到最緊張的時候，立起來「一言而定天下」。

小組會議的第二天，有人提議兩學會的合併或合作辦法。結果是前者被否定而後者保留起來。

宣讀論文共有十一篇，最初是全篇從頭至尾地宣讀，而且在重要的地方還特別提出來加以檢討。後來受了時間的限制，祇是宣讀大意和摘要。計宣讀過的論文有(一)譚炳訓：『建都之工程觀』，(二)朱泰信：『實業計劃中之城市建設』；(三)朱皆平：『衛生工程之國防價值』；(四)過祖源：『戰後建國中之衛生工程』；(五)馬育騏：『南京市下水道計劃』；(六)過守正：『大青島市建設計劃』；(七)譚炳訓：『北平市溝渠建設計劃』；(八)劉景翼：『高架水櫃設計』；(九)過基同：『新橋鎮簡易給水工程』；(十)過祖源：『快性砂濾效能之研究』；(十一)汪德晉：『熱發酵堆肥處理糞便垃圾法之介紹』。

轉載

工程師應以大禹的精神完成建國的使命

——本會凌理事長在卅二年六月六日工程師節講詞——

今天在天水工程界諸同仁舉行「六六」工程師節慶祝會，承各機關首長來賓蒞臨指導，與感莫名，工程師定節紀念這件事，在兩年前決定，當時全國的工程師認爲各部門工程業務在抗戰建國中逐漸加重，各地執業的工程師人數也逐漸增多，爲集中精神力量共趨於建國的同一目標之下加速其業務的開展，對外在喚起各界對於我們工程界的注意，進而使工程師職業者獲得社會上更多的同情與協助起見，工程界已有迅速確定紀念節舉行慶祝宣傳的必要，現在其他各行都有他本行的紀念節，比如勞動節教師節婦女節兒童節等屆時都有盛大的紀念會，藉以促進本身工作，喚起外界注意，我們工程界就在這種需要下，在兩年前確定大禹聖誕六月六日爲「工程師節」。

中國有五千多年的歷史，在這悠久的民族文化的潮流中，具有歷史的偉績，在工程事業上對國家民族有重大貢獻的，首推大禹，據史學家推算國歷六月六日爲大禹聖誕，爲尊崇禹聖在工程上的勳功偉業，就確定六月六日爲工程師節，並由工程師學會請准政府核定並明令公佈施行。這是民國三十年的事，今天是工程師節第三次紀念會，去年今天在重慶曾舉行盛大的慶祝會，大會節目繁多，盛極一時，關於工程獎章的頒發，工程學術教育的獎勵，都在這一天舉行，今年工程師節在重慶舉行慶祝情形雖無確實消息，但據報紙披露，重慶今天除舉行紀念會慶祝外，各大學並敦請名人講演有關國防工業交通等工程建設問題，引起各界對工程事業前途的認識與注意。

今天是中國工程界鼻祖大禹的聖誕，我們紀念工程師節，不僅應學習大禹爲國家民族爲建國工程犧牲的精神，同時對大禹服務的道德人格和他個人的日常生活習慣也應積極倡導倣效，期勉盡戰時工程人員的職責。中國有史以來，在工程方面有所發明的人物，不只大禹一人，何以我們推崇大禹，爲工程界的鼻祖？因爲在大禹以前，據傳有巢氏構木爲巢，燧人氏鑽木取火，這些人可以說是當時的工程師，但在歷史上找不到切實的根據，又如古時用一塊樹木作成一個簡單的小船，以渡過河流，其來源定必極早，這也是工程上的成就，但這種工程，並不是大規模的，大禹能以畢生的精力，完成全國治水的工程，勘定水害，開闢水利，他的成就的遠大，不但影響了當時的國家民族，並奠定幾千年民族生存的基礎，我們知道大禹的父親鯀，治水無功，大禹繼承父志，受命治水之初，就抱定救國救民的宏願，排除萬難，卒底於成，當時全國洪水橫流泛濫於天下，五穀不登，禽獸逼人，大禹隨山刊木，疏通九河，他治水的方法，是因勢利導，因地制宜，所謂「以海爲壑、導山、導水、導津、入河、入江、入海、」系統分明，條理不紊，他治水時胼手胝足，他的兒子啓，才初生呱呱而泣，他並不過問，在外八年，三過其門而不入，這種國而忘家，公而忘私，一往無前大無畏的精神，成就了他不朽的事業。讀「美哉禹功，微禹吾其魚乎」的頌詞，但見聖績彪炳萬世，實在不是偶然的事。

講到大禹的服務道德，與私人生活習慣，在載籍所述：「禹惡旨酒」，「卑宮室」，「惡衣服」，「大禹惜寸陰」，「禹聞善言則拜」，「禹思天下有溺者，猶己溺之也」，此種道德人格與服務精神，拿到現在，無一不足爲現時工程界同仁應奉爲圭臬做效遵行者，我們知道有大禹這種偉大的精神人格，才有大禹治水的豐功偉業，才奠了中國工程建國的基礎，與中國文化政教的階梯，這種基業的艱難締造，歷時愈久，其影響國家民族的價值愈大，我們奉大禹爲工程界的始祖，也不是偶然的事。

中國正逢抗戰建國同時並進的時代，抗戰的主要力量，要靠武裝部隊，但有國防的工程建設，軍用器械的製造，化學戰爭的配備，以及戰地工事運輸工程設計進展的遲速，無一不與工程技術有密切的關係，說到建國更是百端待舉，這裏無論爲輕重工業的建設，農業的改良，交通冶鑛的發展，以及林牧水利墾殖的計劃推動將來成敗利鈍的樞紐，完全要以現在的工程師能否以大禹的精神人格爭取時間，辛勤工作堅定抗戰建國必勝必成的信念來判斷，我們相信在建國的過程中，有許多急待推動的工程事業，期待工程師們能以大禹治水的熱誠毅力接受時代的要求，各盡個人的全力，爲斯人的創造與成就。

工程學與工程教育

——轉載工程第十五卷第五期——

楊繹德

一・引言

中國工程師學會與各專門工程學會第十一屆聯合年會在蘭州舉行，所徵集論文，尤注重者計有十類，而工程教育列為其中之一，其意義深遠可不言而喻。蓋吾國自抗戰軍興以後，各大學多遠道遷移，雖烽煙遍地，而絃歌之聲未嘗或輟。且因國家對於工程建設之需要，各工程學院科系亦屢有增加，近年全國各大學入學試驗，志願習工程者常倍蓰於其他院系，吾政府當局特添設雙班，廣事培植，以備他日抗戰勝利大規模建設展開時各項工程人才之需要。「各國科學之日進千里，吾國工業實瞠乎其後，繼今以往，必須推宏各級工程人才之培育。……蓋立國於當世須有堅實之國防精神，而工程科學與工程事業實為國防之真正基礎」。語 蔣委員長對中國工程師學會第十屆年會訓詞，可知工業建設與工程教育為國家之基本政策，而工程人才之如何培育與工程學術之如何發揚，實為今後工程教育一大問題，故工程教育範圍至廣，歐美各國體制不同，欲詳細研討，端賴鴻製巨著，非愚陋如作者所能盡其什一也。本文之目的，在從歷史觀點以闡明工程學之要義，然後根據工程學之定義以申述工程教育之方針。作者不敏，服務於工程教育，無非本其一得之愚，貢其管窺之見，以就正於海內宏達與工程學術界同志耳。

二・工程制作與工程學術之演進

人類文化之進步與工程學術制作之演進有至密切之關係，此乃公認之事實也。當邃古之世，草昧初開，尚未有文字記載之歷史可資考證，博古家之所憑藉以爲推考之資料者，爲由廢墟古塚所發掘之古物。因古時人類所用武器工具之演進，而分作舊石器新石器銅器鐵器等時代；既有武器工具則必有工程制作殆無疑義。且文明進化之程度以當時所用之工具爲標準，則工程制作意義之重大亦可想見。吾國上古時代歷史傳說所稱有巢氏構木爲巢，燧人氏鑽木取火，構木爲巢乃建築工程之開始，而鑽木取火者，以科學解釋之，即摩擦生熱之理也。易與尚書爲古代典籍之信而足徵者，易繫辭稱神農氏斲木爲耜，揉木爲耒，耒耨之利以教天下；黃帝堯舜刳木爲舟，剡木爲楫，舟楫之利以濟不通，弦木爲弧，剡木爲矢，弧矢之利以威天下，上棟下宇以待風雨。耒耨之利屬於農具，舟楫之利屬於交通，弧矢之利屬於兵工，上棟下宇屬於建築，蓋皆工程製作之演進也。尚書禹貢篇記禹敷土，隨山刊木，奠高山大川，分天下爲九洲，辨壤土之性，誌物產之饒，詳考山脈之系統，江河之源流誠爲吾國古代水利工程與地理學之傑作。遠在四千年前而有如是規模宏大之工程事業與精詳分析之記載，則吾國古時工程學術水準之高可以概見。其在西洋則埃及金字塔之遺跡巍然猶存，菲尼基之造船，巴比倫之河渠水利，亦尚有古碣遺址可尋；凡此皆足以表示古代工程製作之成績已大有可觀。故工程制作演進之歷史雖甚久遠，但在古典籍中，關於工程方面有價值之巨著，如尚書禹貢篇，周禮考工記之類，實未多觀，在西洋則希羅氏（Hero 公元前一五〇年）所著氣流學（Pneumatica）亦爲僅有之工程古籍。蓋工程制作之創始固遠在文字記載歷史以前，而工程學術之發揚則尚待自然科學創明以後。在古簡册中，容或有片段關於工程知識之記載，其所得結論或亦與現代工程學理相暗合，但其所憑者，大抵爲以往之經驗及心得之訣巧，究未可以稱有系統之工程學術也。溯工程制作之演進既如是久遠，而工程學術之發明又須待數理化與生物等科學基礎樹立以後，源流愈遠，憑藉愈厚，則其所發揮之成果愈見充實而精彩。當十七世紀中葉，工程學術之發展已漸露端倪，迨十八世紀以後，各項工程創作蓬勃蔚起，理論研究光彩燦然，在極短時期中而有如是豐富之收穫；所以然者，蓋學術之演進已達成熟之階段，工程技術已積有長時期之經驗，根深蒂固，枝葉繁茂，已能放燦爛之花，而收成功之果也。

三・工程教育發展之過程

考工程教育之發展亦屬晚近期內之事。吾國因受清季政治腐敗之影響，以致在工程教育方面，較之歐美未免落後。在歐美各邦中，發展最早者首推法國，其次爲德，而英美俄諸國稍晚。當羅馬帝國崩潰之後，法人崛起歐西，蔚成大國，羅馬人對於工程上之經驗向稱豐富，而法人承其遺緒，故其軍事與民事工程技術之高，初非歐洲其他國家所能比，但法國工程學校之創始已在十八世紀中葉，最初僅爲製圖成技術學校，且限於土木與採鑛方面。德國工程學校較法國爲後起，最初亦爲建築學校之類，其創立時期則在十八世紀之末。英國則當十九世紀之初，僅有私人組織之學社，以講論科學技術爲宗旨，而正式工程學校之設立則尚在倫敦展覽會(一八五一年)之後。美國歷史最早之藍色樓工科大學(Rensselaer Polytechnic Institute)成立於一八二四年，創辦之初僅授土木工程，名爲藍色樓學校。大抵歐美各國之工程教育，在十九世紀初葉，尚屬技術學校之性質，以土木工程爲主，至於高等工程教育之長足進展，實開始於一八六〇年以後，迄今亦不過八十年而已。

吾國科學研究之進展，在天文醫藥數學各方面，已有久遠之歷史。論工程技術則周禮考工記分攻木攻金攻皮設色刮摩搏埴之工計三十項，觀其對於名詞尺度選料驗工之精密，足見當時工程技術已到達相當程度。文化歷史輾轉演變，迨元明之世，歐亞之交通漸繁，西方教士傳導歐洲之科學技術於中土，配合吾國固有之學術基礎，未始不可以發揚而光大。惟因明末政治不良，滿清入關，兵戎擾攘，乾嘉以降，政治紊亂，以致吾國之學術文化未能適應時代潮流而發展。及同治初年，曾李諸臣提倡西洋學術，設廣方言館與繙譯館於上海，前後出版格致化學製造各書計一百七十八種。迨光緒二十一年(一八九五年)，北洋大學(初名中西學堂)首先成立，設有路礦等工程學部，二十三年(一八九七年)，南洋公學成立，注重算化格致工藝等科，是吾國之高等工程教育亦已有四十餘年之歷史。尤以民國十六年北伐完成之後，國民政府銳意推進大學教育，充實工程科學圖書設備，雖尚未能與歐美負有盛名之高等工程學府媲美，但已追及其第二等大學。自抗戰以來，吾政府宏謀遠算，對於大學教育苦心維護，不遺餘力，而對於工程教育則因時勢之需要，期望更屬殷切，繼往開來，時代之使命甚重，吾國人素富於艱難創造之精神，諒必能發揮其固有之文化力量，以樹立今後學術建設之基礎也。

四·工程學之定義

美國電工學家史篤德氏(H.G.Stott)作工程學之定義，曰「工程學爲組織領導羣工，控制大自然之能力物料，以造福人類之學術」，(見美國電工學會季刊一九〇八年卷)。此定義言簡而意賅，可稱允當，求之吾國古籍，與史篤德氏之定義恰相符者，有尚書大禹謨「正德利用厚生」一語。正德者領導羣工，即論語其身正不令而行之意，利用者控制自然，而厚生者造福人類也。大禹爲工程師之典型，故其言之精當切要如此。觀上述定義，可知工程學爲博大精深之學問，對於自然之澈底研究最關重要，庶能控制大自然之能力物料而充份利用之，若組織領導之才能，雖與個人稟賦有關，但發展而利導之者，亦端賴學識與修養，否則璞玉雖美，未加雕琢，恐尚未能成珪璋之器也。

又美國麻省理工大學校長康濮登氏（K,T,Compton）作工程師之定義，曰「所謂工程師者，乃運用數理化與生物諸科學以及經濟學之知識，更濟之以從觀察實驗研究發明所得之結果，然後利用大自然之質料與能力以造福社會之人才」（參看傑克遜教授著美國工程教育之現狀及趨勢第一章第八頁）。故工程教育實包含自然科學與人文經濟方面，而不僅限於工程技術方面，且工程學之要旨在運用科學知識，而運用之妙則存乎一心。善乎傑克遜（D,G,Jackson）教授之論工程學方法曰「工程學爲科學與藝術之調合，習之者須具精確採集數據之能力，能分析，考索與類別數據；能從數據作成案語；且能運綜合之想像力以審察數據，而利用所作案語以構成工程計劃」（參看同上）。故自然科學爲研究工程學所必需之知識，但工程學則又自有其特殊之色彩與背景者也。

工程學之主要目標爲利用大自然之能力與質料，其意義可得而申述之。蓋宇宙間一切活動均受能與質之支配；就人羣言之，則立國之本首重民生，而民生問題之中心則在如何開發動力，推廣生產；保國之道端賴國防，而國防問題之中心則在如何動員國力，培養資源，要之皆能與質之利用也。是故能盡能與質之用者，國家即不難日進於富強之域。惟欲盡能與質之用者，必先明能與質之理；明汽電之理，然後能創造汽機電機，而盡汽電能力之用，明鋼鐵之理，然後能冶製特種鋼鐵，而盡鋼鐵質料之用。故欲明能與質之理，則自然科學之研究爲重，而欲盡能與質之用，則工程學之研究尚矣。雖然，知其用不可不知其理，苟原理尚未通曉，則應用時感到困難將無從解決；況知其用尚未可以稱盡其用也，欲盡其用則須發揮研究創作之能力，而研究創作所需工具乃基本科學也。科學知識爲工程學之基本工具，而工程學之主要目的，則在運用科學知識以盡能與質之用。故工程學與純粹科學不同，蓋純粹科學之目的在創明能與質之理，而工程學則在盡其用，因欲盡其用乃必先明其理也。工程學又與應用科學之意義不類，蓋工程學之特色在發展分析與創作之能力，並非應用二字所能概其全，惟工程學之目的則在利用厚生耳。

各門學術均有理論與實際兩方面，固不獨工程學爲然，而工程學則理論與實際間之調和尤關重要。蓋工程學所研究之問題屬於物質方面，但研究時之考索推論亦不外乎理想之默運；又因所研究之對象爲實際的，故理想須證之以實驗，而不致蹈於空泛。工程學之最大意義爲務實，凡理論與實際不相符者，不能認爲完滿之理論也，且工程學之要旨，在運用科學知識以盡能與質之用，而造福於社會，科學或稱格致學，故工程學之下手工夫在格物致知，而最高目的則在博施濟衆。其理想至高，而其工夫最切實。吾國以前學術似偏重於誠正修齊治平方面，而對於格物致知方面，則較淡薄。工程學則發揮格致之功能，以得到治平之效果，其研究學理時之精心一志，亦有合乎誠正之義。故工程教育與吾國古代教育思想有契合之處，而其力行務實之精神，更足以祛空談玄理之失。惟理論與實際，理想與物質，貴乎能和諧調劑，庶可以培養研究之興趣而發展創作之能力也。

五、工程教育之一般原則

美國著名電機工程師闌姆氏（B. G. Lamme）論工程人才之訓練（見闌姆氏著電工論文集）其主要原則爲(一)把握基本原理，(二)明白物理觀念，(三)養成想像能力，而更總括之以(四)發展分析才能。所謂分析才能者，即從所能得到之數據與事實，加以分析研究而作正確判斷之謂也。研究工程學理所需要之工具爲數學，精於工程學者，能善用數學工具以解析工程問題，而善於利用數學工具者，則能運用比較基本之數學以解決學理方面之問題，得之於心，應之於手，苟非對於工程學之基本原理確具把握，且對於所研究問題之物理現像澈底了解，更濟之以豐富之想像能力者，未易臻此也。高深數學爲研究工程學之精良工具，而正確數學觀念之養成實爲運用數學工具之前提，此則爲習工程學者所當明瞭，而論工程教育者所宜注意者也。

且所謂工程人才者，若依其學養性質而區別之，可分作研究、設計、製造、測驗、裝建、運用、保全等若干類。大抵研究人才需要淵博之科學工程知識，與超越之創作才能，設計人才須曉暢工程原理，且富有想像推斷之能力；製造人才須熟諳工作機械與工程技術，材料之性質，工廠管理法等；測驗人才須明瞭器材性質，測驗標準等；裝建人才須具豐富之工程知識與技能；運用人才須熟習整個工程系制，與機器之特性；保全人才須精習機器之構造品質與工作技術。各項工程人才均有其特殊之學養經驗，長於研究者未必長於製造，長於設計者未必長於裝建。然就工程教育之立場言之，無論青年學子之稟賦與趣如何，而工程學理之基本訓練實爲首要之圖。蓋教育之目的爲培植青年之學術基礎，以爲他日前進發展之所憑藉，基礎鞏固，則無論何種建築物皆可經營締造於基礎之上，而不虞傾側。故基本原理之訓練固爲工程教育之要義，即專門工程學術之傳授，其主旨亦不外乎基本原理之發揮與推究也。

傑克遜教授論工程教育之趨勢，其最扼要之點爲培養學生自動進修之能力（見美國工程教育之現狀及趨勢第八章第一二八頁），此實不磨之論也。工程學術門類繁多，日新月異，既未可僅賴書本，亦未能限於教室。大學修業時期不過四年，大學教育乃畢生事業之出發點，無論何項事業之成功，端賴不斷之努力進修，苟在大學時期而未能養成自動進修之能力與習慣，則將來一出校門，機會更難，其影響於青年自身前途無量之發展，與夫社會國家之建設與進步者豈可量耶。

六、對於吾國工程教育之管見

無論何種制度莫不與一國之歷史背景與現實環境有密切之關係，工程教育亦然。是故美國之工程教育制度與歐洲不同，而歐洲大陸之制度又與英國異致。簡要言之，在法國則注重理論之研究，入學標準頗高，講學者大多為工程學界知名之士，而學生課業之考績極嚴。在德國則學貴專精，學生在中學校之訓練比較嚴格，入工程學府者大多懷抱工程事業之志趣，而傾向於工作業專題之研究，在英國則認工程學府之主要使命為傳授工程基本學理，而實用技術則當在製造工廠或工程實施之場所訓練之，故英國工學院多採用學院與工廠錯綜教練之制。在美國則工學院中工場實驗室設備豐富，實習科目衆多，除基本科學及工程專門科學外頗注重經濟學；並有若干學院兼施合作科制，採取英國錯綜教練之法。在蘇聯則為完成其經濟建設計劃起見，除設工學院以培養高級工程人才外，並設各級技術學校以訓練大批工程幹部及技工人才；工學院與各工廠間之聯繫既十分密切，且每建新廠即同時成立技術學校，用能在短期之內造就多數之工程技術人員，惟學識技能之是否健全或不無疑問耳。吾國工程教育制度與美國最相近；以國情言之，吾國地大物博，需要大批工程人才作大量之開發，採取美制可稱得宜。惟每個國家各有其歷史傳統之觀念，與社會組織之特點，必須斟酌調劑，因地制宜，方能獲到最高之效益也。

吾國大學教育之歷史傳統精神為注重人格之修養。工程師為領導羣工以開物成務者，必須有開闊之胸襟，與高超之志趣，方能獲得羣工之信仰，而收臂指股肱之效。且國家大規模建設展開之後，各個工業組織勢將成為社會之重心，而工程師則為工業組織之中堅份子，又必賴平素之修養，方足以領導社會。是故存主奴之見，而抱中學為體，西學為用之觀念者，固難免思想陳腐之譏，但吾國之傳統教育精神，所以維繫社會於不墜者，亦未可一概抹煞。且注重人格修養，豈但吾國傳統教育思想如此，即英美諸國之教育趨向亦莫不然。傑克遜教授之言曰，「目下社會人士已更能認識優越之智慧為工程事業成就之因，而優秀之德性則為支持成就之要素；又工程師視品性較知識更重要，而工程科目之傳習正所以訓練誠毅之意志」（參看美國工程教育之現狀及趨勢第八章第一四一頁。）其注重品性有如是者，故吾國大學工程教育，對於人文方面似乎亦有相當重視之需要，以期適合國情也。

工程學除以自然科學為基本外，並以力學科目（包括力學，水力學，靜電與動電力學，熱力學等）為介乎自然科學與專門工程科目間之津梁。吾國學生習工程科學須先習英文或兼習德文，故大學工程學系科目繁重為必然之結果。大學修業期間不過四年，其中斟酌損益實煞費苦心，一國之工程教育，往往受實際環境之支配，而未可純憑理論。吾國目下最堪注意之問題，為大學工科師資之缺乏，與工程科目教材之調整，關於師資缺乏一層，則陳部長在中國工程師學會第八屆年會講中國工程教育問題（工程第十三卷第四號），亦曾慨乎言之；關於教材調整一層，則爲吾國數十年來工程教育之重大問題；苟此兩大主要問題未能合理解決，則其他方面之改革更張，恐終鮮實效，吾國最早工科師資借才異域，現在則多數爲留學歸國之工程學者，恐將來趨勢當爲培養優秀之青年師資，其興趣在於學術研究者，待服務數年之後，在學術研究方面已具相當根據，然後派送國外遊學以求深造，豈但爲大學培養師資，亦卽爲國家造就才俊，又目下工科師資缺乏，與各機關迫切需求工程人才不無關係，如何設法調整，諒吾政府當局已籌之熟矣。吾國大學所用工程教本多數採取美國，一則因吾國工程教育制度近於美國，再則因美國出版事業發達之故。惟吾國社會現狀與美國不同，關於教材方面有應斟酌變通之處。吾政府當局對於此問題關心已久，念茲事體大，影響甚宏，尙有待於全國工程學者之努力與合作者也。

工程學術非但爲國防建設所必需，且其內容包羅宏富，涵蘊精深，理論與實際調和適合，是以能引起學習者之興趣；近年來多數青年之所以志願習工程者，此點亦爲重要之原因。惟工程學之標準頗高，而爲學之道貴乎循序漸進，不可躐等。美國之大學工程教育較吾國更完備，但在彼邦亦頗有主張提高入學標準者。然彼邦出版事業發達，圖書館林立，各大工廠多設立實習訓練班，故大學工程學生卒業以後亦頗多進修之機會。吾國社會現狀，較之美國容有不同，大學教育關係更重。若能利用暑期，酌設講習班之類，以充實基本學識與必要之語文工具，則修習高深科目，更能發生濃厚之興趣，而自動求知之力量可增強不少矣。

一國工程學術之地位，一方面旣與工業建設之發展息息相關，而他方面則創作研究之重要性實堪注意。夫超卓之研究人才至不易得也，但學術之進步則有賴乎少數研究人才努力所獲之成績。當研究某種物質或現象之初，固未能預卜其所研究之結果，對於工業建設將發生若何影響，豈但未能預卜焉，或竟未甚縈懷也，所專心致志者在格物而窮其理耳。一旦研究成功，則能與質之利用更廣，工業建設隨之進步，而社會蒙其福。夫才難之嘆，自古如斯，而創作研究之長才尤屬難得，國家對於此種人才須珍惜之而培育之，此在論工程教育時所不得不鄭重提出者也。

至於工程技術人才之培育，國家正不知費去多少心血；工程學理貫通之後，又須積多年之實地經驗，技術乃能臻純熟之境，故高級工程人才之養成殊非易易。且工程技術類別頗多，門徑互異，因各人性格不同，有宜於製造，有宜於裝建，有宜於設計，有宜於保全，若職務與個性不宜，或所用非所學者，在整個建設事業言之，均屬人才之浪費，亦卽社會之損失，夫工業建設之推進，端賴學識優良，經驗豐需之工程技術人才。以吾國幅員之大，各種工業均待舉辦，而工程人才如果缺乏，一方面固當廣事培植，他方面尤應免除浪費，庶幾人盡其才，而工程技術乃能高速度進步矣。

工業建設與學術研究有互相關聯之勢，工業建設愈發展則學術研究愈需要。吾國對於學術研究方面之基礎甚形薄弱，一旦大規模建設開始後必有許多立待研究之問題；蓋各項工業均有地方性與時代性以及經濟上之種種關係，故研究工作十分重要。且工業技術進步極速，不研究即不能進步，不進步即立刻落伍，故工業建設愈發達則愈感研究工作之需要，蘇聯經濟建設計劃推行時，並成立多數之工業科學研究機關，即其明證。吾國工業落後為有識者所同慨，而研究基礎薄弱實更堪注意；且研究事業之發展較之工業建設或更難立見成效。故吾國工科大學應及早設立研究院以發展研究事業，造就研究人才，蓋非但為工業建設當務之急，抑亦為國家之百年大計也。

又陳部長在第八屆年會講中國工程教育問題，對於建教合作之推進闡述頗詳。蓋大學工科教授所負之使命為發揚工程學術，而工程學則理論與實際間之調和極關重要。工程學府所研究之問題偏重學理方面，但理論須證之以實際而後涵義更明，工業建設所研究之問題偏重實際方面，但實際須本之於學理而後計劃更當，故建教合作實為發揚學術恢宏建設之重要因素。試觀蘇聯歷次經濟建設計劃之實行，工科教授之參與研究設計以底於成者，不在少數，可為明證。且大學工科教授負有培育建設人才之責任，若能熟知國內工業建設實在情況，則學校內學生所習能與國內建設取得聯繫，不致脫節，而愈能激發學生對於學業之興趣。美國哥倫比亞大學工學院長白克氏 (J. W. Barker) 稱傑克遜教授工程教育哲學之精義，為「工程學生須令及早與新問題接觸，以激發其思索之能力。……欲達到此接觸新問題之目的，必須工科教授從事於工程顧問或研究方可」（參看美國電工雜誌一九三九年二月期第六五頁）。可知工科教授之參加工程顧問或研究，不但合於建教合作之義；且對於工程教育之健全發展亦有莫大之裨益也。

七、結論

近來討論大學教育之文屢見報章雜誌，而關於工程教育方面者則未多覯，但大學工程教育之重要，固不待旁徵博引而後明也。國家工業建設之推進，有待於大批工程人才之培育，而大學工程教育則為養成高級工程人才之主要因素，况大學教育為青年畢生事業之出發點，其對於事業前途能發生蓬勃之意與勇往堅毅，百折不撓之精神者，端賴健全之學識為原動能力，故謂吾國現階段之大學工程教育，與今後若干年國運有關者，殆非過甚之辭也。作者不才，以茲問題之重要，尚參考文獻，欲明其究竟，明知管窺蠡測，無當大雅，然千慮必有一得，拋磚可以引玉，爰草此文，略陳鄙見，博雅君子，幸賜教焉。

工與中國文化

潘光旦

（轉載自由論壇一卷一期）

我們相信遠在三千五百年以至於三千年之前，工在中國文化裏是有過他的地位的。周禮冬官考工記便是一個明證。考工記的開頭便說：「國有六職，百工與居一焉。」又說：「知者創物，巧者述之，守之世，謂之工，百工之事，皆聖人之所作也。爍金以爲刃，凝土以爲器，作車以行陸，作舟以行水，此皆聖人之所作也。天有時，地有氣，材有美，工有巧；合此四者，然後可以爲良。」可見當初並沒有貶薄工的意思，更無鄙夷機巧的成見；有之乃是後世的事。再從文字源流上說，工字在六書爲指事，所指爲規矩勾股之事，固然不錯，但同時我疑心他所指的不止於此，他也許是和巫字屬於一類的字，即上下兩畫所指的是天地，而中間一豎有通天地之意，取法乎天，收材於地以成物的人和事，叫做工；巫字我以爲也應當屬於此類，徐鍇對於說文上「與巫同意」一語的解釋是很牽強的，我所不取。古人造字，如果眞有此用意的話，那工的地位就非常之高，至少不在古代巫的地位之下，而根據周官「此皆聖人之所作」一語推之，更可知此種地位的獲得也很在情理之中，不過我不是一個小學專家，以前的小學專家，據我所知，對於工字的來源也從沒有作過這樣的解釋，姑存此一說以供參考。

無論如何，我們如今要把工的文化建設起來，要使工在中國文化中取在其應有的地位，除了把工的地位提高之外，更無第二條途徑；提高工的地位，有條件，也有限制，不講求條件與限制，一味盲目的提倡鼓吹，是非徒無益而又害之的，而被害的對象不止是工的本身，並且是中國文化的全部。

第一個條件是思想的。儒家的人文思想原是相當的完整的，但一變而爲人本論，再變而爲唯人論，結果是終於把三才中的天地兩才擱過一邊，置之不聞不問。擱過了天，是慢忽了哲學和一切形上的東西；擱過了地，是遺忘了科學和一切形下的東西。工和機巧的視同敝屣是思想中擱過了地的必然的結果。關於這一層，我以前在別處已有過比較詳細的論列，在此恕不再贅（說本，二十八年六月十一日昆明益世報）。

惟道家思想一方面，我們不妨再多說幾句話。道家反對人類的故作聰明，妄加創制，固然不利於工的發展，已如上篇所述，但何以對於比較抽象與理論的哲學也不能多有大的貢獻呢？西洋的哲學科學，以至近百年來貫通一切學問的學說如同演化論，不就從自然主義產生出來的麼？只要自然主義能產生科學，那工學與技術的發展豈不是就不至於落空？原來道家的自然主義和西洋的大有不同。西洋的自然主義是宗教的超自然主義的一個反響，富於理智的成分，故其目的是在了解自然，分析自然，中國的自然主義，至少就春秋以後的歷史說，是儒家人爲主義與禮教主義的一個反響，富有感情的色彩；故其目的是在接納自然，順適自然，而所接納順適的自然當然是整個的，不是經由人力而支解了的。在這種自然主義之下，要干涉自然而有所利用，固然事所不行，就是要分析自然而有所了解，也是理有未可。中國人對於自然的態度是只求欣賞，不求認識，只問完整的外形，不問內容的節目。這種的自然主義固然樹立了樂天安命的人生觀，創造了一部分的詩與一部分的畫，在許多人的生活裏養成了種花，飼鳥，遊覽山水，以至於賞玩石塊等習慣，但天命的窺測，花鳥的解剖，以至於地質的控掘檢視，不但在意識裏無此興趣，並且在情緒上可以引起反感。這樣一路的自然主義決不能產生科學的理論，更不能孕育工學的技術，是在邏輯上無可避免的一個結論。

所以第一件我們應努力的事是思想上或民族觀感上的補正。補正的工作可以從好幾個方面下手，根據上文，便已經有兩個方面，一是貫澈儒家的思想，董申通達三才而不蔽於人的人文思想。二是補充道家的自然主義。兩晉六朝以還，一方面爲因佛教的發達，一方面也因爲道家取得了一部分統治階級的信仰，我們本來有過一度重新整理所謂人文思想的機會，因爲二氏是至少不蔽於人的。惟其不蔽於人，所以從晉代以迄宋元，我們多少還有過一些形上與形下的收獲。葛稚川的抱朴子是最好的一例，晚近治科學的人說抱朴子中有不少科學的種子。許多高僧，於發揮佛典而外，往往也做些形下的觀察，例如宋之惠洪，元之冰亭，『冷齋閒話』和『搜采異聞錄』兩種作品裏的一部分的觀察與見解是墨守儒家成說的人所做不到與說不出的。不過就大體說，道家既蔽於一種近乎感傷主義的自然主義，而佛家又蔽於心，客觀與可容分析的物便不在他們注意範圍以內，所以在這時期裏，我們雖不乏有綜合九流的頭腦的大師，有如朱熹，也不能多所裨補了。

近三百年來歐洲學術的輸入當然給了我們第二度的機會。不過這機會我們到如今還沒有能充分與合理的利用這機會引起了一度科學與玄學的論戰，一度本位文化與全盤西化的討論，以及目前對於理工的狂熱的鼓吹；此外還不見有何成績。其實歐西文化也離不了三才的範圍；希臘文化是比較最能兼籌並顧的，比較最通達的，但是早就過去了；文藝復興時代一番重整的努力並沒有成功，晚近一部份英美學者的呼聲也就不啻寒蛩之鳴，無裨實際，實際的局面是，中古時代蔽於天，而文藝復興以還蔽於物，而蔽天蔽物的結果均是以人爲芻狗！唯物論產生了思想的禁錮，其所芻狗的是人；唯貨論也產生了畸形發展的理工的技術，其所芻狗的也是人；三十年來的西洋史，包括蘇俄的革新運動以及第一次與目前第二次的世界大戰在內，可以說是西洋蔽物思想的一個總結算。

中國文化和歐西文化，從三才通論的立場說，是各有所蔽的。就其二千餘年來的發展說，前者是蔽於人，後者初則蔽於天，而終則蔽於物，或蔽於地，不過從本題的立場說，歐西文化表面上佔上一些便宜，底子裏是否便宜，上文已稍加別論。三才通論是一個最較完整的文化觀，是一二比較原始而元氣磅礴的民族所產生的，在歐洲是希臘民族，在中國是代表易經或易經中繫詞的時代的漢民族。自此以降，因為民族的以及文化的種種內在的原因，始則發生了動盪，終則不免於支離破碎。結果在我們是蔽於人，而西洋是先後的蔽於天與蔽於物。說起動盪，最好的比喻而也是物理上所無可避免的一種象徵，是鐘擺的擺動。鐘擺的擺動是弧形的，兩極端是天地兩才，而和地心成直線的一點是人的一才。中國文化蔽於人，就好比鐘擺動得極為微弱，始終離人的一點不遠；西洋文化則好比鐘擺作有力的擺動，所以時常到達極端，而於中間則極少徘徊瞻顧的機會，從這個立場看，蔽於天與蔽於物的文化，表面上雖若南轅北轍，各不相干，而底子裏却並不衝突，因為蔽於天的反動就是蔽於物，而蔽於人的文化則終有一天不知天高地厚，甚至於不知天地為何物，換言之，西洋黑暗時代的宗教文化，對於後來的科學文化和技術文化，不但不是一個阻礙，並且是一個最主要的動力，而在中國則此種動力根本不存在。所以上文說，歐西文化雖同屬偏蔽，而從工的發展的立場看是比我們要佔便宜的。

從上文的討論，就可知歐化東漸以來，我們因為力求位育，而在理論上與行為上的若干努力，到目前為止，不是很徒然的，便是未必有利的。第一，玄學與科學便無須發生論戰，不有蔽於天的因，就不會有蔽於物的果；不有玄學，何來科學？第二，本位文化和全盤西化之爭也是枉費心力，因為兩者所患的偏蔽雖有不同，而其為不免於偏蔽則一。大凡有所偏蔽，即有所廢墜，在西洋所廢墜的是人，而人的廢墜到相當程度以後，天地亦不免於閉塞，目前正進行着的大規模的屠殺如果再延長下去，這天地閉塞，乾坤止息的終局怕也就不遠了。至於中國，直接而先廢墜的是天地，而因為天地廢墜，以致不能講求利用厚生的結果，間接而亦終不免於廢墜的是人。這樣說來，本位文化固有他不能再事維持的理由，而全盤西化又有甚麼可以豔羨的地方呢？至於目前的一味提倡理工，若目的在矯正以前的積習，認為非暫時「過正」，不足以言「矯枉」，則還說得過去，但若非如此不足以重新奠定民族文化的基礎，從而認為國家百年大計的一部分，那即使事實與能力容許我們做到，也無非是甘心於踏上西洋文化的覆轍而已。

所以不說到思想的重新整理則已；否則問題就非常之大。我們想把工向中國文化裏再度配合進去，使他取得他的應有的地位，勢必牽動中國文化的全部結構，勢必參考到西洋文化的內容與演變，而也勢必牽扯到我們對於溝通中西文化的一大問題的應有的態度。但牽涉到的場面雖大。內容雖複雜，如果上文的討論可以成立，而値得做依據的話，那結論倒是相當的單純的。復古，維新，恪守本位，全盤西化，恢復固有精神，銳意於現代化等等，都不成其爲問題，都不是問題，問題是怎樣建立一個通達三才的新文化，使一切的事物，包括工在內，各有他們的地位，使一切的才能，包括機巧的技能在內，各有其用武的場合，而不再發生任何一方面過於偏蔽的弊病。說到這裏，可知思想的整理，無論其爲儒家思想的重新檢討，或道家自然主義的補充，或佛家極端唯心論的糾正，或西洋近代思潮的節取與調和，其實只是一個問題。

第二個條件是教育的。我們把第一個條件打發開以後，這第二個條件是比較簡單了。教育固然所以傳播思想，而國家的教育制度以及時代的教育政策究是建築在思想的基礎上的。思想偏蔽，斯有偏蔽的教育，思想通達，斯有通達的教育。因爲單看重士的行業，所以二千年來，我們只有讀書識字，傳述古人的教育；因爲尊尚法治，提倡民權，才有三四十年來的法政教育；因爲注重技術，銳意建設，我們才有最近幾年來的理工的專門教育。

不過理工的專門教育，特別是近年來提倡與實行的那一種，還不是教育條件的基本的部分，並且如果祇靠他，前途工的地位不但未必能真正提高，並且會有再度淪陷的危險，下文當續有討論。基本的教育條件有三個。一是一般的思想教育，目的在發揮傳播上文所已討論過的民族思想的根本改正。我們如果不能會通三才，不能中用執兩，工的地位是絕對無法提高的，因爲工是兩端中之一的一大部分。第二個可以叫做手藝或技能教育。這種教育我們本來已經有了一些，但是不夠，並且也有錯誤，中小學有手工一課，大抵只學些輕便或美觀的東西，談不上機械的技巧。這是不夠的一方面。我們提倡職業教育，主張手腦並用，也已應有年所，職業學校裏，特別是工業學校裏的機巧的訓練固然比較多，但又僅僅以此爲限，其爲偏蔽，與高一級的理工專門教育相同，並且職業教育的名稱也有弊病，似乎所重的，對社會只是有業人士的增加，生產能力的提高，而對個人只是一種出路的準備，一種吃飯本領的獲得，而不在機巧的訓練本身，這是錯誤的一方面。

如今我們提倡手藝或技能教育，所重的應完全在此種教育所能供給的訓練，至於前途對社會經濟與個人經濟有何利益，至少在辦理與接受教育的人是不問的。這種教育要辦得普遍，即一切學校教育裏應當有一些，而不限於比較低級的普通學校教育和職業教育。這種教育也要辦得相當深刻，即使任何青年對於科學化的工的技巧要直接有些接觸，要親切的獲取一些經驗，不管他們將來進甚麼行業，要使他們對於日常應用的機械能有明白的了解，能自己裝配修理；要讓他們拳拳服膺，物理不比人事，是硬性的，是一是一，二是二的，是無法通融不能違拗的，設有違拗，不但不能成事，說不定對己對人還可以發生很大的危險。要技能教育普遍化與深刻化，我們應當主張各級學校教育，對於理化的課程，以及表證工作，試驗實習，工廠實習等，都得有適度而充分的設備；大學生更應人人修習物理化學和機械大意，也人人應做些最低限度的工廠實習。目的務使民族分子中所有的一些技巧能力，無論剩餘多少，能有充分表見與發揮的機會；其能力過於薄弱，無可表見，或只能胡亂應付，以至於容易出岔子的分子勢必逐漸趨於淪落而歸於淘汰。由反選擇一路而來的一種局面，非改循選擇的路，即非反其道而行之，是無法改變的。

第三我們才數到工的專門教育。在提高工的位置的努力裏，此種專門教育的提倡與專才的獎勵當然有他的地位，不過這裏有一個極端重要的條件，就是此種專門教育，以至於任何專門教育，必須建築在一個比較圓到的通才教育之上。為甚麼必須如此，我們又可以從三個不同的方面加以討論。一是維持健全的教育理想，二是加強工的專才的效用，三是扶植工的地位，使不至於再度驟於淪喪。

教育的理想是在發展整個的人格，我以前在別處討論到過人格有三方面，一是人之所以為人的通性，二是此人所以異於彼人的個性，三是男女的性別。健全的教育是三方面都得充分顧到的，如果捨男女之別不論，則須兼籌並顧的至少尚有兩方面。個人的先天性格儘管不免有所偏倚，教育的鵠的則不能不力求通性與個性的平衡發展。通性是通才教育的對象，而個性是專才教育的對象；一個人應當受的教育是一個通專並重的教育，以至於『通』稍稍重於『專』的教育，因為歸根結蒂，我們必須承認，做人之道重於做事之道；生命的範圍大於事業的範圍；至於一個人究能通到甚麼程度與專到甚麼程度，那自須看他的才力了。這一層理論我相信是古今中外所同的，從前『本末』『博約』『文質』一類的原則所指也未能外此，楊雄的儒伎之分所指的也就是通才與專才之分。

我一向感覺到近年來大學理工教育是不健全的。其所以不健全之故並不在專的過度，而在通的程度遠不足以相副。有充分通識做襯託的專識，無論專到何種程度，是不妨的。至於通識不足，或極端缺乏，即使專的程度不深，也往往可以誤人誤事。所謂誤人，有的是人格的畸形化，成一個偏廢或半身不遂的局面。好像西洋有一位生物學家說過，「專化的代價是死亡」，古生物學界裏此種例子最多，而其中最足以發人深省的是龍類。近代歐西文化的危機，我以為也就在此，歐洲的理工文化，已經發展到一個尾大不掉的程度，其結果是戰爭，屠殺與死亡。族類專化的代價如此，個體大概不會例外。

第二方面指的是對於事業的礙妨。一個專家，如果沒有充分的通識做襯託，其實是等於一個匠人，至多不過比普通的匠人細膩一些罷了。第一，他不大了解他所專擅的學術以外，尚有其他的學術，他不大知道他的專門學術，在整個學術界裏，以至於全部的文化生活裏，究佔多大一個位置，究有多少分量，究應如何配合起來，方才覺得稱當。第二，他不大認識人。他和從前的讀書人似乎恰好相反，懂得「物」是甚麼，聲光電化是甚麼，但人是甚麼，他多少有幾分莫名其妙。因此，不但他的學術事業和別人的不容易配合起來，他和別種學問事業的人，以至於和同學問同一事業中的別人，也容易發生扞格發生齟齬。最近有一位工業界的領袖對我說，而所說事實上等於一種飽嘗痛苦後的呼籲：「我希望工學院的畢業生，於讀畢理工學科以後，再費廿年功夫，來讀些社會學心理學之類，好教他們一旦就業，於應付事工之外，更能應付人，於一己獨力工作之外，更能與人合作！」有組織的工業場合裏，如果人與人不能合作，也就根本沒有事工可言。是不是讀了一些心理社會之學，就可以替這位工程界的領袖解決問題，我們不得而知，不過這一層我們是知道的，人事的處理是屬於通識一方面的，通識不足的人儘管會設計，打樣，以至於發明機械，頭頭是道，件件在行，要他應付人，却是困難極了，而在中國社會裏，必須應付的人事又是特別的多。在中國社會裏，人事不修，事業便根本無法推進。所以上文說，我們為加強工的專才的效用計，也認為必須把工的專門教育安放在一個比較圓到的通乎教育的基礎上。

不過第三個理由終究是最關重要，只有專才，而沒有通識的人，是一個比較健全的社會與文化所瞧不起的人，而此種瞧不起的態度是很有理由的。上文提起過，此種掛一漏萬的專才或只鑽牛角尖而不識大道的專才是畸形的，是殘缺的，也就是不健全的。他只是一個匠人，一個楊子雲所稱的「伎」，一個師乙所自稱的「賤工」。工的所以淪於下賤，實際的中國文化發展，固然要負責任，但及其既經淪落，原有的比較完整的文化標準確乎也很有理由把他看作「不入流品」，「不上場面」！明代的文徵明，最初以單單畫家的資格入侍內廷，曾經受當時科甲出身的人，如狀元姚淶等的白眼，認為畫匠何得廁身士流，直到後來文氏自己也取中甲科，入了詞林，才出得這口氣。姚淶這一類人只用俗眼看人，固然可恨，但此種看法背後所依據的人才理想和文化標準，畢竟是未可厚非的。

如此說來，一個工的專才要教社會瞧得起，必須同時是一個富有通識的人。社會看重這樣一個專才，並不因為他有專識，而是因為他於通識之外，更有專長，於做人之外，又能做事。我認為我們對於民族文化不能控制則已，否則此種對於人才的看法，便應當在積極發揮的範圍之內。惟有這樣一個看法之下，也惟有中智以上的人都能照此看法努力，蔚爲人才，工的地位才有眞正提高的可能，而一經提高，不再有淪於下賤的危險。

但這第三個理由還有另一個方面，就是人才個人的旨趣方面，近年以來，工的專才教育，因爲提倡得力，確乎吸引了不少的智能很高的靑年，如果我們在若干大學裏舉辦一次智力測驗，我相信工學院的平均分數要比其他學院的爲高。工院學生有所寫作，詞理通達的程度往往不在其他院系學生之下，也是一個很好的證明。不過暫時的吸引是一回事，比較永久的維繫是又一回事。就目前的形勢說，我相信一部份習工的高材生遲早會脫離工的一路而別尋發展的方向的，在外國專才教育的歷史裏就有過這種情形。社會學大師斯賓塞爾是以鐵道工程師出身的，德國家位學派的社會學大家勒泊萊原是一位鑛冶工程師。美國優生學的領袖達文包，起初也是專習工程的，後來終於完全放棄工程，而轉入興趣較廣的人文生物學的領域。其他社會與人文科學的歷史裏，也大都可以找到這一類改絃易轍的第一流與第二流的人才。爲甚麼？大凡才能較高的人，學力所及，往往可以求通，也可以求專。其對於通的企求，大抵不在對於專的企求之下，且往往超出專的企求之上。如果所受的教育能通專互重，他自然能安於此種教育，一旦就業，也因爲興趣及準備並未過分受教育的限制之故，而安於其位；他往往能於專業之外，同時從事於一些和專業很不相干的學術上的活動。但如果所受的專門教育只是一種比較高深的技能的訓練，本身的範圍既十分狹窄，又沒有啓迪通識的學科做有力的襯託，則忍耐性到達一個限度的日子，也就是他幡然的變計，去而之他的日子，約言之，前一種教育，表面上不完全爲工的學術設想，實際上則適足以維繫工的人才；後一種的教育，卽目前所施行的教育，表面上十足的爲工的學術設想，實際則適足以驅遣工的人才。而這種人才的陸續引去又勢必教工的行業再度的淪於微賤。

總之，歷史的教訓，我們是不能忘記的。工的所以淪於微賤，就因爲他和通識的教育完全脫了聯繫，因而被擯在文化的洪流之外，音樂也復如此，戲劇的藝術也復如此。繪畫却是一個例外，而其所以成爲例外之故，就因爲凡屬成畫家，而不是畫匠的人，在通才方面，也都有相當的發展；畫家大抵能詩能文，甚至於不乏學術的興趣，經濟的長才。可以當官師，可以辦事業，而同時無害於其畫術的發展，不妨礙他的以畫起家，以畫名世，以至於以畫傳後。上文所提的文徵明就是最好的一個例子。換言之，如果士的教育等於通識的教育，那就任何人應當先受相當於其一般智力的士的教育，然後再從事於相當於其特殊才能的專門教育。士的教育是一個公分母，而農，工，商以至於各種藝術的教育是一些分子；目前的公分母是太不夠大了。我們誠能以這種見地作爲教育興革的張本，則一方面既不怕專才的成爲廢物，而另一方面，對於文、法、理、工等各種學科，卽使因時的需要，而偶有側重，也不至於發生甚麼過於偏蔽的影響。

第三個條件可以說是政治的或社會的。就是國家應積極獎勵多技能與善機巧的人才，這一層我無須多說，一則因理由最爲顯明，再則因實行比較容易，三則因政府及較大的公私企業機關已經逐漸注意及此。不多幾天以前雲南經濟委員會的一個工業組織，不就因爲工作勤愼與對所司機件愛護有加的關係，特別獎勵過一位汽車司機麼？

要提高工的地位，發展工一方面的文化，工的人才固然是最大的要素，而增加這要素的條件，已具如上述不過要素尚不止此，就人事方面說，機巧的才能而外，尙須組織的才能。近代工業的發展，有賴於組織能力的或許比有賴於機巧能力的還要多。中國民族的組織能力是薄弱的，至於如何薄弱，所以薄弱的原因，以及如何改變此種原因，使薄弱者復歸於比較健旺，我以前曾別有論列，無須再贅。還有一個要素是工業的資源。我們的資源不能算多，地大而物則不博，似乎已成爲一般的公論。不過大體說來，我們的資源總還夠我們比較長時期的支配，借了不很充裕的資源來啓發我們不大充裕的工的才力，於事亦尙公允；在短時期內，才力的施展雖不能太多，物資的消耗與狼藉亦宜似乎有限。此次戰爭結束以後，國際大概會規定一種資源上貿遷有無的辦法，此種辦法而成事實，則一旦我們的才力逐漸增加，其所需的更大的資源也就不患無從取給。所以我認爲這一點目前也無須多加討論。

最後還有不能不略加申說的一層。就是上文云云必須和講求品質的人口政策聯繫起來，才會發生効力，尤其應該注意的是，要愼防軒輊生育率的發生。如果我們不能嚴防這一點，而同時居然把工的地位擡得很高，那結果就無異明知故犯，與變本加厲的斷送了民族本質中也許蘊蓄着不太多的技工與機巧的才能。我說明知故犯，因爲我們，不比前人，已經懂得一些選擇的原理；我說變本加厲，因爲前代似乎不曾有過軒輊生育率的現象，而近年以來已大有發生的可能，而變本加厲的方式一種是速率的增加！我們的技巧才能，特別是高級的，本來怕就不多，經不起再度的沙汰，更經不起加速度的沙汰，是很顯然的。

從本篇冗長的討論裏，讀者可以看出，一個文化的問題，究其極也就是一個民族的問題，以至於一個人口的問題。一個民族發展文化，累積文化，及文化發展入某種途徑，累積到相當程度，他就會發生選擇的作用，轉而影響民族的人格，人口的品質，包括繼續創造文化的能力在內，約言之，民族與文化是互爲因果的兩個本體，我們要加以控制的話，決不能舉其一而遺其二，工的一方面如此，其他生活與事業的種種方面，也莫不如此。

工程技術與學理的諧合

——轉載思想與時代——

楊耀德

一　一個工程教育問題

工業建設之推進，有賴乎精深之研究，縝密之設計，與優良之技術，研究屬於學理方面而設計則須兼顧技術方面，故易言之，精深之學理與優秀之技術，乃發展工業之主要因素也，且工程學理與技術二者，譬如車之兩輪，鳥之兩翼，初未可厚此而薄彼也。況工程技術亦不外乎學理之切實運用，而學理研究之成效，正所以促技術之進步。是故學理與技術之調合配合，為工程學之要義，而如何能達到二者之諧合，實為工程教育上之一重要問題也。

吾國興辦工程教育，已歷相當年數，因工業建設之落後，工程學府所造就之專門人才，畢業之後，出而服務，未必能用其實學。又因實業不發達之故，規模宏大，設備新穎之製造工廠，在國內寥若晨星，工程學生缺乏觀摩考察之場所，凡教室之所講授，圖書雜誌之所描寫者，苦無比較印證之機會，以致學理與技術未能充分配合，而難以饜社會之所期望。然此並非吾國單獨所有之問題，即在歐美諸國亦無不感覺此問題之重要。譬如英國各大學工學院多施行錯綜教練制度，(Sandwich System)即學生修業期間，一部份須在工廠實地訓練，其餘部份則在學院攻習，如是錯綜分配，以期學理與技術能互相諧合。高德海博士（Dr. Elmer L. Carthell）在考察工程教育報告書中，引約克廈學院(Yorkshire College Leeds)之工程教育宗旨，謂「學院中之科學訓練，應視為獲得工程基本原理之法門，而工廠中之實地訓練則為學習工程技術之所必要」(The scientific training at the College must be regarded as a means of acquiring principles that underlie the art of engineering and the training in the works as necessary for acquiring the art itself) 在美國則如辛辛那鐵大學工學院(The College of Engineering of the University of Cincinnati)麻省理工大學(Masachusetts Institute of Technology)等亦施行合作制度（Co-operating plan）與各著名工廠合作，有類英國錯綜教練制度。負合作訓練之責者為協調員(Co-ordinator)其任務為協調學生每星期內在工廠中之實習技術與在大學中之理論智識。譬如某日下午，渠將在製造工廠中觀察學生工作，而在次星期內某日，渠將集合學生至教室之內，解釋學生在工廠中所作機件之効能。此種工程教育制度之優點所在，爲訓練青年手腦並用，使與工程實際問題接觸，明瞭工程實施程序與方法，熟悉生產機構之內容狀況，與工友之生活習慣等。要而言之，使教室中之工程知識，與工廠中之實際技術，打成一片，而達到充分諧合之目的也。但在實施此種制度時，必須學院與其附近工廠密切聯繫方可，而協調員所負之責任獨重耳。吾國過去工程教育，因未能達到學理與技術之充分諧合，以致有所學非所用之慨，有心人士乃注重職業教育以補其缺，然職業教育之旨，與工程教育未能盡同，蓋職業教育以訓練應用技能爲主，而工程教育之目的，則在發展分析與創作之能力，而理論與實際間之調和，實爲發展此能力之主要因素。故提倡職業教育，僅足以救社會一時之失，未可以與工程教育相提並論也。近年來吾國政府教育當局選派工學院畢業學生，赴國外各著名工廠實習，以期學理與技術互相印證。經濟部資源委員會與國立大學工學院合作，以研究工程上之實際問題。而大學工學院三年級學生，又往往利用暑期赴各大工廠實習，以補充技術方面之智識。凡此種種，皆足以表示吾國之工程教育，正在向光明之前途邁進，抗戰既勝之後，欲負起建國必成之重大使命，於吾國現階段之工程教育，實不得不寄以無限之期望者也。

二　工程教育與知行合一

工程教育有理論與實際，學理與技術兩方面之訓練，此固不獨現代工程教育如此而已。吾國古時之教育方針，早已如此。孔門心法，始之以博學，終之以力行，知行並重，其道一貫，王陽明承象山之學統，創知行合一之說，謂「行之明覺精察處便是知，知之眞切篤實處便是行」。工程學之理論屬於知的方面，而技術則屬於行的方面，故技術的明覺精察處莫非學理，而學理之眞切篤實處莫非技術，學理技術，打成一片，與知行合一之教，不謀而合。吾國古時六藝之教，理論與實際，學理與技術並重，故斯時所造就之人才，不但能知，而且能行。降及後世，章句之儒，斤斤於句讀之末，而不務實際。故明末學者，若顧亭林輩，講求經濟之學，以致用爲依歸，蓋鑒於當時學術陷於空疏之弊也。工程學之最大意義爲眞切篤實，不但注意於理論方面之知，而更着重於實際方面之行。蓋「知是行的主意，行是知的工夫，知是行之始，行是知之成」。「未有知而不行，知而不行，只是未知」。故工程教育之特色爲訓練手腦並用，不僅用腦，而且動手；即不但能知而且能行也。又工程師作工程計劃之時，必須考慮到實際方面的種種問題，如原料之供給，勞工之需要，機器之設備，交通之便利，成本之估計等，然後斟酌損益，折衷取舍，以制定一切實能行之方案；蓋理論雖深，而不離實際，此乃工程學之特點也。

吾國之工程教育，比較歐美列強，雖未免落後，但其一掃過去空疏玄虛之失，則誠未容忽視。若清季詹天佑氏之造平綏鐵路，穿山越難，工程十分艱巨，卽外邦人士之來觀光者，亦表示異常欽佩，而工程告成之速，乃在規定期限之先，其篤實力行之工夫已足以當第一流工程師而無愧色矣。其他在水利，化工等各方面，吾國工程師所表現之成績亦極可稱道，尤以抗戰以還，吾國工程界人士之服務於交通，兵工，製造等各方面者，莫不固守崗位，堅苦卓絕，在物質條件不利情形之下，埋頭苦幹，不辭勞瘁，而在彈片飛紛之下搶運搶修，效忠邦國，義無反顧，甚至犧牲生命而不惜。其得力於平昔眞切篤實之教育思想者，寧非淺鮮耶。

中庸稱「好學近乎知，力行近乎仁」，好學指學理方面，而力行指實際方面。考工程師之定義，爲「運用數理化與生物諸科學以及經濟學之知識，更濟之以從觀察實驗研究發明所得之結果，然後利用大自然質料與能力，以造福社會之人才」。（美國麻省理工大學校長康潑登氏所說）所謂運用科學知識，研究發明者，豈非近乎知耶。所謂利用質料與能力，以造福社會者，豈非近乎仁耶。世人有批評工程教育爲近乎功利主義者，若根據上述工程師之定義，以論工程教育之旨，則可知其與狹義之功利主義並不相同，惟太過於機械式之工程教育，則恐難免於買櫝還珠之憾，斯則不可不加以注意者也。

三　學理與技術之調和

學理與技術調和爲工程學之要旨，今請論永久磁鐵質料之研究，以明其義。關於永久磁鐵之質料，自居利夫人（Madame Curie）以來，世界各國工程科學人士，從事於研究試驗者，歷數十年而不衰。製磁鐵用之合金鋼料，以及養化物混合料等，種類頻多，日新月異。大抵在第一次歐戰以前，製永久磁鐵多用鎢鋼。歐戰中，鎢價騰貴，不易購致，經多方試驗之結果，乃採用鉻鋼，以代替鎢鋼。在一九一六年間，美國威司丁好司電機製造廠人士，卽試驗鈷鋼之性能，而知其爲優良之磁鐵質料。及一九二〇年（Honda）發表鈷鎢鉻合金鋼之優異磁鐵性能，名之曰KS鋼。此鋼所含鈷之成分甚高，達百分之三十至四十，鎢次之，又鉻更次之。至於不含鎢之鈷鉻合金鋼，亦爲此時歐洲人士所研究，迨一九三〇年前後，鎳鈷合金鋼，鎳鋁合金鋼，鎳鉬合金鋼，新式KS鋼，（含鈷鎳鈦合金元素）。以及養化物與鐵養化物之混合料等，各具特異之磁鐵性能，各國人士，試驗研究，成績昭然。惟關於永久磁鐵質料研究所得之數據，各廠家大都保守祕密，僅見一鱗半爪，難以窺其全豹。且磁性之學理，十分深奧，迄未臻充滿之地步。卽關於永久磁鐵性能之經驗的理論，自S.P.Thompson以後，各家之說不同，亦頗難得一定論。扼要言之，決定磁鐵性能之主要因素，爲頑磁性，矯頑磁力，與磁引度三者；而所謂能量乘積（Energy product）者，又爲判定磁鐵實在效率之要素也。

永久磁鐵之製成，全鋼料所含之合金元素，爲具有決定性之因素外，其加熱處理之方法亦極關重要，且可從加熱處理對於鋼料所發生之影響，而進一步研究合金元素對於磁鐵性能之效應。蓋鋼鐵當加熱或冷却之際，其內部組織發生變化，此種變化，在純鐵，炭素鋼，或合金鋼皆有特異之點。純鐵則結晶組織或作體中心立方，或作面中心立方。炭素鋼則鐵炭化物或溶解在鐵中，或分離而混合。合金鋼則合金元素往往延緩或停止某種內部組織之變化。各種組織皆具特異之性能，以決定磁鐵之特性。故已知鋼料之內部組織，即可預測此鋼料製成磁鐵後之性能如何。鋼料之加熱處理爲屬於技術方面之問題，而其內部組織之研究則屬於學理方面，從加熱處理，對於鋼料組織所發生之影響，以研究永久磁鐵之特性，此工程學理與技術互相調和之一例也。

上述關於永久磁鐵質料之研究，僅舉學理與技術調和之一個顯著的例，以明工程與技術學理相諧合之重要性而已；其他關於工程設計，製造，運用等各方面，亦莫不如是。且即關於純粹學理方面之研究，亦不能絕對脫離實際方面，蓋學理研究之證明，尚有賴於實驗也，況作學理上之研究，必先假定若干條件，以期便於分析，若取研究所得之結果，應用於解決實際問題時，尚須審察斟酌，未能膠柱鼓瑟，蓋實際問題之內容，難免與理論條件有出入故也，且技術方面又有不少經驗與訣竅，未易完全憑學理以得到圓滿之解決。故學理與技術二者，貴乎能和諧配合，庶幾解決各項工程問題時，能作確當之判斷，而定切實之方案也。

四　達到工程技術與學理諧合所經過之階段

欲達到工程技術與學理充分配合之境，誠未可一蹴而幾。大抵初習工程學者，往往對於理論方面容易感覺興趣，而對於技術方面則每較淡薄，但在此時期，對於理論之認識尚未眞切，數學物理之觀念尚未深刻，故關於理論方面之研討．公式之導出運用數據之分析等，亦尙未能眞知灼見也。因其對於理論方面未能切實認識，於是對於技術方面亦覺乾燥乏味，蓋比較機械性之技術，苟非明澈之理論烘托之，難免陷枯燥之感，此在初學者固未可苛求者耳。惟學理之傳授，與技術之訓練，須雙方並進，由淺入深，由近及遠，則學問基礎方能穩固，而濃厚之興趣，亦大半是從困知勉行中得來。故注意於理論方面，而對技術較淡薄，此可謂習工程學之第一階段也。迨經相當期間以後，因學理自學理，而技術自技術，則一套理論公式，數據曲線，變成索索無生氣之書本知識，於是漸起厭倦之意。蓋工程學並非若文學藝術之富有賞欣意味。故習工程學者，大都爲有志於工程事業之青年，初學者以學術基礎未固之故，既未能引起理論上濃厚興趣，而獻身於工程事業之志向則迄未稍衰，於是轉其興趣於實際技術方面。且工程學本有理論與實際二端，苟但講理論，而不務實際，則難免蹈於空虛之感。況各項工程創作，皆由聰明睿智之士，所運其巧思靈慧而發明之成績；若公輸作雲梯以攻城，武侯製木牛以運遠，亞幾米提(Archimedes)造守城之具，希臘(Hero)製蒸汽之輪，此乃中外所稱道之神工奇技也。及現代科學昌明之後，光電熱化各方面之技術創作，更有推陳出新，匠心獨運之妙。習工程學者日與此項新奇機械相接觸，不禁抵掌贊賞，莫能自已。故由學理方面之興趣轉至技術方面之興趣，此為習工程學之第二階段也。神奇之機械技巧，固足以引起學習者之興趣，但此類機械發明，出於創作者之戛戛獨造，非常人所能幾及，所謂創作天才，可望而不可即也。抑機械之發明，初視雖若新奇，究其所本者，亦無非學理上之原則。但若驚其新奇，嘆其神巧，而不務研究其學理上之根源，則於衷心並無所得，於是僅有贊嘆驚異之情，而未能發生真實永久之興趣。惟與此項新奇機械接觸以後，漸悟宇宙之祕造化之妙，理有未窮，知有不盡，恍然於昔日所習之學理，初未能收舉一反三之效。因而觀念漸明，思慮漸周，認識漸昇，見解漸透，而漸起仰之彌高，鑽之彌堅之感想。信學問之無窮，覺自然之偉大，乃復集中力量於理論方面，溫故欲以知新，博學欲以窮理。至於技術方面之興趣則似稍減退，並非真稍減退也，蓋正集中力量於學理方面，未能分心於技術也。因學理與技術尚未達充分諧合之境，欲窺其進一部而達到調和配合，舍更致力於學理方面外，並無他道也。故此時期之專力於學理方面者，非厚於理論而薄於實際也，乃在學理與技術尚未達到諧合以前，所必經之第三階段也。及經相當時期學理上鑽研後，學理與技術，理論與實際，漸能豁然貫通，打成一片，譬如烈火旺盛之餘，漸臻爐火純青之候。於是技術之所表演，無非學理之所旁通，理論之研究，無非實際上之妙諦。凡自然現象之所顯示者，皆研究學問中之資料；運其靈心慧眼，默契於中，目無全牛，胸有成竹，以言創作則水到渠成，以言發明則穎脫而出，若更進而研幾窮理，格物致知，明造化之微，參天人之際，學問至此境界，可謂已登高峯，此則最後成功之階段矣。

五　技而進於道

無論何種技術，苟神而明之，則將進於道。昔庖丁為文惠君解牛，文惠君曰：善哉，技蓋至此乎。庖丁對曰：臣之所好者道也，進乎技矣。臣以神遇，而不以目視，官知止而神欲行，依乎天理，批大郤，導大窾，因其固然，技經肯綮之未嘗，而況大軱乎。臣之刀十九年矣。而刀刃若新發於硎，彼節者有間，而刀刃者無厚，以無厚入有間，恢恢乎其於遊刃必有餘地矣。庖丁之神於解牛，若以現代科學觀念解釋之，蓋其對於牛之生理的知識，與解剖的技術，能充分諧合者也。故外之頭角蹄膊，內之五臟百骸件件有自然之腠理，了然於心目之間，信手所之，迎刃而解。當其始解牛時，所見無非全牛者，因其對於牛之生理組織，尚未明白認識，在眼前只是有一牛而已。三年之後，未皆見全牛，則積多年之解牛經驗，知識漸豐。技術漸進，所見者皆意中之天然節族，因其固然之理，而遊刃乎其間矣。凡初習工程學者，在眼前只見有一部機器。迨積多年之實地經驗，然後對於機器之內部結構，零件配合，功用性質等，一一明瞭，而不僅見機器之外表而已也。且機器之運行，必合乎科學原理，若根據原理則能觀察機器之各種變化，用科學方法分析之，以得出一定不易之因果規律，而工程技術乃愈趨進步。故刀者可喻科學工具，而節者可喻工程問題，以精密愼嚴，滴水不漏之科學工具，解析工程上之各項問題，恢恢乎綽有餘裕矣。又庖丁每至難解之處，則視止行遲，動力甚微，豁然已解，爲之躊躇滿志。蓋各種技術，必有困難之點，固不獨解牛如此，苟細細加以分析，必能得到適當之解決，而至高無上之樂趣，乃在困難問題之圓滿解答也。

莊子一書，善於言道者也，而庖丁解牛一節，描寫技之神妙，已入化境，一片天機，純任自然，工夫到此地步，欲不謂之進於道，不可得也。蓋技而進於道者，不執着於物質而能超然乎迹象之外也。故其胸襟開豁，思想超脫，無所粘滯，無所牽掛，爲學問而學問，爲創作而創作，不問成功順合自然。若是者庶幾近乎道矣。

莊子，天才也，而其文，至文也，若柳子厚之傳記文章，玲瓏剔透，趣味雋永，亦可稱爲能品矣。所作梓人傳一文，描寫梓人技術之精，高出儕輩。量棟宇之任，視木之能舉，度材而制其宜，是精於工程材料者也。指揮刀斤斧鋸之工，使各當其任，是善於領導組織者也。畫宮於堵，盈尺而曲盡其制，計其毫厘而構大廈，無進退焉，是精於打樣設計者也，宮室旣成，書其姓氏於上棟，是尙不得謂之進於道，若論其技，則建築工程之名手也，夫打樣設計，度材制宜，非學理與技術互相調和者，不足以勝任而愉快，若梓人者，雖曰末學，吾必謂之學焉。

歐美各著名製造公司，莫不致力於工程科學之研究，其藏書樓卷帙之豐富，研究所設備之完善，不啻一最高學術研究機關也。禮聘學識優良經驗豐富之專家，俾悉心從事於研究工作，每歲所支出之研究費，恆佔總支出中相當百分數量，蓋技術隨學理而進步，苟非繼續的研究，在製造工業將故步自封，難期發展故也。其研究者，固多屬於各該公司製造方面之各項工程問題，但亦並不限於固定範圍之內。往往在純粹理論方面，似乎與出品製造並無直接關係者，亦不惜費腦力財力以研究之。蓋研究之目的在卽物而窮其理，理爲自然之法則，一定而不易者。自然之法則旣明，則事事物物各得其當，雖與出品製造似無直接關係。而推究其極，則出品製造亦卽在事事物物之中。故製造偏於技術者也，而研究之目的則在窮理，理愈明斯技術愈精。觀歐美各著名工廠對於學術研究之興趣，不禁盟然嘆曰，技而進於道矣。

工務行政

張維翰

——轉載工程第五卷第五期——

抗戰到了今天，已經五年，我們的工程師，在這五年之間走遍了後方的城市和鄉村，荒山和遠水，篳路藍縷，慘淡經營，在艱難辛苦中，負起建設大後方的責任，貢獻之大，可與前方將士媲美，幾年前深山絕澗人跡罕到的地方，現在公路暢通，航行無阻，幾年前的荒山荒地，現在變成滿山的林木，滿地的田畝，我們以前在後方所缺乏的物資，我們的工程師，一一都迅速的經營起來，電燈廠、水泥廠、鋼鐵廠、機器廠、紡織廠、火柴廠、麵粉廠、煤礦廠、油礦廠，一一都辦得成績斐然，這確是值得我們大家慶幸的，因為這是一種有意義的象徵，象徵了中國一切都已見到了曙光，光明的前程，就在眼前。

今天本人獻詞要順便向各位報告的，是關於內務行政中的幾種工務建設問題，簡單說，就是工務行政，工務行政，在歐美各國，早成為一個專門名詞，舉凡都市鄉村，物質環境的改造，例如公私建築、公園、廣場、道路、橋樑、溝渠、堤岸、鐵道、港灣，以及其他公用事業工程，大都包括在內，到過歐美各國的，首先使人注意者，是他們都市的整齊美觀，他們鄉村的清潔秀麗，反觀我國，則城市的建設簡陋，鄉村的穢污不堪，令人感覺十分慚愧，無怪歐美人士，不能了解我們，推原其故，實由於我國對於工務行政，往往未能加以注意，對於市政工程，不能作有計劃的設施，同時這種行政，在我國又是在草創的時期，技術與財力，同樣感到困難，所以近十餘年來的成績，未能令人滿意。

到抗戰已經五年的中國，可說完全改變了敵人對我們的建設，已經破壞無遺，立體式的戰爭，使我們後方的都市城鎮，和交通要地，均同樣遭受摧毀，損失之大，固不待言，但是一切建設的障礙也同時被其消除，未來的建設，差不多寫在白紙上計劃，更容易達到建造理想城市的目的，當然在戰爭猛烈展開的今日，不能像平時一樣，可以撥出巨量的人力物力，從事於都市的建設，然而我們到處所見到的破瓦殘垣，不斷的提醒我們必須要努力準備戰後復興計劃，並且應該利用此次抗戰所得的寶貴教訓。去準備能夠適應未來戰爭條件的復興計劃，亦就是說具備國防條件的復興計劃，總裁在去年致中國工程師學會第十屆年會的訓詞中，曾指示我們「吾人當前努力之二大目標，於抗戰必爭最後勝利，於建國則必須國防絕對安全」，所以一切的建設，必定以國防為先決條件，同時在一面抗戰，一面建國的大原則之下，我們亦應當研究如何用最低限度的物力財力，去完成適應戰時需要的種種建設工程，如此方能顧到現在，準備將來，英國管理公共事業大臣里茲勛爵士，不久以前曾說過，「有計劃有秩序的建設觀念，是對於作戰努力的推動和鼓勵」，這高瞻遠眺的結論，確是由痛苦中體念出來，値得我們效法。

爲求加强工務行政的效率，爲求貫澈上述的主張，爲求業務上的開展，內政部在本年七月一日奉准增設營繕司，掌管全國的建築行政，都市與鄉村的建設計劃，和一般土木與市政工程，這種事業的範圍旣廣，又非短時期所可收效，所以營建司今後的中心工作，可分爲下面列舉之幾項：一、充實主管營建的機構，二、制定都市鄉村復興計劃的方案，三、實施公私建築管理，四、規定建築材料標準，五、推行有效的住宅政策，六、改進公用事業，至其他可舉的業務還很多，上面所說的不過其犖犖大者而已。

本屆工程師聯合年會，遠近各地的會員，都聚會在一堂，所以本人趁此機會，對於內政部主管的工務行政，向各位作一個簡單報告，講到工程行政，兄弟附帶有一個感想，亦可說是一點意見，向各位貢獻，作一種參考。

以前往往有人以爲工程與行政是兩件事，專學工程的人，每每不長於行政，因之遇事常感困難，專門行政的人，對工程又是外行，因之工程行政與工程事業中的利弊得失，與種種問題能不能作一個正確判斷，結果亦是失敗，這種情形現尙存在，不能否認，工程界確於抗戰建國，站在極重要地位的，今天這問題，實在値得我們注意的，並且對於這種情形，今後似應加以補救。

工程事業，當然由工程師來主持，這是天經地義毫無疑問的，也是一般工程先進國家的慣例，因爲工程是專門學識，是多年研究的心得，然而要主持工程事業，單有專門學識，覺得是不够的，還要具有行政的能力，行政並非一種專門學識，而是各種學問的融匯貫通，亦可以說是廣泛的常識，這種常識的求得，本人認爲有兩點必須做到：

一．工程人才的教育，常識之求得，要在學校課程中先植其基，英美教育家，對於廣泛教育的重視，使研究專門學識的人，必須得到一般普通知識，就是着眼在此，所以今後大學與專科學校的必修科，對於工程以外的科學，如文科、法科、商科的課程，以及其他各種有關社會科學，也要加以注重，總裁昭示我們，「必須更推宏各級工程人才之培育」，我們應當作如此的解釋。

二．工程人才的訓練，工程是最系統的科學，所以工程師的頭腦，往往比較一般人淸晰，工程師的毅力，也往往比一般人堅强，然而想綜覽全局，從大處着眼，則必定在事業上，求得不僅一方面，而是各方面的訓練，方能達到目的，以工程師有系統的頭腦，去自行研討，更可具備豐富的常識，工程師的行政能力，必較一般人爲優，應爲公認的事實。

戰後建都問題

傅孟真

——轉載三十二年十一月二十九日大公報——

我一向總覺着我們的都城之在南京，是沒有問題的。因爲遷都是大事，南京又有他的長處，況且他有國民黨建國的象徵意義——總理陵園在那裏。然而近來大公報上常有討論建都地點的文章，而且胡秋原先生之建都長春說，似詭而正，使我心中發出了平日潛伏着的若干思想，現在拉雜寫下，就正於留心此問題的人。

在討論這個問題之前，我們心中先要摒除兩個不自覺的錯誤，第一個是「發懷古之幽情」。我們這個歷史長的民族，有這個「幽情」是很自然的，然而爲建設一個近代化的國家起見。這「幽情」有時是很危險的。文化如此，建都亦然。歷史上的偉大時代，都有他那一個時代的問題，時代變了，問題也變了。即如漢高祖之忽然放棄洛陽而向長安，是應付當時內外問題的一個大手段，決非有所愛於終南渭水的風水。何爲當時的內問題？當時的情形，居關東有不易控制關西的危險，居關西而去函關大梁間布置着兵站與大庫（即敖倉），修好了道路，却可以控制關東。何爲當時的外問題？當時的匈奴可以隨時到渭水北岸，其建牙之所，雖正對大同，然河套陝北在地形上最便於胡騎侵入。當時定都長安的故事，大致是如下面所說的：陳涉起兵後，項羽敗章邯後，關東的「革命軍」都羨慕秦土那塊肥地方，與其所說是羨慕秦土肥沃，毋寧說是羨慕秦土聚積了無量的掠奪品，所謂子女玉帛者。所以義帝才有「先入關者王之」之約。項羽看不到秦土之重要，分給三個月不能自保的將，白白的爲劉邦之資。劉邦也是一個「富貴歸故鄉」的人物，所以才有可笑的大風歌，不過在建都大事上，纔能以理智克服情感。先擇好了洛陽，當時的一個大商業城，正是天下之中心。婁敬忽然提醒了長安之重要，他心中活動了，一問張良，張良雖爲韓世家，却大贊成長安說。劉邦彷彿如夢初醒，立刻駕着車奔赴灰燼的咸陽城，歡喜得結果便把姓婁的改作姓劉（可笑），封他作奉春君。這事張良不先說，大約項羽一死張良的魂魄就便隨赤松子去了！這一幕戲，從美術上說，不少俗氣，從國策上說，極關重要。從此奠定了漢家的天下。如此說來，彼一時代對此一時代，彼時代之內外問題對此時代之內外問題，今日若重演「畢中國襲之長安」那一幕戲，自然應該以長春或瀋陽爲都，至少是北平。胡秋原先生眞是絕頂聰明人，眞是讀史得間者。那些主張遷西安者，大多以歷史爲根據，殊不知當年都西安之根據，即今日都東北之根據也。

第二個要摒除的錯誤，是全國中心說。假如建國必在中心，蘇聯應遷烏拉山東，美國應遷芝加哥，這絕不是必要的。誠然以今天空戰之發達，都城太近邊境，自有大不便處，然必在中心也無必要。決定建都適宜的因素，另有所在。

討論這問題，我們先要認清幾件事實。

第一，中國不是一個工業化的國家，而必須在最近迅速走上工業化之路，若是，則建都最適宜的地方，應當是可為最大工業區域的中心。否則工業化的重心不在國都之四週，便可另成一個經濟重心，這個經濟重心便可變成一個政治重心。以二十餘年前的情形論，北平管不到上海，即以國民政府建置在南京論，上海自有影響南京處。這些年，國家進步多了。金陵控制北方，遠不如從前之難。然而一旦東三省北七省工業化起來，農牧改進起來，經濟的重心自然在北不在南。還有一事，勞工是今後的一個大問題，勞工的力量必隨工業化而進展，於是工業化的區域，自然要產出他的政治力量來，這個趨勢在實行民生主義時更要表現出來的。

第二，中國在文教上確是一個融合統一的民族，然而南北各地之地域性也不算不發達。所謂「省界」一種感覺，仍是多數人下意識中一個原動力。北方諸省人，心中有一個「北人」的自覺，是明顯的事實。我自己是一個北方人，以讀書服務之環境無多地域性，所以朋友以南人為多。但仍常聽到北方朋友或相識對於「北人落伍」之嗟嘆，此為非其罪而落伍也。即我自己，除去對於自己的家鄉時常懸念外，一切地域感覺算是洗刷得夠清楚的了，然而每次蔣夢麟先生開玩笑，說一故事，「你那個賣螃蟹的同鄉說你是南方人」，心中總多多少少發點氣，每試每中。新教育之發達，在南方比在北方先，與外國接觸也如此，加以明清兩代長江流域經濟與人文都比黃河流域發達，遂形成了今天北人稍微落後的事實。這個事實，反而助長北人之地域感覺。如此說來，若把政府放在南方，北人的地域性可以發展下去，若把政府放在北方，南人的地域性不會發展下去，因為今後幾十年中任何政府，總不免南人佔絕大多數。在南則北人或以為「他是他」，在北則南人總以為「他是我」。南方尚如此，何況關內與關外，清朝末年，東三省對內的同心力很強，然自民國初年張氏步公孫氏之後塵，東北同胞之地域自覺，恐怕因政治影響，未必即走衰落的一條路罷。

第三，在空戰發達之前，與國的都城，每每接近邊境，只有苟安的國家才把都城放在中心點。前一項的例子，有漢唐之都長安，明成祖之都北平；後一項的例子，有東漢北宋之都汴洛。空軍發達，這個例子是稍稍改變了，都城不可太接近邊境，然而仍舊需要控制着形勝之地。「都城應建設在全國軍略上最要害之處，」這是我們必須守着的格言。試看今後全國軍略上最要害之處在何一方，這可不問而知。

在防海嗎？誠然，日本問題不徹底解決，我們的國防最要緊的是海防，但是，如果日本問題不徹底解決，而建國的工作亦無從說起，何況建都？所以一切討論，皆以日本消滅其大海軍大陸軍為前提，否則一切不必說了。日本既無大海軍，則試問海波自那一陣風吹來？美國？上帝不許我們想到美國是我們假想的敵人，而不是我們永久的朋友！英國？英國是聰明人，今後決不再作領土冒險了。其實過去一百年之中英糾紛，都是「經濟發展」「與人比賽」兩個觀念為動力，英國從未曾在遠東發展過領土慾。那麼，一百年的「海禍」，將以鴉片始，以「抗戰」終。即此一點，已足證明由南京遷武漢一說之無意義了。然而在陸地上，這問題決不如此簡單，張開歐亞地圖一看，為之駭然！我們若以蘇聯為假想敵，可謂至愚。我們的外交政策，應該是不與鄰邦起任何糾紛的。但是，強大的蘇聯，與我們工業化的基礎地域接壤，這個事實使得我們更該趕快工業化這個區域。我們的頭腦，理當放在與我們接壤最多的友邦之旁，否則有幾為頑冥之慮。大凡兩個國家，接觸近，轉易維持和平，接觸遠，可由忽略而無事生事。遠例如宋金，本是盟國，以不接頭而生侵略。近例如黑龍江之役，假如當時政府在北平，或者對那事注意要多些罷。再就日本說，這亦必須解決他的大海軍大陸軍，固為一切之前提，然而滅亡了他也是做不到，不該做的。他既不滅亡而保持其本土，則以小鬼的脾氣論，二三十年後必又來生事。到那時候，仍舊以朝鮮為跳板，仍以延吉清津一道為從華最方便之路。所以都城在北方，仍是防倭之要着，在南方，則時移世異矣。

如此說來，中華民國首都之應在北平，似乎沒有多問題罷？北平以交通發達之故，可以控制東三省，長城北三省。其地恰當東三省，長城北三省（熱河，察哈爾，綏遠），北四省（冀豫魯晉）共十省（下文簡稱北十省）之大工業農牧圈之中心。

這個十省大工業農牧圈，是中華民國建國的大本錢。有這十省，我們的資源尚不及蘇美與大英帝國，沒有這十省，我們決做不了一等國家，決趕不上法德，只比意大利好些而已。試看下列幾種物品，煤、鐵、棉、麥、大豆、鹽、羊毛，在這十省之出產最及可能出產量，佔全國百分之幾？若說都集中在這十省，也不為過。西北煤油之希望甚大，西南各雜鐵也是國防所必需的，然而比起北十省來，不免如四肢之比本幹了。中國地大而物不盡博，煤在北方充足，鐵則不足，雖不足，猶比南省多得多，鹽之一項，長蘆出塲費比自流井便宜十幾倍，所以西南的鹽鑛，只能以特殊原因保持着，若用自由貿易法，決難得存在。若論農業，則東南西南山區中費力多而成功少，所用方法，與其說是農業，毋寧說是園藝，北十省中，農業百分之八十可以機械化。從飛機上看江浙農田，眞是錦繡山河，若待北十省建設起來，江浙必爲窮省，因爲絲業已無多年之壽命，種稻又以天氣之故，其成本決比不過安南暹羅輕。

中國之資源既集中在北十省，而這北十省又常在危險中，我們便該建都在他的中心點去——北平。

此外還有幾點長處，建都北平然後有之：

一、天氣。中國的都城至少須達一千萬人。這一千萬人，若因天氣有三個月不能工作，則無形減少工作效能四分之一，死亡提早尙不在內。南京武漢兩地，尤其是武漢，夏天太熱，無形中減少工作效能一月至兩月不等。冬天雖比北方暖，然而煤貴，無普遍溫室之習慣，於是走路提着熱水袋；在家帶着手套，無形中減少工作效能不少。北平的冬天是最可愛的，煤既便宜到極度，無論貧賤，家家燒煤，所以在屋子裏如過春日，在屋子外便可活潑了。北平的夏天，有時也熱，但晚上總睡得着覺。若厭北平熱，則一過古北口，不久便是木蘭，即所謂熱河行宮者，火車當日可以來回，汽車也可以昨往今返。在木蘭爲政府夏日辦公處，比牯嶺之於南京，方便何止一百倍？這樣，無形中工作效能大大增加了，即是人命延長了，精神不浪費了。

二、現成都市。戰後建設萬耑待辦，請問我們的資本何自來？借債是有限的，自力更生要吃大苦的。試看蘇聯之建設，老百姓在生活上曾出了多麼高的代價，或者在最初餓死了多少人？我們的物質憑藉，遠不如蘇聯，加以此次戰事殘破範圍之大，我們戰後集中力量在生產上，斷不當以建都城爲第一義。「大興土木」，在歷史上本是亡國條件之一，這原則今天仍然有效。北平是個現成的都會，其可容人口之數，比南京爲多，當作都城目下即可應用。其必要之建設，可在舊城之外，待第一第四次五年計劃辦好後。也許日本人撤退時大毀而去，但修補舊城市總比造新城市容易。

三、有「海口」。這話初看似乎怪了，北平連河都沒有，何以說有海口？我們不要忘，運河在當年本是由杭州直通北京城門下的。所謂二閘，我幼年還常去玩耍。永定河白河問題。是容易早解決的。天津之淤塞，實在人事上太不講了，原不是難辦的。把這一區河渠闢整好，修一條寬渠，小輪船可以到北京。若在冬天，秦皇島距北平不太遠，那是一個不凍港。這話是說，北平接近海運，這也是建都的一要點。

四、練兵方便。這却是極重要的一點。今後工業化和建軍，本來是一件事。以中國社會之形態論，以近代化軍隊之須集中訓練論，當年俾士麥面告李鴻章的那個辦法，就是集中在首都左近練兵，而各地道路修好，仍然是適用的。若以北平四圍各一千五百里爲練兵集中之地，各種地形，幾乎應有盡有。地廣原不必說，山則是眞正的大山，可以演習隘口爭奪戰，可在大山裏面建飛機場，因爲大山與平地，多是直接相連的，戈壁上演習坦克，沙漠中演習進軍，渡永定河之爭奪戰，湖沼戰鬥，（勝芳一帶湖沼甚多）雪中戰鬥，夏秋大雨中戰鬥，皆可出盡其妙。只缺少熱帶森林戰之演習場，與中國南方小丘陵區域之地形而已，至於南京武漢，只有稻田與小山兩種地形，大規模的飛機場已不易尋，若鑿山的飛機場更爲難能。我現在遐想未來的強大中國，其「邦畿千里」之中，大工廠、集體農場、練兵場、飛機廠，錯綜着成一幅錦繡圖案，這樣近代文化的偉大的美麗世界，比起那故宮建築，江南風景，後者算得什麼呢？

有人問我，北平作都城，與建海軍之影響如何？我想這是沒有什麼關係的。近代的國家，不是上古與中世的城國（City-States），不需以海軍基地爲建都之條件。說到這裏，我們要想想我們將來的海軍是怎樣的形態。我想，我們永不以侵略爲主義，永不爭霸大洋，大海軍是不必要的，只是一個輕型護海艦隊，已經夠我們今後一二十年担負的了。這艦隊中，要有兩個大巡洋艦，專爲每年訪問華僑與友邦之用。其主力悉爲潛水艦、驅逐艦、輕快巡洋艦，至於主力艦與航空母艦，我們根本不需要。沿海應該建設些海軍要塞，而旅大與威海建設好了，渤海便是一個中國湖。再加朝鮮必然永遠是我們的盟邦，渤海中也有二三不凍港，北平之「海上安全」，比南京好多了。

北平作都城，只有一個大毛病，就是離內外蒙古交界處，比較還是太近些。這個地方，我指錫林卓布盟與車臣汗交界處——一片大戈壁——而言。我們不要在實際問題中幻想我們的邊疆在外蒙古之北的買賣城呵。不過這話又說回來了，我們的今後外交，應該以協和四鄰爲主義，而且我們既住在強鄰之旁，不更可清醒些麼？

再說對於其他建都說的意見。

武漢，我以爲最無建都價值的是武漢，持此說者，每以天下之中爲言，而建都在天下之中者，總是心中包括着一個對四面八方國內外都害怕的心理，這先要不得。且看造天下之中一說之老祖宗如甸說的，他說：「有德易以昌，無德易以亡。」我們只看到東漢北宋遷洛陽汴京，（即所謂天下之中者）易以亡，未看到他如何易以昌。至於漢唐之都秦，明之改都燕，在當時都是建都邊塞。還有一件事，很可以形容邊塞建都之意義。明成祖改都北平後，在仁宣英三朝，時有改回南京的意思，所以南北各部的印，時有改換，忽而此間加「行在」，忽而彼處加「南京」。但是，這個「復元」主義，到瓦拉之寇，英宗被虜，便無人提了，北京雖然圖得那樣危險，而當時及後來的朝廷，也就從此知道北京必爲京都了。獨怪黃黎洲，他是明朝人，應知本朝事，何以重責明成祖之遷都北平？其理由是「河朔人物，久已不及吳會。」殊不知成祖若不都北平，到英宗時已經半壁江山了，何待努兒哈赤起來？復次，武漢並非天下之中心，於是持武漢說者，又有人口集中說，殊不知天下最無法搬動的是地形和資源，最容易搬動的是人。

西安，西安是個在將來可以發展的都市，天的賦予他很雄麗。不過，按以近代大國國都的條件，還缺少很多。第一•吃水先有問題，渭河在冬天，吸涸了他，也未必夠一個千萬人的大都市之用。至於鑿井之法，本不是大都市所能用，況且在西安必須鑿得極深，即等於用時耗費電力很多。第二•燃料無法解決，米麥尚可運來，若一千萬人的一切燃料都須自遠方來，其生活程度必然在全國經濟中成一毒素了。至於建築資料，也只能靠隴海一條鐵路運，這也是不了之局。其實今日「開發西北」之說，毋寧改為「救甦西北」，除鑛業外，所有農牧的改進，其最大前提是使得山可生樹，地可長草，其辦法則是調理水道，減低冲刷。這麼開發之效，要在數十年之後了。唐代涇渭礬地畝數已遠比漢代為少，到今天，自西安一渡渭橋，面有沙漠之感。所有漢朝陵墓痛地，在當年都是複道蟠閣，在今天我有一次去看，時值五月之末，草還未綠呢。季元鼎先生對我說，他小時，九十兩銀子可以買一所像樣四合房之木料，到今天，這樣木料直無從買去。若一切仰仗隴海路，則我們須知鐵路運輸比起海道運輸來真不知貴多少倍。民國十五年，我有二三十隻書箱自柏林運上海，由柏林至漢堡的鐵路運費，遠比由漢堡至上海的船運費為高！當時雖德國的鐵路有賠款担負，然而德國船走蘇彝士河也有特別担負。所以我們的都城若離水路到達的港口很遠，國民經濟是大受損失的。今日國都之要求，何止漢唐時代之幾十倍？然而今日之西北，又止當漢唐時代幾分之一了•

長春瀋陽。這說本是北平說之偏鋒文字，我既主張北平，可以不論。

南京，其實南京也有不少的好處，他有北平沒有的條件，其地位僅次於北平。國都應該自南京遷北平與否，完全看我們今後立國的決心如何。若照東漢安樂主義的辦法，便在南京住下好了；若有西漢開國的魄力，把都城放有邊塞上，還是到北平去。不過，不求安樂者，子孫有時可以得到安樂；求安樂者，每不得安樂。個人國家，皆是如此的。南京的長處我不多說，因為這篇文本為北平說張軍，而且說得已經太多了。

寫完，友人看了問我，「你想，討論這問題有用嗎？難道你覺得這問題值得最先討論嗎？你不是說，反攻第一，收復失地第一，而笑人家談戰後事嗎？」我只笑而不言。而我這位朋友問得不放鬆，我說，「你把這篇題目改作『如夢令』好了。」

中國之中樞區域與首都

——轉載現代農民第六卷第十一期——

沙學浚

一 引言

抗戰勝利後，中國最大的任務是建國。建國有兩大前提，第一是失地收復，全國統一；第二是全世界尤其太平洋上要有至少三十年的和平。筆者相信這兩大前提都可實現，至少本文是根據這樣的相信而立論。因此，首都之選定應注重便於領導建國，而不是便于平定內亂，更不是準備不久又將爆發的二次抗戰。

在選定國都地點時，視線應射得遠些，空間上要密切注意全世界尤其列強歷史發展之趨向，及其對中國之關係與影響，不要只看中國。時間上，要想像並理想着中國與世界在戰後及三五十年後各是甚麼樣子，如中國工業化成功了多少，友敵關係是否與今日完全相同，不可只看現在。另外還有三個不可。

第一，不可完全根據抗戰教訓，提出海洋可怕，內地安全的退縮政策。抗戰是遠東史與世界史八十年來發展之自由結果，如果日本這次戰敗，很少可能在二三十年內複演這段歷史。抗戰教訓可比病床經驗，建國工作則是運動場上的活動，不能相提並論。

第二，不可用形勢完固的地理觀點考慮首都之安全。安全自然是建都條件之一，但首都之安全繫於全國之安全，而全國之安全繫於國力充沛。如果國力太薄弱，國防無辦法，首都雖深藏於「天下之奧區」，其中恐亦難有「出路」。談首都安全而念念不忘「寇深矣」，兵臨城下，或天險可恃，實在是太軟弱太悲觀的看法。

第三，不可完全根據顧祖禹東控西聯，南阻北接的地理學說及作分作合，大陸發展爲中心的中國歷史來觀察，認識現代的新中國，亦即行將現代化的統一的新中國。歷史是演進的，理解並理想新中國的未來歷史發展，要有新的史地眼光，新的理論根據。

建都問題是一個政治地理的問題。政治地理學的任務，簡單的講在于研究政治權力之分布與地理環境之關係。因此研究建都問題，除開歷史地理（包括氣候地形，經濟交通，聚落民族等項目）兩大因素外，尚須考慮到與權力很有關係的國策與力源兩大因素。

（一）國策　國內外的形勢謂之國勢，根據當前國勢及最近將來的可能變化而確定的立國方針謂之國策。有些國家的首都常隨國策的變化而移徙。俄國彼得大帝建都彼得堡，戰後革命政府遷都莫斯科，均爲國策所決定，前者爲接近海洋接近西歐，後者爲避免威脅（西方領土縮小之結果）建設內部。戰後土耳其遷都安哥拉，與後一點相同。南宋之遷都臨安，亦形勢使然，不得不爾。

（二）力源（Basis 亦譯策源）　借用自克勞什維茲之戰爭論，在本文裏表示一國或一個政治勢力的首都之選定，主要着眼於力量策源地所在之區域，首都建於該區之中央或其不遠之附近，不但感覺安全，而且便於接應與運用。此例甚多，宋之都汴梁，明太祖（起於濠泗）都金陵，孫總統都南京，袁世凱都北平，十七年國府都南京均是，甚至項羽欲東歸都彭城，或亦據此理由。

（三）歷史　美國獨立後建都華盛頓，乃地理所決定，在當時的十三邦，地位適中，以後不他移，可視爲「歷史決定的」，並非華盛頓以外便找不到另一適於建都或更適於建都的地點。秦漢隋唐先後都關中，地理與歷史均有決定的力量。

（四）地理　法都巴黎，英都倫敦，義都羅馬，印度都德里，蒙古高原諸民族先後都和林或庫倫，都是地理決定的，因爲此外便找不到適於建都的地點。無論強弱勝敗絕不遷都。

本文根據歷史與地理兩個因素，確定新首都應在何區域，再就國策與力源兩個因素，確定新首都應在何都市。

二　中樞區域

首都是全國的首都，未講首都，先談全國。全國之範圍不當只指海棠葉狀的領土及與之相聯的領海，而當包括黃海（渤海是其一部分）東海南海北部這三個緣海，共約三百萬方公里的面積。他們是太平洋的緣海（Randmeer）在政治地理上是中國的海疆，中國的緣海。緣海是公海，與中國的生存與安全關係甚密，是中國的生活領域之一部。因爲他們不但是經濟空間，交通空間，聚落空間（緣海中有三千數百島嶼），而且是國防空間。東西沙羣島及台灣琉球收回後，緣海與中國之關係自然更加密切。

海棠葉狀的中國領土分爲兩部，（一）爲邊疆，指蒙古高原，新疆省，青康藏高源以及鄂爾多斯，四川雲南兩省之西北角。餘下來的是（二）腹裏（借用元史「中書省統山東西河北之地謂之腹裏」而擴大其意義與範圍）指東北及舊本部十八省及新設各省之精華區域。兩區之地理環境，民族分布，人民生活不同之點甚多，茲不備述。面積比較，兩區各爲五百餘萬方里，而人口比較，腹裏佔全國百分之九十七，邊疆僅占百分之三。

中國之重心在腹裏，腹裏之存亡即中國之存亡，但欲保障腹裏，必須西北與邊疆爲唇齒，東南控緣海作屏藩，三大區互相依存，不可缺一，而腹裏尤不可缺。

腹裏依地理環境言，分為兩部，南方與北方。

腹裏又依空間價值及地位價值言，分為兩部，即中樞區域與環拱區域，此為本文所特別注意者。

中樞區域即昔日中原之擴大與延長，北邊擴展到宣化盆地（本為河北省之一部，後劃入察哈爾省）至榆關之線，南邊擴展到兩湖與江南，自浙贛路沿線西包長沙常德宜昌襄樊均在其內。再由榆關作一直線到長江江口，由宣化作一直線，經太行山，而到老河口，分別與南邊線相接。此一梯形區域可稱為大中原或現代中原，惟仍以稱為中樞區域（Zentrallandschaft）（原可謂為中央區域，但恐引起幾何學的意義，故改譯，俾合實在意義）為宜，因此乃政治地理學上之專門術語，其範圍與人文地理上所謂心臟地帶或核心區域大致相當，有時不完全符合。

此區面積佔全國十分之一，腹裏五分之一，然量其富什居五六，人口，耕地，農產亦占全國之半，有煤有鐵及其他鑛產，工業則佔百分之八十以上，全國八大都市有五個在此。總之，全國之精華在此，全國之生命力大部在此。

中樞區域好比「中國之大廳」（德國地理學泰斗 Richthofen 語）被八個房闥所環繞，這便是八個區域，就其對中樞區域之地位言，姑稱之為環拱區域，即是（一）東北四省（二）內蒙（三）山西（四）陝甘（五）四川（六）雲貴（七）五嶺南北（指兩廣及湘贛兩省之南半部）（八）東南沿海（指浙東及福建）。八區總面積佔全國十分之四，腹裏五分之四，人口約佔全國或腹裏二分之一。雖有平原，而面積不廣，雖有鑛產，開採較少，雖有工業，不夠發達，尤其重要的是八大區域彼此間之交通，聯絡都不甚便，而各區對中樞區域之交通不但是最重要，而且是很方便，有百川分流，朝宗於海之勢，此乃地位使然，非人力所強制。

中國歷史地理上的重要門戶與通過地帶多數在中樞區域與八大環拱區域之間或其附近，榆關（即山海關）居庸（古稱軍都）東西並峙，其間關口尚多，不遑枚舉。晉冀之間有太行八陘，今日惟平型、井陘、天井三處最稱重要，函谷武關為關中通中原之隘口，襄樊三峽，乃巴蜀漢中往東南之要津，「鎮遠者雲貴之門戶」（明周瑛語），「常德府……滇黔之咽喉」（顧祖禹語），南嶺亦稱五嶺，以桂林摺嶺梅嶺為要隘，衡陽贛州是其哨站，「仙霞嶺兩浙衿束，八閩之咽喉」，（顧祖禹語）附近有六嶺五關之阻，杉關為浙贛往來之要道，亦稱重險。

中國史上多數重要戰爭一部分發生於中樞區域，一部分發生於上述諸門戶之外。中樞區域之戰爭自然是常具決定性的，而爭門戶之戰爭目的在於進窺中樞區域，或由中樞區域佔取環拱區域，亦至重要。

總之，歷史上，地理上，空間價值上，地位價值上，中樞區域爲中國之重要之區域。欲統治中國必先統治腹裏，欲統治腹裏必先統治中樞區域。首都爲國家生活之指導中心政治力量之策動源泉，自必於中樞區域中求之。

英國地理學家柯立西（Vaugtan Cornish）著主要都會（Great Capitals）一書，認爲建都之條件有三：即叉路口（Crossway），堡壘（Fortress），與穀倉（Granary），意圖交通便利，形勢險固，與農產豐饒，或供應充足。就這三個條件言，南京武漢北平均有建都之價值，雖然有質與量的差異。

英國史學家唐比（Toynbee）近著「歷史研究」一書，倡挑戰與反應學說，認爲古今建都常擇外患威脅最大之地如北平，如德里。

美國軍事學家馬罕（Mahan）著「海權及其對於歷史之影響」一書，結論是古今很多國家之興衰存亡繫於海權的強弱得失，海權爲立國之本，在近代歷史表現尤爲顯著。

根據這兩種理論，中國首都須在距海不遠之地，俾便於發展海權，便於應付海上威脅。南京與北平符合此要求，適於建都，武漢深藏內地，適於建立工業中心，但不適於作新中國之首都。筆者主張戰後中國最好遷都北平，其次還都南京。

三　還都南京之理由

曠觀古今中外歷史，凡於戰時遷都者戰勝後必定還都，如不幸而戰敗，或還都或移都，視情形而定。庚子之役，聯軍陷北京，車駕幸西安，和議已定，大臣有獻議遷都西安者，兩江總督劉坤一力爭，遂年車駕還京師。此次同盟國勝利後，蘇聯必將由古比雪夫還都莫斯科（事實上已還都），中國何不敢光榮還都，致爲劉坤一所笑。首都之安全，豈退避政策所能保障！

還都南京之理由除此點似小實大的理由外，尚有其他更重要者。

現代與將來之南京，就地位與地位價值言，與過去之南京根本不同。南京已由邊緣地位變爲中央地位，即是地位適中——是由這樣觀察而得的結論。

先縱分中國爲五帶：（一）邊疆（二）中國弧形斷裂綫（指大興安嶺，太行山，秦嶺斷裂，湖廣斷裂即三峽，貴州斷裂）以西之腹裏部分，（三）弧形斷裂綫以東之腹裏部分，（四）以海岸爲根據之緣海的分環，（五）以沿海島嶼爲根據緣海的外環。南京適居第三帶，即中央地帶。

再縱分腹裏及南洋（其地華僑人口與全部邊疆人口大致相等）爲五帶：（一）東北及內蒙，（二）黃河流域，（三）長江流域，（四）

福建台灣與嶺南，（五）南洋，南京亦居第三帶，即中央地帶。

在縱橫分帶上南京均居第三帶，故爲全國或全民族分布區之中央地位，雖然是中樞區域之東南角。今日之南京，本質上與漢唐時代之西安相若，一方控制中樞區域，一方面對海疆威脅，故今日東南建都，也是以「首都作要塞，以天子守邊疆。」

在交通上，長江大動脈之終點，海外航運之起點，沿海航運之中點，聚會於江南，上海爲中國亦太平洋上最大之世界港，遠東海陸（水運與鐵路）交通最大之焦點（日本任何都市趕不上），其腹地初不限於長江流域，隴海沿線地區亦深受其支配，海州青島甚至天津大連大部分的均爲上海之衛星港，故上海蔚爲中國最大貿易港，遠東最大轉口港，南京控有此港與江浙富區，財賦（包括關稅）之區，交通之便均甲於全國。

在經濟上，長江流域多農產，自給自用之消費農產與國際貿易之實具農產或外匯農產在產量與價值上均超過北方（連東北之大豆計）。中國之工業生產百分之七十在長江流域，其中最大部分集中於上海與江南。甚至畜牧在西北爲原始的，粗放的，而現代高級生產之乳酪業，則以東南爲宜。南京的周地之富，亦甲於全國。

在人口密度上，江南佔全國之首席。長江流域一萬五千萬（不連江浙兩省之六千萬），沿海七省及重要港埠之兩萬萬，作丁字形之分布，其會合點在江南，故爲全國人力重心之所在，同時又爲人才重心之所在，因全國才智之士自來以江浙占主要成份，時至今日，依然如此。（十餘年前東方雜誌有專文論此，一時查不出。）

「東南財賦地，江浙人文藪」一語，在現在又增加了新的意義與重量，故能提高南京之建都價值。

象山港，面臨東海，鄰近日本，支居全國海岸之中央，右聯澎湖，瓊崖，左接青島大連，策應至便，台灣琉球收回，作爲自由出入太平洋之門戶，由象山港控制之最爲方便。象山港之優越的中央地位，增進南京建都之海防價值，爲有史以來所未有。

總之，還都南京爲掌握地理優勢，均衡海陸發展。

四　移都北平之理由

國民政府之建都南京，表面上的理由是遵奉總理遺訓，事實上則爲就近力源——華僑之財力，江浙之財力，南方之人力物力，與準備再進。是時革命力量未達黃河流域，雖欲都北平都西安均不可能，其情形與民元之暫都南京完全相同。

抗戰勝利後，日本崩潰或削弱，朝鮮與正獨立，東北完全（指南滿鐵路及旅順大連租借地）收回，台灣琉球一齊收回，而中國全部統一，這種種如果都照着正義與理想大部或完全實現，筆者主張移都北平。

北伐成功後，北平雖非首都，其地位重要，關係重大等於首都。蘆溝橋事變初起，平津即告淪陷，當時　蔣委員長於告抗戰全體將士第二書說「……平津既是北方政治軍事經濟文化的中心，就是我國家整個命脈之所關……」，北平之淪陷對國民精神上之影響與抗戰進展上之影響，絕不下於首都之淪陷。故移都北平與還都南京，就紀念勝利言意義是完全一致的。此外，尚有種種理由說明北平之適於建都。

首先要講到北平之地位與地位價值。現代北平之地位與歷史上之北平大致相同，而意義加重。北平是北方國防重心，外聯邊塞，內瞰中原，以黃海渤海爲內湖，遼東山東（兩半島）爲門戶。只須有相當力量加以防守，北平便成支哥加。北平是北方最大陸空交通會點，鐵道航空四方輻輳，勢力所達之範圍最廣，全國沿海航運以渤海爲起點。邊疆腹裏緣海三大區在燕山一線最爲接近，見西山之駝隊，遙念朔漠之安危，登津沽之巨輪，遙想太平洋之遼闊。胡人南下牧馬，久成歷史陳跡，但高原隱憂並未消除。登萬里長城，懷漢唐盛業，油然生追奔逐北之思。北平之雄壯形勢，可以概見。

北平之地位價值雖是如此之重要，但其周地之空間價值，因地理環境的稍差，水旱兵災之頻仍，軍閥與敵國之長期統治，在現在（！）却遠不如南方，但發展希望甚大。再就今代之歷史言，國民革命策源於珠江流域，定基於長江流域，其勢力其影響愈北而愈弱，東北亡了十二年，自然最弱。從南京定都到抗戰前夕的十年間，革命勢力雖已「北上」，但不易深入，不易遍布，不易滲透，不易生根。在有些地方雖能生根，却因草莽滋蔓，不易順利成長。

比較的講，南方繁榮，北方衰落，南方活躍，北方凝滯，南方進步，北方保守，成爲明顯之對比。因此在精神上，心理上，物質上，中國處於一種半邊痺痲的狀態，此非僅北方之痛苦，亦南方之不幸。

中國自古以來本是文化統一的國家，經七年抗戰之熔鑄，戰爭意志之集中，政治上常能趨於統一。以文化統一，政治統一爲基礎，今後應以大部分力量加強北方之建設，使北方在精神上，心理上，物質上迅速發展，與南方並駕齊驅。爲達此目的，戰後國都宜建於北方，宜建於東北西北兼顧之北平。

因北方之凝滯，保守與衰落，而重建之工作又十分艱巨，單靠北方民衆之力量自然不夠，必須以南方之有餘（比較而言）補北方之不足，尤其要中央政府移駐北平，領導建設。南方可靠自力而求更生，北方建設須靠中央政府之領導與支持，方能邁進，首都建於北平適合均衡發展之要求。

建設北方之重要問題甚多，這裏只舉八個犖犖大者：

（一）重整東北。（二）建設新疆。（三）治理黃河。黃河之害，大於全國河流之害。（四）組訓移民，東北與蒙新並重，以前爲自由移民，今後爲計劃移民。南洋移民每年只有兩萬人，東北移民，九一八前數年，每年平均百萬人。(五)完成鐵路網，南方水運發達，鐵路網之需要不如北方迫切。(六)振興水利，水利爲中國立國之本，北方因旱災特頻，需要亦較南方爲迫切。(七)建立海權，未來之海軍雖以駐在象山港爲適中，但多數艦艇之建造，則以渤海諸港爲宜，因爲東北既多森林，而晉冀遼寧更多煤鐵。(八)建設重工業，中國煤鑛百分之八十在晉陝河北，鐵鑛百分之八十在遼熱，冀魯與察南，北平居中，有左右逢源之便。北方重工業之發展爲全國工業化之基礎。

這八大重要問題都在北方，其成敗得失之影響，普遍的深刻的達於全國，他們不是北方問題，而是全中國之問題，而是全中國之首要問題，故應用全力求其解決。

總括一句，移都北平之根本理由，是中央重大問題，均衡南北發展。

五 結論

都南京是席豐履厚，都北平是任重道遠。都南京是掌握現在，都北平是創造將來。都南京是守成與創業並重，都北平則爲創業，爲進取。

就海陸並重言，都北平之國防的意義大於都南京，因北方邊務問題未克澄清。都南京之經濟的意義大於都北平，因長江大動脈的雄厚勢力難以搖撼。

都北平，國策因素決定較多；都南京，力源因素關係較大，在歷史背景與地理基礎上各有所據，難作比較。

筆者希望，抗戰勝利後，國內外的形勢容許中國能夠移都北平，否則還都南京。

南京與北平是中樞區域亦即全中國政治地理上的兩大重心。在領導建國的任務上，重要性相等，故其得失榮衰，同爲我國家整個命脈之所關。如果其中一個定爲首都，另一個一定是同首都——具有首都之空間影響，地位價值之大都會。

這兩大都市過去常常決定了中國之分裂，今後因地理機能歷史使命的相輔而成，必能加強中國之鞏固與統一。新中國的建國史，將是一部光輝燦爛的「雙城記」。

建都意見

董時進

戰後都城應設在何處？是時下一般很關心的問題。近來討論此問題者頗多，惟衆議紛歧，距離甚遠，由新疆到東北，各地皆有人主張。一察各方意見，雖不少人以科學的態度，從地理歷史及軍事形勢上着眼，以地點適中，防守安全，氣候佳良，交通便利等爲選擇的標準，無如高談理論，往往昧於實情。至不加思索，盲從附和者，尤其衆多。將來取決之時，當權者，或以私人利益爲前提，或以純粹理論爲依據，大錯鑄成，挽回困難。定都之事，非同小可，研討應不厭其詳，本刊願以常識的觀察，提供參考的意見。

竊以爲國都固然要安全，但最要安全者，尚不是國都，而是國家。都城雖無恙，假使國家精華不能保，又何貴乎有國都？故都城必須能扼守全國之精華地帶，具國亡與亡，國存與存之形勢，而後能保國家之安全。都城之於一國，猶如大本營之於軍隊，大本營雖不可在太前線，亦不可在太後方，如果大本營遠遠躲在後方，只顧本身安全，則前線軍隊覆沒，留此大本營何益？一般着眼於安全之點者，往往於不知不覺中，犯了國都安全，即是國家安全的錯誤心理，而不知其爲截然兩事。現時吾國之都城距敵人甚遠，固不可謂不安全，然而吾人豈可以都城之安全爲滿足，而置精華之喪失於不問？

中國之精華在東部，假使都城躲在西部，必使東部失去保障，而致有都城安全，而國家不安全之危險。

國都地點固以適中爲最佳，但適中並非決定都城地點最重要之條件，尤非惟一之條件。都城固是全國的首腦，但若謂都城必須處全國的中央，就無異說人的腦子必須生長在腹內。中國東部與西部情形迥異，尤不能等量齊觀，而以幾何的中心定都城之地位。世界各國之都城位置適當中央者絕少，尤其地面遼闊，各處氣候土質不一律之國家，此層更不可能。俄京偏在西邊，若要適中，非移至中亞細亞之沙漠不可。美國之地形與中國最爲近似，而華盛頓却在東岸，決無人認爲可以遷至中西部。都城必須在富庶開明之區域及衝要之地點，位置則不妨稍偏。從未有國地點的適中，而設在窮鄉僻野者，不但首都應如此，即省會亦應如此。試看各省省會的位置，也是偏在一方或一隅者居多，其所以偏的原因，自有實際的道理。歷來的首都大鎮，多是自然情勢的產物，絕不是憑理論玄想所造成，故適合實際需要，能歷久而不變。

氣候對於生活的舒適及工作效率都有關係，但其重要決不如許多人所說之甚。國土全部在熱帶或亞熱帶之國家，文化比較低落，誠屬事實，但若以爲都城地位偏南，而國家就必定衰弱，則尙待證明。一個都城的關係，畢竟不至如此之大，即使都城氣候不佳，只要國家大部分的農工商學不甚感受炎熱的妨害，則其國的文化及各種事業，應仍可進步。何況現時禦熱的方法甚多，電扇、冷氣、冷飲、以及良好的建築等，皆可以大減暑熱的威力。謂都城的氣候，足以影響到國家或朝代的壽命，實在於是太偏于理論。

歷代建都南方者國運短，建都北方者國運長，很多人都認爲是南方不宜建都的鐵證。惟有這種似科學而實不科學，似邏輯而實不邏輯的話最容易使人糊塗。一個朝代的興亡，受許多因素的影響，有內在的，有外來的，有屬於各種環境及形勢的。建都在何處，即使有幾許的關係，也必定關係很小。將這偶合而不能證明其有因果關係的事實，認爲是國運久暫的原因，實在是近乎迷信，與堪輿家的風水說一樣的迷信。譬如三國五代，根本是歷史上的混亂時期，其國運之不久，與建都地點何干？近如太平天國之亡，又何嘗是金陵之過。我們與其說建都南方朝代不久長，不如說朝代不久長的恰巧建都在南方。如果說南方氣候不適於生活與工作，何以近代的農工商以及文化事業，都以在南方比較發達。又何以近代的人物也都是出在南方的居多？或謂建都在北方則施政儉樸，國運攸久，此說實亦牽強。歷代施政儉樸與否，根本在人而不在地，開國之君儉樸，嗣後則漸奢侈。首都所在、未有不是比較奢侈的，做都城時的長安情形，絕不是和後來不做都城時的長安一樣。世界上任何國家首都，概是比較繁華奢侈的地方，並不見國家的壽命因此縮短。

細察關於建都地點的言論，大都犯了似是而非的毛病，不僅拋開了事實，純然講理論，而且誤認了因果關係，下了不正確的判斷。實際上都城的位置，並不是很難覓的，只要大體上合乎三個條件，就夠資格。即一．要在衝要地帶，但不必在全國的中心，而且不必在人口或經濟的中心。二．要相當的安全，但不必絕對的安全。三．氣候溫和適宜固佳，但非絕對的必要，不得已時可以犧牲。合乎這三個條件的地點很多，沿平漢綫及其以東的各處都無不適合，至於以西的地方，倒要愼重考慮。不過問題的癥結，不是在地位的適中與否，安全與否，及氣候如何，而是在有無作都城的設備。此點最關重要，反爲一般論者所忽視。建設一個首都，不是一件容易的事體，費了幾百年的時間，及若干專制帝王的威力，與人民的血汗脂膏，才造成了一個北京。南京已經十多年的建設，而去完成的日期還很遠，民窮財盡的中國，戰後必須用其所有的全部人力物力，去做經濟建設及生產工作，雖一絲一毫，也不能浪費去做其他非經濟和不生產的事體。因此，我們戰後的首都，只好儘現成的利用，絕不可再找一塊新地方去建設一個所謂新式的、理想的首都。如果將現成的設備擱在一邊不用，把重要的事業擱在一邊不做，而又耗費無量數的金錢和人力去建設都市，那眞是自取滅亡之道。

可以做都城的地點雖多，而有現成的設備的地方則少，各項條件具備的，只有北平。北平不但有現成的建設，而且那些設備特別具中國的色彩，最能表現中國的個性，絕不是新式的西洋建設，或半中不西的建設所能代替的。北平是中華民族最寶貴的物質遺產，是惟一可以向外人誇耀的物質建設，而同時也確是外人所最重視的，西方的羅馬，東方的北平，是世界上最偉大的兩個都城，無論對於中國人民或外國人，只有這座都城，可以使他們相信中國過去的光榮與偉大，用北平做首都是最適宜而且最方便的，不得已而求其次，也寧可在南京，但決不可在西安或武漢，或任何其他地點。

這類事體必須順應自然，不應空談理論，我想起一個故事可以證明。美國有一位曾任康奈爾大學校長和駐德大使的大教育家，名懷特，當康校的大方場四面的建築完成後，校務會議討論方場上的道途應如何規劃，諸教授紛陳理由，爭執不休，懷特校長才調解道：「諸位先生，你們的理由都對，但是我們最好不忙規劃這些路綫，姑且等候幾個月，讓學生們把路綫走出來之後，再行劃定。」學生走出來的路綫才是最自然而又切合實用的路綫。至今那些道路。仍是學生所走出來的，不是照理論劃定的。還有一個例子：在重慶瓷器口上游數里的對岸，一個地點叫大竹林，數年前有人看中了那塊地方，從各方面推論，認爲可以成立一個繁華的市鎮。於是修建好多房屋，街道，空場等，都計劃得井井有條，儼然一座新式的鎮子，取名興隆場。初開場的一些日子，也還熱鬧，但是過不多天，就逐漸冷靜下來，卒不得不放棄。前幾年曾一度存放兵工材料，後來也沒有用了，現在仍然空閒在那裏，這是建築在純粹理論上的市鎮的下場。

建都不是兒戲，萬不能輕易嘗試，浪費無數金錢，後來又不得不搬遷，那就損失太大了。我敢預言，假使我們定都西安，與洛陽，或武漢，或蘭州，將來都不免蹈興隆場的覆轍。

附錄

修正市組織法

——三十二年五月十九日國民政府公布——

第一條 市之自治，除本法規定外，準用關於縣自治之規定。

第二條 市自治實施辦法，由行政院定之。

第三條 凡人民聚居地方，具有左列情形之一者，設市，受行政院之指揮監督，一、首都，二、人口在百萬以上者，三、在政治經濟文化上有特殊情形者。

第四條 凡人民聚居地方，具有左列情形之一者，得設市，受省政府之指揮監督，一、省會，二、人口在二十萬以上者，三、在政治經濟文化地位重要，其人口在十萬以上者。

第五條 市之設置與廢止，及市區域之劃定或變更，應經國民政府之核准。

第六條 市以下爲區，區內之編制爲保甲，十戶至三十戶爲甲，十甲至三十甲爲保，十保至三十保爲區，其依地方情勢，有酌量變更之必要者，應呈經上級機關之核准。

第七條 中華民國人民，在市區域內，繼續居住六個月以上，或有住所達一年以上，年滿二十歲，經宣誓登記後，爲市公民，有依法行使選舉罷免創制複決之權。有左列情形之一者，不得有公民資格，一、褫奪公權者，二、虧欠公款者，三、曾因贓私處罰有案者，四、禁治產者，五、吸用鴉片或其代用品者。

第八條 市設市政府，其職權如左：一、辦理市自治事項，二、執行上級政府委辦事項。

第九條 市政府於不抵觸中央及上級政府法令範圍內，得發布市令。

第十條 市政府置市長一人，綜理全市事務，並指揮監督所屬機關及職員。

第十一條　市政府設局或科，掌理關於民政、財政、教育、建設、警察、衛生事項，設局或設科，由行政院依其事務之繁簡定之，市政府設局者，置局長科長科員，設科者置科長科員。

第十二條　院轄市市政府置秘書長一人，省轄市市政府置秘書主任一人，掌理文書庶務，及其他不屬於各局科事項。

第十三條　院轄市市政府，必要時得置參事一人或二人，掌理規章之撰擬事項。

第十四條　市政府因事務之需要，得置技術人員及視導人員。

第十五條　院轄市市長秘書長參事局長簡任，秘書科長薦任，科員委任，省轄市市長薦任或簡任，秘書主任局長薦任，秘書科長委任或薦任，科員委任。

第十六條　市政府人員之員額及其職務之分配，按各該市人口之多寡，及事務之繁簡，於各該市市政府組織規程中規定之，前項組織規程，由行政院定之。

第十七條　市政府得酌用雇員。

第十八條　市政府置主辦會計人員主辦統計人員各一人，掌理歲計會計統計事項，受市長之監督指揮，並依國民政府主計處組織法之規定，直接對主計處負責。會計統計需用佐理人員名額，由各該市市政府及主計處就各該市市政府組織規程所定人員名額中，會同決定之。

第十九條　市政府設市政會議，以左列人員組織之，一、市長，二、秘書長或秘書主任，三、參事，四、局長或科長，五、主辦會計人員。

第二十條　左列事項，應經市政會議議決，一、提出於市參議會之要件，二、市政府所屬機構辦事章則，三、市政府所屬機構間，不能解決之事項，四、市長交議事項，五、其他有關市政之重要事項。

第二十一條　市政會議每月至少開會一次，由市長召集之，開會時市長主席。

第二十二條　市政會議議事細則，由該會議定之。

第二十三條　市設市參議會，由市公民及依法成立之職業團體選舉市參議員組織之，但由職業團體選舉之參議員，不得超過總額十分之三。

第二十四條　市參議會議長副議長，由市參議員互選之。

第二十五條　市參議會之組織權及選舉方法，另以法律定之。

第二十六條　市財政依財政收支系統及關係法令之規定。

第二十七條　區設區民代表會，區民代表由保民大會選舉之，每保二人，任期二年，連選得連任，區民代表違法或失職，由保民大會罷免之。

第二十八條　區民代表會之職權如左，一、審議區規約及區與區相互間之公約，二、議決區長交議及本區內公民建議事項，三、選舉或罷免區長副區長，四、聽取區公所報告及向區公所提出詢問事項，五、其他有關本區重要興革事項。

第二十九條　區民代表會置主席一人，由代表互選之，開會時得通知區長保長列席。

第三十條　區民代表會每三個月開會一次，由主席召集之，必要時得舉行臨時會議。

第三十一條　區民代表會非有本區區民代表過半數之出席，不得開會議案之表決，以出席代表過半數之同意行之，可否同數時，取決於主席。

第三十二條　區民代表會決議案，送請區長分別執行，如區長延不執行或執行不當，得請其說明理由，如仍認為不滿意時，得報請市政府核辦。

第三十三條　區長對於區民代表會之決議案，如認為不當，得附理由送請覆議，對於覆議結果，如仍認為不當時，得呈請市政府核辦。

第三十四條　區設區公所置區長一人，副區長一人，由區民代表選舉之，受市政府之監督指揮，辦理本區自治事項，及執行市政府委辦事項，區長副區長任期二年，連選得連任。

第三十五條　區公所得置助理員及雇員。

第三十六條　保設保民大會，由本保每戶推出一人組織之其職權如左，一、審議保甲規約及保與保相互間之公約，二、議決保長交議及本保公民建議事項，三、選舉或罷免保長副保長，四、選舉或罷免區民代表會代表，五、聽取保辦公處工作報告，及向保辦公處提出其他有詢問事項，六、關於本保重要興革事項。

第三十七條　保民大會開會時，保長主席，保長有事故時，副保長主席，保長副保長具有事故或與所議事項有利害關係時，由大會推舉一人主席。

第三十八條　保民大會每三月開會一次，由保長召集之，必要時得召集臨時會議。

第三十九條　第三十一條第三十二條及第三十三條之規定，於保民大會準用之。

第四十條　保設保辦公處，置保長一人，副保長一人，由保民大會選舉之，受區長之監督指揮，辦理本保自治事項及執行市政府委辦事項。

第四十一條　甲設戶長會議，由本甲各戶戶長組織之，戶長有事故不能出席時，應派一人代表出席。

第四十二條　戶長會議之職權如左，一、選舉或罷免甲長，二、本甲應興革事項。

第四十三條　戶長會議由甲長召集之，每月開會一次，必要時經甲長或五戶以上之請求，得舉行臨時會議，開會時，甲長主席，甲長有事故或與所議事項有利害關係時，由出席人推舉一人主席。

第四十四條　戶長會議，非有本甲戶長過半數之出席，不得開會，議案之表決，以出席人過半數之同意行之，可否同數時取決於主席。

第四十五條　戶長會議決議案，由甲長執行之。

第四十六條　甲長認為必要或有本甲居民十人以上之連名請求時，應舉行甲居民會議，討論議決有關本甲興革事項。

第四十七條　在區民代表會未成立之地方區長副區長由市政府委任，在保民大會未成立之地方保長副保長由區公所遴定加倍人數，呈請市政府委任。

第四十八條　區保應辦事項，區民代表會及保民大會議事規則，由市政府定之。

第四十九條　本法施行細則，由行政院定之。

第五十條　本法自公布日施行。

都市計劃法

——二十八年六月八日國民政府公布——

第一條　都市計劃除法律別有規定外依本法之規定定之。

第二條　都市計劃由地方政府依據地方實際情況及其需要擬定之。

第三條　左列各地方應儘先擬定都市計劃。

一、市

二、已開之商埠

三、省會

四、聚居人口在十萬以上者

五、其他經國民政府認為應依本法擬定都市計劃之地方。

第四條　前條規定之地方如因軍事地震水災火災或其他重大事變致受損毀時地方政府認爲有改定都市計劃之必要者應於事變後六個月內重爲都市計劃之擬定。

第五條　就舊城市地方爲都市計劃應依當地情形另闢新市區應就原有市區逐步改造。

第六條　都市計劃擬定後應送由內政部會同關係機關核定轉呈行政院備案交地方政府公布執行。

都市計劃經核定公布後如有變更仍應依前項之規定辦理

第七條　都市計劃公布後其事業分期進行狀況應由地方政府於每年度終編具報告送內政部查核備案。

第八條　地方政府爲擬具都市計劃得遴聘專門人員並指派主管人員組織都市計劃委員會議訂之。

第九條　都市計劃委員會之組織通則由內政部定之。

第十條　都市計劃應表明左列事項：

一、市區現況

二、計劃區域

三、分區使用

四、公用土地

五、道路系統及水道交通

六、公用事業及上下水道

七、實施程序

八、經費

九、其他

前項各款應盡量以圖表表明之其第一款應包括地勢人口氣象交通經濟等狀況並應附具實測地形圖明示山河地勢原有道路村鎮市街及名勝建築等之位置與地名其比例尺不得小於二萬五千分之一。

第十一條　都市計劃區域應依據現在及既往情況並預期至少三十年內發展情形決定之。

第十二條　都市計劃應劃定住宅商業工業等限制使用區必要時並得劃定行政區及文化區。

第十三條　住宅區內土地及建築物之使用不得有礙居住之安寧。

第十四條　商業區內土地及建築物之使用不得有礙商業之便利。

第十五條　具有特殊性質之工廠應就工業區內特別指定地點建築之。

第十六條　行政區應盡可能就市中心地段劃定之。

第十七條　文化區應就幽靜地段劃定之。

第十八條　土地分區使用規定後其土地上原有建築物不合使用規定

著除維修外不得增築但主管地方政府認為必要時得斟酌地方情形限期令其變更使用其因變更使用所受之損害應補償之。

第十九條　市區道路系統應按分區及交通情形與預期之發展佈置之道路佔用土地面積不得少於全市總面積百分之二十。

第二十條　市區道路之縱橫距離應依使用地區分別定之。

第二十一條　市區主要道路交叉處車馬行人集中地點及紀念物建築地段均應設置廣場並應於適當地點設置停車場。

第二十二條　市區公園依天然地勢及人口疏密分別劃定適當地段建設之其佔用土地總面積不得少於全面積百分之十。

第二十三條　市區飲用水以自來水為原則未能設備自來水者其飲用水源應有衛生管理之規定。

第二十四條　市區飲用水源地域不得有排水溝渠之灌注及妨害水源清潔之設置。

第二十五條　市區內市民學校及體育衛生防空消防設備等公用地之設置地點應依市民居住分佈情形適當配置之。

第二十六條　市區內垃圾蓄便利用水運運出者其碼頭應於距市區一公里以外之地位。

第二十七條　市區公墓應於適當地點設置之。

第二十八條　都市計劃得分期分區實施。

第二十九條　新設市區應先完成主要道路及下水溝渠等工程建設。

第三十條　新市區建築地段應優先完成土地重劃。

第三十一條　本法施行細則得由各省政府依當地情形訂定送內政部核轉備案。

第三十二條　本法自公布日施行。

編校後記

三十二年十一月三十日，中國市政工程學會假中央水利實驗處，開第四次理監事聯席會議，守正適因公在渝，得參與出席。會中決定本會出版刊物，分年刊及專刊兩類，先着手編印年刊。當時因本會編審委員會尚未成立，又因黔省紙張印刷，在經濟及美觀上，均較西南各省為優，故經決定關於年刊資料之收集由本會總幹事譚炳訓先生主持，而編印事務則交黔分會，由守正負責。返黔後當即開始編纂，惟因稿件之遲寄，版式之商確，與總會函電往返，不無稍延時日，至四月初始編竣，六月中本會第一次年刊乃得與各界人士相見，於學術界闢一新園地，亦差足自慰，然終以亟於出版，校刊方面，難免有缺漏之處，尚希工程界人士及本會同仁多予賜教，俾作以後編印改進之指針，是所企幸。

過守正三十三年六月於銀坑

市政工程人才登記表

姓名		年齡		像片
別號		籍貫		
通訊處	1.			
	2.			
學歷	國內			
	國外			

普通工程經驗	
	1.……年　月（　年　月至　年　月）
	2.……年　月（　年　月至　年　月）
	3.……年　月（　年　月至　年　月）
	4.……年　月（　年　月至　年　月）
	5.……年　月（　年　月至　年　月）
	計　年　月

市政工程經歷		市政工程負責部份
	1.……年　月（　年　月至　年　月）	年　月
	2.……年　月（　年　月至　年　月）	年　月
	3.……年　月（　年　月至　年　月）	年　月
	4.……年　月（　年　月至　年　月）	年　月
	計　年　月	

普通工程經歷，市政工程經驗連同學歷共　年　月內市政工程負責部份計　年　月
專門著作
擅長技能
現任職務
備考

中華民國　年　月　日填

中國市政工程學會入會審查表

茲有　　　　君熱誠贊成本會宗旨並願遵守工程師信約協同工作照納會費志願加入本會用特檢奉市政工程學會入會申請書　　份請付審查為禱此致

中國市政工程學會

介紹人
1.　　　　住址　　　　簽章
2.　　　　住址　　　　簽章
3.　　　　住址　　　　簽章

時期	學歷（依規定折合經歷）	普通工程經歷	全部市政工程經歷	內市政工程負責部份
原填經歷	年　月	年　月	年　月	年　月
審查合格經歷	年　月	年　月	年　月	年　月

1. 全部工程經歷（連學歷）共　　年　　月
2. 市政工程負責部份計　　年　　月

審查結果	備註：
通知日期	
會證號碼	

本會總分會通訊處及負責人

總會：重慶

總幹事：重慶棗子嵐埡中央設計局公共工程組（電話：二二四七） 譚炳訓

編審委員會：重慶上清寺聚興村十二號（電話：二九六三） 鄭肇經

分會：貴陽 貴州公路局 姚世濂

昆明 西南聯合大學工學院 陶葆楷

桂林 依仁路五十號 袁向華

蘭州 水利林牧公司 沈君怡

贛縣 江西公路處辦事處 過守正

曲江 廣東建設廳 麥蘊瑜

桂陽 湖南建設廳 余籍傳

定價每冊壹百元（郵費另加）

編輯人：中國市政工程學會

發行人：中國市政工程學會理事長 凌鴻勛

承印者：江西印刷廠

出版期：民國三十三年 月

代售處：本會各地分會 全國各大書店

26114